Table of Measurement Abbreviations

U.S. Customary System

Length		Capacity		Weight		Area	
in.	inch	oz	ounce	oz	ounce	in^2	square inch
ft	foot	c	cup	lb	pound	ft^2	square foot
yd	yard	qt	quart				
mi	mile	gal	gallon				

Metric System

Length		Capacity		Weight/Mass		Area	
mm	millimeter (0.001 m)	ml	milliliter (0.001 L)	mg	milligram (0.001 g)	cm^2	square centimeter
cm	centimeter (0.01 m)	cl	centiliter (0.01 L)	cg	centigram (0.01 g)		
dm	decimeter (0.1 m)	dl	deciliter (0.1 L)	dg	decigram (0.1 g)	m^2	square meter
m	meter	L	liter	g	gram		
dam	decameter (10 m)	dal	decaliter (10 L)	dag	decagram (10 g)		
hm	hectometer (100 m)	hl	hectoliter (100 L)	hg	hectogram (100 g)		
km	kilometer (1000 m)	kl	kiloliter (1000 L)	kg	kilogram (1000 g)		

Time

h	hour	min	minute	s	second

Table of Symbols

+	add	<	is less than
−	subtract	≤	is less than or equal to
$\cdot$, $(a)(b)$	multiply	>	is greater than
$\frac{a}{b}$, ÷	divide	≥	is greater than or equal to
( )	parentheses, a grouping symbol	(a, b)	an ordered pair whose first component is a and whose second component is b
[ ]	brackets, a grouping symbol	°	degree (for angles and temperature)
π	pi, a number approximately equal to $\frac{22}{7}$ or 3.14	$\sqrt{a}$	the principal square root of a
$-a$	the opposite, or additive inverse, of a	$|a|$	the absolute value of a
$\frac{1}{a}$	the reciprocal, or multiplicative inverse, of a		reference to the Computer Tutor™
=	is equal to		reference to the Videotapes Library
≈	is approximately equal to		indicates graphing calculator topics
≠	is not equal to		

BEGINNING ALGEBRA

with Applications

FOURTH EDITION

Richard N. Aufmann
Palomar College, California

Vernon C. Barker
Palomar College, California

Joanne S. Lockwood
Plymouth State College, New Hampshire

HOUGHTON MIFFLIN COMPANY   **Boston**   **Toronto**

Geneva, Illinois   Palo Alto   Princeton, New Jersey

Sponsoring Editor: Maureen O'Connor
Associate Editor: Dawn Nuttall
Senior Project Editor: Cynthia Harvey
Editorial Assistant: Adriana Don
Senior Production/Design Coordinator: Carol Merrigan
Senior Manufacturing Coordinator: Marie Barnes
Marketing Manager: Charles Cavaliere

Cover design: Harold Burch, Harold Burch Design, New York City
Cover image: Eiji Yanagi/PHOTONICA

Printed in the U.S.A.

Library of Congress Catalog Card Number: 95-76919

ISBNs:
Text: 0-395-74610-8
Instructor's Annotated Edition: 0-395-74611-6

3456789-VH-   97 98 99

CONTENTS

PREFACE

The fourth edition of *Beginning Algebra with Applications* provides a mathematically sound and comprehensive coverage of the topics considered essential in an introductory algebra course. The text has been designed not only to meet the needs of the traditional college student but also to serve the needs of returning students whose mathematical proficiency may have declined during their years away from formal education.

In this new edition of *Beginning Algebra with Applications,* careful attention has been given to implementing the standards suggested by NCTM and AMATYC. Each chapter begins with a mathematical vignette consisting of an application, historical note, or curiosity related to mathematics. At the end of each section there are Supplemental Exercises that include writing, synthesis, critical thinking, and challenge problems. The chapter ends with a Project in Mathematics and Focus on Problem Solving. The Project in Mathematics feature is an extension or application of a concept that was covered in the chapter. These projects can be used for cooperative learning activities or extra credit. The Focus on Problem Solving feature demonstrates various proven problem-solving strategies and then asks the student to use the strategies to solve problems.

Instructional Features

Interactive Approach

Beginning Algebra with Applications uses an interactive style that gives the student an opportunity to try a skill as it is presented. Each section is divided into objectives, and every objective contains one or more sets of matched-pair examples. The first example in each pair is worked out; the second example, labeled Problem, is for the student to work. By solving this problem, the student practices using concepts as they are presented in the text. There are *complete* worked out solutions to these problems in a Solutions section at the end of the book. By comparing their solutions to model solutions, students can get immediate feedback on and reinforcement of the concepts.

Emphasis on Problem-Solving Strategies

Beginning Algebra with Applications features a carefully sequenced approach to application problems that emphasizes using proven strategies to solve problems. Students are encouraged to develop their own strategies, to draw diagrams, and to write their strategies as part of the solution to each application problem. In each case, model strategies are presented as guides for students to follow as they attempt the matched-pair Problem.

Emphasis on Applications

The traditional approach to teaching algebra covers only the straightforward manipulation of numbers and variables and thereby fails to teach students the practical value of algebra. By contrast, *Beginning Algebra with Applications* contains an extensive collection of contemporary application problems. Wherever appropriate, the last objective of a section presents applications that require the student to use the skills covered in that section to solve practical problems. This carefully integrated, applied approach generates awareness on the student's part of the value of algebra as a real-life tool.

Integrated Learning System Organized by Objectives

Each chapter begins with a list of the learning objectives included within that chapter. Each objective is then restated in the chapter to remind the student of the current topic of discussion. The same objectives that organize the text are reflected in the structure for exercises, for the testing programs, and for the Computer Tutor. Associated with every objective in the text is a corresponding computer tutorial and a corresponding set of test questions.

The Interactive Approach

Instructors have long recognized the need for a text that requires the student to use a skill as it is being taught. *Beginning Algebra with Applications* uses an interactive technique that meets this need. Every objective, including the one shown below, contains at least one pair of examples. One of the examples in the pair is

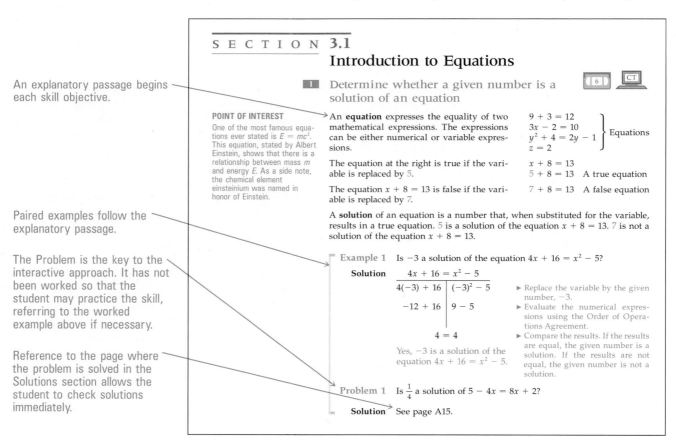

An explanatory passage begins each skill objective.

Paired examples follow the explanatory passage.

The Problem is the key to the interactive approach. It has not been worked so that the student may practice the skill, referring to the worked example above if necessary.

Reference to the page where the problem is solved in the Solutions section allows the student to check solutions immediately.

SECTION 3.1

Introduction to Equations

1 Determine whether a given number is a solution of an equation

POINT OF INTEREST
One of the most famous equations ever stated is $E = mc^2$. This equation, stated by Albert Einstein, shows that there is a relationship between mass m and energy E. As a side note, the chemical element einsteinium was named in honor of Einstein.

An **equation** expresses the equality of two mathematical expressions. The expressions can be either numerical or variable expressions.

$$\left. \begin{array}{l} 9 + 3 = 12 \\ 3x - 2 = 10 \\ y^2 + 4 = 2y - 1 \\ z = 2 \end{array} \right\} \text{Equations}$$

The equation at the right is true if the variable is replaced by 5.

$x + 8 = 13$
$5 + 8 = 13$ A true equation

The equation $x + 8 = 13$ is false if the variable is replaced by 7.

$7 + 8 = 13$ A false equation

A **solution** of an equation is a number that, when substituted for the variable, results in a true equation. 5 is a solution of the equation $x + 8 = 13$. 7 is not a solution of the equation $x + 8 = 13$.

Example 1   Is -3 a solution of the equation $4x + 16 = x^2 - 5$?

Solution

$$4x + 16 = x^2 - 5$$

$4(-3) + 16$	$(-3)^2 - 5$
$-12 + 16$	$9 - 5$

$4 = 4$

Yes, -3 is a solution of the equation $4x + 16 = x^2 - 5$.

▶ Replace the variable by the given number, -3.
▶ Evaluate the numerical expressions using the Order of Operations Agreement.
▶ Compare the results. If the results are equal, the given number is a solution. If the results are not equal, the given number is not a solution.

Problem 1   Is $\frac{1}{4}$ a solution of $5 - 4x = 8x + 2$?

Solution   See page A15.

worked. The second example in the pair (the Problem) is not worked so that the student may "interact" with the text by solving it. In order to provide immediate feedback, a complete solution to this Problem is provided in the Solutions section. The benefit of this interactive style is that students can check whether they have learned the new skill before they attempt a homework assignment.

Emphasis on Applications

The solution of an application problem in *Beginning Algebra with Applications* comprises two parts: **Strategy** and **Solution**. The strategy is a written description of the steps that are necessary to solve the problem; the solution is the implementation of the strategy. Using this format provides students with a structure for problem solving. It also encourages students to write strategies for solving problems, which in turn fosters their organizing problem-solving strategies in a logical way. Having students write strategies is a natural way to incorporate writing into the math curriculum.

A strategy that the student may use in solving an application problem is stated.

This strategy is used in the solution of the worked example.

When students compare their solutions to those in the Solutions section, they will see a complete solution of the problem along with a written strategy for solving the problem.

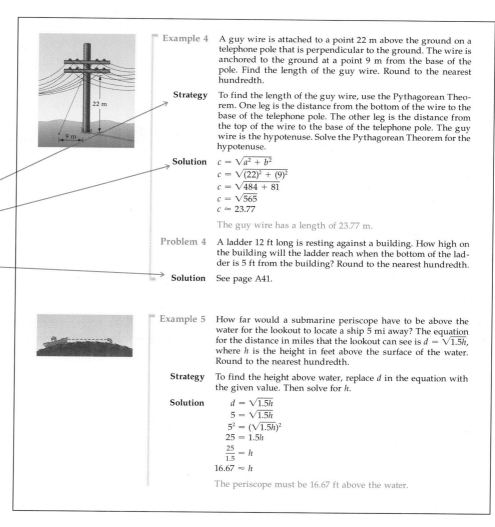

Example 4 A guy wire is attached to a point 22 m above the ground on a telephone pole that is perpendicular to the ground. The wire is anchored to the ground at a point 9 m from the base of the pole. Find the length of the guy wire. Round to the nearest hundredth.

22 m
9 m

Strategy To find the length of the guy wire, use the Pythagorean Theorem. One leg is the distance from the bottom of the wire to the base of the telephone pole. The other leg is the distance from the top of the wire to the base of the telephone pole. The guy wire is the hypotenuse. Solve the Pythagorean Theorem for the hypotenuse.

Solution
$$c = \sqrt{a^2 + b^2}$$
$$c = \sqrt{(22)^2 + (9)^2}$$
$$c = \sqrt{484 + 81}$$
$$c = \sqrt{565}$$
$$c \approx 23.77$$

The guy wire has a length of 23.77 m.

Problem 4 A ladder 12 ft long is resting against a building. How high on the building will the ladder reach when the bottom of the ladder is 5 ft from the building? Round to the nearest hundredth.

Solution See page A41.

Example 5 How far would a submarine periscope have to be above the water for the lookout to locate a ship 5 mi away? The equation for the distance in miles that the lookout can see is $d = \sqrt{1.5h}$, where h is the height in feet above the surface of the water. Round to the nearest hundredth.

Strategy To find the height above water, replace d in the equation with the given value. Then solve for h.

Solution
$$d = \sqrt{1.5h}$$
$$5 = \sqrt{1.5h}$$
$$5^2 = (\sqrt{1.5h})^2$$
$$25 = 1.5h$$
$$\frac{25}{1.5} = h$$
$$16.67 \approx h$$

The periscope must be 16.67 ft above the water.

The Objective-Specific Approach

Many texts in mathematics are not organized in a manner that facilitates management of learning. Typically, students are left to wander through a maze of apparently unrelated lessons, exercise sets, and tests. *Beginning Algebra with Applications* solves this problem by organizing all lessons, exercise sets, computer tutorials, and tests around a carefully constructed hierarchy of objectives. The advantage of this objective-by-objective organization is that it enables the student who is uncertain at any step in the learning process to refer easily to the original presentation and review that material.

The objective-specific approach also gives the instructor greater control over the management of student progress. The computerized testing program and the printed testing program are organized in terms of the same objectives as the text. These references are provided with the answers to the test items. This allows the instructor to identify quickly those objectives for which a student may need additional instruction.

The Computer Tutor is also organized around the objectives of the text. As a result, supplemental instruction is available for any objectives that are troublesome for a student.

A numbered objective statement names the topic of each lesson.

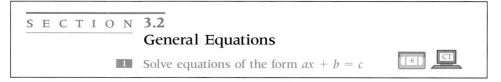

SECTION **3.2**

General Equations

1 Solve equations of the form $ax + b = c$

The exercise sets correspond to the objectives in the text.

EXERCISES 3.2

1 Solve and check.

1. $3x + 1 = 10$ **2.** $4y + 3 = 11$ **3.** $2a - 5 = 7$ **4.** $5m - 6 = 9$

The answers to the Chapter Review Exercises show the objective from which the exercise was taken.

CHAPTER REVIEW EXERCISES   *pages 127–129*

1. No (Obj. 3.1.1)   **2.** 20 (Obj. 3.1.2)   **3.** −7 (Obj. 3.1.3)   **4.** 7 (Obj. 3.2.1)   **5.** 4 (Obj. 3.2.2)
6. $-\frac{1}{5}$ (Obj. 3.2.3)   **7.** 405 (Obj. 3.1.4)   **8.** 25 (Obj. 3.1.4)   **9.** 67.5% (Obj. 3.1.4)
10. ⊢┼┼┼┼┼┼┼┼┼┼⊣ -5-4-3-2-1 0 1 2 3 4 5 (Obj. 3.3.1)   **11.** $x > 2$ ⊢┼┼┼┼┼┼┼┼┼┼⊣ -5-4-3-2-1 0 1 2 3 4 5 (Obj. 3.3.1)

The answers to the Chapter Test exercises show the objective from which the exercise was taken.

CHAPTER TEST   *pages 129–130*

1. −12 (Obj. 3.1.3)   **2.** $-\frac{1}{2}$ (Obj. 3.2.2)   **3.** −3 (Obj. 3.2.1)   **4.** No (Obj. 3.1.1)   **5.** $\frac{1}{8}$ (Obj. 3.1.2)
6. $-\frac{1}{3}$ (Obj. 3.2.3)   **7.** 5 (Obj. 3.2.1)   **8.** $\frac{1}{2}$ (Obj. 3.2.2)   **9.** −5 (Obj. 3.1.2)   **10.** −5 (Obj. 3.2.2)

The answers to the Cumulative Review Exercises also show the objective that relates to each exercise.

CUMULATIVE REVIEW EXERCISES   *pages 299–300*

1. −8, −4 (Obj. 1.1.1)   **2.** {1, 2, 3, 4, 5, 6, 7, 8, 9, 10} (Obj. 1.1.1)   **3.** 10 (Obj. 1.4.2)   **4.** $40a - 28$
(Obj. 2.2.5)   **5.** $\frac{3}{2}$ (Obj. 2.1.1)   **6.** $-\frac{3}{2}$ (Obj. 3.1.3)   **7.** $-\frac{7}{2}$ (Obj. 3.2.3)   **8.** $-\frac{2}{9}$ (Obj. 3.2.3)
9. $x < -5$ (Obj. 3.3.3)   **10.** $x \leq 3$ (Obj. 3.3.3)   **11.** 24% (Obj. 3.1.4)   **12.** (4, 0), (0, −2) (Obj. 5.2.3)

Additional Learning Aids

Project in Mathematics

The Project in Mathematics feature appears near the end of each chapter. These projects can be used for extra credit or as cooperative learning activities. The projects cover various aspects of mathematics and are drawn from fields as diverse as geometry, statistics, science, and business.

Focus on Problem Solving

The Focus on Problem Solving feature introduces students to various proven problem-solving strategies. Some of the topics that are discussed are inductive reasoning, deductive reasoning, finding counterexamples, and using the calculator as a problem-solving tool.

Chapter Summaries

The Chapter Summaries have been rewritten to be a more useful guide to students as they review for a test. The Chapter Summary includes the Key Words and the Essential Rules and Procedures that were covered in the chapter. Each key word and essential rule is accompanied by an example of that concept.

Study Tips

At various places throughout the first few chapters, students are offered suggestions for acquiring good study habits and skills.

Glossary

There is a Glossary that includes definitions of terms used in the text.

Exercises

End-of-Section Exercises

Beginning Algebra with Applications contains more than 6000 exercises. At the end of each section there are exercise sets that are keyed to the corresponding learning objectives. The exercises have been carefully developed to ensure that students can apply the concepts in the section to a variety of problem situations.

Concept Review Exercises

These "Always true, Sometimes true, or Never true" exercises precede every exercise set and are designed to test a student's understanding of the material. Using Concept Review exercises as oral exercises at the end of a class session can lead to interesting class discussions.

Supplemental Exercises

The end-of-section exercises are followed by Supplemental Exercises. These contain a variety of exercise types:

• Challenge problems (designated by [C] in the Instructor's Annotated Edition)
• Problems that ask students to interpret and work with real-world data
• Writing exercises
• Problems that ask students to determine incorrect procedures
• Problems that require a more in-depth analysis

Data Exercises

These exercises, designated by [D] in the Instructor's Annotated Edition, ask students to analyze and solve problems taken from actual situations. They are often required to work with tables, graphs, and charts drawn from a variety of disciplines.

Writing Exercises

Within the Supplemental Exercises, writing exercises are denoted by [W] in the Instructor's Annotated Edition. These exercises ask students to write about a topic in the section or to research and report on a related topic. There are also writing exercises in some of the application problems. These exercises ask students to write a sentence that describes the meaning of their answers in the context of the problem.

Chapter Review Exercises and Chapter Tests

Chapter Review Exercises and a Chapter Test are found at the end of each chapter. These exercises are selected to help the student integrate all the topics presented in the chapter. The answers to all Chapter Review and Chapter Test exercises are given in the Answers section. Along with the answer, there is a reference to the objective that pertains to each exercise.

Cumulative Review Exercises and Final Exam

Cumulative Review Exercises, which appear at the end of each chapter beginning with Chapter 2, help the student maintain skills learned in previous chapters. In addition, a Final Exam appears at the end of Chapter 11. The answers to all Cumulative Review and Final Exam exercises are given in the Answers section. Along with the answer, there is a reference to the objective that pertains to each exercise.

New to This Edition

Topical Coverage

Beginning Algebra with Applications retains its strong commitment to applications of mathematics. However, at the suggestion of users of the book, we have changed the order of the types of application problems presented in Chapter 4 so that there is a better transition from easier to more difficult problems. The new organization also gives more prominence to problems from the field of geometry. Although application problems are featured in this chapter, we have endeavored to provide a balanced approach to application problems that is used consistently throughout the text.

We have added new application problems to the text. Within some of these problems, we have incorporated short writing exercises. These exercises ask students to explain, in the context of the application, the significance of their answer.

Many application problems are now accompanied by diagrams that help students visualize the problem settings. Students are encouraged to draw and use diagrams as a problem-solving strategy.

The material on inequalities in Chapter 10 of the third edition has been dispersed throughout the text in this edition. For instance, solving first-degree inequalities now appears in Chapter 3, Solving Equations and Inequalities. In an analogous manner, the remaining material in Chapter 10 of the third edition has been moved to appropriate sections of the new edition.

Graphing linear equations and inequalities is now covered earlier in the text (Chapter 5). Now that graphing is introduced in an earlier chapter, it is possible to explore application problems not only from an algebraic perspective, but in a graphical setting as well.

So as to maintain the graphing theme started in Chapter 5, Systems of Linear Equations is now Chapter 6.

An introduction to functions and functional notation has been added to Chapter 5.

Scientific notation has been added to Chapter 7.

The pace of Chapter 11, Quadratic Equations, has been altered. There is now more expository material, a greater number of worked examples, and an increased emphasis on checking solutions of equations. The exercise sets have been modified and now offer a smoother transition from easy to difficult exercises.

Margin Notes

Margin notes are interspersed throughout the text. These notes are called Look Closely and Point of Interest. The Look Closely notes warn students that a procedure may be particularly involved or remind students that there are certain checks of their work that should be performed. The Point of Interest notes are interesting sidelights on the topic being discussed. In addition, there are Instructor Notes that are printed only in the Instructor's Annotated Edition.

Study Tips

Effective study skills are an important factor in achieving success in any discipline. This is especially true in mathematics. To help students acquire these skills, we have added Study Tips boxes throughout the first few chapters of the text. Each box contains a suggestion that will lead to improved study habits.

Project in Mathematics

Through the Project in Mathematics feature, some of the standards suggested by the NCTM can be implemented. These projects offer an opportunity for students to explore several topics relating to geometry, statistics, science, and business.

Focus on Problem Solving

Although successful problem solvers use a variety of techniques, it has been well established that there are basic recurring strategies. Among these are trying to solve a related problem, trying to solve an easier problem, working backwards, trial and error, and other techniques. The Focus on Problem Solving features present some of these methods in the context of a problem to be solved. Students are encouraged to apply these strategies in solving similar problems.

Graphing Calculators

Graphing calculators are incorporated as an optional feature at appropriate places throughout the text. In some chapters, there are graphing calculator exercises that are designated by the graphing calculator icon shown at the beginning of this paragraph. The graphing calculator may help some students as they struggle with new concepts. However, all graphing calculator suggestions can be omitted without destroying the integrity of the course. Students can consult the appendix, Guidelines for Using Graphing Calculators, for help with keystroking procedures to use for several models of graphing calculators.

Index of Applications

The Index of Applications that follows the Preface provides a quick reference for application problems from a wide variety of fields.

Computer Tutor

The Computer Tutor, an interactive computer tutorial covering *every* objective in the text, has been completely revised. It is now an algorithmically based tutor that includes color and animation. The algorithmic feature essentially provides an infinite number of practice problems for students to attempt. The algorithms have been carefully crafted to present a variety of problem types from easy to difficult. A complete solution to each problem is available.

An interactive feature of the Computer Tutor requires students to respond to questions about the topic in the current lesson. In this way, students can assess their understanding of concepts as these are presented. There is a Glossary that can be accessed at any time so that students can look up words whose definitions they may have forgotten.

When the student completes a lesson, a printed report is available. This optional report gives the student's name, the objectives studied, the number of problems attempted, the number of problems answered correctly, the percent correct, and the time spent working on the exercises in that objective.

Supplements for the Instructor

Instructor's Annotated Edition

The Instructor's Annotated Edition is an exact replica of the student text except that answers to all exercises are given in the text. Also, there are Instructor Notes in the margins that offer suggestions for presenting the material in that objective.

Instructor's Resource Manual with Solutions Manual

The Instructor's Resource Manual includes suggestions for course sequencing and gives sample answers for the writing exercises. The Solutions Manual contains worked-out solutions for all end-of-section exercises, Concept Review exercises, Chapter Review exercises, Chapter Test exercises, Cumulative Review exercises, and Final Exam exercises.

Computerized Test Generator

The Computerized Test Generator is the first of three sources of testing material. The data base contains more than 2000 test items. These questions are unique to the Test Generator. The Test Generator is designed to provide an unlimited number of chapter tests, cumulative chapter tests, and final exams. It is available for the IBM PC and compatible computers and for the Macintosh. The IBM version provides new on-line testing and also contains algorithms that produce an unlimited number of some types of test questions.

Printed Test Bank with Chapter Tests

The Printed Test Bank, the second component of the testing material, is a printout of all items in the Computerized Test Generator. Instructors who do not have access to a computer can use the test bank to select items to include on tests prepared by hand. The Chapter Tests comprise the printed testing program, which is the third source of testing material. Eight printed tests, four free-response and four multiple-choice, are provided for each chapter. In addition, there are cumulative tests for use after Chapters 4, 8, and 11, and a final exam.

Supplements for the Student

Student Solutions Manual

The *Student Solutions Manual* contains complete solutions to all odd-numbered exercises in the text.

Computer Tutor

The Computer Tutor is an interactive instructional computer program for student use. Each objective of the text is supported by a tutorial in the Computer Tutor. These tutorials contain an interactive lesson that covers the material in the objective. Following the lesson are randomly generated exercises for the student to attempt. A record of the student's progress is available.

The Computer Tutor can be used in several ways: (1) to cover material the student missed because of an absence, (2) to reinforce instruction on a concept that the student has not yet mastered, and (3) to review material in preparation for exams. This tutorial is available for the Macintosh and for IBM PC and compatible computers running Windows.

Within each section of the book, a computer icon appears next to each objective. The icon serves as a reminder that there is a Computer Tutor lesson corresponding to that objective.

Videotapes

The videotape series contains lessons that accompany *Beginning Algebra with Applications*. These lessons follow the format and style of the text and are closely tied to specific sections of the text. Each videotape begins with an application, and then the mathematics needed to solve that application is presented. The tape closes with the solution of the application problem.

Within each section of the book, a videotape icon appears next to each objective that is covered by a videotape lesson. The icon contains the reference number of the appropriate video.

Acknowledgments

The authors would like to thank the people who have reviewed this manuscript and provided many valuable suggestions:

Philip E. Buechner Jr., *Cowley County Community College, KS*
Richard Campbell, *Butte College, CA*
Ronnie L. Foster, *Southern University, LA*
Mary Hartman, *Southwestern Community College, NC*
Georgia Kamvosoulis
William Kincaid, *Wilmington College, OH*
Michael Longfritz, *Rensselear Polytechnic Institute, NY*
Marilyn Massey Moss, *Collin County Community College, TX*
Dorothy Pennington, *Tallahassee Community College, FL*
Claudinna Rowley, *Johnson County Community College, KS*
John Searcy
Mark Stevenson, *Oakland Community College, MI*

INDEX of Applications

1

Real Numbers

Objectives

Early Egyptian Fractions

One of the earliest written documents of mathematics is the Rhind Papyrus. This tablet was found in Egypt in 1858, but it is estimated that the writings date back to 1650 B.C.

The Rhind Papyrus contains over 80 problems. Studying these problems has enabled mathematicians and scientists to understand some of the methods by which the early Egyptians used mathematics.

Evidence gained from the Papyrus shows that the Egyptian method of calculating with fractions was much different from the methods used today. All fractions were represented in terms of what are called unit fractions. A unit fraction is a fraction in which the numerator is 1. This fraction was symbolized with a bar over the number. Examples (using modern numbers) include

$$\overline{3} = \frac{1}{3} \qquad \overline{15} = \frac{1}{15}$$

The early Egyptians also tended to deal with powers of two (2, 4, 8, 16, . . .). As a result, representing fractions with a 2 in the numerator in terms of unit fractions was an important matter. The Rhind Papyrus has a table giving the equivalent unit fractions for all odd denominators from 5 to 101 with 2 as the numerator. Some of these are listed below.

$$\frac{2}{5} = \overline{3}\ \overline{15} \qquad \left(\frac{2}{5} = \frac{1}{3} + \frac{1}{15}\right)$$

$$\frac{2}{7} = \overline{4}\ \overline{28}$$

$$\frac{2}{11} = \overline{6}\ \overline{66}$$

$$\frac{2}{19} = \overline{12}\ \overline{76}\ \overline{114}$$

SECTION 1.1

Introduction to Integers

1 Order relations

It seems to be a human characteristic to group similar items. For instance, a botanist places plants with similar characteristics in groups called species. Nutritionists classify foods according to food groups; for example, pasta, crackers, and rice are among the foods in the bread group.

Mathematicians place objects with similar properties in groups called sets. A **set** is a collection of objects. The objects in a set are called the **elements** of the set.

The **roster method** of writing sets encloses a list of the elements in braces. The set of sections within an orchestra is written {brass, percussion, strings, woodwind}.

The numbers that we use to count objects, such as the number of students in a classroom or the number of people living in an apartment house, are the natural numbers.

Natural numbers = {1, 2, 3, 4, 5, 6, 7, 8, 9, 10, ...}

The three dots mean that the list of natural numbers continues on and on and that there is no largest natural number.

The natural numbers alone do not provide all the numbers that are useful in applications. For instance, a meterologist also needs the number zero and numbers below zero.

Integers = {..., −5, −4, −3, −2, −1, 0, 1, 2, 3, 4, 5, ...}

Each integer can be shown on a number line. The integers to the left of zero on the number line are called **negative integers.** The integers to the right of zero are called **positive integers,** or natural numbers. Zero is neither a positive nor a negative integer.

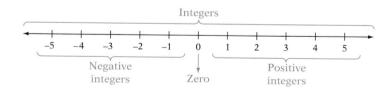

The **graph** of an integer is shown by placing a heavy dot on the number line directly above the number. The graphs of −3 and 4 are shown on the number line below.

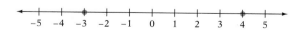

Consider the following sentences.

> The quarterback threw the football, and the receiver caught *it*.
> A student purchased a computer and used *it* to write English and history papers.

In the first sentence, *it* is used to mean the football; in the second sentence, *it* means the computer. In language, the word *it* can stand for many different objects. Similarly, in mathematics, a letter of the alphabet can be used to stand for a number. Such a letter is called a **variable**. Variables are used in the following definition of inequality symbols.

Definition of Inequality Symbols

If a and b are two numbers and a is to the left of b on the number line, then a **is less than** b. This is written $a < b$.

If a and b are two numbers and a is to the right of b on the number line, then a **is greater than** b. This is written $a > b$.

POINT OF INTEREST

The symbols for "is less than" and "is greater than" were introduced by Thomas Harriot around 1630. Before that, ⊏ and ⊐ were used for < and >, respectively.

Negative 4 is less than negative 1.

$-4 < -1$

5 is greater than 0.

$5 > 0$

There are also inequality symbols for **is less than or equal to** ($\leq$) and **is greater than or equal to** ($\geq$).

$7 \leq 15$ 7 is less than or equal to 15. $6 \leq 6$ 6 is less than or equal to 6.
This is true because $7 < 15$.   This is true because $6 = 6$.

Example 1   Use the roster method to write the set of negative integers greater than or equal to -6.

    Solution $A = \{-6, -5, -4, -3, -2, -1\}$ ▶ A set is designated by a capital letter.

Problem 1   Use the roster method to write the set of positive integers less than 5.

    Solution   See page A11.

Example 2   Given $A = \{-6, -2, 0\}$, which elements of set A are less than or equal to -2?

    Solution $-6 < -2$ ▶ Find the order relation between each element of set A and -2.

                  $-2 = -2$

                   $0 > -2$

          The elements -6 and -2 are less than or equal to -2.

Problem 2   Given $B = \{-5, -1, 5\}$, which elements of set B are greater than -1?

Solution   See page A11.

2 Opposites and absolute value

Two numbers that are the same distance from zero on the number line but are on opposite sides of zero are **opposite numbers,** or **opposites.** The opposite of a number is also called its **additive inverse.**

The opposite of 5 is -5.

The opposite of -5 is 5.

The negative sign can be read "the opposite of."

$$-(2) = -2 \qquad \text{The opposite of 2 is } -2.$$
$$-(-2) = 2 \qquad \text{The opposite of } -2 \text{ is 2.}$$

Example 3   Find the opposite number.   A. 6   B. -51

Solution   A. -6 B. 51

Problem 3   Find the opposite number.   A. -9 B. 62

Solution   See page A11.

The **absolute value** of a number is its distance from zero on the number line. Therefore, the absolute value of a number is a positive number or zero. The symbol for absolute value is two vertical bars, $||$.

POINT OF INTEREST

The definition of *absolute value* given in the box is written in what is called rhetorical style. That is, it is written without the use of variables. This is how *all* mathematics was written prior to the Renaissance. During that period from the 14th to the 16th century, the idea of expressing a variable symbolically was developed. In terms of that symbolism, the definition of absolute value is

$$|x| = \begin{cases} x, & x > 0 \\ 0, & x = 0 \\ -x, & x < 0 \end{cases}$$

The distance from 0 to 3 is 3. Therefore, the absolute value of 3 is 3.

$$|3| = 3$$

The distance from 0 to -3 is 3. Therefore, the absolute value of -3 is 3.

$$|-3| = 3$$

Absolute Value

The absolute value of a positive number is the number itself. The absolute value of zero is zero. The absolute value of a negative number is the opposite of the negative number.

Example 4   Evaluate.   A. $|-4|$ B. $-|-10|$

Solution   A. $|-4| = 4$

B. $-|-10| = -10$ ▶ The absolute value sign does not affect the negative sign in front of the absolute value sign. You can read $-|-10|$ as "the opposite of the absolute value of negative 10."

Problem 4   Evaluate.   A. $|-5|$ B. $-|-9|$

Solution   See page A11.

STUDY TIPS

KNOW YOUR INSTRUCTOR'S REQUIREMENTS

To do your best in this course, you must know exactly what your instructor requires. If you don't, you probably will not meet his or her expectations and are not likely to earn a good grade in the course.

Instructors ordinarily explain course requirements during the first few days of class. Course requirements may be stated in a *syllabus,* which is a printed outline of the main topics of the course, or they may be presented orally. When they are listed in a syllabus or on other printed pages, keep them in a safe place. When they are presented orally, make sure to take complete notes. In either case, understand them completely and follow them exactly.

CONCEPT REVIEW 1.1

Determine whether the statement is always true, sometimes true, or never true.

1. The absolute value of a number is positive.

2. The absolute value of a number is negative.

3. If x is an integer, than $|x| < -3$.

4. If x is an integer, then $|x| > -2$.

5. The opposite of a number is a positive number.

6. The set of positive integers is $\{0, 1, 2, 3, 4, \ldots\}$.

7. If a is an integer, then $a \le a$.

8. If a and b are integers and $a > b$, then $a \ge b$.

9. If x is a negative integer, then $|x| = -x$.

10. If a and b are integers and $a < b$, then $|a| < |b|$.

EXERCISES 1.1

1 Place the correct symbol, $<$ or $>$, between the two numbers.

1. $-2 \quad -5$
2. $-6 \quad -1$
3. $-16 \quad 1$
4. $-2 \quad 13$

5. $3 \quad -7$
6. $5 \quad -6$
7. $0 \quad -3$
8. $8 \quad 0$

9. $-42 \quad 27$
10. $-36 \quad 49$
11. $21 \quad -34$
12. $53 \quad -46$

13. $-27 \quad -39$
14. $-51 \quad -20$
15. $-131 \quad 101$
16. $127 \quad -150$

Use the roster method to write the set.

17. the natural numbers less than 9

18. the natural numbers less than or equal to 6

19. the positive integers less than or equal to 8

20. the positive integers less than 4

21. the negative integers greater than -7

22. the negative integers greater than or equal to -5

Solve.

23. Given $A = \{-7,\ 0,\ 2,\ 5\}$, which elements of set A are greater than 2?

24. Given $B = \{-8,\ 0,\ 7,\ 15\}$, which elements of set B are greater than 7?

25. Given $D = \{-23,\ -18,\ -8,\ 0\}$, which elements of set D are less than -8?

26. Given $C = \{-33,\ -24,\ -10,\ 0\}$, which elements of set C are less than -10?

27. Given $E = \{-35,\ -13,\ 21,\ 37\}$, which elements of set E are greater than -10?

28. Given $F = \{-27,\ -14,\ 14,\ 27\}$, which elements of set F are greater than -15?

29. Given $B = \{-52,\ -46, 0, 39, 58\}$, which elements of set B are less than or equal to 0?

30. Given $A = \{-12,\ -9, 0, 12, 34\}$, which elements of set A are greater than or equal to 0?

31. Given $C = \{-23,\ -17, 0, 4, 29\}$, which elements of set C are greater than or equal to -17?

32. Given $D = \{-31,\ -12, 0, 11, 45\}$, which elements of set D are less than or equal to -12?

33. Given that set A is the positive integers less than 10, which elements of set A are greater than or equal to 5?

34. Given that set B is the positive integers less than or equal to 12, which elements of set B are greater than 6?

35. Given that set D is the negative integers greater than or equal to -10, which elements of set D are less than -4?

36. Given that set C is the negative integers greater than -8, which elements of set C are less than or equal to -3?

2 Find the opposite number.

37. 22   **38.** 45   **39.** -31 **40.** -88

41. -168 **42.** -97 **43.** 630   **44.** 450

Evaluate.

45. $-(-18)$ **46.** $-(-30)$ **47.** $-(49)$ **48.** $-(67)$

49. $|16|$ **50.** $|19|$ **51.** $|-12|$ **52.** $|-22|$

53. $-|29|$ **54.** $-|20|$ **55.** $-|-14|$ **56.** $-|-18|$

57. $-|0|$ **58.** $|-30|$ **59.** $-|34|$ **60.** $-|-45|$

Solve.

61. Given $A = \{-8, -5, -2, 1, 3\}$, find
   a. the opposite of each element of set A.
   b. the absolute value of each element of set A.

62. Given $B = \{-11, -7, -3, 1, 5\}$, find
   a. the opposite of each element of set B.
   b. the absolute value of each element of set B.

Place the correct symbol, $<$ or $>$, between the two numbers.

63. $|-83|$ $|58|$ **64.** $|22|$ $|-19|$ **65.** $|43|$ $|-52|$ **66.** $|-71|$ $|-92|$

67. $|-68|$ $|-42|$ **68.** $|12|$ $|-31|$ **69.** $|-45|$ $|-61|$ **70.** $|-28|$ $|43|$

SUPPLEMENTAL EXERCISES 1.1

Write the given numbers in order from least to greatest.

71. $|-5|, 6, -|-8|, -19$ **72.** $-4, |-15|, -|-7|, 0$

73. $-(-3), -22, |-25|, |-14|$ **74.** $-|-26|, -(-8), |-17|, -(5)$

The following table gives equivalent temperatures for combinations of temperature and wind speed. For example, the combination of a temperature of 15°F and a wind blowing at a 10 mph has a cooling power equal to −3°F.

Wind Chill Factors																	
Wind Speed (in mph)	Thermometer Reading (in degrees Fahrenheit)																
	35	30	25	20	15	10	5	0	−5	−10	−15	−20	−25	−30	−35	−40	−45
5	33	27	21	19	12	7	0	−5	−10	−15	−21	−26	−31	−36	−42	−47	−52
10	22	16	10	3	−3	−9	−15	−22	−27	−34	−40	−46	−52	−58	−64	−71	−77
15	16	9	2	−5	−11	−18	−25	−31	−38	−45	−51	−58	−65	−72	−78	−85	−92
20	12	4	−3	−10	−17	−24	−31	−39	−46	−53	−60	−67	−74	−81	−88	−95	−103
25	8	1	−7	−15	−22	−29	−36	−44	−51	−59	−66	−74	−81	−88	−96	−103	−110
30	6	−2	−10	−18	−25	−33	−41	−49	−56	−64	−71	−79	−86	−93	−101	−109	−116
35	4	−4	−12	−20	−27	−35	−43	−52	−58	−67	−74	−82	−89	−97	−105	−113	−120
40	3	−5	−13	−21	−29	−37	−45	−53	−60	−69	−76	−84	−92	−100	−107	−115	−123
45	2	−6	−14	−22	−30	−38	−46	−54	−62	−70	−78	−85	−93	−102	−109	−117	−125

75. Use the table. Which of the following weather conditions feels colder?

   a. a temperature of 5°F with a 20 mph wind or a temperature of 10°F with a 15 mph wind

   b. a temperature of −25°F with a 10 mph wind or a temperature of −15°F with a 20 mph wind

76. Use the table. Which of the following weather conditions feels warmer?

   a. a temperature of 25°F with a 25 mph wind or a temperature of 10°F with a 10 mph wind

   b. a temperature of −5°F with a 10 mph wind or a temperature of −15°F with a 5 mph wind

Complete.

77. On the number line, the two points that are four units from 0 are _____ and _____ .

78. On the number line, the two points that are six units from 0 are _____ and _____ .

79. On the number line, the two points that are seven units from 4 are _____ and _____ .

80. On the number line, the two points that are five units from −3 are _____ and _____ .

81. If a is a positive number, then $-a$ is a _____ number.

82. If a is a negative number, then $-a$ is a _____ number.

83. An integer that is its own opposite is _____ .

84. True or False. If $a \geq 0$, then $|a| = a$.

85. True or False. If $a \leq 0$, then $|a| = -a$.

86. Give examples of some games that use negative integers in the scoring.

87. In your own words, explain the meaning of the absolute value of a number and the opposite of a number.

88. Give some examples of English words that are used as variables.

89. Write an essay describing your interpretation of the data presented in the table at the right.

The Twenty-First Century: An Easier Ladder to Climb Number of 30- to 34-year-olds competing for jobs			
	1992	**1999**	**2005**
Men	10.3 million	9.2 million	8.5 million
Women	8.3 million	7.9 million	7.8 million
Total	**18.6 million**	**17.1 million**	**16.3 million**

S E C T I O N **1.2**

Operations with Integers

1 Add integers

A number can be represented anywhere along the number line by an arrow. A positive number is represented by an arrow pointing to the right, and a negative number is represented by an arrow pointing to the left. The size of the number is represented by the length of the arrow.

Addition is the process of finding the total of two numbers. The numbers being added are called **addends**. The total is called the **sum**. Addition of integers can be shown on the number line. To add integers, find the point on the number line corresponding to the first addend. At that point, draw an arrow representing the second addend. The sum is the number directly below the tip of the arrow.

$4 + 2 = 6$

$-4 + (-2) = -6$

$-4 + 2 = -2$

$4 + (-2) = 2$

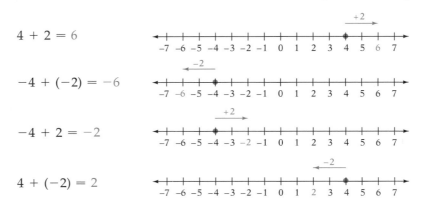

The pattern for addition shown on the number lines on page 10 is summarized in the following rules for adding integers.

Addition of Integers

Integers with the same sign
To add two numbers with the same sign, add the absolute values of the numbers. Then attach the sign of the addends.

$2 + 8 = 10$

$-2 + (-8) = -10$

Integers with different signs
To add two numbers with different signs, find the absolute value of each number. Then subtract the smaller of these absolute values from the larger one. Attach the sign of the number with the larger absolute value.

$-2 + 8 = 6$

$2 + (-8) = -6$

Example 1   Add.   A. $162 + (-247)$

B. $-14 + (-47)$

C. $-4 + (-6) + (-8) + 9$

Solution   A. $162 + (-247) = -85$ ▶ The signs are different. Subtract the absolute values of the numbers $(247 - 162)$. Attach the sign of the number with the larger absolute value.

B. $-14 + (-47) = -61$ ▶ The signs are the same. Add the absolute values of the numbers $(14 + 47)$. Attach the sign of the addends.

C. $-4 + (-6) + (-8) + 9$ ▶ To add more than two numbers, add the first two numbers. Then add the sum to the third number. Continue until all the numbers have been added.
$= -10 + (-8) + 9$
$= -18 + 9$
$= -9$

Problem 1   Add.

A. $-162 + 98$ B. $-154 + (-37)$ C. $-36 + 17 + (-21)$

Solution   See page A11.

2 Subtract integers

Subtraction is the process of finding the difference between two numbers. Subtraction of an integer is defined as addition of the opposite integer.

Subtract $8 - 3$ by using addition of the opposite.

$$\text{Subtraction} \longrightarrow \text{Addition of the Opposite}$$

$$8 \;\; - \;\; (+3) \;\; = \;\; 8 \;\; + \;\; (-3) \;\; = \;\; 5$$

$$\underline{\quad\text{Opposites}\quad}$$

To subtract one integer from another, add the opposite of the second integer to the first number.

first number	−	second number	=	first number	+	the opposite of the second number		

$$40 \;\; - \;\; 60 \;\; = \;\; 40 \;\; + \;\; (-60) \;\; = \;\; -20$$

$$-40 \;\; - \;\; 60 \;\; = \;\; -40 \;\; + \;\; (-60) \;\; = \;\; -100$$

$$-40 \;\; - \;\; (-60) \;\; = \;\; -40 \;\; + \;\; 60 \;\; = \;\; 20$$

$$40 \;\; - \;\; (-60) \;\; = \;\; 40 \;\; + \;\; 60 \;\; = \;\; 100$$

Example 2   Subtract: $-12 - 8$

Solution $-12 - 8 = -12 + (-8)$ ▶ Rewrite subtraction as addition of the opposite.

   $= -20$ ▶ Add.

Problem 2   Subtract: $-8 - 14$

Solution   See page A11.

When subtraction occurs several times in a problem, rewrite each subtraction as addition of the opposite. Then add.

Example 3   Subtract: $-8 - 30 - (-12) - 7 - (-14)$

Solution $-8 - 30 - (-12) - 7 - (-14)$

   $= -8 + (-30) + 12 + (-7) + 14$ ▶ Rewrite each subtraction as addition of the opposite.

   $= -38 + 12 + (-7) + 14$ ▶ Add the first two numbers. Then add the sum to the third number. Continue until all the numbers have been added.
   $= -26 + (-7) + 14$
   $= -33 + 14$
   $= -19$

Problem 3   Subtract: $4 - (-3) - 12 - (-7) - 20$

Solution   See page A11.

3   ## Multiply integers

Multiplication is the process of finding the product of two numbers.

Several different symbols are used to indicate multiplication. The numbers being multiplied are called **factors**; for instance, 3 and 2 are factors in each of the examples at the right. The result is called the **product**. Note that when parentheses are used and there is no arithmetic operation symbol, the operation is multiplication.

$$3 \times 2 = 6$$
$$3 \cdot 2 = 6$$
$$(3)(2) = 6$$
$$3(2) = 6$$
$$(3)2 = 6$$

When 5 is multiplied by a sequence of decreasing integers, each product decreases by 5.

$$(5)(3) = 15$$
$$(5)(2) = 10$$
$$(5)(1) = 5$$
$$(5)(0) = 0$$

The pattern developed can be continued so that 5 is multiplied by a sequence of negative numbers. The resulting products must be negative in order to maintain the pattern of decreasing by 5.

$$(5)(-1) = -5$$
$$(5)(-2) = -10$$
$$(5)(-3) = -15$$
$$(5)(-4) = -20$$

This illustrates that the product of a positive number and a negative number is negative.

When −5 is multiplied by a sequence of decreasing integers, each product increases by 5.

$$(-5)(3) = -15$$
$$(-5)(2) = -10$$
$$(-5)(1) = -5$$
$$(-5)(0) = 0$$

The pattern developed can be continued so that −5 is multiplied by a sequence of negative numbers. The resulting products must be positive in order to maintain the pattern of increasing by 5.

$$(-5)(-1) = 5$$
$$(-5)(-2) = 10$$
$$(-5)(-3) = 15$$
$$(-5)(-4) = 20$$

This illustrates that the product of two negative numbers is positive.

This pattern for multiplication is summarized in the following rules for multiplying integers.

Multiplication of Integers

Integers with the same sign
To multiply two numbers with the same sign, multiply the absolute values of the numbers. The product is positive.

$$4 \cdot 8 = 32$$

$$(-4)(-8) = 32$$

Integers with different signs
To multiply two numbers with different signs, multiply the absolute values of the numbers. The product is negative.

$$-4 \cdot 8 = -32$$

$$(4)(-8) = -32$$

Example 4   Multiply.   A. $-42 \cdot 62$ B. $2(-3)(-5)(-7)$

Solution   A. $-42 \cdot 62$ ▸ The signs are different. The product is
                 $= -2604$ negative.

B. $2(-3)(-5)(-7)$ ▸ To multiply more than two numbers, mul-
   $= -6(-5)(-7)$ tiply the first two numbers. Then multiply
   $= 30(-7)$ the product by the third number. Con-
   $= -210$ tinue until all the numbers have been
                            multiplied.

Problem 4   Multiply.   A. $-38 \cdot 51$ B. $-7(-8)(9)(-2)$

Solution   See page A11.

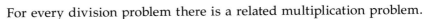

4 ▪ **Divide integers**

For every division problem there is a related multiplication problem.

$$\text{Division: } \frac{8}{2} = 4 \qquad \text{Related multiplication: } 4 \cdot 2 = 8$$

This fact can be used to illustrate the rules for dividing signed numbers.

The quotient of two numbers with the $\dfrac{12}{3} = 4$ because $4 \cdot 3 = 12$.
same sign is positive.

$\dfrac{-12}{-3} = 4$ because $4(-3) = -12$.

The quotient of two numbers with dif- $\dfrac{12}{-3} = -4$ because $-4(-3) = 12$.
ferent signs is negative.

$\dfrac{-12}{3} = -4$ because $-4 \cdot 3 = -12$.

Division of Integers

> **Integers with the same sign**
> To divide two numbers with the same sign, divide the absolute values of
> the numbers. The quotient is positive.
>
> **Integers with different signs**
> To divide two numbers with different signs, divide the absolute values of
> the numbers. The quotient is negative.

Note that $\dfrac{-12}{3} = -4$, $\dfrac{12}{-3} = -4$, and $-\dfrac{12}{3} = -4$. This suggests the following rule.

$$\text{If } a \text{ and } b \text{ are two integers, } b \neq 0, \text{ then } \frac{a}{-b} = \frac{-a}{b} = -\frac{a}{b}.$$

Read $b \neq 0$ as "b is not equal to 0." The reason why the denominator must not be
equal to 0 is explained in the following discussion of zero and one in division.

Zero and One in Division

Zero divided by any number other than zero is zero.	$\dfrac{0}{a} = 0, a \neq 0$	because $0 \cdot a = 0.$
Division by zero is not defined.	$\dfrac{4}{0} = ?$	$? \times 0 = 4$ There is no number whose product with zero is 4.
Any number other than zero divided by itself is 1.	$\dfrac{a}{a} = 1, a \neq 0$	because $1 \cdot a = a.$
Any number divided by 1 is the number.	$\dfrac{a}{1} = a$	because $a \cdot 1 = a.$

Example 5   Divide.   A. $(-120) \div (-8)$ B. $\dfrac{95}{-5}$ C. $-\dfrac{-81}{3}$

Solution   A. $(-120) \div (-8) = 15$

B. $\dfrac{95}{-5} = -19$

C. $-\dfrac{-81}{3} = -(-27) = 27$

Problem 5   Divide.   A. $(-135) \div (-9)$ B. $\dfrac{84}{-6}$ C. $-\dfrac{36}{-12}$

Solution   See page A11.

5 Application problems

To solve an application problem, first read the problem carefully. The **Strategy** involves identifying the quantity to be found and planning the steps needed to find that quantity. The **Solution** involves performing each operation stated in the Strategy and writing the answer.

Positive and negative numbers are used to express the profitability of a company. A profit is recorded as a positive number, a loss as a negative number.

The bar graph at the right shows the annual profit and loss, to the nearest ten million dollars, for a corporation for the years 1993, 1994, and 1995. Determine the total profit or loss for the three years.

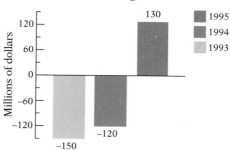

Strategy
To determine the total profit or loss, add the profits and losses for the three years.

Solution
$-150 + (-120) + 130 = -270 + 130 = -140$

The company had a loss of $140 million.

Example 6   The daily low temperatures, in degrees Celsius, during one week were recorded as follows: $-8°, 2°, 0°, -7°, 1°, 6°, -1°$. Find the average daily low temperature for the week.

Strategy   To find the average daily low temperature:
- Add the seven temperature readings.
- Divide by 7.

Solution $-8 + 2 + 0 + (-7) + 1 + 6 + (-1)$
$= -6 + 0 + (-7) + 1 + 6 + (-1)$
$= -6 + (-7) + 1 + 6 + (-1)$
$= -13 + 1 + 6 + (-1)$
$= -12 + 6 + (-1)$
$= -6 + (-1)$
$= -7$

$-7 \div 7 = -1$

The average daily low temperature was $-1°$C.

Problem 6   The daily high temperatures, in degrees Celsius, during one week were recorded as follows: $-5°, -6°, 3°, 0°, -4°, -7°, -2°$. Find the average daily high temperature for the week.

Solution   See page A11.

STUDY TIPS

SURVEY THE CHAPTER

Before you begin reading a chapter, take a few minutes to survey it. Glancing through the chapter will give you an overview of its content and help you see how the pieces fit together as you read.

Begin by reading the chapter title. The title summarizes what the chapter is about. Next read the section headings. The section headings summarize the major topics presented in the chapter. Then read the objectives under each section heading. The objective headings describe the learning goals for that section. Keep these headings in mind as you work through the material. They provide direction as you study.

CONCEPT REVIEW 1.2

Determine whether the statement is always true, sometimes true, or never true.

1. The sum of two integers is larger than either of the integers being added.

2. The sum of two nonzero integers with the same sign is positive.

3. The quotient of two integers with different signs is negative.

4. To find the opposite of a number, multiply the number by -1.

5. To find the opposite of an integer, divide the integer by -1.

6. To subtract $a - b$, add the opposite of a to b.

7. $4(-8) = -4$

8. If x and y are two integers and $y = 0$, then $\dfrac{x}{y} = 0$.

9. If x and y are two different integers, then $x - y = y - x$.

10. If x is an integer and $4x = 0$, then $x = 0$.

EXERCISES 1.2

1 Add.

1. $-3 + (-8)$
2. $-12 + (-1)$

3. $-4 + (-5)$
4. $-12 + (-12)$

5. $6 + (-9)$
6. $4 + (-9)$

7. $-6 + 7$
8. $-12 + 6$

9. $2 + (-3) + (-4)$
10. $7 + (-2) + (-8)$

11. $-3 + (-12) + (-15)$
12. $9 + (-6) + (-16)$

13. $-17 + (-3) + 29$
14. $13 + 62 + (-38)$

15. $-3 + (-8) + 12$
16. $-27 + (-42) + (-18)$

17. $13 + (-22) + 4 + (-5)$
18. $-14 + (-3) + 7 + (-6)$

19. $-22 + 10 + 2 + (-18)$
20. $-6 + (-8) + 13 + (-4)$

21. $-126 + (-247) + (-358) + 339$
22. $-651 + (-239) + 524 + 487$

2   Subtract.

23. $16 - 8$ **24.** $12 - 3$ **25.** $7 - 14$

26. $7 - (-2)$ **27.** $3 - (-4)$ **28.** $-6 - (-3)$

29. $-4 - (-2)$ **30.** $6 - (-12)$ **31.** $-12 - 16$

32. $-4 - 3 - 2$ **33.** $4 - 5 - 12$ **34.** $12 - (-7) - 8$

35. $-12 - (-3) - (-15)$ **36.** $4 - 12 - (-8)$ **37.** $13 - 7 - 15$

38. $-6 + 19 - (-31)$ **39.** $-30 - (-65) - 29 - 4$ **40.** $42 - (-82) - 65 - 7$

41. $-16 - 47 - 63 - 12$ **42.** $42 - (-30) - 65 - (-11)$

43. $-47 - (-67) - 13 - 15$ **44.** $-18 - 49 - (-84) - 27$

3   Multiply.

45. $14 \cdot 3$ **46.** $62 \cdot 9$ **47.** $5(-4)$ **48.** $4(-7)$ **49.** $-8(2)$

50. $-9(3)$ **51.** $(-5)(-5)$ **52.** $(-3)(-6)$ **53.** $(-7)(0)$ **54.** $-32 \cdot 4$

55. $-24 \cdot 3$ **56.** $19(-7)$ **57.** $6(-17)$ **58.** $-8(-26)$ **59.** $-4(-35)$

60. $-5(23)$ **61.** $5 \cdot 7(-2)$ **62.** $8(-6)(-1)$ **63.** $(-9)(-9)(2)$ **64.** $-8(-7)(-4)$

65. $-5(8)(-3)$ **66.** $(-6)(5)(7)$ **67.** $-1(4)(-9)$ **68.** $6(-3)(-2)$ **69.** $4(-4) \cdot 6(-2)$

70. $-5 \cdot 9(-7) \cdot 3$ **71.** $-9(4) \cdot 3(1)$ **72.** $8(8)(-5)(-4)$ **73.** $-6(-5)(12)(0)$ **74.** $7(9) \cdot 10(-1)$

4   Divide.

75. $12 \div (-6)$ **76.** $18 \div (-3)$ **77.** $(-72) \div (-9)$ **78.** $(-64) \div (-8)$

79. $0 \div (-6)$ **80.** $-49 \div 0$ **81.** $45 \div (-5)$ **82.** $-24 \div 4$

83. $-36 \div 4$ **84.** $-56 \div 7$ **85.** $-81 \div (-9)$ **86.** $-40 \div (-5)$

87. $72 \div (-3)$ **88.** $44 \div (-4)$ **89.** $-60 \div 5$ **90.** $144 \div 9$

91. $78 \div (-6)$ **92.** $84 \div (-7)$ **93.** $-72 \div 4$ **94.** $-80 \div 5$

95. $-114 \div (-6)$ **96.** $-128 \div 4$ **97.** $-130 \div (-5)$ **98.** $(-280) \div 8$

99. $-132 \div (-12)$ **100.** $-156 \div (-13)$ **101.** $-182 \div 14$ **102.** $-144 \div 12$

103. $143 \div 11$ **104.** $168 \div 14$

5   Solve.

105. Find the temperature after a rise of 9°C from −6°C.

106. Find the temperature after a rise of 7°C from −18°C.

107. The high temperature for the day was 10°C. The low temperature was −4°C. Find the difference between the high and low temperatures for the day.

108. The low temperature for the day was −2°C. The high temperature was 11°C. Find the difference between the high and low temperatures for the day.

109. The temperature at which mercury boils is 360°C. Mercury freezes at −39°C. Find the difference between the temperature at which mercury boils and the temperature at which it freezes.

110. The temperature at which radon boils is −62°C. Radon freezes at −71°C. Find the difference between the temperature at which radon boils and the temperature at which it freezes.

The elevation, or height, of places on the earth is measured in relation to sea level, or the average level of the ocean's surface. The following table shows height above sea level as a positive number and depth below sea level as a negative number.

Continent	Highest Elevation (in meters)		Lowest Elevation (in meters)	
Africa	Mt. Kilimanjaro	5895	Qattara Depression	−133
Asia	Mt. Everest	8848	Dead Sea	−400
Europe	Mt. Elbrus	5634	Caspian Sea	−28
North America	Mt. McKinley	6194	Death Valley	−86
South America	Mt. Aconcagua	6960	Salinas Grandes	−40

111. Use the table to find the difference in elevation between Mt. Elbrus and the Caspian Sea.

112. Use the table to find the difference in elevation between Mt. Aconcagua and Salinas Grandes.

113. Use the table to find the difference in elevation between Mt. Kilimanjaro and the Qattara Depression.

114. Use the table to find the difference in elevation between Mt. McKinley and Death Valley.

115. Use the table to find the difference in elevation between Mt. Everest and the Dead Sea.

116. The daily low temperatures, in degrees Celsius, during one week were recorded as follows: 4°, −5°, 8°, 0°, −9°, −11°, −8°. Find the average daily low temperature for the week.

117. The daily high temperatures, in degrees Celsius, during one week were recorded as follows: −8°, −9°, 6°, 7°, −2°, −14°, −1°. Find the average daily high temperature for the week.

118. On January 22, 1943, the temperature at Spearfish, South Dakota, rose from −4°F to 45°F in two minutes. How many degrees did the temperature rise during those two minutes?

119. In a 24 h period in January of 1916, the temperature in Browning, Montana, dropped from 44°F to −56°F. How many degrees did the temperature drop during that time?

The bar graph at the right below shows the profit and loss, in millions of dollars, for a major company for the years 1992 through 1995. Use this chart for Exercises 120–123.

120. What was the total profit or loss for the four years shown on the graph?

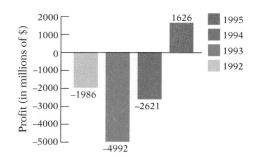

121. What was the difference between the profit or loss in 1995 and that in 1994?

122. What was the difference between the profit or loss in 1992 and that in 1993?

123. What was the difference between the profit or loss in 1995 and that in 1993?

124. To discourage random guessing on a multiple-choice exam, a professor assigns 5 points for a correct answer, −2 points for an incorrect answer, and 0 points for leaving the question blank. What is the score for a student who had 20 correct answers, had 13 incorrect answers, and left 7 questions blank?

125. To discourage random guessing on a multiple-choice exam, a professor assigns 7 points for a correct answer, −3 points for an incorrect answer, and −1 point for leaving the question blank. What is the score for a student who had 17 correct answers, had 8 incorrect answers, and left 2 questions blank?

SUPPLEMENTAL EXERCISES 1.2

Simplify.

126. $|-7 + 12|$

127. $|13 - (-4)|$

128. $|-13 - (-2)|$

129. $|18 - 21|$

Assuming the pattern is continued, find the next three numbers in the pattern.

130. $-7, -11, -15, -19, \ldots$

131. $16, 11, 6, 1, \ldots$

132. $7, -14, 28, -56, \ldots$

133. $256, -64, 16, -4, \ldots$

Solve.

134. 32,844 is divisible by 3. By rearranging the digits, find the largest possible number that is still divisible by 3.

135. 4563 is not divisible by 4. By rearranging the digits, find the largest possible number that is divisible by 4.

136. How many three-digit numbers of the form 8__4 are divisible by 3?

In each exercise, determine which statement is false.

137. **a.** $|3 + 4| = |3| + |4|$ **b.** $|3 - 4| = |3| - |4|$ **c.** $|4 + 3| = |4| + |3|$ **d.** $|4 - 3| = |4| - |3|$

138. **a.** $|5 + 2| = |5| + |2|$ **b.** $|5 - 2| = |5| - |2|$ **c.** $|2 + 5| = |2| + |5|$ **d.** $|2 - 5| = |2| - |5|$

Determine which statement is true for all real numbers.

139. **a.** $|x + y| \leq |x| + |y|$ **b.** $|x + y| = |x| + |y|$ **c.** $|x + y| \geq |x| + |y|$

140. **a.** $||x| - |y|| \leq |x| - |y|$ **b.** $||x| - |y|| = |x| - |y|$ **c.** $||x| - |y|| \geq |x| - |y|$

141. Is the difference between two integers always smaller than either one of the numbers in the difference? If not, give an example for which the difference between two integers is greater than either integer.

142. If $-4x$ equals a positive integer, is x a positive or a negative integer? Explain your answer.

143. Explain why division by zero is not allowed.

S E C T I O N **1.3**

Rational Numbers

1 Write rational numbers as decimals

POINT OF INTEREST

As early as A.D. 630 the Hindu mathematician Brahmagupta wrote a fraction as one number over another separated by a space. The Arab mathematician al Hassar (around A.D. 1050) was the first to show a fraction with the horizontal bar separating the numerator and denominator.

A **rational number** is the quotient of two integers. Therefore, a rational number is a number that can be written in the form $\frac{a}{b}$, where a and b are integers, and b is not zero. A rational number written in this way is commonly called a **fraction**.

$\dfrac{a}{b}$ ←an integer
←a nonzero integer

$\left. \begin{array}{c} \\ \\ \end{array} \right\}$ Rational numbers

$\dfrac{2}{3}, \dfrac{-4}{9}, \dfrac{18}{-5}, \dfrac{4}{1}$

Because an integer can be written as the quotient of the integer and 1, every integer is a rational number.

$5 = \dfrac{5}{1} \qquad -3 = \dfrac{-3}{1}$

A number written in **decimal notation** is also a rational number.

three-tenths $0.3 = \dfrac{3}{10}$

thirty-five hundredths $0.35 = \dfrac{35}{100}$

negative four-tenths $-0.4 = \dfrac{-4}{10}$

A rational number written as a fraction can be written in decimal notation.

LOOK CLOSELY

The fraction bar can be read "divided by."

$\dfrac{5}{8} = 5 \div 8$

Example 1   Write $\dfrac{5}{8}$ as a decimal.

Solution

$$\begin{array}{r} 0.625 \\ 8\overline{)5.000} \\ -4\,8 \\ \hline 20 \\ -16 \\ \hline 40 \\ -40 \\ \hline 0 \end{array}$$ ←This is called a **terminating decimal.**

←The remainder is zero.

$\dfrac{5}{8} = 0.625$

LOOK CLOSELY

Dividing the numerator by the denominator results in a remainder of 0. The decimal 0.625 is a terminating decimal.

Problem 1   Write $\dfrac{4}{25}$ as a decimal.

Solution   See page A12.

Example 2   Write $\frac{4}{11}$ as a decimal.

Solution   0.3636 . . . ← This is called a **repeating decimal**.

$$11)\overline{4.0000}$$
$$\underline{-3\,3}$$
$$70$$
$$\underline{-66}$$
$$40$$
$$\underline{-33}$$
$$70$$
$$\underline{-66}$$
$$4 \leftarrow \text{The remainder is never zero.}$$

$\frac{4}{11} = 0.\overline{36}$ ← The bar over the digits 3 and 6 is used to show that these digits repeat.

Problem 2   Write $\frac{4}{9}$ as a decimal. Place a bar over the repeating digits of the decimal.

Solution   See page A12.

Rational numbers can be written as fractions, such as $-\frac{6}{7}$ or $\frac{8}{3}$, where the numerator and denominator are integers. But every rational number also can be written as a repeating decimal (such as 0.25767676 . . .) or a terminating decimal (such as 1.73). This is illustrated in Examples 1 and 2.

Some numbers cannot be written either as a repeating decimal or as a terminating decimal. These numbers are called **irrational numbers**. For example, 2.45445444544445 . . . is an irrational number. Two other examples are $\sqrt{2}$ and π.

$$\sqrt{2} = 1.414213562 \ldots \qquad \pi = 3.141592654 \ldots$$

The three dots mean that the digits continue on and on without ever repeating or terminating. Although we cannot write a decimal that is exactly equal to $\sqrt{2}$ or to π, we can give approximations of these numbers. The symbol $\approx$ is read "is approximately equal to." Shown below is $\sqrt{2}$ rounded to the nearest thousandth and π rounded to the nearest hundredth.

$$\sqrt{2} \approx 1.414 \qquad \pi \approx 3.14$$

The rational numbers and the irrational numbers taken together are called the **real numbers**.

2   Convert among percents, fractions, and decimals

"A population growth rate of 3%," "a manufacturer's discount of 25%," and "an 8% increase in pay" are typical examples of the many ways in which percent is used in applied problems. **Percent** means "parts of 100." Thus 27% means 27 parts of 100.

In applied problems involving a percent, it is usually necessary either to rewrite the percent as a fraction or a decimal or to rewrite a fraction or a decimal as a percent.

To write 27% as a fraction, remove the percent sign and multiply by $\frac{1}{100}$.

$$27\% = 27\left(\frac{1}{100}\right) = \frac{27}{100}$$

To write a percent as a decimal, remove the percent sign and multiply by 0.01.

To write 33% as a decimal, remove the percent sign and multiply by 0.01.

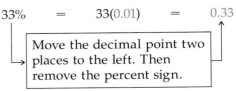

$$33\% = 33(0.01) = 0.33$$

Move the decimal point two places to the left. Then remove the percent sign.

Write 100% as a decimal. $100\% = 100(0.01) = 1$

Example 3   Write 130% as a fraction and as a decimal.

Solution $130\% = 130\left(\frac{1}{100}\right) = \frac{130}{100} = 1\frac{3}{10}$ ▶ To write a percent as a fraction, remove the percent sign and multiply by $\frac{1}{100}$.

$130\% = 130(0.01) = 1.30$ ▶ To write a percent as a decimal, remove the percent sign and multiply by 0.01.

Problem 3   Write 125% as a fraction and as a decimal.

Solution   See page A12.

Example 4   Write $33\frac{1}{3}\%$ as a fraction.

Solution $33\frac{1}{3}\% = 33\frac{1}{3}\left(\frac{1}{100}\right) = \frac{100}{3}\left(\frac{1}{100}\right)$ ▶ Write the mixed number $33\frac{1}{3}$
$= \frac{1}{3}$ as the improper fraction $\frac{100}{3}$.

Problem 4   Write $16\frac{2}{3}\%$ as a fraction.

Solution   See page A12.

Example 5   Write 0.25% as a decimal.

Solution $0.25\% = 0.25(0.01) = 0.0025$ ▶ Remove the percent sign and multiply by 0.01.

Problem 5   Write 6.08% as a decimal.

Solution   See page A12.

A fraction or decimal can be written as a percent by multiplying by 100%. Recall that 100% = 1, and multiplying a number by 1 does not change the number.

To write $\frac{5}{8}$ as a percent, multiply by 100%.

$$\frac{5}{8} = \frac{5}{8}(100\%) = \frac{500}{8}\% = 62.5\% \text{ or } 62\frac{1}{2}\%$$

To write 0.82 as a percent, multiply by 100%.

$$0.82 \quad = \quad 0.82(100\%) \quad = \quad 82\%$$

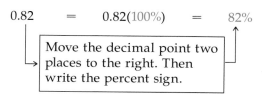

Move the decimal point two places to the right. Then write the percent sign.

Example 6 Write as a percent. A. 0.027 B. 1.34

Solution A. $0.027 = 0.027(100\%) = 2.7\%$ ▶ To write a decimal as a percent, multiply by 100%.

B. $1.34 = 1.34(100\%) = 134\%$

Problem 6 Write as a percent. A. 0.043 B. 2.57

Solution See page A12.

Example 7 Write $\frac{5}{6}$ as a percent. Round to the nearest tenth of a percent.

Solution $\frac{5}{6} = \frac{5}{6}(100\%) = \frac{500}{6}\% \approx 83.3\%$ ▶ To write a fraction as a percent, multiply by 100%.

Problem 7 Write $\frac{5}{9}$ as a percent. Round to the nearest tenth of a percent.

Solution See page A12.

Example 8 Write $\frac{7}{16}$ as a percent. Write the remainder in fractional form.

Solution $\frac{7}{16} = \frac{7}{16}(100\%) = \frac{700}{16}\% = 43\frac{3}{4}\%$ ▶ Multiply the fraction by 100%.

Problem 8 Write $\frac{9}{16}$ as a percent. Write the remainder in fractional form.

Solution See page A12.

3 Add and subtract rational numbers

To add or subtract fractions, first rewrite each fraction as an equivalent fraction with a common denominator, using the least common multiple (LCM) of the denominators as the common denominator. Then add the numerators, and place the sum over the common denominator. Write the answer in simplest form.

The sign rules for adding integers apply to addition of rational numbers.

Example 9   Add: $-\dfrac{5}{6} + \dfrac{3}{10}$

Solution   Prime factorization of 6 and 10:

$6 = 2 \cdot 3 \qquad 10 = 2 \cdot 5$

$LCM = 2 \cdot 3 \cdot 5 = 30$

▶ Find the LCM of the denominators 6 and 10. The LCM of denominators is sometimes called the **least common denominator** (LCD).

$$-\dfrac{5}{6} + \dfrac{3}{10} = -\dfrac{25}{30} + \dfrac{9}{30}$$

$$= \dfrac{-25 + 9}{30}$$

$$= \dfrac{-16}{30}$$

$$= -\dfrac{8}{15}$$

▶ Rewrite the fractions as equivalent fractions, using the LCM of the denominators as the common denominator.

▶ Add the numerators, and place the sum over the common denominator.

▶ Write the answer in simplest form.

Problem 9   Subtract: $\dfrac{5}{9} - \dfrac{11}{12}$

Solution   See page A12.

Example 10   Simplify: $-\dfrac{3}{4} + \dfrac{1}{6} - \dfrac{5}{8}$

Solution $-\dfrac{3}{4} + \dfrac{1}{6} - \dfrac{5}{8} = -\dfrac{18}{24} + \dfrac{4}{24} - \dfrac{15}{24}$

▶ The LCM of 4, 6, and 8 is 24.

$$= \dfrac{-18}{24} + \dfrac{4}{24} + \dfrac{-15}{24}$$

$$= \dfrac{-18 + 4 + (-15)}{24}$$

$$= \dfrac{-29}{24}$$

$$= -\dfrac{29}{24}$$

Problem 10   Simplify: $-\dfrac{7}{8} - \dfrac{5}{6} + \dfrac{1}{2}$

Solution   See page A12.

To add or subtract decimals, write the numbers so that the decimal points are in a vertical line. Then proceed as in the addition or subtraction of integers. Write the decimal point in the answer directly below the decimal points in the problem.

Example 11   Add: 14.02 + 137.6 + 9.852

Solution

$$\begin{array}{r} 14.02 \\ 137.6 \\ +\quad 9.852 \\ \hline 161.472 \end{array}$$

▸ Write the decimals so that the decimal points are in a vertical line.

▸ Write the decimal point in the sum directly below the decimal points in the problem.

Problem 11   Add: 3.097 + 4.9 + 3.09

Solution   See page A12.

Example 12   Add: −114.039 + 84.76

Solution

$$\begin{array}{r} 114.039 \\ -\quad 84.76 \\ \hline 29.279 \end{array}$$

▸ The signs are different. Subtract the absolute value of the number with the smaller absolute value from the absolute value of the number with the larger absolute value.

$$\begin{array}{l} -114.039 + 84.76 \\ = -29.279 \end{array}$$

▸ Attach the sign of the number with the larger absolute value.

Problem 12   Subtract: 16.127 − 67.91

Solution   See page A12.

---

4   ## Multiply and divide rational numbers

The sign rules for multiplying and dividing integers apply to multiplication and division of rational numbers. The product of two fractions is the product of the numerators divided by the product of the denominators.

Example 13   Multiply: $\dfrac{3}{8} \cdot \dfrac{12}{17}$

Solution

$$\dfrac{3}{8} \cdot \dfrac{12}{17} = \dfrac{3 \cdot 12}{8 \cdot 17}$$

▸ Multiply the numerators. Multiply the denominators.

$$= \dfrac{3 \cdot \overset{1}{\cancel{2}} \cdot \overset{1}{\cancel{2}} \cdot 3}{2 \cdot \underset{1}{\cancel{2}} \cdot \underset{1}{\cancel{2}} \cdot 17}$$

▸ Write the prime factorization of each factor. Divide by the common factors.

$$= \dfrac{9}{34}$$

▸ Multiply the numbers remaining in the numerator. Multiply the numbers remaining in the denominator.

Problem 13   Multiply: $-\dfrac{7}{12} \cdot \dfrac{9}{14}$

Solution   See page A12.

To divide fractions, invert the divisor. Then proceed as in the multiplication of fractions.

Example 14 Divide: $\dfrac{3}{10} \div \left(-\dfrac{18}{25}\right)$

Solution $\dfrac{3}{10} \div \left(-\dfrac{18}{25}\right) = -\left(\dfrac{3}{10} \div \dfrac{18}{25}\right)$ ▶ The signs are different. The quotient is negative.

$= -\left(\dfrac{3}{10} \cdot \dfrac{25}{18}\right)$ ▶ Change division to multiplication. Invert the divisor.

$= -\left(\dfrac{3 \cdot 25}{10 \cdot 18}\right)$ ▶ Multiply the numerators. Multiply the denominators.

$= -\left(\dfrac{\overset{1}{\cancel{3}} \cdot \overset{1}{\cancel{5}} \cdot 5}{2 \cdot \underset{1}{\cancel{5}} \cdot 2 \cdot \underset{1}{\cancel{3}} \cdot 3}\right)$

$= -\dfrac{5}{12}$

Problem 14 Divide: $-\dfrac{3}{8} \div \left(-\dfrac{5}{12}\right)$

Solution See page A12.

To multiply decimals, multiply as in the multiplication of whole numbers. Write the decimal point in the product so that the number of decimal places in the product equals the sum of the decimal places in the factors.

Example 15 Multiply: $(-6.89)(0.00035)$

Solution

$$
\begin{array}{rl}
6.89 & \text{2 decimal places} \\
\times \quad 0.00035 & \text{5 decimal places} \\
\hline
3445 & \\
2067 \quad\;\; & \\
\hline
0.0024115 & \text{7 decimal places}
\end{array}
$$
▶ Multiply the absolute values.

$(-6.89)(0.00035) = -0.0024115$ ▶ The signs are different. The product is negative.

Problem 15 Multiply: $(-5.44)(3.8)$

Solution See page A13.

To divide decimals, move the decimal point in the divisor to make it a whole number. Move the decimal point in the dividend the same number of places to the right. Place the decimal point in the quotient directly over the decimal point in the dividend. Then divide as in the division of whole numbers.

Example 16   Divide $1.32 \div 0.27$. Round to the nearest tenth.

Solution $0.27{\overline{)1.32}}$

▶ Move the decimal point two places to the right in the divisor and in the dividend. Place the decimal point in the quotient.

$$
\begin{array}{r}
4.88 \approx 4.9 \\
27{\overline{)132.00}} \\
-108 \\
\hline
24\,0 \\
-21\,6 \\
\hline
240 \\
-216 \\
\hline
24
\end{array}
$$

▶ The symbol $\approx$ is used to indicate that the quotient is an approximate value that has been rounded off.

Problem 16   Divide $-0.394 \div 1.7$. Round to the nearest hundredth.

Solution   See page A13.

STUDY TIPS

USE THE TEXTBOOK TO LEARN THE MATERIAL

For each objective studied, read very carefully all of the material from the objective statement to the examples provided for that objective. As you read, note carefully the formulas and words printed in **boldface** type. It is important for you to know these formulas and the definitions of these words.

You will note that each example is paired with a problem. The example is worked out for you; the problem is left for you to do. After studying the example, do the problem. Immediately look up the answer to this problem in the Solutions section at the back of the text. The page number on which the solution appears is printed on the solution line below the problem statement. If your answer is correct, continue. If your answer is incorrect, check your solution against the one given in the Solutions section. It may be helpful to review the worked-out example also. Determine where you made your mistakes.

Next, do the problems in the exercise set that correspond to the objective just studied. The answers to all the odd-numbered exercises appear in the Answers section in the back of the book. Check your answers to the exercises against these.

If you have difficulty solving problems in the exercise set, review the material in the text. Many examples are solved within the text material. Review the solutions to these problems. Reread the examples provided for the objective. If, after checking these sources and trying to find your mistakes, you are still unable to solve a problem correctly, make a note of the exercise number so that you can ask someone for help with that problem.

CONCEPT REVIEW 1.3

Determine whether the statement is always true, sometimes true, or never true.

1. To multiply two fractions, you must first rewrite the fractions as equivalent fractions with a common denominator.

2. The number $\frac{\pi}{3}$ is an example of a rational number.

3. A rational number can be written as a terminating decimal.

4. The number 2.585585558... is an example of a repeating decimal.

5. An irrational number is a real number.

6. 37%, 0.37, and $\frac{37}{100}$ are three numbers that have the same value.

7. To write a decimal as a percent, multiply the decimal by $\frac{1}{100}$.

8. If $a, b, c,$ and d are natural numbers, then $\frac{a}{b} + \frac{c}{d} = \frac{a+c}{b+d}$.

9. -12 is an example of a number that is both an integer and a rational number.

10. $\frac{7}{9}$ is an example of a number that is both a rational number and an irrational number.

EXERCISES 1.3

1   Write as a decimal. Place a bar over the repeating digits of a repeating decimal.

1. $\frac{1}{3}$ 2. $\frac{2}{3}$ 3. $\frac{1}{4}$ 4. $\frac{3}{4}$

5. $\frac{2}{5}$ 6. $\frac{4}{5}$ 7. $\frac{1}{6}$ 8. $\frac{5}{6}$

9. $\frac{1}{8}$ 10. $\frac{7}{8}$ 11. $\frac{2}{9}$ 12. $\frac{8}{9}$

13. $\frac{5}{11}$ 14. $\frac{10}{11}$ 15. $\frac{7}{12}$ 16. $\frac{11}{12}$

17. $\frac{4}{15}$ 18. $\frac{8}{15}$ 19. $\frac{9}{16}$ 20. $\frac{15}{16}$

21. $\frac{6}{25}$ 22. $\frac{14}{25}$ 23. $\frac{9}{40}$ 24. $\frac{21}{40}$

25. $\frac{15}{22}$ 26. $\frac{19}{22}$ 27. $\frac{11}{24}$ 28. $\frac{19}{24}$

29. $\frac{5}{33}$ 30. $\frac{25}{33}$ 31. $\frac{3}{37}$ 32. $\frac{14}{37}$

2   Write as a fraction and as a decimal.

33. 75% **34.** 40% **35.** 50% **36.** 10%

37. 64% **38.** 88% **39.** 175% **40.** 160%

41. 19% **42.** 87% **43.** 5% **44.** 2%

45. 450% **46.** 380% **47.** 8% **48.** 4%

Write as a fraction.

49. $11\frac{1}{9}\%$ **50.** $37\frac{1}{2}\%$ **51.** $31\frac{1}{4}\%$ **52.** $66\frac{2}{3}\%$

53. $\frac{1}{2}\%$ **54.** $5\frac{3}{4}\%$ **55.** $6\frac{1}{4}\%$ **56.** $83\frac{1}{3}\%$

Write as a decimal.

57. 7.3% **58.** 9.1% **59.** 15.8% **60.** 0.3%

61. 9.15% **62.** 121.2% **63.** 18.23% **64.** 0.15%

Write as a percent.

65. 0.15 **66.** 0.37 **67.** 0.05 **68.** 0.02

69. 0.175 **70.** 0.125 **71.** 1.15 **72.** 2.142

73. 0.008 **74.** 0.004 **75.** 0.065 **76.** 0.083

Write as a percent. Round to the nearest tenth of a percent.

77. $\frac{27}{50}$ **78.** $\frac{83}{100}$ **79.** $\frac{1}{3}$ **80.** $\frac{3}{8}$

81. $\frac{4}{9}$ **82.** $\frac{9}{20}$ **83.** $2\frac{1}{2}$ **84.** $1\frac{2}{7}$

Write as a percent. Write the remainder in fractional form.

85. $\frac{3}{8}$ **86.** $\frac{7}{16}$ **87.** $\frac{5}{14}$ **88.** $\frac{4}{7}$

89. $1\frac{1}{4}$ **90.** $2\frac{5}{8}$ **91.** $1\frac{5}{9}$ **92.** $1\frac{13}{16}$

3   Simplify.

93. $\dfrac{2}{3} + \dfrac{5}{12}$

94. $\dfrac{1}{2} + \dfrac{3}{8}$

95. $\dfrac{5}{8} - \dfrac{5}{6}$

96. $\dfrac{1}{9} - \dfrac{5}{27}$

97. $-\dfrac{5}{12} - \dfrac{3}{8}$

98. $-\dfrac{5}{6} - \dfrac{5}{9}$

99. $-\dfrac{6}{13} + \dfrac{17}{26}$

100. $-\dfrac{7}{12} + \dfrac{5}{8}$

101. $-\dfrac{5}{8} - \left(-\dfrac{11}{12}\right)$

102. $\dfrac{1}{3} + \dfrac{5}{6} - \dfrac{2}{9}$

103. $\dfrac{1}{2} - \dfrac{2}{3} + \dfrac{1}{6}$

104. $-\dfrac{3}{8} - \dfrac{5}{12} - \dfrac{3}{16}$

105. $-\dfrac{5}{16} + \dfrac{3}{4} - \dfrac{7}{8}$

106. $\dfrac{1}{2} - \dfrac{3}{8} - \left(-\dfrac{1}{4}\right)$

107. $\dfrac{3}{4} - \left(-\dfrac{7}{12}\right) - \dfrac{7}{8}$

108. $\dfrac{1}{3} - \dfrac{1}{4} - \dfrac{1}{5}$

109. $\dfrac{2}{3} - \dfrac{1}{2} + \dfrac{5}{6}$

110. $\dfrac{5}{16} + \dfrac{1}{8} - \dfrac{1}{2}$

111. $\dfrac{5}{8} - \left(-\dfrac{5}{12}\right) + \dfrac{1}{3}$

112. $\dfrac{1}{8} - \dfrac{11}{12} + \dfrac{1}{2}$

113. $-\dfrac{7}{9} + \dfrac{14}{15} + \dfrac{8}{21}$

114. $1.09 + 6.2$

115. $-32.1 - 6.7$

116. $5.13 - 8.179$

117. $-13.092 + 6.9$

118. $2.54 - 3.6$

119. $5.43 + 7.925$

120. $-16.92 - 6.925$

121. $-3.87 + 8.546$

122. $6.9027 - 17.692$

123. $2.09 - 6.72 - 5.4$

124. $16.4 + 3.09 - 7.93$

125. $-18.39 + 4.9 - 23.7$

126. $19 - (-3.72) - 82.75$

127. $-3.07 - (-2.97) - 17.4$

128. $-3.09 - 4.6 - 27.3$

129. $317.09 - 46.902 + 583.0714$

130. $71.0235 - 86.0974 + 254.309$

4   Simplify.

131. $\dfrac{1}{2}\left(-\dfrac{3}{4}\right)$

132. $-\dfrac{2}{9}\left(-\dfrac{3}{14}\right)$

133. $\left(-\dfrac{3}{8}\right)\left(-\dfrac{4}{15}\right)$

134. $\dfrac{5}{8}\left(-\dfrac{7}{12}\right)\dfrac{16}{25}$

135. $\left(\dfrac{1}{2}\right)\left(-\dfrac{3}{4}\right)\left(-\dfrac{5}{8}\right)$

136. $\left(\dfrac{5}{12}\right)\left(-\dfrac{8}{15}\right)\left(-\dfrac{1}{3}\right)$

137. $\dfrac{3}{8} \div \dfrac{1}{4}$

138. $\dfrac{5}{6} \div \left(-\dfrac{3}{4}\right)$

139. $-\dfrac{5}{12} \div \dfrac{15}{32}$

140. $\dfrac{1}{8} \div \left(-\dfrac{5}{12}\right)$

141. $-\dfrac{4}{9} \div \left(-\dfrac{2}{3}\right)$

142. $-\dfrac{6}{11} \div \dfrac{4}{9}$

143. $(1.2)(3.47)$

144. $(-0.8)(6.2)$

145. $(-1.89)(-2.3)$

146. $(6.9)(-4.2)$

147. $(1.06)(-3.8)$

148. $(-2.7)(-3.5)$

149. $(1.2)(-0.5)(3.7)$

150. $(-2.4)(6.1)(0.9)$

151. $(-0.8)(3.006)(-5.1)$

Simplify. Round to the nearest hundredth.

152. $-24.7 \div 0.09$

153. $-1.27 \div (-1.7)$

154. $9.07 \div (-3.5)$

155. $-354.2086 \div 0.1719$

156. $-2658.3109 \div (-0.0473)$

157. $(-3.92) \div (-45.008)$

SUPPLEMENTAL EXERCISES 1.3

Classify each of the following numbers as a natural number, an integer, a positive integer, a negative integer, a rational number, an irrational number, and a real number. List all that apply.

158. -1

159. 28

160. $-\dfrac{9}{34}$

161. -7.707

162. $5.2\overline{6}$

163. $0.171771777\ldots$

Solve.

164. Find the average of $\dfrac{5}{8}$ and $\dfrac{3}{4}$.

165. Postage for first-class mail is $.32 for the first ounce or fraction of an ounce and $.23 for each additional ounce or fraction of an ounce. Find the cost of mailing a $4\dfrac{1}{2}$ oz letter by first-class mail.

166. The graph at the right shows the profit and loss of Hartmarx for the years 1990 through 1994. What was the difference between the profit for 1994 and the loss for 1990?

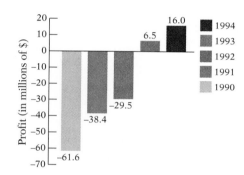

167. Use a calculator to determine the decimal representations of $\dfrac{17}{99}, \dfrac{45}{99}$, and $\dfrac{73}{99}$. Make a conjecture as to the decimal representation of $\dfrac{83}{99}$. Does your conjecture work for $\dfrac{33}{99}$? What about $\dfrac{1}{99}$?

168. If the same positive integer is added to both the numerator and the denominator of $\dfrac{2}{5}$, is the new fraction less than, equal to, or greater than $\dfrac{2}{5}$?

169. A magic square is one in which the numbers in every row, column, and diagonal sum to the same number. Complete the magic square at the right.

$\dfrac{2}{3}$		
	$\dfrac{1}{6}$	$\dfrac{5}{6}$
		$-\dfrac{1}{3}$

170. Let x represent the price of a car. If the sales tax is 6% of the price, express the total of the price of the car and the sales tax in terms of x.

171. Let x represent the price of a suit. If the suit is on sale at a discount rate of 30%, express the price of the suit after the discount in terms of x.

172. The price of a pen was 60¢. During a storewide sale, the price was reduced to a different whole number of cents, and the entire stock was sold for $54.59. Find the price of the pen during the sale.

173. Find three natural numbers $a, b,$ and c such that $\dfrac{1}{a} + \dfrac{1}{b} + \dfrac{1}{c}$ is a natural number.

174. If a and b are rational numbers and $a < b$, is it always possible to find a rational number c such that $a < c < b$? If not, explain why. If so, show how to find one.

175. In your own words, define **a.** a rational number, **b.** an irrational number, and **c.** a real number.

176. Explain why you "invert and multiply" when dividing a fraction by a fraction.

177. Explain why you need a common denominator when adding two fractions and why you don't need a common denominator when multiplying two fractions.

178. Write an essay describing your interpretation of the data presented in the chart shown below.

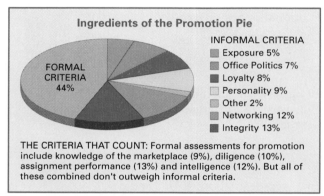

Ingredients of the Promotion Pie

FORMAL CRITERIA 44%

INFORMAL CRITERIA
- Exposure 5%
- Office Politics 7%
- Loyalty 8%
- Personality 9%
- Other 2%
- Networking 12%
- Integrity 13%

THE CRITERIA THAT COUNT: Formal assessments for promotion include knowledge of the marketplace (9%), diligence (10%), assignment performance (13%) and intelligence (12%). But all of these combined don't outweigh informal criteria.

First appeared in WORKING WOMAN in July 1994. Source: STROOK & STROOK & LAVAN. Reprinted with the permission of WORKING WOMAN Magazine. Copyright © 1994 by WORKING WOMAN, Inc.

SECTION 1.4

Exponents and the Order of Operations Agreement

1 Exponential expressions

POINT OF INTEREST

René Descartes (1596–1650) was the first mathematician to use exponential notation extensively as it is used today. However, for some unknown reason, he always used xx for x^2.

Repeated multiplication of the same factor can be written using an exponent.

$$2 \cdot 2 \cdot 2 \cdot 2 \cdot 2 = 2^5 \longleftarrow \text{exponent}$$
$$\uparrow \text{base}$$

$$a \cdot a \cdot a \cdot a = a^4 \longleftarrow \text{exponent}$$
$$\uparrow \text{base}$$

The **exponent** indicates how many times the factor, called the **base**, occurs in the multiplication. The multiplication $2 \cdot 2 \cdot 2 \cdot 2 \cdot 2$ is in **factored form**. The exponential expression 2^5 is in **exponential form**.

2^1 is read "the first power of two" or just "two." ⟶ Usually the exponent 1 is not written.

2^2 is read "the second power of two" or "two squared."

2^3 is read "the third power of two" or "two cubed."

2^4 is read "the fourth power of two."

2^5 is read "the fifth power of two."

a^5 is read "the fifth power of a."

There is a geometric interpretation of the first three natural-number powers.

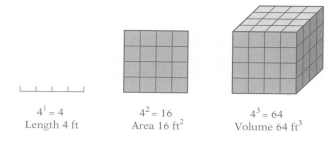

$4^1 = 4$
Length 4 ft

$4^2 = 16$
Area 16 ft^2

$4^3 = 64$
Volume 64 ft^3

To evaluate an exponential expression, write each factor as many times as indicated by the exponent. Then multiply.

$$3^5 = 3 \cdot 3 \cdot 3 \cdot 3 \cdot 3 = 243$$

$$2^3 \cdot 3^2 = (2 \cdot 2 \cdot 2) \cdot (3 \cdot 3) = 8 \cdot 9 = 72$$

LOOK CLOSELY

The -4 is squared only when the negative sign is *inside* the parentheses. In $(-4)^2$, we are squaring -4; in -4^2, we are finding the opposite of 4^2.

Example 1   Evaluate $(-4)^2$ and -4^2.

Solution $(-4)^2 = (-4)(-4) = 16$

$-4^2 = -(4 \cdot 4) = -16$

Problem 1   Evaluate $(-5)^3$ and -5^3.

Solution   See page A13.

LOOK CLOSELY

The product of an even number of negative factors is positive. The product of an odd number of negative factors is negative.

Example 2   Evaluate $(-2)^4$ and $(-2)^5$.

Solution
$(-2)^4 = (-2)(-2)(-2)(-2)$
$= 4(-2)(-2)$
$= -8(-2)$
$= 16$

$(-2)^5 = (-2)(-2)(-2)(-2)(-2)$
$= 4(-2)(-2)(-2)$
$= -8(-2)(-2)$
$= 16(-2)$
$= -32$

Problem 2   Evaluate $(-3)^3$ and $(-3)^4$.

Solution   See page A13.

Example 3   Evaluate $(-3)^2 \cdot 2^3$ and $\left(-\dfrac{2}{3}\right)^3$.

Solution $(-3)^2 \cdot 2^3 = (-3)(-3) \cdot (2)(2)(2) = 9 \cdot 8 = 72$

$\left(-\dfrac{2}{3}\right)^3 = \left(-\dfrac{2}{3}\right)\left(-\dfrac{2}{3}\right)\left(-\dfrac{2}{3}\right) = -\dfrac{2 \cdot 2 \cdot 2}{3 \cdot 3 \cdot 3} = -\dfrac{8}{27}$

Problem 3   Evaluate $(3^3)(-2)^3$ and $\left(-\dfrac{2}{5}\right)^2$.

Solution   See page A13.

2   The Order of Operations Agreement

Evaluate $2 + 3 \cdot 5$.

There are two arithmetic operations, addition and multiplication, in this problem. The operations could be performed in different orders.

Add first.	$2 + 3 \cdot 5$	Multiply first.	$2 + 3 \cdot 5$
Then multiply.	$5 \quad \cdot 5$	Then add.	$2 + \quad 15$
	25		17

In order to prevent there being more than one answer to the same problem, an Order of Operations Agreement has been established.

The Order of Operations Agreement

Step 1   Perform operations inside grouping symbols. Grouping symbols include parentheses ( ), brackets [ ], absolute value symbols | |, and the fraction bar.

Step 2   Simplify exponential expressions.

Step 3   Do multiplication and division as they occur from left to right.

Step 4   Do addition and subtraction as they occur from left to right.

Example 4   Simplify: $12 - 24(8 - 5) \div 2^2$

Solution $12 - 24(8 - 5) \div 2^2$
$= 12 - 24(3) \div 2^2$ ▶ Perform operations inside grouping symbols.
$= 12 - 24(3) \div 4$ ▶ Simplify exponential expressions.
$= 12 - 72 \div 4$ ▶ Do multiplication and division as they occur from left to right.
$= 12 - 18$
$= -6$ ▶ Do addition and subtraction as they occur from left to right.

Problem 4   Simplify: $36 \div (8 - 5)^2 - (-3)^2 \cdot 2$

Solution   See page A13.

One or more of the steps shown in Example 4 may not be needed to simplify an expression. In that case, proceed to the next step in the Order of Operations Agreement.

Example 5   Simplify: $\frac{4 + 8}{2 + 1} - |3 - 1| + 2$

Solution $\frac{4 + 8}{2 + 1} - |3 - 1| + 2$

$= \frac{12}{3} - |2| + 2$ ▶ Perform operations inside grouping symbols (above and below the fraction bar and inside the absolute value symbols).

$= \frac{12}{3} - 2 + 2$ ▶ Find the absolute value of 2.

$= 4 - 2 + 2$ ▶ Do multiplication and division as they occur from left to right.

$= 2 + 2$ ▶ Do addition and subtraction as they occur
$= 4$ from left to right.

Problem 5   Simplify: $27 \div 3^2 + (-3)^2 \cdot 4$

Solution   See page A13.

When an expression has grouping symbols inside grouping symbols, first perform the operations inside the *inner* grouping symbols by following Steps 2, 3, and 4 of the Order of Operations Agreement. Then perform the operations inside the *outer* grouping symbols by following Steps 2, 3, and 4 in sequence.

Example 6   Simplify: $6 \div [4 - (6 - 8)] + 2^2$

Solution $6 \div [4 - (6 - 8)] + 2^2$

$= 6 \div [4 - (-2)] + 2^2$ ▶ Perform operations inside inner grouping symbols.

$= 6 \div 6 + 2^2$ ▶ Perform operations inside outer grouping symbols.

$= 6 \div 6 + 4$ ▶ Simplify exponential expressions.
$= 1 + 4$ ▶ Do multiplication and division.
$= 5$ ▶ Do addition and subtraction.

Problem 6   Simplify: $4 - 3[4 - 2(6 - 3)] \div 2$

Solution   See page A13.

Using your calculator to simplify numerical expressions sometimes requires use of the $\boxed{+/-}$ key or, on some calculators, the negative key, which is frequently shown as $\boxed{(-)}$. These keys change the sign of the number currently in the display. To enter -4:

• For those calculators with $\boxed{+/-}$, press **4** and then $\boxed{+/-}$.

• For those calculators with $\boxed{(-)}$, press $\boxed{(-)}$ and then **4**.

Here are the keystrokes for evaluating the expression $3(-4) - (-5)$.

Calculators with $\boxed{+/-}$ key: $3\ \boxed{\times}\ 4\ \boxed{+/-}\ \boxed{-}\ 5\ \boxed{+/-}\ \boxed{=}$

Calculators with $\boxed{(-)}$ key: $3\ \boxed{\times}\ \boxed{(-)}\ 4\ \boxed{-}\ \boxed{(-)}\ 5\ \boxed{=}$

This example illustrates that calculators make the distinction between a negative sign and a minus sign.

STUDY TIPS

USE THE END-OF-CHAPTER MATERIAL

To help you review the material presented within a chapter, a Chapter Summary appears at the end of each chapter. In the Chapter Summary, definitions of the important terms and concepts introduced in the chapter are provided under "Key Words." Listed under "Essential Rules and Procedures" are the formulas and procedures presented in the chapter. After completing a chapter, be sure to read the Chapter Summary. Use it to check your understanding of the material presented and to determine what concepts you need to review.

Each chapter ends with Chapter Review Exercises and a Chapter Test. The problems these contain summarize what you should have learned when you have finished the chapter. Do these exercises as you prepare for an examination. Check your answers against those in the back of the book. Answers to all Chapter Review and Chapter Test exercises are provided there. The objective being reviewed by any particular problem is written in parentheses following the answer. For any problem you answer incorrectly, review the material corresponding to that objective in the textbook. Determine *why* your answer was wrong.

CONCEPT REVIEW 1.4

Determine whether the statement is always true, sometimes true, or never true.

1. $(-5)^2$, -5^2, and $-(5)^2$ all represent the same number.

2. By the Order of Operations Agreement, addition is performed before division.

3. In the expression 3^8, 8 is the base.

4. The expression 9^4 is in exponential form.

5. To evaluate the expression $6 + 7 \cdot 10$ means to determine what one number it is equal to.

6. When using the Order of Operations Agreement, we consider the absolute value symbol a grouping symbol.

7. Using the Order of Operations Agreement, it is possible to get more than one correct answer to a problem.

8. The Order of Operations Agreement is used for natural numbers, integers, rational numbers, and real numbers.

9. The expression $(2^3)^2$ equals $2^{(3^2)}$.

10. If a and b are positive integers, then $a^b = b^a$.

EXERCISES 1.4

1 Evaluate.

1. 6^2
2. 7^4
3. -7^2
4. -4^3
5. $(-3)^2$
6. $(-2)^3$
7. $(-3)^4$
8. $(-5)^3$
9. $\left(\frac{1}{2}\right)^2$
10. $\left(-\frac{3}{4}\right)^3$
11. $(0.3)^2$
12. $(1.5)^3$
13. $\left(\frac{2}{3}\right)^2 \cdot 3^3$
14. $\left(-\frac{1}{2}\right)^3 \cdot 8$
15. $(0.3)^3 \cdot 2^3$
16. $(0.5)^2 \cdot 3^3$
17. $(-3) \cdot 2^2$
18. $(-5) \cdot 3^4$
19. $(-2) \cdot (-2)^3$
20. $(-2) \cdot (-2)^2$
21. $2^3 \cdot 3^3 \cdot (-4)$
22. $(-3)^3 \cdot 5^2 \cdot 10$
23. $(-7) \cdot 4^2 \cdot 3^2$
24. $(-2) \cdot 2^3 \cdot (-3)^2$
25. $\left(\frac{2}{3}\right)^2 \cdot \frac{1}{4} \cdot 3^3$
26. $\left(\frac{3}{4}\right)^2 \cdot (-4) \cdot 2^3$
27. $8^2 \cdot (-3)^5 \cdot 5$

2 Simplify by using the Order of Operations Agreement.

28. $4 - 8 \div 2$
29. $2^2 \cdot 3 - 3$
30. $2(3 - 4) - (-3)^2$
31. $16 - 32 \div 2^3$
32. $24 - 18 \div 3 + 2$
33. $8 - (-3)^2 - (-2)$
34. $16 + 15 \div (-5) - 2$
35. $14 - 2^2 - |4 - 7|$
36. $3 - 2[8 - (3 - 2)]$
37. $-2^2 + 4[16 \div (3 - 5)]$

38. $6 + \dfrac{16 - 4}{2^2 + 2} - 2$

39. $24 \div \dfrac{3^2}{8 - 5} - (-5)$

40. $96 \div 2[12 + (6 - 2)] - 3^3$

41. $4 \cdot [16 - (7 - 1)] \div 10$

42. $16 \div 2 - 4^2 - (-3)^2$

43. $18 \div |9 - 2^3| + (-3)$

44. $16 - 3(8 - 3)^2 \div 5$

45. $4(-8) \div [2(7 - 3)^2]$

46. $\dfrac{(-10) + (-2)}{6^2 - 30} \div |2 - 4|$

47. $16 - 4 \cdot \dfrac{3^3 - 7}{2^3 + 2} - (-2)^2$

48. $(0.2)^2 \cdot (-0.5) + 1.72$

49. $0.3(1.7 - 4.8) + (1.2)^2$

50. $(1.8)^2 - 2.52 \div (1.8)$

51. $(1.65 - 1.05)^2 \div 0.4 + 0.8$

52. $\dfrac{3}{8} \div \left|\dfrac{5}{6} + \dfrac{2}{3}\right|$

53. $\left(\dfrac{3}{4}\right)^2 - \left(\dfrac{1}{2}\right)^3 \div \dfrac{3}{5}$

SUPPLEMENTAL EXERCISES 1.4

54. Using the Order of Operations Agreement, describe how to simplify Exercise 44.

Place the correct symbol, $<$ or $>$, between the two numbers.

55. $(0.9)^3$ 1^5

56. $(-3)^3$ $(-2)^5$

57. $(-1.1)^2$ $(0.9)^2$

Simplify.

58. $1^2 + 2^2 + 3^2 + 4^2$

59. $1^3 + 2^3 + 3^3 + 4^3$

60. $(-1)^3 + (-2)^3 + (-3)^3 + (-4)^3$

61. $(-2)^2 + (-4)^2 + (-6)^2 + (-8)^2$

Solve.

62. Find a rational number, r, that satisfies the condition.
   a. $r^2 < r$ **b.** $r^2 = r$ **c.** $r^2 > r$

63. A computer can do 600,000 additions in 1 s. To the nearest second, how many seconds will it take the computer to do 10^7 additions?

64. The sum of two natural numbers is 41. Each of the two numbers is the square of a natural number. Find the two numbers.

Determine the ones digit when the expression is evaluated.

65. 34^{202}

66. 23^{502}

67. 27^{622}

68. Prepare a report on the Kelvin scale. Include in your report an explanation of how to convert between the Kelvin scale and the Celsius scale.

Project in Mathematics

Moving Averages

In many courses, your course grade depends on the average of all your test scores. You compute the average by calculating the sum of all your test scores and then dividing that result by the number of tests. Statisticians call this average an **arithmetic mean.** Besides its application to finding the average of your test scores, the arithmetic mean is used in many other situations. For instance, in Objective 1.2.5, Example 6 and Problem 6 on page 16, you saw it applied to finding the average daily low temperature and the average daily high temperature.

Stock market analysts calculate the **moving average** of a stock's price changes. This is the arithmetic mean of the changes in the value of a stock for a given number of days. To illustrate this procedure, we will calculate the five-day moving average of a stock's price changes. In actual practice, a stock market analyst may use 15 days, 30 days, or some other number of days.

The following table shows the amount of increase or decrease, in cents, from the previous closing price of a stock for a 10-day period.

Day 1	Day 2	Day 3	Day 4	Day 5	Day 6	Day 7	Day 8	Day 9	Day 10
+50	−175	+225	0	−275	−75	−50	+50	−475	−50

To calculate the five-day moving average of this stock's price changes, determine the average of the stock's price changes for days 1 through 5, days 2 through 6, days 3 through 7, and so on.

Days 1–5	Days 2–6	Days 3–7	Days 4–8	Days 5–9	Days 6–10
+50	−175	+225	0	−275	−75
−175	+225	0	−275	−75	−50
+225	0	−275	−75	−50	+50
0	−275	−75	−50	+50	−475
−275	−75	−50	+50	−475	−50
Sum = −175	Sum = −300	Sum = −175	Sum = −350	Sum = −825	Sum = −600

$$\text{Ave} = \frac{-175}{5} = -35 \quad \text{Ave} = \frac{-300}{5} = -60 \quad \text{Ave} = \frac{-175}{5} = -35 \quad \text{Ave} = \frac{-350}{5} = -70 \quad \text{Ave} = \frac{-825}{5} = -165 \quad \text{Ave} = \frac{-600}{5} = -120$$

The five-day moving average is the list of means: −35, −60, −35, −70, −165, and −120. If the list tends to increase, the price of the stock is showing an upward trend; if it decreases, the price of the stock is showing a downward trend. These trends help an analyst recommend stocks.

Calculate the five-day moving average for at least three different stocks. Discuss and compare the results for the different stocks.

For this project, you will need to use stock tables that are printed in the business section of major newspapers. Your college library should have copies of these publications. In a stock table, the column headed "Chg." provides the change in the price of a share of the stock; that is, it gives the difference between the closing price for the day shown and the closing price for the previous day. The symbol + indicates that the change was an increase in price; the symbol − indicates that the change was a decrease in price.

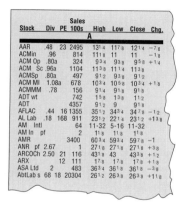

Stock	Div	PE	Sales 100s	High	Low	Close	Chg.
A							
AAR	.48	23	2495	13 1/4	11 7/8	12 1/4	−7/8
ACMin	.96		814	11 1/8	11	11	−1/8
ACM Op	.80a		324	9 3/4	9 3/8	9 5/8	+1/4
ACM Sc	.96a		1104	11 3/8	11 1/4	11 3/8	
ACMSp	.80a		497	9 1/2	9 3/8	9 1/2	
ACM MI	1.08a		678	10 3/4	10 5/8	10 3/4	+1/8
ACMMM	.78		156	9 1/4	9 1/8	9 1/8	
ADT wt			742	15 5/8	13 5/8	11 1/2	
ADT			4357	9 1/2	9	9 1/8	
AFLAC	.44	16	1355	35 1/2	34 3/4	34 7/8	−1/2
AL Lab	.18	168	911	23 1/2	22 1/4	23 1/2	+13/8
AM Intl			64	11-32	5-16	11-32	
AM In pf			2	1 1/8	1 1/8	1 1/8	
AMR			3400	60 3/4	59 3/4	59 7/8	−1
ANR pf	2.67		1	27 1/8	27 1/8	27 1/8	+3/8
ARCOCh	2.50	21	116	43 1/8	43	43 3/8	+1/2
ARX		12	111	17 7/8	17 7/8	17 7/8	+1/8
ASA Ltd	2		483	36 3/4	36 1/8	36 1/8	−3/8
AbtLab s	68	18	20304	26 1/2	26 3/8	26 3/8	+11/8

Focus on Problem Solving

Inductive Reasoning

Suppose you save $200 each month. The total amount you have saved at the end of each month can be described in a list of numbers.

$$200, 400, 600, 800, 1000, 1200, 1400, 1600, \ldots$$

The list of numbers that indicates your total savings is an ordered list of numbers called a **sequence**. Each of the numbers in a sequence is called a **term** of the sequence. The list is ordered because the position of a number in the list indicates the month in which that total amount has been saved. For example, the 7th term of the sequence is 1400 because a total of $1400 has been saved by the 7th month.

On page 21 of this text, the direction line for Exercises 130–133 reads, "Assuming the pattern is continued, find the next three numbers in the pattern." Exercise 130 is

$$-7, -11, -15, -19, \ldots$$

This list of numbers is a sequence. The first step in solving this problem is to observe the pattern in the list of numbers. In this case, each number in the list is 4 less than the previous number.

This process of discovering the pattern in the list of numbers is inductive reasoning. **Inductive reasoning** involves making generalizations from specific examples; in other words, we reach a conclusion by making observations about particular facts or cases.

On page 33, Exercise 167 asks that you determine the decimal representations of $\frac{17}{99}, \frac{45}{99}$, and $\frac{73}{99}$ and then make a conjecture as to the decimal representation of $\frac{83}{99}$. This is another example of using inductive reasoning to solve a problem.

Try each of the following exercises. Each exercise requires inductive reasoning.

Name the next two terms in the sequence.

1. 1, 3, 5, 7, 1, 3, 5, 7, 1, ...

2. 1, 4, 2, 5, 3, 6, 4, ...

3. 1, 2, 4, 7, 11, 16, ...

4. A, B, E, F, I, J, ...

Draw the next shape in the sequence.

5.

6. |• ||• ||•• |||•• |||•••

Solve.

7. Convert $\frac{1}{11}$, $\frac{2}{11}$, $\frac{3}{11}$, $\frac{4}{11}$, and $\frac{5}{11}$ to decimals. Then use the pattern you observe to convert $\frac{6}{11}$, $\frac{7}{11}$, and $\frac{9}{11}$ to decimals.

8. Convert $\frac{1}{33}$, $\frac{2}{33}$, $\frac{4}{33}$, $\frac{5}{33}$, and $\frac{7}{33}$ to decimals. Then use the pattern you observe to convert $\frac{8}{33}$, $\frac{13}{33}$, and $\frac{19}{33}$ to decimals.

Chapter Summary

Key Words

A *set* is a collection of objects. The objects in the set are called the *elements* of the set. The *roster method* of writing sets encloses a list of the elements in braces.   (Objective 1.1.1)

The set of even natural numbers less than 10 is written $A = \{2, 4, 6, 8\}$. The elements in this set are the numbers 2, 4, 6, and 8.

The set of *natural numbers* is {1, 2, 3, 4, 5, 6, 7, ...}. The set of *integers* is {... −4, −3, −2, −1, 0, 1, 2, 3, 4, ...}. (Objective 1.1.1)

A number *a is less than* another number *b*, written $a < b$, if *a* is to the left of *b* on the number line. A number *a is greater than* another number *b*, written $a > b$, if *a* is to the right of *b* on the number line. The symbol $\leq$ means *is less than or equal to.* The symbol $\geq$ means *is greater than or equal to.* (Objective 1.1.1)

$-9 < 7$

$4 > -5$

Two numbers that are the same distance from zero on the number line but on opposite sides of zero are *opposite numbers,* or *opposites.* The opposite of a number is also called its *additive inverse.*   (Objective 1.1.2)

5 and −5 are opposites.

The *absolute value* of a number is its distance from zero on the number line. The absolute value of a number is positive or zero.   (Objective 1.1.2)

$|-12| = 12$

$-|7| = -7$

A *rational number* is a number that can be written in the form $\frac{a}{b}$, where *a* and *b* are integers and $b \neq 0$. A rational number written in this form is commonly called a *fraction.* (Objective 1.3.1)

$\frac{3}{8}$, $\frac{-9}{11}$, and $\frac{4}{1}$ are rational numbers.

An *irrational number* is a number that has a decimal representation that never terminates or repeats.   (Objective 1.3.1)

π, $\sqrt{2}$, and 2.17117111711117... are irrational numbers.

The rational numbers and the irrational numbers taken together are called the *real numbers*.   (Objective 1.3.1)

$\dfrac{3}{8}, \dfrac{-9}{11}, \dfrac{4}{1}\pi,$ $\sqrt{2},$ and $2.17117111711117\ldots$ are real numbers.

Percent means "parts of 100."   (Objective 1.3.2)

63% means 63 of 100 equal parts.

An expression of the form a^n is in *exponential form*, where a is the *base* and n is the *exponent*.   (Objective 1.4.1)

6^3 is an exponential expression in which 6 is the base and 3 is the exponent.

Essential Rules and Procedures

To add two integers with the same sign, add the absolute values of the numbers. Then attach the sign of the addends. (Objective 1.2.1)

$9 + 3 = 12$
$-9 + (-3) = -12$

To add two integers with different signs, find the absolute value of each number. Then subtract the smaller of these absolute values from the larger one. Attach the sign of the number with the greater absolute value.   (Objective 1.2.1)

$9 + (-3) = 6$
$-9 + 3 = -6$

To subtract one integer from another, add the opposite of the second integer to the first integer.   (Objective 1.2.2)

$6 - 11 = 6 + (-11) = -5$
$-2 - (-8) = -2 + 8 = 6$

To multiply two integers with the same sign, multiply the absolute values of the numbers. The product is positive. (Objective 1.2.3)

$3(5) = 15$
$(-3)(-5) = 15$

To multiply two integers with different signs, multiply the absolute values of the numbers. The product is negative. (Objective 1.2.3)

$-3(5) = -15$
$3(-5) = -15$

To divide two integers with the same sign, divide the absolute values of the numbers. The quotient is positive. (Objective 1.2.4)

$14 \div 2 = 7$
$-14 \div (-2) = 7$

To divide two integers with different signs, divide the absolute values of the numbers. The quotient is negative. (Objective 1.2.4)

$14 \div (-2) = -7$
$-14 \div 2 = -7$

Zero and One in Division   (Objective 1.2.4)

$\dfrac{0}{a} = 0, a \neq 0$

$\dfrac{0}{3} = 0$

Division by zero is undefined.

$\dfrac{9}{0}$ is undefined.

$\dfrac{a}{a} = 1, a \neq 0$

$\dfrac{-2}{-2} = 1$

$\dfrac{a}{1} = a$

$\dfrac{-6}{1} = -6$

To convert a percent to a decimal, remove the percent sign and multiply by 0.01.   (Objective 1.3.2)

$85\% = 85(0.01) = 0.85$

To convert a percent to a fraction, remove the percent sign and multiply by $\frac{1}{100}$.   (Objective 1.3.2)

$$25\% = 25 \cdot \frac{1}{100} = \frac{25}{100} = \frac{1}{4}$$

To convert a decimal or a fraction to a percent, multiply by 100%.   (Objective 1.3.2)

$$0.5 = 0.5(100\%) = 50\%$$

$$\frac{3}{8} = \frac{3}{8} \cdot 100\% = \frac{300}{8}\% = 37.5\%$$

Order of Operations Agreement   (Objective 1.4.2)

Step 1   Perform operations inside grouping symbols.
Step 2   Simplify exponential expressions.
Step 3   Do multiplication and division as they occur from left to right.
Step 4   Do addition and subtraction as they occur from left to right.

$$16 \div (-2)^3 + (7 - 12)$$
$$= 16 \div (-2)^3 + (-5)$$
$$= 16 \div (-8) + (-5)$$
$$= -2 + (-5)$$

$$= -7$$

Chapter Review Exercises

1. Use the roster method to write the set of natural numbers less than 7.

2. Write $\frac{5}{8}$ as a percent.

3. Evaluate: $-|-4|$

4. Subtract: $16 - (-30) - 42$

5. Divide: $-561 \div (-33)$

6. Write $\frac{7}{9}$ as a decimal. Place a bar over the repeating digits of the decimal.

7. Simplify: $(6.02)(-0.89)$

8. Simplify: $\frac{-10 + 2}{2 + (-4)} \div 2 + 6$

9. Find the opposite of -4.

10. Subtract: $16 - 30$

11. Write 0.672 as a percent.

12. Write $79\frac{1}{2}\%$ as a fraction.

13. Divide: $-72 \div 8$

14. Write $\frac{17}{20}$ as a decimal.

15. Divide: $\frac{5}{12} \div \left(-\frac{5}{6}\right)$

16. Simplify: $3^2 - 4 + 20 \div 5$

17. Place the correct symbol, $<$ or $>$, between the two numbers.
$$-1 \quad 0$$

18. Add: $-22 + 14 + (-8)$

19. Multiply: $(-5)(-6)(3)$

20. Subtract: $6.039 - 12.92$

21. Given $A = \{-5, -3, 0\}$, which elements of set A are less than or equal to -3?

22. Write 7% as a decimal.

23. Evaluate: $\frac{3}{4} \cdot (4)^2$

24. Place the correct symbol, $<$ or $>$, between the two numbers.
$$-2 \quad -40$$

25. Add: $13 + (-16)$

26. Multiply: $(-4)(12)$

27. Add: $-\frac{2}{5} + \frac{7}{15}$

28. Evaluate: $(-3^3) \cdot 2^2$

29. Write $2\frac{7}{9}$ as a percent. Round to the nearest tenth of a percent.

30. Write 240% as a decimal.

31. Find the opposite of -2.

32. Subtract: $7 - 21$

33. Divide: $96 \div (-12)$

34. Write $\frac{7}{20}$ as a decimal.

35. Divide: $-\frac{7}{16} \div \frac{3}{8}$

36. Simplify: $2^3 \div 4 - 2(2 - 7)$

37. Evaluate: $|-3|$

38. Subtract: $12 - (-10) - 4$

39. Write $2\frac{8}{9}$ as a percent. Write the remainder in fractional form.

40. Given $C = \{-12, -8, -1, 7\}$, find
   a. the opposite of each element of set C.
   b. the absolute value of each element of set C.

41. Divide: $(-204) \div (-17)$

42. Write $\frac{7}{11}$ as a decimal. Place a bar over the repeating digits of the decimal.

43. Divide: $0.2654 \div (-0.023)$
   Round to the nearest tenth.

44. Simplify: $(7 - 2)^2 - 5 - 3 \cdot 4$

45. Place the correct symbol, $<$ or $>$, between the two numbers.
$$8 \quad -10$$

46. Add: $-12 + 8 + (-4)$

47. Multiply: $2(-3)(-12)$

48. Add: $-\frac{5}{8} + \frac{1}{6}$

49. Given $D = \{-24, -17, -9, 0, 4\}$, which elements of set D are greater than -19?

50. Write 0.002 as a percent.

51. Evaluate: $-4^2 \cdot \left(\frac{1}{2}\right)^2$

52. Add: $-1.329 + 4.89$

53. Evaluate: $-|17|$

54. Subtract: $-5 - 22 - (-13) - 19 - (-6)$

55. Multiply: $\left(\frac{1}{3}\right)\left(-\frac{4}{5}\right)\left(\frac{3}{8}\right)$

56. Place the correct symbol, $<$ or $>$, between the two numbers.
$$-43 \quad -34$$

57. Write $\frac{18}{25}$ as a decimal.

58. Evaluate $(-2)^3 \cdot 4^2$.

59. Write 0.075 as a percent.

60. Write $\frac{19}{35}$ as a percent. Write the remainder in fractional form.

61. Add: $14 + (-18) + 6 + (-20)$

62. Multiply: $-4(-8)(12)(0)$

63. Simplify: $2^3 - 7 + 16 \div (-3 + 5)$

64. Simplify: $\frac{3}{4} + \frac{1}{2} - \frac{3}{8}$

65. Divide: $-128 \div (-8)$

66. Place the correct symbol, $<$ or $>$, between the two numbers.
$$-57 \quad 28$$

67. Evaluate: $\left(-\frac{1}{3}\right)^3 \cdot 9^2$

68. Add: $-7 + (-3) + (-12) + 16$

69. Multiply: $5(-2)(10)(-3)$

70. Use the roster method to write the set of negative integers greater than -4.

71. Find the temperature after a rise of 14°C from −6°C.

72. The daily low temperatures, in degrees Celsius, for a three-day period were recorded as follows: −8°, 7°, −5°. Find the average low temperature for the three-day period.

73. The high temperature for the day was 8°C. The low temperature was −5°C. Find the difference between the high and low temperatures for the day.

74. Find the temperature after a rise of 7°C from −13°C.

75. The temperature on the surface of the planet Venus is 480°C. The temperature on the surface of the planet Pluto is −234°C. Find the difference between the surface temperatures on Venus and Pluto.

Chapter Test

1. Write 55% as a fraction.

2. Given $B = \{-8, -6, -4, -2\}$, which elements of set B are less than -5?

3. Subtract: $-9 - (-6)$

4. Write $\frac{17}{20}$ as a decimal.

5. Multiply: $\frac{3}{4}\left(-\frac{2}{21}\right)$

6. Divide: $-75 \div 5$

7. Evaluate: $\left(-\frac{2}{3}\right)^3 \cdot 3^2$

8. Add: $-7 + (-3) + 12$

9. Use the roster method to write the set of positive integers less than or equal to 6.

10. Write 1.59 as a percent.

11. Evaluate: $|-29|$

12. Place the correct symbol, $<$ or $>$, between the two numbers.
$$-47 \quad -68$$

13. Subtract: $-\dfrac{4}{9} - \dfrac{5}{6}$

14. Multiply: $-6(-43)$

15. Simplify: $8 + \dfrac{12-4}{3^2-1} - 6$

16. Divide: $-\dfrac{5}{8} \div \left(-\dfrac{3}{4}\right)$

17. Write $\dfrac{3}{13}$ as a percent. Round to the nearest tenth of a percent.

18. Write 6.2% as a decimal.

19. Subtract: $13 - (-5) - 4$

20. Write $\dfrac{13}{30}$ as a decimal. Place a bar over the repeating digits of the decimal.

21. Multiply: $(-0.9)(2.7)$

22. Divide: $-180 \div (-12)$

23. Evaluate: $2^2 \cdot (-4)^2 \cdot 10$

24. Add: $15 + (-8) + (-19)$

25. Evaluate: $-|-34|$

26. Place the correct symbol, $<$ or $>$, between the two numbers.
$$53 \quad -92$$

27. Given $A = \{-17, -6, 5, 9\}$, find
   a. the opposite of each element of set A.
   b. the absolute value of each element of set A.

28. Write $\dfrac{16}{23}$ as a percent. Write the remainder in fractional form.

29. Add: $-18.354 + 6.97$

30. Multiply: $-4(8)(-5)$

31. Simplify: $9(-4) \div [2(8-5)^2]$

32. Find the temperature after a rise of 12°C from −8°C.

33. The daily high temperatures, in degrees Celsius, for a four-day period were recorded as follows: −8°, −6°, 3°, −5°. Find the average high temperature for the four-day period.

2

Variable Expressions

Objectives

History of Variables

Before the 16th century, unknown quantities were represented by words. In Latin, the language in which most scholarly works were written, the word *res*, meaning "thing," was used. In Germany the word *zahl*, meaning "number," was used. In Italy the word *cosa*, also meaning "thing," was used.

Then in 1637, René Descartes, a French mathematician, began using the letters *x*, *y*, and *z* to represent variables. It is interesting to note, upon examining Descartes's work, that toward the end of the book the letters *y* and *z* were no longer used and *x* became the choice for a variable.

One explanation of why the letters *y* and *z* appeared less frequently has to do with the nature of printing presses during Descartes's time. A printer had a large tray that contained all the letters of the alphabet. There were many copies of each letter, especially those letters used frequently. For example, there were more *e*'s than *q*'s. Because the letters *y* and *z* do not occur frequently in French, a printer would have few of these letters on hand. Consequently, when Descartes started using these letters as variables, it quickly depleted the printer's supply and *x*'s had to be used instead.

Today, most nations use *x* as the standard letter for a single unknown. In fact, x-rays were so named because the scientists who discovered them did not know what they were and thus labeled them the "unknown rays," or x-rays.

S E C T I O N **2.1**

Evaluating Variable Expressions

1 Evaluate variable expressions

Often we discuss a quantity without knowing its exact value, such as the price of gold next month, the cost of a new automobile next year, or the tuition for next semester. In algebra, a letter of the alphabet is used to stand for a quantity that is unknown or one that can change, or *vary*. The letter is called a **variable**. An expression that contains one or more variables is a **variable expression.**

A variable expression is shown at the right. The expression can be rewritten by writing subtraction as the addition of the opposite.

$$3x^2 - 5y + 2xy - x - 7$$

$$3x^2 + (-5y) + 2xy + (-x) + (-7)$$

Note that the expression has five addends. The **terms** of a variable expression are the addends of the expression. The expression has five terms.

5 terms

$$\underbrace{3x^2 \quad - \quad 5y \quad + \quad 2xy \quad - \quad x}_{\text{Variable terms}} \quad \underbrace{- \quad 7}_{\substack{\text{Constant} \\ \text{term}}}$$

The terms $3x^2$, $-5y$, $2xy$, and $-x$ are **variable terms.**

The term -7 is a **constant term,** or simply a **constant**.

Each variable term is composed of a **numerical coefficient** and a **variable part** (the variable or variables and their exponents).

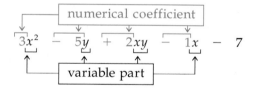

When the numerical coefficient is 1 or -1, the 1 is usually not written ($x = 1x$ and $-x = -1x$).

Example 1 Name the variable terms of the expression $2a^2 - 5a + 7$.

Solution $2a^2$, $-5a$

Problem 1 Name the constant term of the expression $6n^2 + 3n - 4$.

Solution See page A13.

Variable expressions occur naturally in science. In a physics lab, a student may discover that a weight of 1 pound will stretch a spring $\frac{1}{2}$ inch. A weight of 2 pounds will stretch the spring 1 inch. By experimenting, the student can discover that the distance the spring will stretch is found by multiplying the weight by $\frac{1}{2}$. By letting W represent the weight attached to the spring, the distance the spring stretches can be represented by the variable expression $\frac{1}{2}W$.

With a weight of W pounds, the spring will stretch $\frac{1}{2} \cdot W = \frac{1}{2}W$ inches.

With a weight of 10 pounds, the spring will stretch $\frac{1}{2} \cdot 10 = 5$ inches.

With a weight of 3 pounds, the spring will stretch $\frac{1}{2} \cdot 3 = 1\frac{1}{2}$ inches.

Replacing the variable or variables in a variable expression with numbers and then simplifying the resulting numerical expression is called **evaluating the variable expression.**

Example 2   Evaluate $ab - b^2$ when $a = 2$ and $b = -3$.

Solution $ab - b^2$

$2(-3) - (-3)^2$ ▶ Replace each variable in the expression with the number it represents.

$= 2(-3) - 9$ ▶ Use the Order of Operations Agreement to simplify the resulting numerical expression.
$= -6 - 9$
$= -15$

Problem 2   Evaluate $2xy + y^2$ when $x = -4$ and $y = 2$.

Solution   See page A13.

Example 3   Evaluate $\dfrac{a^2 - b^2}{a - b}$ when $a = 3$ and $b = -4$.

Solution $\dfrac{a^2 - b^2}{a - b}$

$\dfrac{(3)^2 - (-4)^2}{3 - (-4)}$ ▶ Replace each variable in the expression with the number it represents.

$= \dfrac{9 - 16}{3 - (-4)}$ ▶ Use the Order of Operations Agreement to simplify the resulting numerical expression.

$= \dfrac{-7}{7}$

$= -1$

Problem 3   Evaluate $\dfrac{a^2 + b^2}{a + b}$ when $a = 5$ and $b = -3$.

Solution   See page A14.

Example 4 Evaluate $x^2 - 3(x - y) - z^2$ when $x = 2$, $y = -1$, and $z = 3$.

Solution $x^2 - 3(x - y) - z^2$

$(2)^2 - 3[2 - (-1)] - (3)^2$ ▶ Replace each variable in the expression with the number it represents.

$= (2)^2 - 3(3) - (3)^2$ ▶ Use the Order of Operations Agreement to simplify the resulting numerical expression.
$= 4 - 3(3) - 9$
$= 4 - 9 - 9$
$= -5 - 9$
$= -14$

Problem 4 Evaluate $x^3 - 2(x + y) + z^2$ when $x = 2$, $y = -4$, and $z = -3$.

Solution See page A14.

Example 5 The diameter of the base of a right circular cylinder is 5 cm. The height of the cylinder is 8.5 cm. Find the volume of the cylinder. Round to the nearest tenth.

8.5 cm

5 cm

Solution $V = \pi r^2 h$ ▶ Use the formula for the volume of a right circular cylinder.

$V = \pi (2.5)^2 (8.5)$ ▶ $r = \dfrac{1}{2}d = \dfrac{1}{2}(5) = 2.5$

$V = \pi (6.25)(8.5)$ ▶ Use the π key on your calculator to enter the value of π.
$V \approx 166.9$

The volume is approximately 166.9 cm^3.

Problem 5 The diameter of the base of a right circular cone is 9 cm. The height of the cone is 9.5 cm. Find the volume of the cone. Round to the nearest tenth.

9.5 cm

9 cm

Solution See page A14.

A graphing calculator can be used to evaluate variable expressions. When the value of each variable is stored in the calculator's memory and a variable expression is then entered on the calculator, the calculator evaluates that variable expression for the values of the variables stored in its memory. See the Appendix on page A1 for a description of keystroking procedures for different models of graphing calculators.

CONCEPT REVIEW 2.1

Determine whether the statement is always true, sometimes true, or never true.

1. The expression $3x^2$ is a variable expression.

2. In the expression $8y^3 - 4y$, the terms are $8y^3$ and $4y$.

3. For the expression x^5, the value of x is 1.

4. For the expression $6a + 7b$, 7 is a constant term.

5. The Order of Operations Agreement is used in evaluating a variable expression.

6. The result of evaluating a variable expression is a single number.

EXERCISES 2.1

1   Name the terms of the variable expression. Then underline the constant term.

1. $2x^2 + 5x - 8$ 2. $-3n^2 - 4n + 7$ 3. $6 - a^4$

Name the variable terms of the expression. Then underline the variable part of each term.

4. $9b^2 - 4ab + a^2$ 5. $7x^2y + 6xy^2 + 10$ 6. $5 - 8n - 3n^2$

Name the coefficients of the variable terms.

7. $x^2 - 9x + 2$ 8. $12a^2 - 8ab - b^2$ 9. $n^3 - 4n^2 - n + 9$

Evaluate the variable expression when $a = 2$, $b = 3$, and $c = -4$.

10. $3a + 2b$ **11.** $a - 2c$ **12.** $-a^2$

13. $2c^2$ **14.** $-3a + 4b$ **15.** $3b - 3c$

16. $b^2 - 3$ **17.** $-3c + 4$ **18.** $16 \div (2c)$

19. $6b \div (-a)$ **20.** $bc \div (2a)$ **21.** $-2ab \div c$

22. $a^2 - b^2$ **23.** $b^2 - c^2$ **24.** $(a + b)^2$

25. $a^2 + b^2$ **26.** $2a - (c + a)^2$ **27.** $(b - a)^2 + 4c$

28. $b^2 - \dfrac{ac}{8}$ **29.** $\dfrac{5ab}{6} - 3cb$ **30.** $(b - 2a)^2 + bc$

Evaluate the variable expression when $a = -2$, $b = 4$, $c = -1$, and $d = 3$.

31. $\dfrac{b + c}{d}$ **32.** $\dfrac{d - b}{c}$ **33.** $\dfrac{2d + b}{-a}$

34. $\dfrac{b + 2d}{b}$ **35.** $\dfrac{b - d}{c - a}$ **36.** $\dfrac{2c - d}{-ad}$

37. $(b + d)^2 - 4a$ **38.** $(d - a)^2 - 3c$ **39.** $(d - a)^2 \div 5$

40. $(b - c)^2 \div 5$ **41.** $b^2 - 2b + 4$ **42.** $a^2 - 5a - 6$

43. $\dfrac{bd}{a} \div c$ **44.** $\dfrac{2ac}{b} \div (-c)$ **45.** $2(b + c) - 2a$

46. $3(b - a) - bc$ **47.** $\dfrac{b - 2a}{bc^2 - d}$ **48.** $\dfrac{b^2 - a}{ad + 3c}$

49. $\dfrac{1}{3}d^2 - \dfrac{3}{8}b^2$ **50.** $\dfrac{5}{8}a^4 - c^2$ **51.** $\dfrac{-4bc}{2a - b}$

52. $\dfrac{abc}{b - d}$ **53.** $a^3 - 3a^2 + a$ **54.** $d^3 - 3d - 9$

55. $-\dfrac{3}{4}b + \dfrac{1}{2}(ac + bd)$ **56.** $-\dfrac{2}{3}d - \dfrac{1}{5}(bd - ac)$ **57.** $(b - a)^2 - (d - c)^2$

58. $(b + c)^2 + (a + d)^2$ **59.** $4ac + (2a)^2$ **60.** $3dc - (4c)^2$

Evaluate the variable expression when $a = 2.7$, $b = -1.6$, and $c = -0.8$.

61. $c^2 - ab$ **62.** $(a + b)^2 - c$ **63.** $\dfrac{b^3}{c} - 4a$

Solve. Round to the nearest tenth.

64. Find the volume of a sphere that has a radius of 8.5 cm. See Figure 1.

65. Find the volume of a right circular cylinder that has a radius of 1.25 in. and a height of 5.25 in.

66. The radius of the base of a right circular cylinder is 3.75 ft. The height of the cylinder is 9.5 ft. Find the surface area of the cylinder.

67. The length of one base of a trapezoid is 17.5 cm, and the length of the other base is 10.25 cm. The height is 6.75 cm. What is the area of the trapezoid? See Figure 2.

8.5 cm

Figure 1

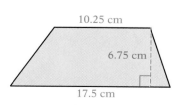

10.25 cm

6.75 cm

17.5 cm

Figure 2

68. A right circular cone has a height of 2.75 in. The diameter of the base is 1 in. Find the volume of the cone.

69. A right circular cylinder has a height of 12.6 m. The diameter of the base is 7 m. Find the volume of the cylinder. See Figure 3.

12.6 m

7 m

Figure 3

SUPPLEMENTAL EXERCISES 2.1

Evaluate the variable expression when $a = -2$ and $b = -3$.

70. $|2a + 3b|$ **71.** $|-4ab|$ **72.** $|5a - b|$

Evaluate the variable expression when $a = \dfrac{2}{3}$ and $b = -\dfrac{3}{2}$.

73. $\dfrac{1}{3} a^5 b^6$ **74.** $\dfrac{(2ab)^3}{2a^3 b^3}$ **75.** $|5ab - 8a^2 b^2|$

Evaluate the following expressions for $x = 2$, $y = 3$, and $z = -2$.

76. $3^x - x^3$ **77.** $2^y - y^2$ **78.** z^y **79.** z^x **80.** $x^x - y^y$ **81.** $y^{(x^2)}$

82. For each of the following, determine the first natural number x, greater than 2, for which the second expression is larger than the first. On the basis of your answers, make a conjecture that appears to be true about the expressions x^n and n^x, where $n = 3, 4, 5, 6, 7, \ldots$ and x is a natural number greater than 2.
    a. $x^3, 3^x$ **b.** $x^4, 4^x$ **c.** $x^5, 5^x$ **d.** $x^6, 6^x$

83. Explain in your own words the meaning of "evaluate an algebraic expression."

S E C T I O N 2.2

Simplifying Variable Expressions

1 The Properties of the Real Numbers

The Properties of the Real Numbers describe the way operations on numbers can be performed. Here are some of the Properties of the Real Numbers and an example of each.

The Commutative Property of Addition

If a and b are real numbers, then $a + b = b + a$.

Two terms can be added in either order; the sum is the same.

$$4 + 3 = 7 \quad \text{and} \quad 3 + 4 = 7$$

The Commutative Property of Multiplication

If a and b are real numbers, then $a \cdot b = b \cdot a$.

Two factors can be multiplied in either order; the product is the same.

$$(5)(-2) = -10 \quad \text{and} \quad (-2)(5) = -10$$

The Associative Property of Addition

If a, b, and c are real numbers, then $(a + b) + c = a + (b + c)$.

When three or more terms are added, the terms can be grouped (with parentheses, for example) in any order; the sum is the same.

$$2 + (3 + 4) = 2 + 7 = 9 \quad \text{and}$$
$$(2 + 3) + 4 = 5 + 4 = 9$$

The Associative Property of Multiplication

If a, b, and c are real numbers, then $(a \cdot b) \cdot c = a \cdot (b \cdot c)$.

When three or more factors are multiplied, the factors can be grouped in any order; the product is the same.

$$(2 \cdot 3) \cdot 4 = 6 \cdot 4 = 24 \quad \text{and}$$
$$2 \cdot (3 \cdot 4) = 2 \cdot 12 = 24$$

The Addition Property of Zero

If a is a real number, then $a + 0 = 0 + a = a$.

The sum of a term and zero is the term.

$$4 + 0 = 4 \qquad 0 + 4 = 4$$

The Multiplication Property of Zero

If a is a real number, then $a \cdot 0 = 0 \cdot a = 0$.

The product of a term and zero is zero.

$$(5)(0) = 0 \qquad (0)(5) = 0$$

The Multiplication Property of One

If a is a real number, then $a \cdot 1 = 1 \cdot a = a$.

The product of a term and 1 is the term.

$$6 \cdot 1 = 6 \qquad 1 \cdot 6 = 6$$

The Inverse Property of Addition

If a is a real number, then $a + (-a) = (-a) + a = 0$.

The sum of a number and its opposite is zero.
The opposite of a number is called its **additive inverse.**

$$8 + (-8) = 0 \qquad (-8) + 8 = 0$$

The Inverse Property of Multiplication

If a is a real number and $a \neq 0$, then $a \cdot \frac{1}{a} = \frac{1}{a} \cdot a = 1$.

The product of a number and its reciprocal is 1.
$\frac{1}{a}$ is the **reciprocal** of a. $\frac{1}{a}$ is also called the **multiplicative inverse** of a.

$$7 \cdot \frac{1}{7} = 1 \qquad \frac{1}{7} \cdot 7 = 1$$

The Distributive Property

If a, b, and c are real numbers, then $a(b + c) = ab + ac$ or $(b + c)a = ba + ca$.

By the Distributive Property, the term outside the parentheses is multiplied by each term inside the parentheses.

$$2(3 + 4) = 2 \cdot 3 + 2 \cdot 4 \qquad (4 + 5)2 = 4 \cdot 2 + 5 \cdot 2$$
$$2 \cdot 7 = 6 + 8 \qquad 9 \cdot 2 = 8 + 10$$
$$14 = 14 \qquad 18 = 18$$

Example 1 Complete the statement by using the Commutative Property of Multiplication.

$(6)(5) = (?)(6)$

Solution $(6)(5) = (5)(6)$ ▶ The Commutative Property of Multiplication states that $a \cdot b = b \cdot a$.

Problem 1   Complete the statement by using the Inverse Property of Addition.

$7 + ? = 0$

Solution   See page A14.

Example 2   Identify the property that justifies the statement.

$2(8 + 5) = 16 + 10$

Solution   The Distributive Property   ▶ The Distributive Property states that $a(b + c) = ab + ac$.

Problem 2   Identify the property that justifies the statement.

$5 + (13 + 7) = (5 + 13) + 7$

Solution   See page A14.

2   Simplify variable expressions using the Properties of Addition

Like terms of a variable expression are the terms with the same variable part. (Because $x^2 = x \cdot x$, x^2 and x are not like terms.)

like terms

$\overbrace{}^{} $ like terms

$3x \;+\; 4 \;-\; 7x \;+\; 9 \;-\; x^2$

like terms

Constant terms are like terms. 4 and 9 are like terms.

To **combine** like terms, use the Distributive Property $ba + ca = (b + c)a$ to add the coefficients.

$2x + 3x = (2 + 3)x$
$= 5x$

Example 3   Simplify.   A. $-2y + 3y$ B. $5x - 11x$

Solution   A. $-2y + 3y$
$= (-2 + 3)y$ ▶ Use the Distributive Property $ba + ca = (b + c)a$.
$= 1y$ ▶ Add the coefficients.
$= y$ ▶ Use the Multiplication Property of One.

B. $5x - 11x$
$= [5 + (-11)]x$ ▶ Use the Distributive Property $ba + ca = (b + c)a$.
$= -6x$ ▶ Add the coefficients.

Problem 3   Simplify.   A. $9x + 6x$ B. $-4y - 7y$

Solution   See page A14.

In simplifying more complicated expressions, use the Properties of Addition.

The Commutative Property of Addition can be used when adding two like terms. The terms can be added in either order. The sum is the same.

$$2x + (-4x) = -4x + 2x$$
$$[2 + (-4)]x = (-4 + 2)x$$
$$-2x = -2x$$

The Associative Property of Addition is used when adding three or more terms. The terms can be grouped in any order. The sum is the same.

$$3x + 5x + 9x = (3x + 5x) + 9x = 3x + (5x + 9x)$$
$$8x + 9x = 3x + 14x$$
$$17x = 17x$$

By the Addition Property of Zero, the sum of a term and zero is the term.

$$5x + 0 = 0 + 5x = 5x$$

By the Inverse Property of Addition, the sum of a term and its additive inverse is zero.

$$7x + (-7x) = -7x + 7x = 0$$

Example 4   Simplify.   A. $8x + 3y - 8x$ B. $4x^2 + 5x - 6x^2 - 2x$

Solution   A. $8x + 3y - 8x$
    $= 3y + 8x - 8x$ ▶ Use the Commutative Property of Addition to rearrange the terms.

    $= 3y + (8x - 8x)$ ▶ Use the Associative Property of Addition to group like terms.

    $= 3y + 0$ ▶ Use the Inverse Property of Addition.

    $= 3y$ ▶ Use the Addition Property of Zero.

    B. $4x^2 + 5x - 6x^2 - 2x$
    $= 4x^2 - 6x^2 + 5x - 2x$ ▶ Use the Commutative Property of Addition to rearrange the terms.

    $= (4x^2 - 6x^2) + (5x - 2x)$ ▶ Use the Associative Property of Addition to group like terms.

    $= -2x^2 + 3x$ ▶ Combine like terms.

Problem 4   Simplify.   A. $3a - 2b + 5a$ B. $x^2 - 7 + 9x^2 - 14$

Solution   See page A14.

3 Simplify variable expressions using the Properties of Multiplication

The Properties of Multiplication are used in simplifying variable expressions.

The Associative Property is used when multiplying three or more factors. $2(3x) = (2 \cdot 3)x = 6x$

The Commutative Property can be used to change the order in which factors are multiplied.

$(3x) \cdot 2 = 2 \cdot (3x) = 6x$

By the Multiplication Property of One, the product of a term and 1 is the term.

$(8x)(1) = (1)(8x) = 8x$

By the Inverse Property of Multiplication, the product of a term and its reciprocal is 1.

$5x \cdot \dfrac{1}{5x} = \dfrac{1}{5x} \cdot 5x = 1, \ x \neq 0$

Example 5 Simplify.   A. $2(-x)$ B. $\dfrac{3}{2}\left(\dfrac{2x}{3}\right)$ C. $(16x)2$

Solution A. $2(-x) = 2(-1 \cdot x)$ ▶ $-x = -1x = -1 \cdot x$
   $= [2 \cdot (-1)]x$ ▶ Use the Associative Property of Multiplication to group factors.
   $= -2x$ ▶ Multiply.

B. $\dfrac{3}{2}\left(\dfrac{2x}{3}\right) = \dfrac{3}{2}\left(\dfrac{2}{3}x\right)$ ▶ Note that $\dfrac{2x}{3} = \dfrac{2}{3} \cdot \dfrac{x}{1} = \dfrac{2}{3}x$.

   $= \left(\dfrac{3}{2} \cdot \dfrac{2}{3}\right)x$ ▶ Use the Associative Property of Multiplication to group factors.

   $= 1x$ ▶ Use the Inverse Property of Multiplication.

   $= x$ ▶ Use the Multiplication Property of One.

C. $(16x)2 = 2(16x)$ ▶ Use the Commutative Property of Multiplication to rearrange factors.

   $= (2 \cdot 16)x$ ▶ Use the Associative Property of Multiplication to group factors.

   $= 32x$ ▶ Multiply.

Problem 5 Simplify.   A. $-7(-2a)$ B. $-\dfrac{5}{6}(-30y^2)$ C. $(-5x)(-2)$

Solution See page A14.

---

4 ## Simplify variable expressions using the Distributive Property

The Distributive Property is used to remove parentheses from a variable expression.

$3(2x - 5)$
$= 3(2x) - 3(5)$
$= 6x - 15$

An extension of the Distributive Property is used when an expression contains more than two terms.

$4(x^2 + 6x - 1)$
$= 4(x^2) + 4(6x) - 4(1)$
$= 4x^2 + 24x - 4$

Example 6 Simplify.   A. $-3(5 + x)$ B. $-(2x - 4)$
   C. $(2y - 6)2$ D. $5(3x + 7y - z)$

Solution   A. $-3(5 + x)$
$$= -3(5) + (-3)x \qquad \blacktriangleright \text{ Use the Distributive Property.}$$
$$= -15 - 3x \qquad\qquad \blacktriangleright \text{ Multiply.}$$

B. $-(2x - 4)$
$$= -1(2x - 4) \qquad\qquad \blacktriangleright \text{ Just as } -x = -1x,$$
$$\qquad\qquad\qquad\qquad\qquad -(2x - 4) = -1(2x - 4).$$
$$= -1(2x) - (-1)(4) \qquad \blacktriangleright \text{ Use the Distributive Property.}$$
$$= -2x + 4 \qquad\qquad\qquad \blacktriangleright \text{ Note: When a negative sign im-}$$
mediately precedes the parenthe-
ses, remove the parentheses and
change the sign of *each* term inside the parentheses.

C. $(2y - 6)2$
$$= (2y)(2) - (6)(2) \qquad \blacktriangleright \text{ Use the Distributive Property}$$
$$= 4y - 12 \qquad\qquad\qquad (b + c)a = ba + ca.$$

D. $5(3x + 7y - z)$
$$= 5(3x) + 5(7y) - 5(z) \qquad \blacktriangleright \text{ Use the Distributive Property.}$$
$$= 15x + 35y - 5z$$

Problem 6   Simplify.   A. $7(4 + 2y)$ B. $-(5x - 12)$
C. $(3a - 1)5$ D. $-3(6a^2 - 8a + 9)$

Solution   See page A14.

---

5 Simplify general variable expressions

When simplifying variable expressions, use the Distributive Property to remove
parentheses and brackets used as grouping symbols.

Example 7   Simplify: $4(x - y) - 2(-3x + 6y)$

Solution $4(x - y) - 2(-3x + 6y)$
$$= 4x - 4y + 6x - 12y \qquad \blacktriangleright \text{ Use the Distributive Property to re-}$$
move parentheses.
$$= 10x - 16y \qquad\qquad\qquad \blacktriangleright \text{ Combine like terms.}$$

Problem 7   Simplify: $7(x - 2y) - 3(-x - 2y)$

Solution   See page A14.

Example 8   Simplify: $2x - 3[2x - 3(x + 7)]$

Solution $2x - 3[2x - 3(x + 7)]$
$$= 2x - 3[2x - 3x - 21] \qquad \blacktriangleright \text{ Use the Distributive Property to re-}$$
move the inner grouping symbols.
$$= 2x - 3[-x - 21] \qquad\qquad \blacktriangleright \text{ Combine like terms inside the}$$
grouping symbols.
$$= 2x + 3x + 63 \qquad\qquad\qquad \blacktriangleright \text{ Use the Distributive Property to re-}$$
move the brackets.
$$= 5x + 63 \qquad\qquad\qquad\qquad \blacktriangleright \text{ Combine like terms.}$$

Problem 8   Simplify: $3y - 2[x - 4(2 - 3y)]$

Solution   See page A14.

CONCEPT REVIEW 2.2

Determine whether the statement is always true, sometimes true, or never true.

1. The Associative Property of Addition states that two terms can be added in either order and the sum will be the same.

2. When three numbers are multiplied, the numbers can be grouped in any order and the product will be the same.

3. The Multiplication Property of One states that multiplying a number by one does not change the number.

4. The sum of a number and its additive inverse is zero.

5. The product of a number and its multiplicative inverse is one.

6. The terms x and x^2 are like terms because both have a coefficient of 1.

7. $3(x + 4) = 3x + 4$ is an example of the correct application of the Distributive Property.

8. To add like terms, add the coefficients; the variable part remains unchanged.

EXERCISES 2.2

1   Use the given property to complete the statement.

1. The Commutative Property of Multiplication
$2 \cdot 5 = 5 \cdot ?$

2. The Commutative Property of Addition
$9 + 17 = ? + 9$

3. The Associative Property of Multiplication
$(4 \cdot 5) \cdot 6 = 4 \cdot (? \cdot 6)$

4. The Associative Property of Addition
$(4 + 5) + 6 = ? + (5 + 6)$

5. The Distributive Property
$2(4 + 3) = 8 + ?$

6. The Addition Property of Zero
$? + 0 = -7$

7. The Inverse Property of Addition
$8 + ? = 0$

8. The Inverse Property of Multiplication
$\frac{1}{-5}(-5) = ?$

9. The Multiplication Property of One
$? \cdot 1 = -4$

10. The Multiplication Property of Zero
$12 \cdot ? = 0$

Identify the property that justifies the statement.

11. $-7 + 7 = 0$

12. $(-8)\left(-\frac{1}{8}\right) = 1$

13. $23 + 19 = 19 + 23$

14. $-21 + 0 = -21$

15. $2 + (6 + 14) = (2 + 6) + 14$

16. $(-3 + 9)8 = -24 + 72$

17. $3 \cdot 5 = 5 \cdot 3$

18. $-32(0) = 0$

19. $(4 \cdot 3) \cdot 5 = 4 \cdot (3 \cdot 5)$

20. $\frac{1}{4}(1) = \frac{1}{4}$

2   Simplify.

21. $6x + 8x$ **22.** $12x + 13x$ **23.** $9a - 4a$ **24.** $12a - 3a$

25. $4y - 10y$ **26.** $8y - 6y$ **27.** $-3b - 7$ **28.** $-12y - 3$

29. $-12a + 17a$ **30.** $-3a + 12a$ **31.** $5ab - 7ab$ **32.** $9ab - 3ab$

33. $-12xy + 17xy$ **34.** $-15xy + 3xy$ **35.** $-3ab + 3ab$ **36.** $-7ab + 7ab$

37. $-\frac{1}{2}x - \frac{1}{3}x$ **38.** $-\frac{2}{5}y + \frac{3}{10}y$ **39.** $\frac{3}{8}x^2 - \frac{5}{12}x^2$ **40.** $\frac{2}{3}y^2 - \frac{4}{9}y^2$

41. $3x + 5x + 3x$ **42.** $8x + 5x + 7x$ **43.** $5a - 3a + 5a$ **44.** $10a - 17a + 3a$

45. $-5x^2 - 12x^2 + 3x^2$ **46.** $-y^2 - 8y^2 + 7y^2$ **47.** $7x - 8x + 3y$

48. $8y - 10x + 8x$ **49.** $7x - 3y + 10x$ **50.** $8y + 8x - 8y$

51. $3a - 7b - 5a + b$

52. $-5b + 7a - 7b + 12a$

53. $3x - 8y - 10x + 4x$

54. $3y - 12x - 7y + 2y$

55. $x^2 - 7x - 5x^2 + 5x$

56. $3x^2 + 5x - 10x^2 - 10x$

3   Simplify.

57. $4(3x)$

58. $12(5x)$

59. $-3(7a)$

60. $-2(5a)$

61. $-2(-3y)$

62. $-5(-6y)$

63. $(4x)2$

64. $(6x)12$

65. $(3a)(-2)$

66. $(7a)(-4)$

67. $(-3b)(-4)$

68. $(-12b)(-9)$

69. $-5(3x^2)$

70. $-8(7x^2)$

71. $\frac{1}{3}(3x^2)$

72. $\frac{1}{5}(5a)$

73. $\frac{1}{8}(8x)$

74. $-\frac{1}{4}(-4a)$

75. $-\frac{1}{7}(-7n)$

76. $\left(\frac{3x}{4}\right)\left(\frac{4}{3}\right)$

77. $\frac{12x}{5}\left(\frac{5}{12}\right)$

78. $(-6y)\left(-\frac{1}{6}\right)$

79. $(-10n)\left(-\frac{1}{10}\right)$

80. $\frac{1}{3}(9x)$

81. $\frac{1}{7}(14x)$

82. $-\frac{1}{5}(10x)$

83. $-\frac{1}{8}(16x)$

84. $-\frac{2}{3}(12a^2)$

85. $-\frac{5}{8}(24a^2)$

86. $-\frac{1}{2}(-16y)$

87. $-\frac{3}{4}(-8y)$

88. $(16y)\left(\frac{1}{4}\right)$

89. $(33y)\left(\frac{1}{11}\right)$

90. $(-6x)\left(\frac{1}{3}\right)$

91. $(-10x)\left(\frac{1}{5}\right)$

92. $(-8a)\left(-\frac{3}{4}\right)$

4   Simplify.

93. $-(x + 2)$

94. $-(x + 7)$

95. $2(4x - 3)$

96. $5(2x - 7)$

97. $-2(a + 7)$

98. $-5(a + 16)$

99. $-3(2y - 8)$

100. $-5(3y - 7)$

101. $(5 - 3b)7$

102. $(10 - 7b)2$

103. $-3(3 - 5x)$

104. $-5(7 - 10x)$

105. $3(5x^2 + 2x)$

106. $6(3x^2 + 2x)$

107. $-2(-y + 9)$

108. $-5(-2x + 7)$

109. $(-3x - 6)5$

110. $(-2x + 7)7$

111. $2(-3x^2 - 14)$

112. $5(-6x^2 - 3)$

113. $-3(2y^2 - 7)$

114. $-8(3y^2 - 12)$

115. $-(6a^2 - 7b^2)$

116. $3(x^2 + 2x - 6)$

117. $4(x^2 - 3x + 5)$

118. $-2(y^2 - 2y + 4)$

119. $-3(y^2 - 3y - 7)$

120. $2(-a^2 - 2a + 3)$

121. $4(-3a^2 - 5a + 7)$

122. $-5(-2x^2 - 3x + 7)$

123. $-3(-4x^2 + 3x - 4)$

124. $3(2x^2 + xy - 3y^2)$

125. $5(2x^2 - 4xy - y^2)$

126. $-(3a^2 + 5a - 4)$

127. $-(8b^2 - 6b + 9)$

5   Simplify.

128. $4x - 2(3x + 8)$ **129.** $6a - (5a + 7)$ **130.** $9 - 3(4y + 6)$

131. $10 - (11x - 3)$ **132.** $5n - (7 - 2n)$ **133.** $8 - (12 + 4y)$

134. $3(x + 2) - 5(x - 7)$ **135.** $2(x - 4) - 4(x + 2)$ **136.** $12(y - 2) + 3(7 - 3y)$

137. $6(2y - 7) - 3(3 - 2y)$ **138.** $3(a - b) - 4(a + b)$ **139.** $2(a + 2b) - (a - 3b)$

140. $4[x - 2(x - 3)]$ **141.** $2[x + 2(x + 7)]$ **142.** $-2[3x + 2(4 - x)]$

143. $-5[2x + 3(5 - x)]$ **144.** $-3[2x - (x + 7)]$ **145.** $-2[3x - (5x - 2)]$

146. $2x - 3[x - 2(4 - x)]$ **147.** $-7x + 3[x - 7(3 - 2x)]$

148. $-5x - 2[2x - 4(x + 7)] - 6$ **149.** $4a - 2[2b - (b - 2a)] + 3b$

150. $2x + 3(x - 2y) + 5(3x - 7y)$ **151.** $5y - 2(y - 3x) + 2(7x - y)$

SUPPLEMENTAL EXERCISES 2.2

Simplify.

152. $C - 0.7C$ **153.** $\frac{1}{3}(3x + y) - \frac{2}{3}(6x - y)$ **154.** $-\frac{1}{4}[2x + 2(y - 6y)]$

Complete.

155. A number that has no reciprocal is _____ .

156. A number that is its own reciprocal is _____ .

157. The additive inverse of $a - b$ is _____ .

158. Determine whether the statement is true or false. If the statement is false, give an example that illustrates that it is false.
   a. Division is a commutative operation.
   b. Division is an associative operation.
   c. Subtraction is an associative operation.
   d. Subtraction is a commutative operation.
   e. Addition is a commutative operation.

159. Define an operation $\otimes$ as $a \otimes b = (a \cdot b) - (a + b)$.
   For example, $7 \otimes 5 = (7 \cdot 5) - (7 + 5) = 35 - 12 = 23$.
   a. Is $\otimes$ a commutative operation? Support your answer.
   b. Is $\otimes$ an associative operation? Support your answer.

160. Give examples of two operations that occur in everyday experience that are not commutative (for example, putting on socks and then shoes).

161. In your own words, explain the Distributive Property.

S E C T I O N **2.3**

Translating Verbal Expressions into Variable Expressions

1 Translate a verbal expression into a variable expression given the variable

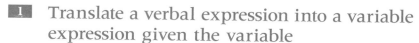

POINT OF INTEREST

The way in which expressions are symbolized has changed over time. Here are some expressions as they may have appeared in the early 16th century.

R p. 9 for $x + 9$. The symbol **R** was used for a variable to the first power. The symbol **p.** was used for plus.

R m. 3 for $x - 3$. The symbol **R** is still used for the variable. The symbol **m.** is used for minus.

The square of a variable was designated by **Q** and the cube was designated by **C**. The expression $x^3 + x^2$ was written **C p. Q.**

One of the major skills required in applied mathematics is the ability to translate a verbal expression into a variable expression. This requires recognizing the verbal phrases that translate into mathematical operations. Here is a partial list of the phrases used to indicate the different mathematical operations.

Addition	added to	6 added to y	$y + 6$
	more than	8 more than x	$x + 8$
	the sum of	the sum of x and z	$x + z$
	increased by	t increased by 9	$t + 9$
	the total of	the total of 5 and y	$5 + y$
	plus	b plus seventeen	$b + 17$
Subtraction	minus	x minus 2	$x - 2$
	less than	7 less than t	$t - 7$
	less	7 less t	$7 - t$
	subtracted from	5 subtracted from d	$d - 5$
	decreased by	m decreased by 3	$m - 3$
	the difference between	the difference between y and 4	$y - 4$
Multiplication	times	10 times t	$10t$
	of	one-half of x	$\frac{1}{2}x$
	the product of	the product of y and z	yz
	multiplied by	y multiplied by 11	$11y$
	twice	twice n	$2n$
Division	divided by	x divided by 12	$\frac{x}{12}$
	the quotient of	the quotient of y and z	$\frac{y}{z}$
	the ratio of	the ratio of t to 9	$\frac{t}{9}$
Power	the square of	the square of x	x^2
	the cube of	the cube of a	a^3

Translating a phrase that contains the word *sum, difference, product,* or *quotient* can sometimes cause a problem. In the examples at the right, note where the operation symbol is placed.

the *sum* of x and y $\qquad$ $x + y$

the *difference* between x and y $\qquad$ $x - y$

the *product* of x and y $\qquad$ $x \cdot y$

the *quotient* of x and y $\qquad$ $\dfrac{x}{y}$

Example 1 Translate into a variable expression.

A. the total of 5 times b and c

B. the quotient of 8 less than n and 14

C. 13 more than the sum of 7 and the square of x

Solution A. the total of 5 times b and c $\qquad$ ▸ Identify words that indicate mathematical operations.

$5b + c$ $\qquad$ ▸ Use the operations to write the variable expression.

B. the quotient of 8 less than n and 14 $\qquad$ ▸ Identify words that indicate mathematical operations.

$\dfrac{n - 8}{14}$ $\qquad$ ▸ Use the operations to write the variable expression.

C. 13 more than the sum of 7 and the square of x

$(7 + x^2) + 13$

Problem 1 Translate into a variable expression.

A. 18 less than the cube of x

B. y decreased by the sum of z and 9

C. the difference between the square of q and the sum of r and t

Solution See page A14.

2 Translate a verbal expression into a variable expression by assigning the variable

In most applications that involve translating phrases into variable expressions, the variable to be used is not given. To translate these phrases, a variable must be assigned to an unknown quantity before the variable expression can be written.

Example 2 Translate "a number multiplied by the total of six and the cube of the number" into a variable expression.

Solution the unknown number: n $\qquad$ ▸ Assign a variable to one of the unknown quantities.

the cube of the number: n^3 $\qquad$ ▸ Use the assigned variable to write an expression for any the total of six and the cube of the number: $6 + n^3$ $\qquad$ other unknown quantity.

$n(6 + n^3)$ $\qquad$ ▸ Use the assigned variable to write the variable expression.

LOOK CLOSELY

The expression $n(6 + n^3)$ must have parentheses. If we write $n \cdot 6 + n^3$, then by the Order of Operations Agreement, only the 6 is multiplied by n, but we want the *total* of 6 and n^3 to be multiplied by n.

Problem 2   Translate "a number added to the product of five and the square of the number" into a variable expression.

Solution   See page A15.

Example 3   Translate "the quotient of twice a number and the difference between the number and twenty" into a variable expression.

Solution   the unknown number: n
twice the number: $2n$
the difference between the number and twenty: $n - 20$

$$\frac{2n}{n - 20}$$

Problem 3   Translate "the product of three and the sum of seven and twice a number" into a variable expression.

Solution   See page A15.

3 Translate a verbal expression into a variable  expression and then simplify the resulting expression

After translating a verbal expression into a variable expression, simplify the variable expression by using the Properties of the Real Numbers.

Example 4   Translate and simplify "the total of four times an unknown number and twice the difference between the number and eight."

Solution   the unknown number: n ▶ Assign a variable to one of the unknown quantities.

four times the unknown number: $4n$ ▶ Use the assigned vari-
twice the difference between   able to write an ex-
   the number and eight: $2(n - 8)$ pression for any other
                                                                            unknown quantity.
$4n + 2(n - 8)$ ▶ Use the assigned vari-
                                                                            able to write the vari-
                                                                            able expression.
$= 4n + 2n - 16$ ▶ Simplify the variable
$= 6n - 16$ expression.

Problem 4   Translate and simplify "a number minus the difference between twice the number and seventeen."

Solution   See page A15.

Example 5   Translate and simplify "the difference between five-eighths of a number and two-thirds of the same number."

Solution   the unknown number: n ▶ Assign a variable to one of the unknown quantities.

five-eighths of the number: $\dfrac{5}{8}n$ ▶ Use the assigned variable to write an expression for any other unknown quantity.

two-thirds of the number: $\dfrac{2}{3}n$

$$\dfrac{5}{8}n - \dfrac{2}{3}n$$ ▶ Use the assigned variable to write the variable expression.

$$= \dfrac{15}{24}n - \dfrac{16}{24}n$$ ▶ Simplify the variable expression.

$$= -\dfrac{1}{24}n$$

Problem 5   Translate and simplify "the sum of three-fourths of a number and one-fifth of the same number."

Solution   See page A15.

---

4   ## Translate application problems

Many of the applications of mathematics require that you identify an unknown quantity, assign a variable to that quantity, and then attempt to express another unknown quantity in terms of that variable.

Suppose we know that the sum of two numbers is 10 and that one of the two numbers is 4. We can find the other number by subtracting 4 from 10.

one number: 4
other number: $10 - 4 = 6$
The two numbers are 4 and 6.

Now suppose we know that the sum of two numbers is 10, we don't know either number, and we want to express *both* numbers in terms of the *same* variable. Let one number be x. Again, we can find the other number by subtracting x from 10.

one number: x
other number: $10 - x$
The two numbers are x and $10 - x$.

Note that the sum of x and $10 - x$ is 10. $x + (10 - x) = x + 10 - x = 10$

Example 6   The length of a swimming pool is 20 ft longer than the width. Express the length of the pool in terms of the width.

LOOK CLOSELY
Any variable can be used. For example, if the width is y, then the length is $y + 20$.

Solution   the width of the pool: W ▶ Assign a variable to the width of the pool.

the length is 20 more than the width: $W + 20$ ▶ Express the length of the pool in terms of W.

Problem 6   An older computer takes twice as long to process a set of data as does a newer model. Express the amount of time it takes the older computer to process the data in terms of the amount of time it takes the newer model.

Solution   See page A15.

Example 7   An investor divided $5000 between two accounts, one a mutual fund and the other a money market fund. Use one variable to express the amounts invested in each account.

Solution   the amount invested in the mutual fund: x ▶ Assign a variable to the amount invested in one account.

the amount invested in the money market fund: $5000 - x$ ▶ Express the amount invested in the other account in terms of x.

Problem 7   A guitar string 6 ft long was cut into two pieces. Use one variable to express the lengths of the two pieces.

Solution   See page A15.

LOOK CLOSELY

It is also correct to assign the variable to the amount in the money market fund. Then the amount in the mutual fund is $5000 - x$.

STUDY TIPS

TAKE CAREFUL NOTES IN CLASS

You need a notebook in which to keep class notes and records about assignments and tests. Make sure to take complete and well-organized notes. Your instructor will explain text material that may be difficult for you to understand on your own and may provide important information that is not provided in the textbook. Be sure to include in your notes everything that is written on the chalkboard.

Information recorded in your notes about assignments should explain exactly what they are, how they are to be done, and when they are due. Information about tests should include exactly what text material and topics will be covered on each test and the dates on which the tests will be given.

CONCEPT REVIEW 2.3

Determine whether the statement is always true, sometimes true, or never true.

1. "Five less than n" can be translated as "$5 - n$."

2. A variable expression contains an equals sign.

3. If the sum of two numbers is 12 and one of the two numbers is x, then the other number can be expressed as $x - 12$.

4. The words *total* and *times* both indicate multiplication.

5. The words *quotient* and *ratio* both indicate division.

6. The expressions $7y - 8$ and $(7y) - 8$ are equivalent.

7. The phrase "five times the sum of x and y" and the phrase "the sum of five times x and y" yield the same variable expression.

EXERCISES 2.3

1 Translate into a variable expression.

1. d less than 19

2. the sum of 6 and c

3. r decreased by 12

4. w increased by 55

5. a multiplied by 28

6. y added to 16

7. 5 times the difference between n and 7

8. 30 less than the square of b

9. y less the product of 3 and y

10. the sum of four-fifths of m and 18

11. the product of -6 and b

12. 9 increased by the quotient of t and 5

13. 4 divided by the difference between p and 6

14. the product of 7 and the total of r and 8

15. the quotient of 9 less than x and twice x

16. the product of a and the sum of a and 13

17. 21 less than the product of s and negative 4

18. 14 more than one-half of the square of z

19. the ratio of 8 more than d to d

20. the total of 9 times the cube of m and the square of m

21. three-eighths of the sum of t and 15

22. s decreased by the quotient of s and 2

23. w increased by the quotient of 7 and w

24. the difference between the square of c and the total of c and 14

25. d increased by the difference between 16 times d and 3

26. the product of 8 and the total of b and 5

2   Translate into a variable expression.

27. a number divided by nineteen

28. thirteen less a number

29. forty more than a number

30. three-sevenths of a number

31. the square of the difference between a number and ninety

32. the quotient of twice a number and five

33. the sum of four-ninths of a number and twenty

34. eight subtracted from the product of fifteen and a number

35. the product of a number and ten more than the number

36. six less than the total of a number and the cube of the number

37. fourteen added to the product of seven and a number

38. the quotient of three and the total of four and a number

39. the quotient of twelve and the sum of a number and two

40. eleven plus one-half of a number

41. the ratio of two and the sum of a number and one

42. a number multiplied by the difference between twice the number and nine

43. the difference between sixty and the quotient of a number and fifty

44. the product of nine less than a number and the number

45. the sum of the square of a number and three times the number

46. the quotient of seven more than twice a number and the number

47. the sum of three more than a number and the cube of the number

48. a number decreased by the difference between the cube of the number and ten

49. the square of a number decreased by one-fourth of the number

50. four less than seven times the square of a number

51. twice a number decreased by the quotient of seven and the number

52. eighty decreased by the product of thirteen and a number

53. the cube of a number decreased by the product of twelve and the number

54. the quotient of five and the sum of a number and nineteen

3   Translate into a variable expression. Then simplify the expression.

55. a number increased by the total of the number and ten

56. a number added to the product of five and the number

57. a number decreased by the difference between nine and the number

58. eight more than the sum of a number and eleven

59. the difference between one-fifth of a number and three-eighths of the number

60. a number minus the sum of the number and fourteen

61. four more than the total of a number and nine

62. the sum of one-eighth of a number and one-twelfth of the number

63. twice the sum of three times a number and forty

64. the sum of a number divided by two and the number

65. seven times the product of five and a number

66. sixteen multiplied by one-fourth of a number

67. the total of seventeen times a number and twice the number

68. the difference between nine times a number and twice the number

69. a number plus the product of the number and twelve

70. nineteen more than the difference between a number and five

71. three times the sum of the square of a number and four

72. a number subtracted from the product of the number and seven

73. three-fourths of the sum of sixteen times a number and four

74. the difference between fourteen times a number and the product of the number and seven

75. sixteen decreased by the sum of a number and nine

76. eleven subtracted from the difference between eight and a number

77. five more than the quotient of four times a number and two

78. twenty minus the sum of four-ninths of a number and three

79. six times the total of a number and eight

80. four times the sum of a number and twenty

81. seven minus the sum of a number and two

82. three less than the sum of a number and ten

83. one-third of the sum of a number and six times the number

84. twice the quotient of four times a number and eight

85. the total of eight increased by the cube of a number and twice the cube of the number

86. the sum of five more than the square of a number and twice the square of the number

87. twelve more than a number added to the difference between the number and six

88. a number plus four added to the difference between three and twice the number

89. the sum of a number and nine added to the difference between the number and twenty

90. seven increased by a number added to twice the difference between the number and two

91. fourteen plus the product of three less than a number and ten

92. a number plus the product of the number minus five and seven

4   Write a variable expression.

93. A propeller-driven plane flies at a rate that is one-half the rate of a jet plane. Express the rate of the propeller plane in terms of the rate of the jet plane.

94. In football, the number of points awarded for a touchdown is three times the number of points awarded for a safety. Express the number of points awarded for a touchdown in terms of the number of points awarded for a safety.

95. A mixture contains four times as many peanuts as cashews. Express the amount of peanuts in the mixture in terms of the amount of cashews.

96. In a coin bank, there are ten more dimes than quarters. Express the number of dimes in the coin bank in terms of the number of quarters.

97. A 5¢ stamp in a stamp collection is 25 years older than an 8¢ stamp in the collection. Express the age of the 5¢ stamp in terms of the age of the 8¢ stamp.

98. The length of a rectangle is five meters more than twice the width. Express the length of the rectangle in terms of the width.

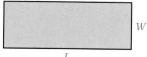

99. In a triangle, the measure of the smallest angle is three degrees less than one-half the measure of the largest angle. Express the measure of the smallest angle in terms of the measure of the largest angle.

100. The cost of renting a car for a day is $39.95 plus 15¢ per mile driven. Express the cost of renting the car in terms of the number of miles driven. Use decimals for constants and coefficients.

101. First-class mail costs 32¢ for the first ounce and 23¢ for each additional ounce. Express the cost of mailing a package first class in terms of the weight of the package. Use decimals for constants and coefficients.

102. The sum of two numbers is twenty-three. Use one variable to represent the two numbers.

103. A rope twelve feet long was cut into two pieces. Use one variable to express the lengths of the two pieces. See Figure 1.

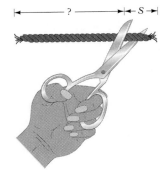

Figure 1

104. A coin purse contains thirty-five coins in nickels and dimes. Use one variable to express the number of nickels and the number of dimes in the coin purse.

105. Twenty gallons of oil were poured into two containers of different size. Use one variable to express the amount of oil poured into each container.

106. Two cars are traveling in opposite directions and at different rates. Two hours later the cars are two hundred miles apart. Express the distance traveled by the slower car in terms of the distance traveled by the faster car. See Figure 2.

107. An employee is paid $640 per week plus $24 for each hour of overtime worked. Express the employee's weekly pay in terms of the number of hours of overtime worked.

108. An auto repair bill is $92 for parts and $25 for each hour of labor. Express the amount of the repair bill in terms of the number of hours of labor.

Figure 2

SUPPLEMENTAL EXERCISES 2.3

Complete each statement with the word *even* or *odd*.

109. If k is an odd integer, then $k + 1$ is an _____ integer.

110. If k is an odd integer, then $k - 2$ is an _____ integer.

111. If n is an integer, then $2n$ is an _____ integer.

112. If m and n are even integers, then $m - n$ is an _____ integer.

113. If m and n are even integers, then mn is an _____ integer.

114. If m and n are odd integers, then $m + n$ is an _____ integer.

115. If m and n are odd integers, then $m - n$ is an _____ integer.

116. If m and n are odd integers, then mn is an _____ integer.

117. If m is an even integer and n is an odd integer, then $m - n$ is an _____ integer.

118. If m is an even integer and n is an odd integer, then $m + n$ is an _____ integer.

119. The chemical formula for glucose (sugar) is $C_6H_{12}O_6$. This formula means that there are twelve hydrogen atoms, six carbon atoms, and six oxygen atoms in each molecule of glucose. If x represents the number of atoms of oxygen in a pound of sugar, express the number of hydrogen atoms in the pound of sugar in terms of x.

$$
\begin{array}{c}
\text{H} \diagdown \ \diagup \text{O} \\
\text{C} \\
| \\
\text{H} - \text{C} - \text{OH} \\
| \\
\text{HO} - \text{C} - \text{H} \\
| \\
\text{H} - \text{C} - \text{OH} \\
| \\
\text{H} - \text{C} - \text{OH} \\
| \\
\text{CH}_2\text{OH}
\end{array}
$$

120. A wire whose length is given as x inches is bent into a square. Express the length of a side of the square in terms of x.

121. A block-and-tackle system is designed so that pulling one end of a rope five feet will move a weight on the other end a distance of three feet. If x represents the distance the rope is pulled, express the distance the weight moves in terms of x.

122. Translate the expressions $5x + 8$ and $5(x + 8)$ into phrases.

123. In your own words, explain how variables are used.

124. Explain the similarities and the differences between the expressions "the difference between x and 5" and "5 less than x."

Project in Mathematics

Applications of Patterns in Mathematics

For the circle shown at the left, use a straight line to connect each dot on the circle with every other dot on the circle. How many different straight lines are there?

Follow the same procedure for each of the circles shown below. How many different straight lines are there in each?

1. Find a pattern to describe the number of dots on a circle and the corresponding number of different lines drawn.

2. Use the pattern from Exercise 1 to determine the number of different lines that would be drawn in a circle with 7 dots and in a circle with 8 dots.

3. You are arranging a tennis tournament with nine players. How many singles matches will be played among the nine players if each player plays each of the other players once?

Now consider **triangular numbers,** which were studied by ancient Greek mathematicians. The numbers 1, 3, 6, 10, 15, 21 are the first six triangular numbers.

Note in the following diagram that a triangle can be formed using the number of dots that correspond to a triangular number.

Observe that the number of dots in a row is one more than the number in the row above. The total number of dots can be found by addition.

$1 = 1$

$1 + 2 + 3 = 6$

$1 + 2 + 3 + 4 + 5 = 15$

$1 + 2 = 3$

$1 + 2 + 3 + 4 = 10$

$1 + 2 + 3 + 4 + 5 + 6 = 21$

4. Use this pattern to find the 7th triangular number and the 8th triangular number. Check your answers by drawing the corresponding triangles of dots.

5. Discuss the relationship between triangular numbers and the pattern describing the correspondence between the number of dots on a circle and the number of different lines drawn (see Exercise 1).

6. Suppose you are in charge of scheduling softball games for a league. There are 10 teams in the league. Use the pattern of triangular numbers to determine the number of games that must be scheduled.

Focus on Problem Solving

Variables in Abstract Problems

As you progress in your study of algebra, you will find that the problems become less concrete and more abstract. Problems that are concrete provide information pertaining to a specific instance. Abstract problems are theoretical; they are stated without reference to a specific instance. Let's look at an example of an abstract problem.

> How many cents are in d dollars?

How can you solve this problem? Are you able to solve the same problem if the information given is concrete?

> How many cents are in 5 dollars?

You know that there are 100 cents in 1 dollar. To find the number of cents in 5 dollars, multiply 5 by 100.

$100 \cdot 5 = 500$ There are 500 cents in 5 dollars.

Use the same procedure to find the number of cents in d dollars: multiply d by 100.

$100 \cdot d = 100d$ There are $100d$ cents in d dollars.

This problem might be taken a step further:

If one pen costs c cents, how many pens can be purchased with d dollars?

Consider the same problem using numbers in place of the variables.

If one pen costs 25 cents, how many pens can be purchased with 2 dollars?

To solve this problem, you need to calculate the number of cents in 2 dollars (multiply 2 by 100) and divide the result by the cost per pen (25 cents).

$$\frac{100 \cdot 2}{25} = \frac{200}{25} = 8$$ If one pen costs 25 cents, 8 pens can be purchased with 2 dollars.

Use the same procedure to solve the related abstract problem. Calculate the number of cents in d dollars (multiply d by 100), and divide the result by the cost per pen (c cents).

$$\frac{100 \cdot d}{c} = \frac{100d}{c}$$ If one pen costs c cents, $\frac{100d}{c}$ pens can be purchased with d dollars.

At the heart of the study of algebra is the use of variables. It is the variables in the problems above that make them abstract. Variables enable us to generalize situations and state rules about mathematics.

Try each of the following problems.

1. If you travel m miles on one gallon of gasoline, how far can you travel on g gallons of gasoline?

2. If you walk a mile in x minutes, how far can you walk in h hours?

3. If one photocopy costs n nickels, how many photocopies can you make for q quarters?

Chapter Summary

Key Words

A *variable* is a letter that is used to stand for a quantity that is unknown. A *variable expression* is an expression that contains one or more variables.   (Objective 2.1.1)

$5x - 4y + 7z$ is a variable expression. It contains the variables x, y, and z.

The *terms of a variable expression* are the addends of the expression. Each term is a *variable term* or a *constant term*. (Objective 2.1.1)

The expression $4a^2 - 6b^3 + 7$ has three terms: $4a^2$, $-6b^3$, and 7. $4a^2$ and $-6b^3$ are variable terms. 7 is a constant term.

A variable term is composed of a *numerical coefficient* and a *variable part*.   (Objective 2.1.1)

For the expression $8p^4r$, 8 is the coefficient and p^4r is the variable part.

Like terms of a variable expression are the terms with the same variable part.   (Objective 2.2.2)

For the expression $2st - 3t + 9s - 11st$, the terms $2st$ and $-11st$ are like terms.

The *additive inverse* of a number is the opposite of the number. (Objective 2.2.1)

The additive inverse of 8 is -8.
The additive inverse of -15 is 15.

The *multiplicative inverse* of a number is the reciprocal of the number.   (Objective 2.2.1)

The multiplicative inverse of $\frac{3}{8}$ is $\frac{8}{3}$.

Essential Rules and Procedures

The Commutative Property of Addition
If a and b are real numbers, then $a + b = b + a$.
(Objective 2.2.1)

$5 + 2 = 7$ and $2 + 5 = 7$

The Commutative Property of Multiplication
If a and b are real numbers, then $ab = ba$.   (Objective 2.2.1)

$6(-3) = -18$ and $-3(6) = -18$

The Associative Property of Addition
If a, b, and c are real numbers, then $(a + b) + c = a + (b + c)$.
(Objective 2.2.1)

$-1 + (4 + 7) = -1 + 11 = 10$
$(-1 + 4) + 7 = 3 + 7 = 10$

The Associative Property of Multiplication
If a, b, and c are real numbers, then $(ab)c = a(bc)$.
(Objective 2.2.1)

$(-2 \cdot 5) \cdot 3 = -10 \cdot 3 = -30$
$-2 \cdot (5 \cdot 3) = -2 \cdot 15 = -30$

The Addition Property of Zero
If a is a real number, then $a + 0 = 0 + a = a$.
(Objective 2.2.1)

$9 + 0 = 9$ $0 + 9 = 9$

The Multiplication Property of Zero
If a is a real number, then $a \cdot 0 = 0 \cdot a = 0$.   (Objective 2.2.1)

$-8(0) = 0$ $0(-8) = 0$

The Multiplication Property of One
If a is a real number, then $1 \cdot a = a \cdot 1 = a$.   (Objective 2.2.1)

$7 \cdot 1 = 7$ $1 \cdot 7 = 7$

The Inverse Property of Addition
If a is a real number, then $a + (-a) = (-a) + a = 0$.
(Objective 2.2.1)

$4 + (-4) = 0$ $-4 + 4 = 0$

The Inverse Property of Multiplication

If a is a real number and $a \neq 0$, then $a \cdot \frac{1}{a} = \frac{1}{a} \cdot a = 1$.
(Objective 2.2.1)

$6 \cdot \frac{1}{6} = 1$ $\frac{1}{6} \cdot 6 = 1$

The Distributive Property
If a, b, and c are real numbers, then $a(b + c) = ab + ac$.
(Objective 2.2.1)

$5(2 + 3) = 5 \cdot 2 + 5 \cdot 3$
$\qquad = 10 + 15$
$\qquad = 25$

Chapter Review Exercises

1. Simplify: $-7y^2 + 6y^2 - (-2y^2)$

2. Simplify: $(12x)\left(\frac{1}{4}\right)$

3. Simplify: $\frac{2}{3}(-15a)$

4. Simplify: $-2(2x - 4)$

5. Simplify: $5(2x + 4) - 3(x - 6)$

6. Evaluate $a^2 - 3b$ when $a = 2$ and $b = -4$.

7. Complete the statement by using the Inverse Property of Addition.

    $-9 + ? = 0$

8. Simplify: $-4(-9y)$

9. Simplify: $-2(-3y + 9)$

10. Simplify: $3[2x - 3(x - 2y)] + 3y$

11. Simplify: $-4(2x^2 - 3y^2)$

12. Simplify: $3x - 5x + 7x$

13. Evaluate $b^2 - 3ab$ when $a = 3$ and $b = -2$.

14. Simplify: $\frac{1}{5}(10x)$

15. Simplify: $5(3 - 7b)$

16. Simplify: $2x + 3[4 - (3x - 7]$

17. Identify the property that justifies the statement.

    $-4(3) = 3(-4)$

18. Simplify: $3(8 - 2x)$

19. Simplify: $-2x^2 - (-3x^2) + 4x^2$

20. Simplify: $-3x - 2(2x - 7)$

21. Simplify: $-3(3y^2 - 3y - 7)$

22. Simplify: $-2[x - 2(x - y)] + 5y$

23. Evaluate $\frac{-2ab}{2b - a}$ when $a = -4$ and $b = 6$.

24. Simplify: $(-3)(-12y)$

25. Simplify: $4(3x - 2) - 7(x + 5)$

26. Simplify: $(16x)\left(\frac{1}{8}\right)$

27. Simplify: $-3(2x^2 - 7y^2)$

28. Evaluate $3(a - c) - 2ab$ when $a = 2$, $b = 3$, and $c = -4$.

29. Simplify: $2x - 3(x - 2)$

30. Simplify: $2a - (-3b) - 7a - 5b$

31. Simplify: $-5(2x^2 - 3x + 6)$

32. Simplify: $3x - 7y - 12x$

33. Simplify: $\frac{1}{2}(12a)$

34. Simplify: $2x + 3[x - 2(4 - 2x)]$

35. Simplify: $3x + (-12y) - 5x - (-7y)$

36. Simplify: $\left(-\frac{5}{6}\right)(-36b)$

37. Complete the statement by using the Distributive Property.

    $(6 + 3)7 = 42 + ?$

38. Simplify: $4x^2 + 9x - 6x^2 - 5x$

39. Simplify: $-\frac{3}{8}(16x^2)$

40. Simplify: $-3[2x - (7x - 9)]$

41. Simplify: $-(8a^2 - 3b^2)$

42. Identify the property that justifies the statement.

    $-32(0) = 0$

43. Translate "b decreased by the product of 7 and b" into a variable expression.

44. Translate "the sum of a number and twice the square of the number" into a variable expression.

45. Translate "three less than the quotient of six and a number" into a variable expression.

46. Translate "10 divided by the difference between y and 2" into a variable expression.

47. Translate and simplify "eight times the quotient of twice a number and sixteen."

48. Translate and simplify "the product of four and the sum of two and five times a number."

49. One car was driven 15 mph faster than a second car. Express the speed of the first car in terms of the speed of the second car.

50. A coffee merchant made twenty pounds of a blend of coffee using only mocha java beans and expresso beans. Use one variable to express the amounts of mocha java beans and expresso beans in the coffee blend.

C hapter Test

1. Simplify: $(9y)4$

2. Simplify: $7x + 5y - 3x - 8y$

3. Simplify: $8n - (6 - 2n)$

4. Evaluate $3ab - (2a)^2$ when $a = -2$ and $b = -3$.

5. Identify the property that justifies the statement.

$$\frac{1}{4}(1) = \frac{1}{4}$$

6. Simplify: $-4(-x + 10)$

7. Simplify: $\frac{2}{3}x^2 - \frac{7}{12}x^2$

8. Simplify: $(-10x)\left(-\frac{2}{5}\right)$

9. Simplify: $(-4y^2 + 8)6$

10. Complete the statement by using the Inverse Property of Addition.

$$-19 + ? = 0$$

11. Evaluate $\frac{-3ab}{2a + b}$ when $a = -1$ and $b = 4$.

12. Simplify: $5(x + y) - 8(x - y)$

13. Simplify: $6b - 9b + 4b$

14. Simplify: $13(6a)$

15. Simplify: $3(x^2 - 5x + 4)$

16. Evaluate $4(b - a) + bc$ when $a = 2$, $b = -3$, and $c = 4$.

17. Simplify: $6x - 3(y - 7x) + 2(5x - y)$

18. Translate "the quotient of 8 more than n and 17" into a variable expression.

19. Translate "the difference between the sum of a and b and the square of b" into a variable expression.

20. Translate "the sum of the square of a number and the product of the number and eleven" into a variable expression.

21. Translate and simplify "twenty times the sum of a number and nine."

22. Translate and simplify "two more than a number added to the difference between the number and three."

23. Translate and simplify "a number minus the product of one-fourth and twice the number."

24. The distance from Neptune to the Sun is 30 times the distance from Earth to the Sun. Express the distance from Neptune to the Sun in terms of the distance from Earth to the Sun.

25. A nine-foot board is cut into two pieces of different lengths. Use one variable to express the lengths of the two pieces.

Cumulative Review Exercises

1. Add: $-4 + 7 + (-10)$

2. Subtract: $-16 - (-25) - 4$

3. Multiply: $(-2)(3)(-4)$

4. Divide: $(-60) \div 12$

5. Write $1\frac{1}{4}$ as a decimal.

6. Write 60% as a fraction and as a decimal.

7. Use the roster method to write the set of negative integers greater than or equal to -4.

8. Write $\frac{2}{25}$ as a percent.

9. Subtract: $\frac{7}{12} - \frac{11}{16} - \left(-\frac{1}{3}\right)$

10. Divide: $\frac{5}{12} \div \left(\frac{3}{2}\right)$

11. Multiply: $\left(-\frac{9}{16}\right)\left(\frac{8}{27}\right)\left(-\frac{3}{2}\right)$

12. Simplify: $-3^2 \cdot \left(-\frac{2}{3}\right)^3$

13. Simplify: $-2^5 \div (3 - 5)^2 - (-3)$

14. Simplify: $\left(-\frac{3}{4}\right)^2 - \left(\frac{3}{8} - \frac{11}{12}\right)$

15. Evaluate $a - 3b^2$ when $a = 4$ and $b = -2$.

16. Simplify: $-2x^2 - (-3x^2) + 4x^2$

17. Simplify: $8a - 12b - 9a$

18. Simplify: $\frac{1}{3}(9a)$

19. Simplify: $\left(-\dfrac{5}{8}\right)(-32b)$

20. Simplify: $5(4 - 2x)$

21. Simplify: $-3(-2y + 7)$

22. Simplify: $-2(3x^2 - 4y^2)$

23. Simplify: $-4(2y^2 - 5y - 8)$

24. Simplify: $-4x - 3(2x - 5)$

25. Simplify: $3(4x - 1) - 7(x + 2)$

26. Simplify: $3x + 2[x - 4(2 - x)]$

27. Simplify: $3[4x - 2(x - 4y)] + 5y$

28. Translate "the difference between six and the product of a number and twelve" into a variable expression.

29. Translate and simplify "the total of five and the difference between a number and seven."

30. A turboprop plane can fly at one-half the speed of sound. Express the speed of the turboprop plane in terms of the speed of sound.

3

Solving Equations and Inequalities

Objectives

Mersenne Primes

A prime number that can be written in the form $2^n - 1$, where n is also prime, is called a Mersenne prime. The table at the right shows some Mersenne primes.

$$3 = 2^2 - 1$$
$$7 = 2^3 - 1$$
$$31 = 2^5 - 1$$
$$127 = 2^7 - 1$$

Not every prime number is a Mersenne prime. For example, 5 is a prime number but not a Mersenne prime. Also, not all numbers in the form $2^n - 1$, where n is prime, yield a prime number. For example, $2^{11} - 1 = 2047$, which is not a prime number.

The search for Mersenne primes has been quite extensive, especially since the advent of the computer. One reason for the extensive research into large prime numbers (not only Mersenne primes) involves cryptology.

Cryptology is the study of making or breaking secret codes. One method of making a code that is difficult to break is called public key cryptology. For this method to work, it is necessary to use very large prime numbers. To keep anyone from breaking the code, each prime should have at least 200 digits.

Today, the largest known Mersenne prime is $2^{216091} - 1$. This number has 65,050 digits in its representation.

Another Mersenne prime got special recognition in a postage-meter stamp. It is the number $2^{11213} - 1$. This number has 3276 digits in its representation.

SECTION **3.1**

Introduction to Equations

1 Determine whether a given number is a solution of an equation

POINT OF INTEREST

One of the most famous equations ever stated is $E = mc^2$. This equation, stated by Albert Einstein, shows that there is a relationship between mass m and energy E. As a side note, the chemical element einsteinium was named in honor of Einstein.

An **equation** expresses the equality of two mathematical expressions. The expressions can be either numerical or variable expressions.

$$\left.\begin{array}{l} 9 + 3 = 12 \\ 3x - 2 = 10 \\ y^2 + 4 = 2y - 1 \\ z = 2 \end{array}\right\} \text{Equations}$$

The equation at the right is true if the variable is replaced by 5.

$x + 8 = 13$
$5 + 8 = 13$ A true equation

The equation $x + 8 = 13$ is false if the variable is replaced by 7.

$7 + 8 = 13$ A false equation

A **solution** of an equation is a number that, when substituted for the variable, results in a true equation. 5 is a solution of the equation $x + 8 = 13$. 7 is not a solution of the equation $x + 8 = 13$.

Example 1   Is -3 a solution of the equation $4x + 16 = x^2 - 5$?

Solution
$$4x + 16 = x^2 - 5$$

$4(-3) + 16$	$(-3)^2 - 5$
$-12 + 16$	$9 - 5$

$4 = 4$

Yes, -3 is a solution of the equation $4x + 16 = x^2 - 5$.

▶ Replace the variable by the given number, -3.
▶ Evaluate the numerical expressions using the Order of Operations Agreement.
▶ Compare the results. If the results are equal, the given number is a solution. If the results are not equal, the given number is not a solution.

Problem 1   Is $\frac{1}{4}$ a solution of $5 - 4x = 8x + 2$?

Solution   See page A15.

Example 2   Is -4 a solution of $4 + 5x = x^2 - 2x$?

Solution
$$4 + 5x = x^2 - 2x$$

$4 + 5(-4)$	$(-4)^2 - 2(-4)$
$4 + (-20)$	$16 - (-8)$

$-16 \neq 24$

No, -4 is not a solution of the equation $4 + 5x = x^2 - 2x$.

▶ Replace the variable by the given number, -4.
▶ Evaluate the numerical expressions using the Order of Operations Agreement.
▶ Compare the results. If the results are equal, the given number is a solution. If the results are not equal, the given number is not a solution.

Problem 2 Is 5 a solution of $10x - x^2 = 3x - 10$?

Solution See page A15.

2 **Solve equations of the form $x + a = b$**

To **solve** an equation means to find a solution of the equation. The simplest equation to solve is an equation of the form **variable = constant,** because the constant is the solution.

If $x = 5$, then 5 is the solution of the equation because $5 = 5$ is a true equation.

The solution of the equation shown at the right is 7.

$$x + 2 = 9 \qquad 7 + 2 = 9$$

Note that if 4 is added to each side of the equation, the solution is still 7.

$$x + 2 + 4 = 9 + 4$$
$$x + 6 = 13 \qquad 7 + 6 = 13$$

If -5 is added to each side of the equation, the solution is still 7.

$$x + 2 + (-5) = 9 + (-5)$$
$$x - 3 = 4 \qquad 7 - 3 = 4$$

This illustrates the Addition Property of Equations.

Addition Property of Equations

The same number or variable term can be added to each side of an equation without changing the solution of the equation.

This property is used in solving equations. Note the effect of adding, to each side of the equation $x + 2 = 9$, the opposite of the constant term 2. After each side of the equation is simplified, the equation is in the form variable = constant. The solution is the constant.

$$x + 2 = 9$$
$$x + 2 + (-2) = 9 + (-2)$$
$$x + 0 = 7$$
$$x = 7$$

variable = constant

The solution is 7.

In solving an equation, the goal is to rewrite the given equation in the form variable = constant. The Addition Property of Equations can be used to rewrite an equation in this form. The Addition Property of Equations is used to **remove a term** from one side of an equation **by adding the opposite of that term** to each side of the equation.

Example 3 Solve and check: $y - 6 = 9$

Solution $y - 6 = 9$ ► The goal is to rewrite the equation in the form *variable = constant*.

$y - 6 + 6 = 9 + 6$ ► Add the opposite of the constant term -6 to each side of the equation (the Addition Property of Equations).

$y + 0 = 15$ ► Simplify using the Inverse Property of Addition.

$$y = 15$$ ▶ Simplify using the Addition Property of Zero. Now the equation is in the form *variable = constant*.

Check $\dfrac{y - 6 = 9}{15 - 6 \mid 9}$

$9 = 9$ ▶ This is a true equation. The solution checks.

The solution is 15.   ▶ Write the solution.

Problem 3   Solve and check: $x - \dfrac{1}{3} = -\dfrac{3}{4}$

Solution   See page A15.

Because subtraction is defined in terms of addition, the Addition Property of Equations makes it possible to subtract the same number from each side of an equation without changing the solution of the equation.

Solve: $y + \dfrac{1}{2} = \dfrac{5}{4}$

The goal is to rewrite the equation in the form *variable = constant*.

$$y + \dfrac{1}{2} = \dfrac{5}{4}$$

Add the opposite of the constant term $\dfrac{1}{2}$ to each side of the equation. This is equivalent to subtracting $\dfrac{1}{2}$ from each side of the equation.

$$y + \dfrac{1}{2} - \dfrac{1}{2} = \dfrac{5}{4} - \dfrac{1}{2}$$

$$y + 0 = \dfrac{5}{4} - \dfrac{2}{4}$$

$$y = \dfrac{3}{4}$$

The solution $\dfrac{3}{4}$ checks.   The solution is $\dfrac{3}{4}$.

Example 4   Solve: $\dfrac{1}{2} = x + \dfrac{2}{3}$

Solution $\dfrac{1}{2} = x + \dfrac{2}{3}$

$\dfrac{1}{2} - \dfrac{2}{3} = x + \dfrac{2}{3} - \dfrac{2}{3}$ ▶ Subtract $\dfrac{2}{3}$ from each side of the equation.

$-\dfrac{1}{6} = x$ ▶ Simplify each side of the equation.

The solution is $-\dfrac{1}{6}$.

Problem 4   Solve: $-8 = 5 + x$

Solution   See page A15.

Note from the solution to Example 4 above that an equation can be rewritten in the form *constant = variable*. Whether the equation is written in the form *variable = constant* or in the form *constant = variable*, the solution is the constant.

3   Solve equations of the form $ax = b$

The solution of the equation shown at the right is 3.	$2x = 6$	$2 \cdot 3 = 6$

Note that if each side of the equation is multiplied by 5, the solution is still 3.

$$5 \cdot 2x = 5 \cdot 6$$
$$10x = 30 \qquad 10 \cdot 3 = 30$$

If each side is multiplied by -4, the solution is still 3.

$$(-4) \cdot 2x = (-4) \cdot 6$$
$$-8x = -24 \qquad -8 \cdot 3 = -24$$

This illustrates the Multiplication Property of Equations.

Multiplication Property of Equations

Each side of an equation can be multiplied by the same nonzero number without changing the solution of the equation.

This property is used in solving equations. Note the effect of multiplying each side of the equation $2x = 6$ by the reciprocal of the coefficient 2. After each side of the equation is simplified, the equation is in the form variable = constant. The solution is the constant.

$$2x = 6$$
$$\frac{1}{2} \cdot 2x = \frac{1}{2} \cdot 6$$
$$1x = 3$$
$$x = 3$$

$$\boxed{\text{variable}} = \boxed{\text{constant}}$$

The solution is 3.

In solving an equation, the goal is to rewrite the given equation in the form variable = constant. The Multiplication Property of Equations can be used to rewrite an equation in this form. The Multiplication Property of Equations is used to **remove a coefficient** from a variable term in an equation **by multiplying each side of the equation by the reciprocal of the coefficient.**

Example 5   Solve: $\dfrac{3x}{4} = -9$

Solution

$$\frac{3x}{4} = -9 \qquad\qquad \blacktriangleright \frac{3x}{4} = \frac{3}{4}x$$

$$\frac{4}{3} \cdot \frac{3}{4}x = \frac{4}{3}(-9) \qquad \blacktriangleright \text{Multiply each side of the equation by the}$$
reciprocal of the coefficient $\dfrac{3}{4}$ (the Multiplication Property of Equations).

$$1x = -12 \qquad\qquad \blacktriangleright \text{Simplify using the Inverse Property of Multiplication.}$$

$$x = -12 \qquad\qquad \blacktriangleright \text{Simplify using the Multiplication Property of One. Now the equation is in the form } variable = constant.$$

The solution is -12. $\blacktriangleright$ Write the solution.

Problem 5   Solve: $-\dfrac{2x}{5} = 6$

Solution   See page A16.

Because division is defined in terms of multiplication, the Multiplication Property of Equations makes it possible to divide each side of an equation by the same number without changing the solution of the equation.

Solve: $8x = 16$

The goal is to rewrite the equation in the form *variable* = *constant*.	$8x = 16$

Multiply each side of the equation by the reciprocal of 8. This is equivalent to dividing each side by 8.

$$\frac{8x}{8} = \frac{16}{8}$$
$$x = 2$$

The solution 2 checks.   The solution is 2.

When using the Multiplication Property of Equations to solve an equation, multiply each side of the equation by the reciprocal of the coefficient when the coefficient is a fraction. Divide each side of the equation by the coefficient when the coefficient is an integer or a decimal.

Example 6   Solve and check: $4x = 6$

Solution $4x = 6$

$\dfrac{4x}{4} = \dfrac{6}{4}$ ▶ Divide each side of the equation by 4, the coefficient of x.

$x = \dfrac{3}{2}$ ▶ Simplify each side of the equation.

Check $4x = 6$

$4\left(\dfrac{3}{2}\right) \;\Big|\; 6$

$6 = 6$ ▶ This is a true equation. The solution checks.

The solution is $\dfrac{3}{2}$.

Problem 6   Solve and check: $6x = 10$

Solution   See page A16.

Before using one of the Properties of Equations, check to see if one or both sides of the equation can be simplified. In Example 7 below, like terms appear on the left side of the equation. The first step in solving this equation is to combine the like terms so that there is only one variable term on the left side of the equation.

Example 7   Solve: $5x - 9x = 12$

Solution $5x - 9x = 12$

$-4x = 12$ ▶ Combine like terms.

$\dfrac{-4x}{-4} = \dfrac{12}{-4}$ ▶ Divide each side of the equation by -4.

$x = -3$

The solution is -3.

> **Problem 7**   Solve: $4x - 8x = 16$
>
> **Solution**   See page A16.

4   Application problems: the basic percent equation

The solution of a problem that involves a percent requires solving the basic percent equation shown below.

The Basic Percent Equation

$$\text{Percent} \cdot \text{Base} = \text{Amount}$$
$$P \quad \cdot \quad B \quad = \quad A$$

To translate a problem involving a percent into an equation, remember that the word *of* translates to "multiply" and the word *is* translates to "=". The base usually follows the word *of*.

20% of what number is 30?

Given: $P = 20\% = 0.20$
           $A = 30$
Unknown: Base

$$PB = A$$
$$(0.20)B = 30$$
$$\frac{0.20B}{0.20} = \frac{30}{0.20}$$
$$B = 150$$

The number is 150.

What percent of 40 is 30?

Given: $B = 40$
           $A = 30$
Unknown: Percent

$$PB = A$$
$$P(40) = 30$$
$$40P = 30$$
$$\frac{40P}{40} = \frac{30}{40}$$

The fraction must be written as a percent in order to answer the question.

$$P = \frac{3}{4}$$
$$P = 75\%$$

30 is 75% of 40.

Find 25% of 200.

Given: $P = 25\% = 0.25$
           $B = 200$
Unknown: Amount

$$PB = A$$
$$0.25(200) = A$$
$$50 = A$$

25% of 200 is 50.

In most cases, the percent is written as a decimal before solving the basic percent equation. However, some percents are more easily written as a fraction. For example,

$$33\frac{1}{3}\% = \frac{1}{3} \qquad 66\frac{2}{3}\% = \frac{2}{3} \qquad 16\frac{2}{3}\% = \frac{1}{6} \qquad 83\frac{1}{3}\% = \frac{5}{6}$$

Example 8   12 is $33\frac{1}{3}\%$ of what number?

Solution

$$PB = A$$
$$\frac{1}{3}B = 12 \qquad \blacktriangleright 33\frac{1}{3}\% = \frac{1}{3}$$
$$3 \cdot \frac{1}{3}B = 3 \cdot 12$$
$$B = 36$$

The number is 36.

Problem 8   27 is what percent of 60?

Solution   See page A16.

Example 9   A fundraiser has raised $3460. The goal is to raise $5000. What percent of the goal has been raised?

Strategy   To find the percent of the goal raised, solve the basic percent equation using $B = 5000$ and $A = 3460$. The percent is unknown.

Solution

$$PB = A$$
$$P(5000) = 3460$$
$$5000P = 3460$$
$$\frac{5000P}{5000} = \frac{3460}{5000}$$
$$P = 0.692$$

69.2% of the goal has been raised.

Problem 9   A student correctly answered 72 of the 80 questions on an exam. What percent of the questions were answered correctly?

Solution   See page A16.

Example 10   In a survey of 1500 registered voters, 27% responded that they believed that the economy in this country was improving. How many people believed the country's economy was improving?

Strategy   To find the number of people, solve the basic percent equation using $B = 1500$ and $P = 27\% = 0.27$. The amount is unknown.

Solution $PB = A$
$$0.27(1500) = A$$
$$405 = A$$

Of those surveyed, 405 people believed the country's economy was improving.

Problem 10   An engineer estimates that 32% of the gasoline used by a car is used efficiently. Using this estimate, determine how many gallons out of 15 gal of gasoline are used efficiently?

Solution   See page A16.

STUDY TIPS

GET HELP FOR ACADEMIC DIFFICULTIES

If you do have trouble in this course, teachers, counselors, and advisers can help. They usually know of study groups, tutors, or other sources of help that are available. They may suggest visiting an office of academic skills, a learning center, a tutorial service, or some other department or service on campus.

Students who have already taken the course and who have done well in it may be a source of assistance. If they have a good understanding of the material, they may be able to help by explaining it to you.

CONCEPT REVIEW 3.1

Determine whether the statement is always true, sometimes true, or never true.

1. Both sides of an equation can be multiplied by the same number without changing the solution of the equation.

2. Both sides of an equation can be divided by the same number without changing the solution of the equation.

3. For an equation of the form $ax = b$, $a \neq 0$, multiplying both sides of the equation by the reciprocal of a will result in an equation of the form $x = constant$.

4. Use the Multiplication Property of Equations to remove a term from one side of an equation.

5. Adding a negative 3 to each side of an equation yields the same result as subtracting 3 from each side of the equation.

6. In the basic percent equation, the amount follows the word *of*.

7. An equation contains an equals sign.

EXERCISES 3.1

1

1. Is 4 a solution of
   $2x = 8$?

2. Is 3 a solution of
   $y + 4 = 7$?

3. Is -1 a solution of
   $2b - 1 = 3$?

4. Is -2 a solution of
   $3a - 4 = 10$?

5. Is 1 a solution of
   $4 - 2m = 3$?

6. Is 2 a solution of
   $7 - 3n = 2$?

7. Is 5 a solution of
   $2x + 5 = 3x$?

8. Is 4 a solution of
   $3y - 4 = 2y$?

9. Is 0 a solution of
   $4a + 5 = 3a + 5$?

10. Is 0 a solution of
    $4 - 3b = 4 - 5b$?

11. Is -2 a solution of
    $4 - 2n = n + 10$?

12. Is -3 a solution of
    $5 - m = 2 - 2m$?

13. Is 3 a solution of
    $z^2 + 1 = 4 + 3z$?

14. Is 2 a solution of
    $2x^2 - 1 = 4x - 1$?

15. Is -1 a solution of
    $y^2 - 1 = 4y + 3$?

16. Is -2 a solution of
    $m^2 - 4 = m + 3$?

17. Is 4 a solution of
    $x(x + 1) = x^2 + 5$?

18. Is 3 a solution of
    $2a(a - 1) = 3a + 3$?

19. Is $-\frac{1}{4}$ a solution of
    $8t + 1 = -1$?

20. Is $\frac{1}{2}$ a solution of
    $4y + 1 = 3$?

21. Is $\frac{2}{5}$ a solution of
    $5m + 1 = 10m - 3$?

22. Is $\frac{3}{4}$ a solution of
    $8x - 1 = 12x + 3$?

23. Is 2.1 a solution of
    $x^2 - 4x = x + 1.89$?

24. Is 1.5 a solution of
    $c^2 - 3c = 4c - 8.25$?

2   Solve and check.

25. $x + 5 = 7$

26. $y + 3 = 9$

27. $b - 4 = 11$

28. $z - 6 = 10$

29. $2 + a = 8$

30. $5 + x = 12$

31. $m + 9 = 3$

32. $t + 12 = 10$

33. $n - 5 = -2$

34. $x - 6 = -5$

35. $b + 7 = 7$

36. $y - 5 = -5$

37. $a - 3 = -5$

38. $x - 6 = -3$

39. $z + 9 = 2$

40. $n + 11 = 1$

41. $10 + m = 3$

42. $8 + x = 5$

43. $9 + x = -3$

44. $10 + y = -4$

45. $b - 5 = -3$

46. $t - 6 = -4$

47. $2 = x + 7$

48. $-8 = n + 1$

49. $4 = m - 11$

50. $-6 = y - 5$

51. $12 = 3 + w$

52. $-9 = 5 + x$

53. $4 = -10 + b$

54. $-7 = -2 + x$

55. $m + \frac{2}{3} = -\frac{1}{3}$

56. $c + \frac{3}{4} = -\frac{1}{4}$

57. $x - \frac{1}{2} = \frac{1}{2}$

58. $x - \frac{2}{5} = \frac{3}{5}$

59. $\frac{5}{8} + y = \frac{1}{8}$

60. $\frac{4}{9} + a = -\frac{2}{9}$

61. $m + \frac{1}{2} = -\frac{1}{4}$

62. $b + \frac{1}{6} = -\frac{1}{3}$

63. $x + \frac{2}{3} = \frac{3}{4}$

64. $n + \frac{2}{5} = \frac{2}{3}$

65. $-\dfrac{5}{6} = x - \dfrac{1}{4}$

66. $-\dfrac{1}{4} = c - \dfrac{2}{3}$

67. $-\dfrac{1}{21} = m + \dfrac{2}{3}$

68. $\dfrac{5}{9} = b - \dfrac{1}{3}$

69. $\dfrac{5}{12} = n + \dfrac{3}{4}$

70. $d + 1.3619 = 2.0148$

71. $w + 2.932 = 4.801$

72. $-0.813 + x = -1.096$

73. $-1.926 + t = -1.042$

3 Solve and check.

74. $5x = 15$

75. $4y = 28$

76. $3b = -12$

77. $2a = -14$

78. $-3x = 6$

79. $-5m = 20$

80. $-3x = -27$

81. $-6n = -30$

82. $20 = 4c$

83. $18 = 2t$

84. $-32 = 8w$

85. $-56 = 7x$

86. $8d = 0$

87. $-5x = 0$

88. $36 = 9z$

89. $35 = -5x$

90. $-64 = 8a$

91. $-32 = -4y$

92. $-42 = 6t$

93. $-12m = -144$

94. $\dfrac{x}{3} = 2$

95. $\dfrac{x}{4} = 3$

96. $-\dfrac{y}{2} = 5$

97. $-\dfrac{b}{3} = 6$

98. $\dfrac{n}{7} = -4$

99. $\dfrac{t}{6} = -3$

100. $\dfrac{2}{5}x = 12$

101. $-\dfrac{4}{3}c = -8$

102. $\dfrac{5}{6}y = -20$

103. $-\dfrac{2}{3}d = 8$

104. $-\dfrac{3}{5}m = 12$

105. $\dfrac{2n}{3} = 2$

106. $\dfrac{5x}{6} = -10$

107. $\dfrac{-3z}{8} = 9$

108. $\dfrac{-4x}{5} = -12$

109. $-6 = -\dfrac{2}{3}y$

110. $-15 = -\dfrac{3}{5}x$

111. $\dfrac{2}{9} = \dfrac{2}{3}y$

112. $-\dfrac{6}{7} = -\dfrac{3}{4}b$

113. $-\dfrac{2}{5}m = -\dfrac{6}{7}$

114. $5x + 2x = 14$

115. $3n + 2n = 20$

116. $7d - 4d = 9$

117. $10y - 3y = 21$

118. $2x - 5x = 9$

119. $\dfrac{x}{1.4} = 3.2$

120. $\dfrac{z}{2.9} = -7.8$

121. $3.4a = 7.004$

122. $2.3m = 2.415$

123. $-3.7x = 7.77$

124. $-1.6m = 5.44$

4 Solve.

125. 12 is what percent of 50?

126. What percent of 125 is 50?

127. Find 18% of 40.

128. What is 25% of 60?

129. 12% of what is 48?

130. 45% of what is 9?

131. What is $33\dfrac{1}{3}\%$ of 27?

132. Find $16\dfrac{2}{3}\%$ of 30.

133. What percent of 12 is 3?

134. 10 is what percent of 15?

135. 60% of what is 3?

136. 75% of what is 6?

137. 12 is what percent of 6?

138. 20 is what percent of 16?

139. $5\dfrac{1}{4}\%$ of what is 21?

140. $37\dfrac{1}{2}\%$ of what is 15?

141. Find 15.4% of 50.

142. What is 18.5% of 46?

143. 1 is 0.5% of what?

144. 3 is 1.5% of what?

145. $\frac{3}{4}$% of what is 3?

146. $\frac{1}{2}$% of what is 3?

147. Find 125% of 16.

148. What is 250% of 12?

149. 16.4 is what percent of 20.4? Round to the nearest percent.

150. Find 18.3% of 625. Round to the nearest tenth.

151. A large grocery store chain expects to make a profit of 1.8% of total sales. What is the expected profit for a day in which total sales are $1,000,000?

152. Approximately 21% of air is oxygen. Using this estimate, determine how many liters of oxygen are in a room containing 21,600 L of air.

153. The cost of a 30 s television commercial during the 1967 Super Bowl was $175,000 (in 1994 dollars). In 1994, a 30 s commercial for the Super Bowl cost $900,000. What percent of the 1967 commercial cost is the 1994 commercial cost? Round to the nearest percent.

154. A ski vacation package regularly costs $1450 for one week, but people who make an early reservation receive a 15% discount. Find the amount of the early-reservation discount.

155. A football stadium increased its 60,000-seat capacity by 15%. How many seats were added to the stadium?

156. To override a presidential veto, at least $66\frac{2}{3}$% of the Senate must vote to override the veto. There are 100 senators in the Senate. What is the minimum number of votes needed to override a veto?

157. To receive a B grade in a history course, a student must correctly answer 75 of the 90 questions on an exam. What percent of the questions must a student answer correctly to receive a B grade?

158. In a recent city election, 16,400 out of 80,000 registered voters voted. What percent of the people voted in the election?

159. The number of women serving as president of an accredited college or university during the 1994–95 school year was approximately 425. This represented 14% of the schools nationwide. To the nearest thousand, how many accredited colleges and universities are in the United States?

160. The circle graph at the right shows the sales of full-size pickup trucks for different manufacturers during a recent year. What percent of the total number of pickups sold by these companies were Ford F-series pickups? Round to the nearest tenth of a percent.

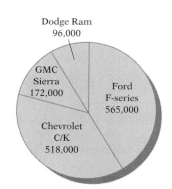

Dodge Ram
96,000

GMC
Sierra
172,000

Ford
F-series
565,000

Chevrolet
C/K
518,000

161. A university consists of three colleges: business, engineering, and fine arts. There are 2900 students in the business college, 1500 students in the engineering college, and 1000 students in the fine arts college. What percent of the total number of students in the university are in the fine arts college? Round to the nearest percent.

162. An airline knowingly overbooks flights by selling 18% more tickets than there are seats available. How many tickets would this airline sell for an airplane that has 150 seats?

163. A survey of 250 people was conducted to determine soft drink preferences. 100 people preferred cola flavor, 60 people preferred lemon/lime flavor, 50 people preferred orange flavor, and 40 people preferred cherry flavor. What percent of the people surveyed preferred a drink that was not cola flavored?

164. There are approximately 8760 h in one year. It is estimated that children aged 2 to 5 watch 25 h of television each week. What percent of the year does a child aged 2 to 5 spend watching TV? Round to the nearest tenth of a percent.

165. The total sales for December for a hobby shop were $25,000. For January, total sales showed an 11% decrease from December's sales. What were the total sales for January?

SUPPLEMENTAL EXERCISES 3.1

Solve.

166. $\dfrac{2m + m}{5} = -9$

167. $\dfrac{3y - 8y}{7} = 15$

168. $\dfrac{1}{\dfrac{1}{x}} = 5$

169. $\dfrac{1}{\dfrac{1}{x}} + 8 = -19$

170. $\dfrac{4}{\dfrac{3}{b}} = 8$

171. $\dfrac{5}{\dfrac{7}{a}} - \dfrac{3}{\dfrac{7}{a}} = 6$

172. Solve for x: $x \div 28 = 1481$ remainder 25

173. Find the value of x in the diagram shown at the right.

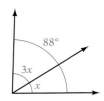

88°

3x

x

174. Your bill for dinner, including a 7.25% sales tax, was $62.74. You want to leave a 15% tip on the cost of the dinner before the sales tax. Find the amount of the tip to the nearest dollar.

175. The total cost for a dinner was $54.86. This included a 15% tip calculated on the cost of the dinner after a 6% sales tax. Find the cost of dinner.

176. A retailer decides to increase the original price of each item in the store by 10%. After the price increase, the retailer notices a significant drop in sales and so decides to reduce the current price of each item in the store by 10%. Are the prices back to the original prices? If not, are the prices lower or higher than the original prices?

177. If a quantity increases by 100%, how many times its original value is the new value?

178. Make up an equation of the form $x + a = b$ that has 2 as a solution.

179. Make up an equation of the form $ax = b$ that has -2 as a solution.

180. Suppose you have a $100,000 30-year mortgage. What is the difference between the monthly payment if the interest rate on your loan is 9.75% and the monthly payment if the interest rate is 9.25%? (*Hint:* You will need to discuss the question with an agent at a bank that provides mortgage services, discuss the question with a real estate agent, or find and use the formula for determining the monthly payment on a mortgage.)

181. Explain the difference between the word *expression* and the word *equation*.

182. The following problem does not contain enough information: "How many hours does it take to fly from Los Angeles to New York?" What additional information do we need in order to answer the question?

183. Write an explanation of the data presented in the table below. Include an explanation of how the percent gaps were calculated. (An MBA is a Masters degree in Business Administration.)

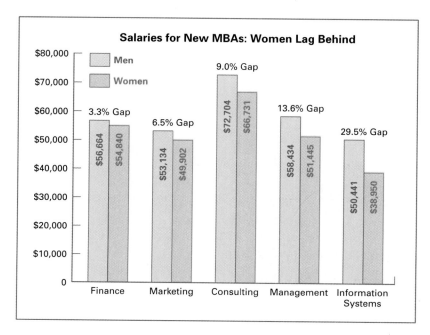

Salaries for New MBAs: Women Lag Behind

S E C T I O N **3.2**

General Equations

1 Solve equations of the form $ax + b = c$

In solving an equation of the form $ax + b = c$, the goal is to rewrite the equation in the form *variable = constant*. This requires applying both the Addition and the Multiplication Properties of Equations.

To solve the equation $\frac{2}{5}x - 3 = -7$, add 3 to each side of the equation.

$$\frac{2}{5}x - 3 = -7$$

$$\frac{2}{5}x - 3 + 3 = -7 + 3$$

Simplify.

$$\frac{2}{5}x = -4$$

Multiply each side of the equation by the reciprocal of the coefficient $\frac{2}{5}$.

$$\frac{5}{2} \cdot \frac{2}{5}x = \frac{5}{2}(-4)$$

Simplify. Now the equation is in the form *variable = constant*.

$$x = -10$$

Check: $\frac{2}{5}x - 3 = -7$

$$\begin{array}{c|c} \frac{2}{5}(-10) - 3 & -7 \\ -4 - 3 & -7 \\ -7 = -7 & \end{array}$$

Write the solution.

The solution is -10.

Example 1   Solve: $3x - 7 = -5$

Solution
$$3x - 7 = -5$$
$$3x - 7 + 7 = -5 + 7 \qquad \blacktriangleright \text{ Add 7 to each side of the equation.}$$
$$3x = 2 \qquad \blacktriangleright \text{ Simplify.}$$
$$\frac{3x}{3} = \frac{2}{3} \qquad \blacktriangleright \text{ Divide each side of the equation by 3.}$$
$$x = \frac{2}{3} \qquad \blacktriangleright \text{ Simplify. Now the equation is in the form } \textit{variable = constant}.$$

The solution is $\frac{2}{3}$. $\blacktriangleright$ Write the solution.

Problem 1   Solve: $5x + 7 = 10$

Solution   See page A16.

Example 2   Solve: $5 = 9 - 2x$

Solution
$$5 = 9 - 2x$$
$$5 - 9 = 9 - 9 - 2x$$ ▶ Subtract 9 from each side of the equation.
$$-4 = -2x$$ ▶ Simplify.
$$\frac{-4}{-2} = \frac{-2x}{-2}$$ ▶ Divide each side of the equation by -2.
$$2 = x$$ ▶ Simplify.

The solution is 2. ▶ Write the solution.

Problem 2   Solve: $11 = 11 + 3x$

Solution   See page A16.

Note in Example 4 on page 89 that the original equation, $\frac{1}{2} = x + \frac{2}{3}$, contained fractions with denominators of 2 and 3. The least common multiple of 2 and 3 is 6. The least common multiple has the property that both 2 and 3 will divide evenly into it. Therefore, if both sides of the equation are multiplied by 6, both of the denominators will divide evenly into 6. The result is an equation that does not contain any fractions. Multiplying an equation that contains fractions by the LCM of the denominators is called **clearing denominators**. It is an alternative method of solving an equation that contains fractions.

$$\frac{1}{2} = x + \frac{2}{3}$$

Multiply both sides of the equation by 6, the LCM of the denominators.
$$6\left(\frac{1}{2}\right) = 6\left(x + \frac{2}{3}\right)$$

Simplify each side of the equation. Use the Distributive Property on the right side of the equation.
$$3 = 6(x) + 6\left(\frac{2}{3}\right)$$

Note that multiplying both sides of the equation by the LCM of the denominators eliminated the fractions.
$$3 = 6x + 4$$

Solve the resulting equation.
$$3 - 4 = 6x + 4 - 4$$
$$-1 = 6x$$
$$\frac{-1}{6} = \frac{6x}{6}$$
$$-\frac{1}{6} = x$$

The solution is $-\frac{1}{6}$.

Example 3   Solve: $\dfrac{2}{3} + \dfrac{1}{4}x = -\dfrac{1}{3}$

Solution

$$\dfrac{2}{3} + \dfrac{1}{4}x = -\dfrac{1}{3}$$

▶ The equation contains fractions. Find the LCM of the denominators.

$$12\left(\dfrac{2}{3} + \dfrac{1}{4}x\right) = 12\left(-\dfrac{1}{3}\right)$$

▶ The LCM of 3 and 4 is 12. Multiply each side of the equation by 12.

$$12\left(\dfrac{2}{3}\right) + 12\left(\dfrac{1}{4}x\right) = -4$$

▶ Use the Distributive Property to multiply the left side of the equation by 12.

$$8 + 3x = -4$$
$$8 - 8 + 3x = -4 - 8$$

▶ The equation now contains no fractions.

$$3x = -12$$
$$\dfrac{3x}{3} = \dfrac{-12}{3}$$
$$x = -4$$

The solution is −4.

Problem 3   Solve: $\dfrac{5}{2} - \dfrac{2}{3}x = \dfrac{1}{2}$

Solution   See page A17.

---

2   Solve equations of the form $ax + b = cx + d$

In solving an equation of the form $ax + b = cx + d$, the goal is to rewrite the equation in the form *variable = constant*. Begin by rewriting the equation so that there is only one variable term in the equation. Then rewrite the equation so that there is only one constant term.

Solve: $4x - 5 = 6x + 11$

	$4x - 5 = 6x + 11$
Subtract 6x from each side of the equation.	$4x - 6x - 5 = 6x - 6x + 11$
Simplify. Now there is only one variable term in the equation.	$-2x - 5 = 11$
Add 5 to each side of the equation.	$-2x - 5 + 5 = 11 + 5$
Simplify. Now there is only one constant term in the equation.	$-2x = 16$
Divide each side of the equation by −2.	$\dfrac{-2x}{-2} = \dfrac{16}{-2}$
Simplify. Now the equation is in the form *variable = constant*.	$x = -8$

Check: $4x - 5 = 6x + 11$

$4(-8) - 5$	$6(-8) + 11$
$-32 - 5$	$-48 + 11$
$-37 =$	-37

Write the solution.

The solution is −8.

Example 4   Solve: $4x - 3 = 8x - 7$

Solution

$$4x - 3 = 8x - 7$$
$$4x - 8x - 3 = 8x - 8x - 7 \quad \blacktriangleright \text{ Subtract } 8x \text{ from each side of the equation.}$$
$$-4x - 3 = -7 \quad \blacktriangleright \text{ Simplify. Now there is only one variable term in the equation.}$$
$$-4x - 3 + 3 = -7 + 3 \quad \blacktriangleright \text{ Add 3 to each side of the equation.}$$
$$-4x = -4 \quad \blacktriangleright \text{ Simplify. Now there is only one constant term in the equation.}$$
$$\frac{-4x}{-4} = \frac{-4}{-4} \quad \blacktriangleright \text{ Divide each side of the equation by } -4.$$
$$x = 1 \quad \blacktriangleright \text{ Simplify. Now the equation is in the form } variable = constant.$$

The solution is 1. $\blacktriangleright$ Write the solution.

Problem 4   Solve: $5x + 4 = 6 + 10x$

Solution   See page A17.

---

3   ## Solve equations containing parentheses

When an equation contains parentheses, one of the steps in solving the equation requires the use of the Distributive Property. The Distributive Property is used to remove parentheses from a variable expression.

$a(b + c) = ab + ac$

Solve: $4 + 5(2x - 3) = 3(4x - 1)$

Use the Distributive Property to remove parentheses.

$$4 + 5(2x - 3) = 3(4x - 1)$$
$$4 + 10x - 15 = 12x - 3$$

Simplify.

$$10x - 11 = 12x - 3$$

Subtract $12x$ from each side of the equation.

$$10x - 12x - 11 = 12x - 12x - 3$$

Simplify. Now there is only one variable term in the equation.

$$-2x - 11 = -3$$

Add 11 to each side of the equation.

$$-2x - 11 + 11 = -3 + 11$$

Simplify. Now there is only one constant term in the equation.

$$-2x = 8$$

Divide each side of the equation by -2.

$$\frac{-2x}{-2} = \frac{8}{-2}$$

Simplify. Now the equation is in the form $variable = constant.$

$$x = -4$$

Check:
$$\frac{4 + 5(2x - 3) = 3(4x - 1)}{}$$

$4 + 5[2(-4) - 3]$	$3[4(-4) - 1]$
$4 + 5(-8 - 3)$	$3(-16 - 1)$
$4 + 5(-11)$	$3(-17)$
$4 - 55$	-51
$-51 = -51$	

Write the solution.　　　　　　　　The solution is -4.

In Chapter 2, we discussed the use of a graphing calculator to evaluate variable expressions. The same procedure can be used to check the solution of an equation. Consider the example above. After we divide both sides of the equation by -2, the solution appears to be -4. To check this solution, store the value of x, -4, in the calculator. Evaluate the expression on the left side of the original equation: $4 + 5(2x - 3)$. The result is -51. Now evaluate the expression on the right side of the original equation: $3(4x - 1)$. The result is -51. Because the results are equal, the solution -4 checks. See the Appendix on page A1 for a description of keystroking procedures for different models of graphing calculators.

Example 5　Solve: $3x - 4(2 - x) = 3(x - 2) - 4$

Solution　$3x - 4(2 - x) = 3(x - 2) - 4$
　　　　　$3x - 8 + 4x = 3x - 6 - 4$　　▶ Use the Distributive Property to remove parentheses.

　　　　　　$7x - 8 = 3x - 10$　　▶ Simplify.
　　　$7x - 3x - 8 = 3x - 3x - 10$　　▶ Subtract $3x$ from each side of the equation.
　　　　　　$4x - 8 = -10$
　　　$4x - 8 + 8 = -10 + 8$　　▶ Add 8 to each side of the equation.
　　　　　　　$4x = -2$
　　　　　　$\dfrac{4x}{4} = \dfrac{-2}{4}$　　▶ Divide each side of the equation by 4.

　　　　　　　$x = -\dfrac{1}{2}$　　▶ The equation is in the form *variable = constant*.

　　　　The solution is $-\dfrac{1}{2}$.

Problem 5　Solve: $5x - 4(3 - 2x) = 2(3x - 2) + 6$

Solution　See page A17.

Example 6　Solve: $3[2 - 4(2x - 1)] = 4x - 10$

Solution　$3[2 - 4(2x - 1)] = 4x - 10$
　　　　　$3[2 - 8x + 4] = 4x - 10$　　▶ Use the Distributive Property to remove the parentheses.

　　　　　　$3[6 - 8x] = 4x - 10$　　▶ Simplify inside the brackets.
　　　　　$18 - 24x = 4x - 10$　　▶ Use the Distributive Property to remove the brackets.

$$18 - 24x - 4x = 4x - 4x - 10$$ ▸ Subtract $4x$ from each side
$$18 - 28x = -10$$ of the equation.
$$18 - 18 - 28x = -10 - 18$$ ▸ Subtract 18 from each side
$$-28x = -28$$ of the equation.
$$\frac{-28x}{-28} = \frac{-28}{-28}$$ ▸ Divide each side of the
$$x = 1$$ equation by -28.

The solution is 1.

Problem 6 Solve: $-2[3x - 5(2x - 3)] = 3x - 8$

Solution See page A17.

4 Application problems

Example 7 The pressure at a certain depth in the ocean can be approximated by the equation $P = 15 + \frac{1}{2}D$, where P is the pressure in pounds per square inch, and D is the depth in feet. Use this equation to find the depth when the pressure is 35 lb/in².

Strategy The pressure is 35 lb/in². Therefore, $P = 35$. To find the depth, replace P in the given equation by 35, and solve for D.

Solution
$$P = 15 + \frac{1}{2}D$$

$$35 = 15 + \frac{1}{2}D$$

$$35 - 15 = 15 - 15 + \frac{1}{2}D$$

$$20 = \frac{1}{2}D$$

$$2 \cdot 20 = 2 \cdot \frac{1}{2}D$$

$$40 = D$$

The depth is 40 ft.

Problem 7 To determine the cost of production, an economist uses the equation $T = U \cdot N + F$, where T is the total cost, U is the unit cost, N is the number of units made, and F is the fixed cost. Use this equation to find the number of units made during a month when the total cost was $8000, the unit cost was $15, and the fixed costs were $2000.

Solution See page A17.

A lever system is shown below. It consists of a lever, or bar; a fulcrum; and two forces, F_1 and F_2. The distance d represents the length of the lever, x represents the distance from F_1 to the fulcrum, and $d - x$ represents the distance from F_2 to the fulcrum.

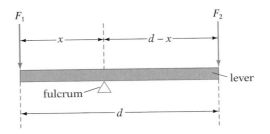

A principle of physics states that when the lever system balances,

$$F_1x = F_2(d - x)$$

Example 8   A lever is 10 ft long. A force of 100 lb is applied to one end of the lever, and a force of 400 lb is applied to the other end. What is the location of the fulcrum when the system balances?

Strategy   Draw a diagram of the situation.

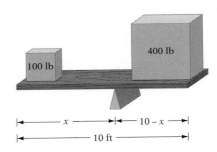

The lever is 10 ft long, so $d = 10$. One force is 100 lb, so $F_1 = 100$. The other force is 400 lb, so $F_2 = 400$. To find the location of the fulcrum when the system balances, replace the variables F_1, F_2, and d in the lever system equation with the given values, and solve for x.

Solution
$$F_1x = F_2(d - x)$$
$$100x = 400(10 - x)$$
$$100x = 4000 - 400x$$
$$100x + 400x = 4000 - 400x + 400x$$
$$500x = 4000$$
$$\frac{500x}{500} = \frac{4000}{500}$$
$$x = 8$$

The fulcrum is 8 ft from the 100 lb force.

Problem 8   A lever is 14 ft long. At a distance of 6 ft from the fulcrum, a force of 40 lb is applied. How large a force must be applied to the other end of the lever so that the system will balance?

Solution   See page A17.

STUDY TIPS

REVIEWING MATERIAL

Reviewing material is the repetition that is essential for learning. Much of what we learn is soon forgotten unless we review it. If you find that you do not remember information that you studied previously, you probably have not reviewed it sufficiently. *You will remember best what you review most.*

One method of reviewing material is to begin a study session by reviewing a concept you have studied previously. For example, before trying to solve a new type of problem, spend a few minutes solving a kind of problem you already know how to solve. Not only will you provide yourself with the review practice you need, but you are also likely to put yourself in the right frame of mind for learning how to solve the new type of problem.

CONCEPT REVIEW 3.2

Determine whether the statement is always true, sometimes true, or never true.

1. The same variable term can be added to both sides of an equation without changing the solution of the equation.

2. The same variable term can be subtracted from both sides of an equation without changing the solution of the equation.

3. An equation of the form $ax + b = c$ cannot be solved if a is a negative number.

4. The solution of the equation $\frac{x}{3} = 0$ is 0.

5. The solution of the equation $\frac{x}{0} = 3$ is 0.

6. In solving an equation of the form $ax + b = cx + d$, the goal is to rewrite the equation in the form *variable = constant*.

7. In solving an equation of the form $ax + b = cx + d$, subtracting cx from each side of the equation results in an equation with only one variable term in it.

EXERCISES 3.2

1   Solve and check.

1. $3x + 1 = 10$

2. $4y + 3 = 11$

3. $2a - 5 = 7$

4. $5m - 6 = 9$

5. $5 = 4x + 9$

6. $2 = 5b + 12$

7. $13 = 9 + 4z$

8. $7 - c = 9$

9. $2 - x = 11$

10. $4 - 3w = -2$

11. $5 - 6x = -13$

12. $8 - 3t = 2$

13. $-5d + 3 = -12$

14. $-8x - 3 = -19$

15. $-7n - 4 = -25$

16. $-12x + 30 = -6$

17. $-13 = -11y + 9$

18. $2 = 7 - 5a$

19. $3 = 11 - 4n$

20. $-35 = -6b + 1$

21. $-8x + 3 = -29$

22. $-3m - 21 = 0$

23. $7x - 3 = 3$

24. $8y + 3 = 7$

25. $6a + 5 = 9$

26. $3m + 4 = 11$

27. $11 = 15 + 4n$

28. $4 = 2 - 3c$

29. $9 - 4x = 6$

30. $7 - 8z = 0$

31. $1 - 3x = 0$

32. $6y - 5 = -7$

33. $8b - 3 = -9$

34. $5 - 6m = 2$

35. $7 - 9a = 4$

36. $9 = -12c + 5$

37. $10 = -18x + 7$

38. $2y + \frac{1}{3} = \frac{7}{3}$

39. $3x - \frac{5}{6} = \frac{13}{6}$

40. $5y + \frac{3}{7} = \frac{3}{7}$

41. $9x + \frac{4}{5} = \frac{4}{5}$

42. $8 = 7d - 1$

43. $8 = 10x - 5$

44. $4 = 7 - 2w$

45. $7 = 9 - 5a$

46. $-6y + 5 = 13$

47. $-4x + 3 = 9$

48. $\frac{1}{2}a - 3 = 1$

49. $\frac{1}{3}m - 1 = 5$

50. $\frac{2}{5}y + 4 = 6$

51. $\frac{3}{4}n + 7 = 13$

52. $-\frac{2}{3}x + 1 = 7$

53. $-\frac{3}{8}b + 4 = 10$

54. $\frac{x}{4} - 6 = 1$

55. $\frac{y}{5} - 2 = 3$

56. $\frac{2x}{3} - 1 = 5$

57. $\frac{3c}{7} - 1 = 8$

58. $4 - \frac{3}{4}z = -2$

59. $3 - \frac{4}{5}w = -9$

60. $5 + \frac{2}{3}y = 3$

61. $17 + \frac{5}{8}x = 7$

62. $17 = 7 - \frac{5}{6}t$

63. $9 = 3 - \frac{2x}{7}$

64. $3 = \frac{3a}{4} + 1$

65. $7 = \frac{2x}{5} + 4$

66. $5 - \frac{4c}{7} = 8$

67. $7 - \frac{5}{9}y = 9$

68. $6a + 3 + 2a = 11$

69. $5y + 9 + 2y = 23$

70. $7x - 4 - 2x = 6$

71. $11z - 3 - 7z = 9$

72. $2x - 6x + 1 = 9$

73. $b - 8b + 1 = -6$

74. $-1 = 5m + 7 - m$

75. $8 = 4n - 6 + 3n$

76. $0.15y + 0.025 = -0.074$

77. $1.2x - 3.44 = 1.3$

78. $3.5 = 3.5 + 0.076x$

79. $-6.5 = 4.3y - 3.06$

Solve.

80. If $2x - 3 = 7$, evaluate $3x + 4$.

81. If $3x + 5 = -4$, evaluate $2x - 5$.

82. If $4 - 5x = -1$, evaluate $x^2 - 3x + 1$.

83. If $2 - 3x = 11$, evaluate $x^2 + 2x - 3$.

2   Solve and check.

84. $8x + 5 = 4x + 13$ 85. $6y + 2 = y + 17$ 86. $7m + 4 = 6m + 7$

87. $11n + 3 = 10n + 11$ 88. $5x - 4 = 2x + 5$ 89. $9a - 10 = 3a + 2$

90. $12y - 4 = 9y - 7$ 91. $13b - 1 = 4b - 19$ 92. $15x - 2 = 4x - 13$

93. $7a - 5 = 2a - 20$ 94. $3x + 1 = 11 - 2x$ 95. $n - 2 = 6 - 3n$

96. $2x - 3 = -11 - 2x$ 97. $4y - 2 = -16 - 3y$ 98. $2b + 3 = 5b + 12$

99. $m + 4 = 3m + 8$ 100. $4x - 7 = 5x + 1$ 101. $6d - 2 = 7d + 5$

102. $4y - 8 = y - 8$ 103. $5a + 7 = 2a + 7$ 104. $6 - 5x = 8 - 3x$

105. $10 - 4n = 16 - n$ 106. $5 + 7x = 11 + 9x$ 107. $3 - 2y = 15 + 4y$

108. $2x - 4 = 6x$ 109. $2b - 10 = 7b$ 110. $8m = 3m + 20$

111. $9y = 5y + 16$ 112. $-3x - 4 = 2x + 6$ 113. $-5a - 3 = 2a + 18$

114. $-8n + 3 = -5n - 6$ 115. $-10x + 4 = -x - 14$ 116. $-x - 4 = -3x - 16$

117. $8 - 4x = 18 - 5x$ 118. $6 - 10a = 8 - 9a$ 119. $5 - 7m = 2 - 6m$

120. $8b + 5 = 5b + 7$ 121. $6y - 1 = 2y + 2$ 122. $7x - 8 = x - 3$

123. $10x - 3 = 3x - 1$ 124. $5n + 3 = 2n + 1$ 125. $8a - 2 = 4a - 5$

126. $8.7y = 3.9y + 9.6$ 127. $4.5x - 5.4 = 2.7x$ 128. $5.6x = 7.2x - 6.4$

Solve.

129. If $5x = 3x - 8$, evaluate $4x + 2$.   130. If $7x + 3 = 5x - 7$, evaluate $3x - 2$.

131. If $2 - 6a = 5 - 3a$, evaluate $4a^2 - 2a + 1$.   132. If $1 - 5c = 4 - 4c$, evaluate $3c^2 - 4c + 2$.

3   Solve and check.

133. $5x + 2(x + 1) = 23$ 134. $6y + 2(2y + 3) = 16$ 135. $9n - 3(2n - 1) = 15$

136. $12x - 2(4x - 6) = 28$ 137. $7a - (3a - 4) = 12$ 138. $9m - 4(2m - 3) = 11$

139. $5(3 - 2y) + 4y = 3$ 140. $4(1 - 3x) + 7x = 9$ 141. $10x + 1 = 2(3x + 5) - 1$

142. $5y - 3 = 7 + 4(y - 2)$ 143. $4 - 3a = 7 - 2(2a + 5)$ 144. $9 - 5x = 12 - (6x + 7)$

145. $3y - 7 = 5(2y - 3) + 4$

146. $2a - 5 = 4(3a + 1) - 2$

147. $5 - (9 - 6x) = 2x - 2$

148. $7 - (5 - 8x) = 4x + 3$

149. $3[2 - 4(y - 1)] = 3(2y + 8)$

150. $5[2 - (2x - 4)] = 2(5 - 3x)$

151. $3a + 2[2 + 3(a - 1)] = 2(3a + 4)$

152. $5 + 3[1 + 2(2x - 3)] = 6(x + 5)$

153. $-2[4 - (3b + 2)] = 5 - 2(3b + 6)$

154. $-4[x - 2(2x - 3)] + 1 = 2x - 3$

155. $0.3x - 2(1.6x) - 8 = 3(1.9x - 4.1)$

156. $0.56 - 0.4(2.1y + 3) = 0.2(2y + 6.1)$

Solve.

157. If $4 - 3a = 7 - 2(2a + 5)$, evaluate $a^2 + 7a$.

158. If $9 - 5x = 12 - (6x + 7)$, evaluate $x^2 - 3x - 2$.

159. If $2z - 5 = 3(4z + 5)$, evaluate $\dfrac{z^2}{z - 2}$.

160. If $3n - 7 = 5(2n + 7)$, evaluate $\dfrac{n^2}{2n - 6}$.

4   Solve.

The distance s that an object will fall in t seconds is given by $s = 16t^2 + vt$, where v is the initial velocity of the object.

161.   Find the initial velocity of an object that falls 80 ft in 2 s.

162.   Find the initial velocity of an object that falls 144 ft in 3 s.

A company uses the equation $V = C - 6000t$ to determine the depreciated value V, after t years, of a milling machine that originally cost C dollars. Equations such as this are used in accounting for straight-line depreciation.

163.   A milling machine originally cost $50,000. In how many years will the depreciated value be $38,000?

164.   A milling machine originally cost $78,000. In how many years will the depreciated value be $48,000?

When two lines intersect, four angles are formed, as shown at the right. Two angles that are on opposite sides of the intersection of two lines are called **vertical angles.** Angle w and angle y are vertical angles. Angle x and angle z are vertical angles. Vertical angles have the same measure.

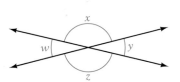

165.   Find the value of x.

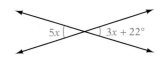

166.   Find the value of y.

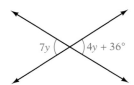

Anthropologists can approximate the height of a primate from the size of its humerus (the bone extending from the shoulder to the elbow) by using the equation $H = 1.2L + 27.8$, where L is the length of the humerus and H is the height of the primate.

167. An anthropologist estimates the height of a primate to be 66 in. What is the approximate length of the humerus of this primate? Round to the nearest tenth.

168. The height of a primate is estimated to be 62 in. Find the approximate length of the humerus of this primate.

The world record time for a 1 mi race can be approximated by the equation $t = 17.08 - 0.0067y$, where t is the time and y is the year of the race.

169. Approximate the year in which the first "4 min mile" was run. (The actual year was 1954.) Round to the nearest whole number.

170. In 1985, the world record for a 1 mi race was 3.77 min. For what year does the equation predict this record time? Round to the nearest whole number.

A telephone company estimates that the number N of phone calls made per day between two cities of population P_1 and P_2 that are d miles apart is given by the equation $N = \frac{2.51P_1P_2}{d^2}$.

171. Estimate the population P_2, given that P_1 is 48,000, the number of phone calls is 1,100,000, and the distance between the cities is 75 mi. Round to the nearest thousand.

172. Estimate the population P_1, given that P_2 is 125,000, the number of phone calls is 2,500,000, and the distance between the cities is 50 mi. Round to the nearest thousand.

To determine the break-even point, or the number of units that must be sold so that no profit or loss occurs, an economist uses the equation $Px = Cx + F$, where P is the selling price per unit, x is the number of units sold, C is the cost to make each unit, and F is the fixed cost.

173. An economist has determined that the selling price per unit for a television is $250. The cost to make one television is $165, and the fixed cost is $29,750. Find the break-even point.

174. A business analyst has determined that the selling price per unit for a cellular telephone is $98. The cost to make one cellular telephone is $67, and the fixed cost is $5890. Find the break-even point.

175. A manufacturing engineer determines that the cost per unit for a compact disc is $3.35 and that the fixed cost is $6180. The selling price for the compact disc is $8.50. Find the break-even point.

176. The manufacture of a softball bat requires two steps: (1) cut the rough shape and (2) sand the bat to its final form. The cost to rough-shape the bat is $.45, and the cost to sand the bat to final form is $1.05. Find the break-even point if the selling price per bat is $7 and the fixed cost is $16,500.

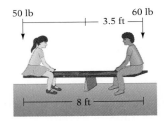

Figure 1

Use the lever-system equation $F_1x = F_2(d - x)$.

177. Two children are sitting 8 ft apart on a see-saw. One child weighs 60 lb and the second child weighs 50 lb. The fulcrum is 3.5 ft from the child weighing 60 lb. Is the see-saw balanced? (See Figure 1.)

178. An adult and a child are on a see-saw 14 ft long. The adult weighs 175 lb and the child weighs 70 lb. How many feet from the child must the fulcrum be placed so that the see-saw balances?

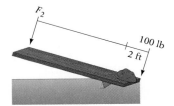

Figure 2

179. A lever 10 ft long is used to balance a 100 lb rock. The fulcrum is placed 2 ft from the rock. What force must be applied to the other end of the lever to balance the rock? (See Figure 2.)

180. A 50 lb weight is applied to the left end of a see-saw 10 ft long. The fulcrum is 4 ft from the 50 lb weight. A weight of 30 lb is applied to the right end of the see-saw. Is the 30 lb weight adequate to balance the see-saw?

181. In preparation for a stunt, two acrobats are standing on a plank 18 ft long. One acrobat weighs 128 lb and the second acrobat weighs 160 lb. How far from the 128 lb acrobat must the fulcrum be placed so that the acrobats are balanced on the plank?

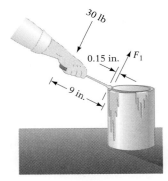

182. A screwdriver 9 in. long is used as a lever to open a can of paint. The tip of the screwdriver is placed under the lip of the lid with the fulcrum 0.15 in. from the lip. A force of 30 lb is applied to the other end of the screwdriver. Find the force on the lip of the lid. (See Figure 3.)

Figure 3

SUPPLEMENTAL EXERCISES 3.2

Solve. If the equation has no solution, write "No solution."

183. $3(2x - 1) - (6x - 4) = -9$

184. $\frac{1}{5}(25 - 10b) + 4 = \frac{1}{3}(9b - 15) - 6$

185. $3[4(w + 2) - (w + 1)] = 5(2 + w)$

186. $\frac{2(5x - 6) - 3(x - 4)}{7} = x + 2$

187.   Solve for x: $32{,}166 \div x = 518$ remainder 50

188.   One-half of a certain number equals two-thirds of the same number. Find the number.

189.   Write an equation of the form $ax + b = cx + d$ that has 4 as a solution.

190.   If a, b, and c are real numbers, is it always possible to solve the equation $ax + b = c$? If not, what values of a, b, and/or c must be excluded?

191.   Does the sentence "Solve $3x - 4(x - 1)$" make sense? Why or why not?

192.   The equation $x = x + 1$ has no solution, whereas the solution of the equation $2x + 1 = 1$ is zero. Is there a difference between no solution and a solution of zero? Explain your answer.

193.   Explain in your own words the steps you would take to solve the equation $\frac{3}{4}x - 7 = 2$. State the Property of Real Numbers or the Property of Equations that you used at each step.

194.   Explain what is wrong with the following demonstration, which suggests that $2 = 3$.

$$2x + 5 = 3x + 5$$
$$2x + 5 - 5 = 3x + 5 - 5 \quad \blacktriangleright \text{Subtract 5 from each side of the equation.}$$
$$2x = 3x$$
$$\frac{2x}{x} = \frac{3x}{x} \qquad\qquad \blacktriangleright \text{Divide each side of the equation by } x.$$
$$2 = 3$$

195.   Archimedes supposedly said, "Give me a long enough lever and I can move the world." Explain what Archimedes meant by this statement.

S E C T I O N   **3.3**

Inequalities

1   Solve inequalities using the Addition Property of Inequalities

An expression that contains the symbol $>$, $<$, $\geq$, or $\leq$ is called an **inequality**. An inequality expresses the relative order of two mathematical expressions. The expressions can be either numerical or variable expressions.

$$\left.\begin{array}{l} 4 > 2 \\ 3x \leq 7 \\ x^2 - 2x > y + 4 \end{array}\right\} \text{Inequalities}$$

The **solution set of an inequality** is a set of numbers, each element of which, when substituted for the variable, results in a true inequality. The solution set of an inequality can be graphed on the number line.

The graph of the solution set of $x > 1$ is shown at the right. The solution set is the real numbers greater than 1. The parenthesis on the graph indicates that 1 is not included in the solution set.

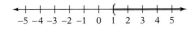

The graph of the solution set of $x \geq 1$ is shown at the right. The bracket at 1 indicates that 1 is included in the solution set.

The graph of the solution set of $x < -1$ is shown at the right. The numbers less than -1 are to the left of -1 on the number line.

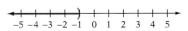

Example 1   Graph: $x < 3$

Solution   ▶ The solution set is the numbers less than 3.

Problem 1   Graph: $x > -2$

Solution   See page A18.

The inequality at the right is true if the variable is replaced by 7, 9.3, or $\frac{15}{2}$.

$$x + 3 > 8$$

$\left. \begin{array}{l} 7 + 3 > 8 \\ 9.3 + 3 > 8 \\ \frac{15}{2} + 3 > 8 \end{array} \right\}$ True inequalities

The inequality $x + 3 > 8$ is false if the variable is replaced by 4, 1.5, or $-\frac{1}{2}$.

$\left. \begin{array}{l} 4 + 3 > 8 \\ 1.5 + 3 > 8 \\ -\frac{1}{2} + 3 > 8 \end{array} \right\}$ False inequalities

There are many values of the variable x that make the inequality $x + 3 > 8$ true. The solution set of $x + 3 > 8$ is any number greater than 5.

The graph of the solution set of $x + 3 > 8$ is shown at the right.

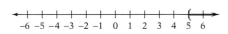

In solving an inequality, the goal is to rewrite the given inequality in the form *variable > constant* or *variable < constant*. The Addition Property of Inequalities is used to rewrite an inequality in this form.

Addition Property of Inequalities

> The same number can be added to each side of an inequality without changing the solution set of the inequality.
>
> $$\text{If } a > b, \text{ then } a + c > b + c.$$
> $$\text{If } a < b, \text{ then } a + c < b + c.$$

The Addition Property of Inequalities also holds true for an inequality containing the symbol $\geq$ or $\leq$.

The Addition Property of Inequalities is used when, in order to rewrite an inequality in the form *variable* $>$ *constant* or *variable* $<$ *constant*, a term must be removed from one side of the inequality. Add the opposite of the term to each side of the inequality.

Because subtraction is defined in terms of addition, the Addition Property of Inequalities makes it possible to subtract the same number from each side of an inequality without changing the solution set of the inequality.

To rewrite the inequality at the right, subtract 4 from each side of the inequality.

$$x + 4 < 5$$
$$x + 4 - 4 < 5 - 4$$

Simplify.

$$x < 1$$

The graph of the solution set of $x + 4 < 5$ is shown at the right.

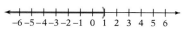

$$-6\ -5\ -4\ -3\ -2\ -1\ \ 0\ \ 1\ \ 2\ \ 3\ \ 4\ \ 5\ \ 6$$

To solve $5x - 6 \leq 4x - 4$, subtract $4x$ from each side of the inequality.

$$5x - 6 \leq 4x - 4$$
$$5x - 4x - 6 \leq 4x - 4x - 4$$

Simplify.

$$x - 6 \leq -4$$

Add 6 to each side of the inequality.

$$x - 6 + 6 \leq -4 + 6$$

Simplify.

$$x \leq 2$$

Example 2   Solve and graph the solution set of $x + 5 > 3$.

Solution
$$x + 5 > 3$$
$$x + 5 - 5 > 3 - 5 \quad \blacktriangleright \text{ Subtract 5 from each side of the inequality.}$$
$$x > -2$$

$$-5\ -4\ -3\ -2\ -1\ \ 0\ \ 1\ \ 2\ \ 3\ \ 4\ \ 5$$

Problem 2   Solve and graph the solution set of $x + 2 < -2$.

Solution   See page A18.

Example 3   Solve: $7x - 14 \leq 6x - 16$

Solution

$$7x - 14 \leq 6x - 16$$
$$7x - 6x - 14 \leq 6x - 6x - 16$$ ▸ Subtract $6x$ from each side of the inequality.
$$x - 14 \leq -16$$
$$x - 14 + 14 \leq -16 + 14$$ ▸ Add 14 to each side of the inequality.
$$x \leq -2$$

Problem 3   Solve: $5x + 3 > 4x + 5$

Solution   See page A18.

2 Solve inequalities using the Multiplication Property of Inequalities

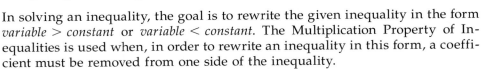

In solving an inequality, the goal is to rewrite the given inequality in the form *variable > constant* or *variable < constant*. The Multiplication Property of Inequalities is used when, in order to rewrite an inequality in this form, a coefficient must be removed from one side of the inequality.

Multiplication Property of Inequalities

Rule 1
Each side of an inequality can be multiplied by the same positive number without changing the solution set of the inequality.

If $a > b$ and $c > 0$, then $ac > bc$.   If $a < b$ and $c > 0$, then $ac < bc$.

LOOK CLOSELY
$c > 0$ means c is a positive number.

LOOK CLOSELY
The inequality symbols are not changed.

$$5 > 4$$ $$6 < 9$$
$$5(2) > 4(2)$$ $$6(3) < 9(3)$$
$$10 > 8$$ A true inequality $$18 < 27$$ A true inequality

Multiplication Property of Inequalities

Rule 2
If each side of an inequality is multiplied by the same negative number and the inequality symbol is reversed, then the solution set of the inequality is not changed.

If $a > b$ and $c < 0$, then $ac < bc$.   If $a < b$ and $c < 0$, then $ac > bc$.

LOOK CLOSELY
$c < 0$ means c is a negative number.

LOOK CLOSELY
The inequality symbol is reversed in each case.

$$5 > 4$$ $$6 < 9$$
$$5(-2) < 4(-2)$$ $$6(-3) > 9(-3)$$
$$-10 < -8$$ A true inequality $$-18 > -27$$ A true inequality

The Multiplication Property of Inequalities also holds true for an inequality containing the symbol $\geq$ or $\leq$.

To rewrite the inequality at the right, multiply each side of the inequality by the reciprocal of the coefficient $-\frac{3}{2}$. Because $-\frac{2}{3}$ is a negative number, the inequality symbol must be reversed.

Simplify.

$$-\frac{3}{2}x \leq 6$$

$$-\frac{2}{3}\left(-\frac{3}{2}x\right) \geq -\frac{2}{3}(6)$$

$$x \geq -4$$

The graph of the solution set of $-\frac{3}{2}x \leq 6$ is shown at the right.

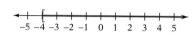

Recall that division is defined in terms of multiplication. Therefore, the Multiplication Property of Inequalities allows each side of an inequality to be divided by the same number. When each side of an inequality is divided by a positive number, the inequality symbol remains the same. When each side of an inequality is divided by a negative number, the inequality symbol must be reversed.

LOOK CLOSELY

Any time an inequality is multiplied or divided by a negative number, the inequality symbol must be reversed. Compare these two examples:

$2x < -4$ Divide each side
$\frac{2x}{2} < \frac{-4}{2}$ by *positive* 2. The
$x < -2$ inequality symbol *is not* reversed.

$-2x < 4$ Divide each side
$\frac{-2x}{-2} > \frac{4}{-2}$ by *negative* 2. The
$x > -2$ inequality symbol *is reversed.*

Solve: $-5x > 8$

Divide each side of the inequality by -5. Because -5 is a negative number, the inequality symbol must be reversed.

Simplify.

$$-5x > 8$$

$$\frac{-5x}{-5} < \frac{8}{-5}$$

$$x < -\frac{8}{5}$$

Example 4   Solve and graph the solution set of $7x < -14$.

Solution $7x < -14$

$\frac{7x}{7} < \frac{-14}{7}$ ▶ Divide each side of the inequality by 7. Because 7 is a
$x < -2$ *positive* number, do not change the inequality symbol.

Problem 4   Solve and graph the solution set of $3x < 9$.

Solution   See page A18.

Example 5   Solve: $-\frac{5}{8}x \leq \frac{5}{12}$

Solution $-\frac{5}{8}x \leq \frac{5}{12}$

$-\frac{8}{5}\left(-\frac{5}{8}x\right) \geq -\frac{8}{5}\left(\frac{5}{12}\right)$ ▶ Multiply each side of the inequality by the reciprocal of $-\frac{5}{8}$. $-\frac{8}{5}$ is a negative number. Reverse the inequality symbol.

$x \geq -\frac{2}{3}$

> **Problem 5**   Solve: $-\frac{3}{4}x \geq 18$
>
> **Solution**   See page A18.

3   Solve general inequalities

In solving an inequality, it is often necessary to apply both the Addition and the Multiplication Properties of Inequalities.

LOOK CLOSELY

Solving these inequalities is similar to solving the equations solved in Section 3.2 *except* that when you multiply or divide the inequality by a negative number, you must reverse the inequality symbol.

Solve: $3x - 2 < 5x + 4$

	$3x - 2 < 5x + 4$
Subtract $5x$ from each side of the inequality. Simplify.	$3x - 5x - 2 < 5x - 5x + 4$ $-2x - 2 < 4$
Add 2 to each side of the inequality. Simplify.	$-2x - 2 + 2 < 4 + 2$ $-2x < 6$
Divide each side of the inequality by -2. Because -2 is a negative number, the inequality symbol must be reversed. Simplify.	$\dfrac{-2x}{-2} > \dfrac{6}{-2}$ $x > -3$

> **Example 6**   Solve: $7x - 3 \leq 3x + 17$
>
> **Solution**
> $$7x - 3 \leq 3x + 17$$
> $$7x - 3x - 3 \leq 3x - 3x + 17 \qquad \blacktriangleright \text{Subtract } 3x \text{ from each side of the inequality.}$$
> $$4x - 3 \leq 17$$
> $$4x - 3 + 3 \leq 17 + 3 \qquad \blacktriangleright \text{Add 3 to each side of the inequality.}$$
> $$4x \leq 20$$
> $$\frac{4x}{4} \leq \frac{20}{4} \qquad \blacktriangleright \text{Divide each side of the inequality by 4.}$$
> $$x \leq 5$$
>
> **Problem 6**   Solve: $5 - 4x > 9 - 8x$
>
> **Solution**   See page A18.

When an inequality contains parentheses, one of the steps in solving the inequality requires the use of the Distributive Property.

To solve $-2(x - 7) > 3 - 4(2x - 3)$, use the Distributive Property to remove parentheses. Simplify.	$-2(x - 7) > 3 - 4(2x - 3)$ $-2x + 14 > 3 - 8x + 12$ $-2x + 14 > 15 - 8x$
Add $8x$ to each side of the inequality. Simplify.	$-2x + 8x + 14 > 15 - 8x + 8x$ $6x + 14 > 15$
Subtract 14 from each side of the inequality. Simplify.	$6x + 14 - 14 > 15 - 14$ $6x > 1$
Divide each side of the inequality by 6.	$\dfrac{6x}{6} > \dfrac{1}{6}$
Simplify.	$x > \dfrac{1}{6}$

Example 7   Solve: $3(3 - 2x) \geq -5x - 2(3 - x)$

Solution
$$3(3 - 2x) \geq -5x - 2(3 - x)$$
$$9 - 6x \geq -5x - 6 + 2x \qquad \blacktriangleright \text{Use the Distributive Property.}$$
$$9 - 6x \geq -3x - 6$$
$$9 - 6x + 3x \geq -3x + 3x - 6 \qquad \blacktriangleright \text{Add } 3x \text{ to each side of the in-}$$
$$9 - 3x \geq -6 \qquad\qquad\qquad\quad \text{equality.}$$
$$9 - 9 - 3x \geq -6 - 9 \qquad\quad \blacktriangleright \text{Subtract 9 from each side of}$$
$$-3x \geq -15 \qquad\qquad\qquad \text{the inequality.}$$
$$\frac{-3x}{-3} \leq \frac{-15}{-3} \qquad\qquad \blacktriangleright \text{Divide each side of the in-}$$
$$x \leq 5 \qquad\qquad\qquad\qquad \text{equality by } -3. \text{ Reverse the}$$
$$\text{inequality symbol.}$$

Problem 7   Solve: $8 - 4(3x + 5) \leq 6(x - 8)$

Solution   See page A18.

STUDY TIPS

BE PREPARED FOR TESTS

The Chapter Test at the end of a chapter should be used to prepare for an examination. We suggest that you try the Chapter Test a few days before your actual exam. Do these exercises in a quiet place and try to complete the exercises in the same amount of time as you will be allowed for your exam. When completing the exercises, practice the strategies of successful test takers: 1) look over the entire test before you begin to solve any problem; 2) write down any rules or formulas you may need so they are readily available; 3) read the directions carefully; 4) work the problems that are easiest for you first; 5) check your work, looking particularly for careless errors.

When you have completed the exercises in the Chapter Test, check your answers. If you missed a question, review the material in that objective and rework some of the exercises from that objective. This will strengthen your ability to perform the skills in that objective.

CONCEPT REVIEW 3.3

Determine whether the statement is always true, sometimes true, or never true.

1. The same variable term can be added to both sides of an inequality without changing the solution set of the inequality.

2. The same variable term can be subtracted from both sides of an inequality without changing the solution set of the inequality.

3. Both sides of an inequality can be multiplied by the same number without changing the solution set of the inequality.

4. Both sides of an inequality can be divided by the same number without changing the solution set of the inequality.

5. If $a > 0$ and $b < 0$, then $ab > 0$.

6. If $a < 0$, then $a^2 > 0$.

7. If $a > 0$ and $b < 0$, then $a^2 > b$.

8. If $a > b$, then $-a > -b$.

9. If $a < b$, then $ac < bc$.

10. If $a \neq 0$, $b \neq 0$, and $a > b$, then $\dfrac{1}{a} > \dfrac{1}{b}$.

EXERCISES 3.3

1   Graph.

1. $x > 2$ 2. $x \geq -1$ 3. $x \leq 0$ 4. $x < 4$

Solve and graph the solution set.

5. $x + 1 < 3$ 6. $y + 2 < 2$

7. $x - 5 > -2$ 8. $x - 3 > -2$

9. $n + 4 \geq 7$ 10. $x + 5 \geq 3$

11. $x - 6 \leq -10$ 12. $y - 8 \leq -11$

13. $5 + x \geq 4$ 14. $-2 + n \geq 0$

Solve.

15. $y - 3 \geq -12$ 16. $x + 8 \geq -14$ 17. $3x - 5 < 2x + 7$

18. $5x + 4 < 4x - 10$ 19. $8x - 7 \geq 7x - 2$ 20. $3n - 9 \geq 2n - 8$

21. $2x + 4 < x - 7$ 22. $9x + 7 < 8x - 7$ 23. $4x - 8 \leq 2 + 3x$

24. $5b - 9 < 3 + 4b$ 25. $6x + 4 \geq 5x - 2$ 26. $7x - 3 \geq 6x - 2$

27. $2x - 12 > x - 10$ 28. $3x + 9 > 2x + 7$ 29. $d + \dfrac{1}{2} < \dfrac{1}{3}$

30. $x - \dfrac{3}{8} < \dfrac{5}{6}$ 31. $x + \dfrac{5}{8} \geq -\dfrac{2}{3}$ 32. $y + \dfrac{5}{12} \geq -\dfrac{3}{4}$

33. $2x - \dfrac{1}{2} < x + \dfrac{3}{4}$ 34. $6x - \dfrac{1}{3} \leq 5x - \dfrac{1}{2}$ 35. $3x + \dfrac{5}{8} > 2x + \dfrac{5}{6}$

36. $4b - \frac{7}{12} \geq 3b - \frac{9}{16}$

37. $x + 5.8 \leq 4.6$

38. $n - 3.82 \leq 3.95$

39. $x - 0.23 \leq 0.47$

40. $3.8x < 2.8x - 3.8$

41. $1.2x < 0.2x - 7.3$

2 Solve and graph the solution set.

42. $3x < 12$

43. $8x \leq -24$

44. $5y \geq 15$

45. $24x > -48$

46. $16x \leq 16$

47. $3x > 0$

48. $-8x > 8$

49. $-2n \leq -8$

50. $-6b > 24$

51. $-4x < 8$

Solve.

52. $-5y \geq 20$

53. $3x < 5$

54. $7x > 2$

55. $-8x \leq -40$

56. $-6x \leq -40$

57. $10x > -25$

58. $-3x \geq \frac{6}{7}$

59. $-5x \geq \frac{10}{3}$

60. $\frac{5}{6}n < 15$

61. $\frac{2}{3}x < -12$

62. $\frac{5}{6}x < -20$

63. $-\frac{3}{8}x < 6$

64. $\frac{3}{4}x < 12$

65. $\frac{2}{3}y \geq 4$

66. $\frac{5}{8}x \geq 10$

67. $-\frac{2}{3}x \leq 4$

68. $-\frac{3}{7}x \leq 6$

69. $-\frac{2}{11}b \geq -6$

70. $-\frac{4}{7}x \geq -12$

71. $\frac{2}{3}n < \frac{1}{2}$

72. $\frac{3}{5}x > \frac{7}{10}$

73. $-\frac{2}{3}x \geq \frac{4}{7}$

74. $-\frac{3}{8}x \geq \frac{9}{14}$

75. $-\frac{3}{4}y \geq -\frac{5}{8}$

76. $-\frac{8}{9}x \geq -\frac{16}{27}$

77. $\frac{2}{3}x \leq \frac{9}{14}$

78. $\frac{7}{12}x \leq \frac{9}{14}$

79. $-\frac{3}{5}y < \frac{9}{10}$

80. $-\frac{5}{12}y < \frac{1}{6}$

81. $-0.27x < 0.135$

82. $-0.63x < 4.41$

83. $8.4y \geq -6.72$

84. $3.7y \geq -1.48$

85. $1.5x \leq 6.30$

86. $2.3x \leq 5.29$

87. $-3.9x \geq -19.5$

88. $0.035x < -0.0735$

89. $0.07x < -0.378$

90. $-11.7x \leq 4.68$

3 Solve.

91. $4x - 8 < 2x$

92. $7x - 4 < 3x$

93. $2x - 8 > 4x$

94. $3y + 2 > 7y$

95. $8 - 3x \leq 5x$

96. $10 - 3x \leq 7x$

97. $3x + 2 \geq 5x - 8$ **98.** $2n - 9 \geq 5n + 4$ **99.** $5x - 2 < 3x - 2$

100. $8x - 9 > 3x - 9$ **101.** $0.1(180 + x) > x$ **102.** $x > 0.2(50 + x)$

103. $0.15x + 55 > 0.10x + 80$ **104.** $-3.6b + 16 < 2.8b + 25.6$ **105.** $2(3x - 1) > 3x + 4$

106. $5(2x + 7) > -4x - 7$ **107.** $3(2x - 5) \geq 8x - 5$ **108.** $5x - 8 \geq 7x - 9$

109. $2(2y - 5) \leq 3(5 - 2y)$ **110.** $2(5x - 8) \leq 7(x - 3)$

111. $5(2 - x) > 3(2x - 5)$ **112.** $4(3d - 1) > 3(2 - 5d)$

113. $5(x - 2) > 9x - 3(2x - 4)$ **114.** $3x - 2(3x - 5) > 4(2x - 1)$

115. $4 - 3(3 - n) \leq 3(2 - 5n)$ **116.** $15 - 5(3 - 2x) \leq 4(x - 3)$

117. $2x - 3(x - 4) \geq 4 - 2(x - 7)$ **118.** $4 + 2(3 - 2y) \leq 4(3y - 5) - 6y$

119. $\frac{1}{2}(9x - 10) \leq -\frac{1}{3}(12 - 6x)$ **120.** $\frac{1}{4}(8 - 12d) < \frac{2}{5}(10d + 15)$

121. $\frac{2}{3}(9t - 15) + 4 < 6 + \frac{3}{4}(4 - 12t)$ **122.** $\frac{3}{8}(16 - 8c) - 9 \geq \frac{3}{5}(10c - 15) + 7$

123. $3[4(n - 2) - (1 - n)] > 5(n - 4)$ **124.** $2(m + 7) \leq 4[3(m - 2) - 5(1 + m)]$

SUPPLEMENTAL EXERCISES 3.3

Use the roster method to list the set of positive integers that are solutions of the inequality.

125. $7 - 2b \leq 15 - 5b$ **126.** $13 - 8a \geq 2 - 6a$

127. $2(2c - 3) < 5(6 - c)$ **128.** $-6(2 - d) \geq 4(4d - 9)$

Use the roster method to list the set of integers that are common to the solution sets of the two inequalities.

129. $5x - 12 \leq x + 8$ **130.** $6x - 5 > 9x - 2$
$3x - 4 \geq 2 + x$ $5x - 6 < 8x + 9$

131. $4(x - 2) \leq 3x + 5$ **132.** $3(x + 2) < 2(x + 4)$
$7(x - 3) \geq 5x - 1$ $4(x + 5) > 3(x + 6)$

Graph.

133. $|x| < 3$ **134.** $|x| < 4$

135. $|x| > 2$ **136.** $|x| > 1$

137. In your own words, state the Addition Property of Inequalities and the Multiplication Property of Inequalities.

138. What differentiates solving linear equations from solving linear inequalities?

Project in Mathematics

Measurements as Approximations

From arithmetic, you know the rules for rounding decimals.

If the digit to the right of the given place value is less than 5, that digit and all digits to the right are dropped.

6.31 rounded to the nearest tenth is 6.3.

If the digit to the right of the given place value is greater than or equal to 5, increase the given place value by one and drop all digits to its right.

6.28 rounded to the nearest tenth is 6.3.

Given the rules for rounding numbers, what range of values can the number 6.3 represent? The smallest possible number that would be rounded to 6.3 is 6.25; any number smaller than that would not be rounded up to 6.3.

What about the largest possible number that would be rounded to 6.3? The number 6.34 would be rounded down to 6.3. So would the numbers 6.349, 6.3499, 6.349999, and so on. Therefore, we cannot name the largest possible value of 6.3. We can say that the number must be less than 6.35. Any number less than 6.35 would be rounded down to 6.3.

The exact value of 6.3 is greater than or equal to 6.25 and less than 6.35.

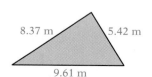

The dimensions of a rectangle are given as 4.3 cm by 3.2 cm. Using the smallest and the largest possible values of the length and of the width, we can represent the possible values of the area, A, of the rectangle as follows:

$$4.25(3.15) \le A < 4.35(3.25)$$
$$13.3875 \le A < 14.1375$$

The area is greater than or equal to 13.3875 cm^2 and less than 14.1375 cm^2.

Solve.

1. The measurements of the three sides of a triangle are given as 8.37 m, 5.42 m, and 9.61 m. Find the possible lengths of the perimeter of the triangle.

2. The length of a side of a square is given as 4.7 cm. Find the possible values for the area of the square.

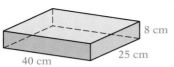

8 cm

40 cm   25 cm

3. What are the possible values for the area of a rectangle whose dimensions are 6.5 cm by 7.8 cm?

4. The length of a box is 40 cm, the width is 25 cm, and the height is 8 cm. What are the possible values of the volume of the box?

Focus on Problem Solving

Counterexamples

Some of the exercises in this text ask you to determine whether a statement is true or false. For instance, the statement "every real number has a reciprocal" is false because 0 is a real number and 0 does not have a reciprocal.

Finding an example, such as 0 has no reciprocal, to show that a statement is not always true is called *finding a counterexample*. A counterexample is an example that shows that a statement is not always true.

Consider the statement "the product of two numbers is greater than either factor." A counterexample to this statement is the factors $\frac{2}{3}$ and $\frac{3}{4}$. The product of these numbers is $\frac{1}{2}$, and $\frac{1}{2}$ is *smaller* than $\frac{2}{3}$ or $\frac{3}{4}$. There are many other counterexamples to the given statement.

Here are some counterexamples to the statement "the square of a number is always larger than the number."

$$\left(\frac{1}{2}\right)^2 = \frac{1}{4} \quad \text{but} \quad \frac{1}{4} < \frac{1}{2} \qquad\qquad 1^2 = 1 \quad \text{but} \quad 1 = 1$$

For each of the next five statements, find at least one counterexample to show that the statement, or conjecture, is false.

1. The product of two integers is always a positive number.

2. The sum of two prime numbers is never a prime number.

3. For all real numbers, $|x + y| = |x| + |y|$.

4. If x and y are nonzero real numbers and $x > y$, then $x^2 > y^2$.

5. The quotient of any two nonzero real numbers is less than either one of the numbers.

When a problem is posed, it may not be known whether the problem statement is true or false. For instance, Christian Goldbach (1690–1764) stated that every even integer greater than 2 can be written as the sum of two prime numbers. No one has been able to find a counterexample to this statement, but neither has anyone been able to prove that it is always true.

In the next set of problems, answer true if the statement is always true. If there is an instance when the statement is false, give a counterexample.

6. The reciprocal of a positive number is always smaller than the number.

7. If $x < 0$, then $|x| = -x$.

8. For any two real numbers x and y, $x + y > x - y$.

9. For any positive integer n, $n^2 + n + 17$ is a prime number.

10. The list of numbers 1, 11, 111, 1111, 11111, ... contains infinitely many composite numbers. (*Hint:* A number is divisible by 3 if the sum of the digits of the number is divisible by 3.)

Chapter Summary

Key Words

An *equation* expresses the equality of two mathematical expressions.   (Objective 3.1.1)

$5(3x - 2) = 4x + 7$ is an equation.

A *solution of an equation* is a number that, when substituted for the variable, results in a true equation.   (Objective 3.1.1)

1 is the solution of the equation $6x - 4 = 2$ because $6(1) - 4 = 2$ is a true equation.

To *solve* an equation means to find a solution of the equation. The goal is to rewrite the equation in the form *variable = constant*, because the constant is the solution. (Objective 3.1.2)

The equation $x = -9$ is in the form *variable = constant*. The constant, -9, is the solution of the equation.

An *inequality* is an expression that contains the symbol $<$, $>$, $\leq$, or $\geq$.   (Objective 3.3.1)

$8x - 1 \geq 5x + 23$ is an inequality.

The *solution set of an inequality* is a set of numbers, each element of which, when substituted for the variable, results in a true inequality.   (Objective 3.3.1)

The solution set of $8x - 1 \geq 5x + 23$ is $x \geq 8$ because every number greater than or equal to 8, when substituted for the variable, results in a true inequality.

Essential Rules and Procedures

The Addition Property of Equations
The same number or variable term can be added to each side of an equation without changing the solution of the equation. (Objective 3.1.2)

$$x + 12 = -19$$
$$x + 12 - 12 = -19 - 12$$
$$x = -31$$

The Multiplication Property of Equations
Each side of an equation can be multiplied by the same nonzero number without changing the solution of the equation.   (Objective 3.1.3)

$$-6x = 24$$
$$\frac{-6x}{-6} = \frac{24}{-6}$$
$$x = -4$$

To solve an equation of the form $ax + b = c$ or an equation of the form $ax + b = cx + d$, use both the Addition and the Multiplication Properties of Equations.   (Objectives 3.2.1 and 3.2.2)

$$7x - 4 = 3x + 16$$
$$7x - 3x - 4 = 3x - 3x + 16$$
$$4x - 4 = 16$$
$$4x - 4 + 4 = 16 + 4$$
$$4x = 20$$
$$\frac{4x}{4} = \frac{20}{4}$$
$$x = 5$$

To solve an equation containing parentheses, use the Distributive Property to remove parentheses.   (Objective 3.2.3)

$$3(2x - 1) = 15$$
$$6x - 3 = 15$$
$$6x - 3 + 3 = 15 + 3$$
$$6x = 18$$
$$\frac{6x}{6} = \frac{18}{6}$$
$$x = 3$$

The Addition Property of Inequalities
The same number or variable term can be added to each side of an inequality without changing the solution set of the inequality.

If $a > b$, then $a + c > b + c$.
If $a < b$, then $a + c < b + c$.   (Objective 3.3.1)

$$x - 9 \le 14$$
$$x - 9 + 9 \le 14 + 9$$
$$x \le 23$$

The Multiplication Property of Inequalities
(Objective 3.3.2)

Rule 1 Each side of an inequality can be multiplied by the same **positive number** without changing the solution set of the inequality.

If $a > b$ and $c > 0$, then $ac > bc$.
If $a < b$ and $c > 0$, then $ac < bc$.

$$7x < -21$$
$$\frac{7x}{7} < \frac{-21}{7}$$
$$x < -3$$

Rule 2 If each side of an inequality is multiplied by the same **negative number** and the inequality symbol is reversed, then the solution set of the inequality is not changed.

If $a > b$ and $c < 0$, then $ac < bc$.
If $a < b$ and $c < 0$, then $ac > bc$.

$$-7x \ge 21$$
$$\frac{-7x}{-7} \le \frac{21}{-7}$$
$$x \le -3$$

The Basic Percent Equation
Percent · Base = Amount
$$P \cdot B = A$$
(Objective 3.1.4)

40% of what number is 16?
$$PB = A$$
$$0.40B = 16$$
$$\frac{0.40B}{0.40} = \frac{16}{0.40}$$
$$B = 40$$
The number is 40.

Chapter Review Exercises

1. Is 3 a solution of $5x - 2 = 4x + 5$?

2. Solve: $x - 4 = 16$

3. Solve: $8x = -56$

4. Solve: $5x - 6 = 29$

5. Solve: $5x + 3 = 10x - 17$

6. Solve: $3(5x + 2) + 2 = 10x + 5[x - (3x - 1)]$

7. What is 81% of 500?

8. 18 is 72% of what number?

9. 27 is what percent of 40?

10. Graph: $x \leq -2$

11. Solve and graph the solution set of $x - 3 > -1$.

12. Solve and graph the solution set of $3x > -12$.

13. Solve: $3x + 4 \geq -8$

14. Solve: $7x - 2(x + 3) \geq x + 10$

15. Is 2 a solution of $x^2 + 4x + 1 = 3x + 7$?

16. Solve: $4.6 = 2.1 + x$

17. Solve: $\frac{x}{7} = -7$

18. Solve: $14 + 6x = 17$

19. Solve: $12y - 1 = 3y + 2$

20. Solve: $x + 5(3x - 20) = 10(x - 4)$

21. What is $66\frac{2}{3}\%$ of 24?

22. 60 is 48% of what number?

23. 0.5 is what percent of 3?

24. Solve and graph the solution set of $2 + x < -2$.

25. Solve and graph the solution set of $5x \leq -10$.

26. Solve: $6x + 3(2x - 1) = -27$

27. Solve: $a - \dfrac{1}{6} = \dfrac{2}{3}$

28. Solve: $\dfrac{3}{5}a = 12$

29. Solve: $32 = 9x - 4 - 3x$

30. Solve: $-4[x + 3(x - 5)] = 3(8x + 20)$

31. Solve: $4x - 12 < x + 24$

32. What is $\dfrac{1}{2}\%$ of 3000?

33. Solve: $3x + 7 + 4x = 42$

34. Solve: $5x - 6 > 19$

35. 8 is what percent of 200?

36. Solve: $6x - 9 < 4x + 3(x + 3)$

37. Solve: $5 - 4(x + 9) > 11(12x - 9)$

38. Find the measure of the third angle of a triangle if the first angle is $20°$ and the second angle is $50°$. Use the equation $A + B + C = 180°$, where A, B, and C are the measures of the angles of a triangle.

39. A lever is 12 ft long. At a distance of 2 ft from the fulcrum, a force of 120 lb is applied. How large a force must be applied to the other end of the lever so that the system will balance? Use the lever system equation $F_1 x = F_2(d - x)$.

40. Find the time it takes for the velocity of a falling object to increase from 4 ft/s to 100 ft/s. Use the equation $v = v_0 + 32t$, where v is the final velocity of a falling object, v_0 is the initial velocity, and t is the time it takes for the object to fall.

41. Find the discount on a mattress set if the sale price is $220.75 and the regular price is $300. Use the equation $S = R - D$, where S is the sale price, R is the regular price, and D is the discount.

42. A baseball playoff series was increased from 5 games to 7 games. What percent increase does this represent?

43. The pressure at a certain depth in the ocean can be approximated by the equation $P = 15 + \dfrac{1}{2}D$, where P is the pressure in pounds per square inch and D is the depth in feet. Use this equation to find the depth when the pressure is 55 lb/in².

44. A lever is 8 ft long. A force of 25 lb is applied to one end of the lever, and a force of 15 lb is applied to the other end. Find the location of the fulcrum when the system balances. Use the lever system equation $F_1 x = F_2(d - x)$.

45. Find the length of a rectangle when the perimeter is 84 ft and the width is 18 ft. Use the equation $P = 2L + 2W$, where P is the perimeter of a rectangle, L is the length, and W is the width.

Chapter Test

1. Solve: $\frac{3}{4}x = -9$

2. Solve: $6 - 5x = 5x + 11$

3. Solve: $3x - 5 = -14$

4. Is -2 a solution of $x^2 - 3x = 2x - 6$?

5. Solve: $x + \frac{1}{2} = \frac{5}{8}$

6. Solve: $5x - 2(4x - 3) = 6x + 9$

7. Solve: $7 - 4x = -13$

8. Solve: $11 - 4x = 2x + 8$

9. Solve: $x - 3 = -8$

10. Solve: $3x - 2 = 5x + 8$

11. Solve: $-\frac{3}{8}x = 5$

12. Solve: $6x - 3(2 - 3x) = 4(2x - 7)$

13. Solve: $6 - 2(5x - 8) = 3x - 4$

14. Solve: $9 - 3(2x - 5) = 12 + 5x$

15. Solve: $3(2x - 5) = 8x - 9$

16. 20 is what percent of 16?

17. 30% of what is 12?

18. Graph: $x > -2$

19. Solve and graph the solution set of $-2 + x \le -3$.

20. Solve and graph the solution set of $\frac{3}{8}x > -\frac{3}{4}$.

21. Solve: $x + \frac{1}{3} > \frac{5}{6}$

22. Solve: $3(x - 7) \ge 5x - 12$

23. Solve: $-\dfrac{3}{8}x \le 6$

24. Solve: $6x - 3(2 - 3x) \le 4(2x - 7)$

25. Solve: $3(2x - 5) \ge 8x - 9$

26. Solve: $15 - 3(5x - 7) < 2(7 - 2x)$

27. Solve: $-6x + 16 = -2x$

28. 20 is $83\dfrac{1}{3}\%$ of what number?

29. Solve and graph the solution set of $\dfrac{2}{3}x \ge 2$.

30. Is 5 a solution of $x^2 + 2x + 1 = (x + 1)^2$?

31. A person's weight on the moon is $16\dfrac{2}{3}\%$ of the person's weight on Earth. If an astronaut weighs 180 lb on Earth, how much would the astronaut weigh on the moon?

32. A chemist mixes 100 g of water at 80°C with 50 g of water at 20°C. Use the equation $m_1 \cdot (T_1 - T) = m_2 \cdot (T - T_2)$ to find the final temperature of the water after mixing. In this equation, m_1 is the quantity of water at the hotter temperature, T_1 is the temperature of the hotter water, m_2 is the quantity of water at the cooler temperature, T_2 is the temperature of the cooler water, and T is the final temperature of the water after mixing.

33. A financial manager has determined that the cost per unit for a calculator is $15 and that the fixed costs per month are $2000. Find the number of calculators produced during a month in which the total cost was $5000. Use the equation $T = U \cdot N + F$, where T is the total cost, U is the cost per unit, N is the number of units produced, and F is the fixed cost.

Cumulative Review Exercises

1. Subtract: $-6 - (-20) - 8$

2. Multiply: $(-2)(-6)(-4)$

3. Subtract: $-\dfrac{5}{6} - \left(-\dfrac{7}{16}\right)$

4. Divide: $-2\dfrac{1}{3} \div 1\dfrac{1}{6}$

5. Simplify: $-4^2 \cdot \left(-\dfrac{3}{2}\right)^3$

6. Simplify: $25 - 3 \cdot \dfrac{(5 - 2)^2}{2^3 + 1} - (-2)$

7. Evaluate $3(a - c) - 2ab$, when $a = 2$, $b = 3$, and $c = -4$.

8. Simplify: $3x - 8x + (-12x)$

9. Simplify: $2a - (-3b) - 7a - 5b$

10. Simplify: $(16x)\left(\dfrac{1}{8}\right)$

11. Simplify: $-4(-9y)$

12. Simplify: $-2(-x^2 - 3x + 2)$

13. Simplify: $-2(x - 3) + 2(4 - x)$

14. Simplify: $-3[2x - 4(x - 3)] + 2$

15. Use the roster method to write the set of negative integers greater than -8.

16. Write $\dfrac{7}{8}$ as a percent. Write the remainder in fractional form.

17. Write 342% as a decimal.

18. Write $62\dfrac{1}{2}\%$ as a fraction.

19. Is -3 a solution of $x^2 + 6x + 9 = x + 3$?

20. Solve: $x - 4 = -9$

21. Solve: $\dfrac{3}{5}x = -15$

22. Solve: $13 - 9x = -14$

23. Solve: $5x - 8 = 12x + 13$

24. Solve: $8x - 3(4x - 5) = -2x - 11$

25. Solve: $-\dfrac{3}{4}x > \dfrac{2}{3}$

26. Solve: $5x - 4 \geq 4x + 8$

27. Solve: $3x + 17 < 5x - 1$

28. Translate "the difference between eight and the quotient of a number and twelve" into a variable expression.

29. Translate and simplify "the sum of a number and two more than the number."

30. A club treasurer has some five-dollar bills and some ten-dollar bills. The treasurer has a total of 35 bills. Use one variable to express the number of each denomination of bill.

31. A wire is cut into two pieces. The length of the longer piece is four inches less than three times the length of the shorter piece. Express the length of the longer piece in terms of the length of the shorter piece.

32. A computer programmer receives a weekly wage of $650, and $110.50 is deducted for income tax. Find the percent of the computer programmer's salary deducted for income tax.

33. A business manager has determined that the cost per unit for a camera is $70 and that the fixed costs per month are $3500. Find the number of cameras produced during a month in which the total cost was $21,000. Use the equation $T = U \cdot N + F$, where T is the total cost, U is the cost per unit, N is the number of units produced, and F is the fixed cost.

34. A chemist mixes 300 g of water at 75°C with 100 g of water at 15°C. Use the equation $m_1 \cdot (T_1 - T) = m_2 \cdot (T - T_2)$ to find the final temperature of the water. In this equation, m_1 is the quantity of water at the hotter temperature, T_1 is the temperature of the hotter water, m_2 is the quantity of water at the cooler temperature, T_2 is the temperature of the cooler water, and T is the final temperature of the water after mixing.

35. A lever is 25 ft long. At a distance of 12 ft from the fulcrum, a force of 26 lb is applied. How large a force must be applied to the other end of the lever so that the system will balance?

4

Solving Equations and Inequalities: Applications

Objectives

Word Problems

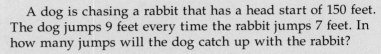

Word problems have been challenging students of mathematics for a long time. Here are two types of problems you may have seen before.

A number added to $\frac{1}{7}$ of the number is 19. What is the number?

A dog is chasing a rabbit that has a head start of 150 feet. The dog jumps 9 feet every time the rabbit jumps 7 feet. In how many jumps will the dog catch up with the rabbit?

What is unusual about these problems is their age. The first one is about 4000 years old and occurred as Problem 1 in the Rhind Papyrus. The second problem is about 1500 years old and comes from a Latin algebra book written in 450 A.D.

These examples illustrate that word problems have been around for a long time. The long history of word problems also reflects the importance that each generation has placed on solving these problems. It is through word problems that the initial steps of applying mathematics are taken.

The answer to the first problem is $16\frac{5}{8}$.

The answer to the second problem is 75 jumps.

SECTION 4.1
Translating Sentences into Equations

1 Translate a sentence into an equation and solve

An equation states that two mathematical expressions are equal. Therefore, to translate a sentence into an equation requires recognizing the words or phrases that mean "equals." Besides "equals," some of these phrases are "is," "is equal to," "amounts to," and "represents."

Once the sentence is translated into an equation, the equation can be solved by rewriting the equation in the form *variable = constant*.

Example 1 Translate "two times the sum of a number and eight equals the sum of four times the number and six" into an equation and solve.

Solution the unknown number: *n*

▶ Assign a variable to the unknown quantity.

two times the sum of a number and eight	equals	the sum of four times the number and six

▶ Find two verbal expressions for the same value.

$$2(n + 8) = 4n + 6$$

▶ Write a mathematical expression for each verbal expression. Write the equals sign.
▶ Solve the equation.

$$2n + 16 = 4n + 6$$
$$2n - 4n + 16 = 4n - 4n + 6$$
$$-2n + 16 = 6$$
$$-2n + 16 - 16 = 6 - 16$$
$$-2n = -10$$
$$\frac{-2n}{-2} = \frac{-10}{-2}$$
$$n = 5$$

The number is 5.

Problem 1 Translate "nine less than twice a number is five times the sum of the number and twelve" into an equation and solve.

Solution See page A18.

2 Application problems

Example 2 The temperature of the sun on the Kelvin scale is 6500 K. This is 4740° less than the temperature on the Fahrenheit scale. Find the Fahrenheit temperature.

Strategy   To find the Fahrenheit temperature, write and solve an equation using F to represent the Fahrenheit temperature.

Solution

6500	is	4740° less than the Fahrenheit temperature

$$6500 = F - 4740$$
$$6500 + 4740 = F - 4740 + 4740$$
$$11{,}240 = F$$

The temperature of the sun is 11,240°F.

Problem 2   A molecule of octane gas has eight carbon atoms. This represents twice the number of carbon atoms in a butane gas molecule. Find the number of carbon atoms in a butane molecule.

Solution   See page A18.

Example 3   A board 10 ft long is cut into two pieces. Three times the length of the shorter piece is twice the length of the longer piece. Find the length of each piece.

LOOK CLOSELY

We could also let x represent the length of the longer piece and $10 - x$ represent the length of the shorter piece. Then our equation would be

$$3(10 - x) = 2x$$

Show that this equation results in the same solutions.

Strategy   Draw a diagram. To find the length of each piece, write and solve an equation using x to represent the length of the shorter piece and $10 - x$ to represent the length of the longer piece.

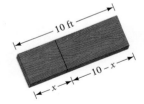

Solution

three times the shorter piece	is	twice the longer piece

$$3x = 2(10 - x)$$
$$3x = 20 - 2x$$
$$3x + 2x = 20 - 2x + 2x$$
$$5x = 20$$
$$\frac{5x}{5} = \frac{20}{5}$$
$$x = 4$$

▶ The length of the shorter piece is 4 ft.

$$10 - x = 10 - 4 = 6$$

▶ Substitute the value of x into the variable expression for the longer piece and evaluate.

The length of the shorter piece is 4 ft. The length of the longer piece is 6 ft.

Problem 3   A company manufactures 160 bicycles per day. Four times the number of 3-speed bicycles made equals 30 less than the number of 10-speed bicycles made. Find the number of 10-speed bicycles manufactured each day.

Solution   See page A19.

STUDY TIPS

FIND GOOD STUDY AREAS

Find a place to study where you are comfortable and can concentrate well. Many students find the campus library to be a good place. You might select two or three places at the college library where you like to study. Or there may be a small, quiet lounge on the third floor of a building where you find you can study well. Take the time to find places that promote good study habits.

CONCEPT REVIEW 4.1

Determine whether the statement is always true, sometimes true, or never true.

1. When two expressions represent the same value, we say that the expressions are equal to each other.

2. Both "four less than a number" and "the difference between four and a number" are translated as "$4 - n$."

3. When translating a sentence into an equation, we can use any variable to represent an unknown number.

4. When translating a sentence into an equation, translate the word "is" as "=."

5. If a rope 19 in. long is cut into two pieces and one of the pieces has length x in., then the length of the other piece can be expressed as $(x - 19)$ in.

6. In addition to a number, the answer to an application problem must have units, such as meters, dollars, minutes, or miles per hour.

EXERCISES 4.1

1   Translate into an equation and solve.

1. The difference between a number and fifteen is seven. Find the number.

2. The sum of five and a number is three. Find the number.

3. The product of seven and a number is negative twenty-one. Find the number.

4. The quotient of a number and four is two. Find the number.

5. Four less than three times a number is five. Find the number.

6. The difference between five and twice a number is one. Find the number.

7. Four times the sum of twice a number and three is twelve. Find the number.

8. Twenty-one is three times the difference between four times a number and five. Find the number.

9. Twelve is six times the difference between a number and three. Find the number.

10. The difference between six times a number and four times the number is negative fourteen. Find the number.

11. Twenty-two is two less than six times a number. Find the number.

12. Negative fifteen is three more than twice a number. Find the number.

13. Seven more than four times a number is three more than two times the number. Find the number.

14. The difference between three times a number and four is five times the number. Find the number.

15. Eight less than five times a number is four more than eight times the number. Find the number.

16. The sum of a number and six is four less than six times the number. Find the number.

17. Twice the difference between a number and twenty-five is three times the number. Find the number.

18. Four times a number is three times the difference between thirty-five and the number. Find the number.

19. The sum of two numbers is twenty. Three times the smaller is equal to two times the larger. Find the two numbers.

20. The sum of two numbers is fifteen. One less than three times the smaller is equal to the larger. Find the two numbers.

21. The sum of two numbers is eighteen. The total of three times the smaller and twice the larger is forty-four. Find the two numbers.

22. The sum of two numbers is two. The difference between eight and twice the smaller number is two less than four times the larger. Find the two numbers.

2   Write an equation and solve.

23. As a result of depreciation, the value of a car is now $9600. This is three-fifths of its original value. Find the original value of the car.

24. The operating speed of a personal computer is 8 megahertz. This is one-fourth the speed of a newer model. Find the speed of the newer personal computer.

25. One measure of computer speed is mips (*millions* of *instructions* per second). One computer has a rating of 10 mips, which is two-thirds the speed of a second computer. Find the mips rating of the second computer.

26. A team scored 26 points during a football game. This team scored twice as many field goals (3 points each) as it did touchdowns (7 points each). Find the number of field goals scored by the team.

27. A basketball team scored 105 points during one game. There were as many two-point baskets as free throws (1 point each), and the number of three-point baskets was five less than the number of free throws. Find the number of three-point baskets.

28. A university employs a total of 600 teaching assistants and research assistants. There are three times as many teaching assistants as research assistants. Find the number of research assistants employed by the university.

29. A 24 lb soil supplement contains iron, potassium, and a mulch. There is five times as much mulch as iron and twice as much potassium as iron. Find the amount of mulch in the soil supplement.

30. A real estate agent sold two homes and received commissions totaling $6000. The agent's commission on one home was one and one-half times the commission on the second home. Find the agent's commission on each home.

31. The purchase price of a new big-screen TV, including finance charges, was $3276. A down payment of $450 was made. The remainder was paid in 24 equal monthly installments. Find the monthly payment.

32. The purchase price of a new computer system, including finance charges, was $6350. A down payment of $350 was made. The remainder was paid in 24 equal monthly installments. Find the monthly payment.

33. The cost to replace a water pump in a sports car was $600. This included $375 for the water pump and $45 per hour for labor. How many hours of labor were required to replace the water pump?

34. The cost of electricity in a certain city is $.08 for each of the first 300 kWh (kilowatt-hours) and $.13 for each kWh over 300 kWh. Find the number of kilowatt-hours used by a family that receives a $51.95 electric bill.

35. An investor deposited $5000 into two accounts. Two times the smaller deposit is $1000 more than the larger deposit. Find the amount deposited into each account.

36. The length of a rectangular cement patio is 3 ft less than three times the width. The sum of the length and width of the patio is 21 ft. Find the length of the patio.

37. Greek architects considered a rectangle whose length was approximately 1.6 times its width to be the most visually appealing. Find the length and width of a rectangle constructed in this manner if the sum of the length and width is 130 ft.

38. A computer screen consists of tiny dots of light called pixels. In a certain graphics mode, there are 640 horizontal pixels. This is 40 more than three times the number of vertical pixels. Find the number of vertical pixels.

39. A wire 12 ft long is cut into two pieces. Each piece is bent into the shape of a square. The perimeter of the larger square is twice the perimeter of the smaller square. Find the perimeter of the larger square.

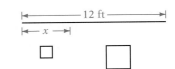

40. Five thousand dollars is divided between two scholarships. Three times the smaller scholarship is equal to twice the larger. Find the amount of the larger scholarship.

41. A carpenter is building a wood door frame. The height of the frame is 1 ft less than three times the width. What is the width of the largest door frame that can be constructed from a board 19 ft long? (*Hint:* A door frame consists of only three sides; there is no frame below a door.)

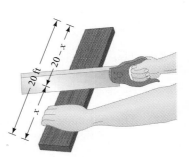

42. A 20 ft board is cut into two pieces. Twice the length of the shorter piece is 4 ft longer than the length of the longer piece. Find the length of the shorter piece.

Two angles are **complementary angles** if the sum of the measures of the angles is 90°. Two angles are **supplementary angles** if the sum of the measures of the angles is 180°. In the diagrams at the right, angles a and b are complementary angles; angles x and y are supplementary angles.

43. Find two complementary angles such that the larger angle is 6 degrees more than twice the smaller angle.

44. Find two complementary angles such that the smaller angle is 15 degrees less than one-half the larger angle.

45. Find two supplementary angles such that the larger angle is 12 degrees less than three times the smaller angle.

46. Find two supplementary angles such that the smaller angle is 10 degrees less than two-thirds the larger angle.

SUPPLEMENTAL EXERCISES 4.1

Solve.

47. The amount of liquid in a container triples every minute. The container becomes completely filled at 3:40 P.M. What fractional part of the container is filled at 3:39 P.M.?

48. A cyclist traveling at a constant speed completes $\frac{3}{5}$ of a trip in $1\frac{1}{2}$ h. In how many additional hours will the cylist complete the entire trip?

49. The charges for a long-distance telephone call are $1.21 for the first three minutes and $.42 for each additional minute or fraction of a minute. If the charges for a call were $6.25, how many minutes did the phone call last?

50. Four employees are paid at four consecutive levels on a wage scale. The difference between any two consecutive levels is $320 per month. The average of the four employees' monthly wages is $2880. What is the monthly wage of the highest-paid employee?

51. A coin bank contains nickels, dimes, and quarters. There are 14 nickels in the bank, $\frac{1}{6}$ of the coins are dimes, and $\frac{3}{5}$ of the coins are quarters. How many coins are in the coin bank?

52. During one day at an office, one-half of the amount of money in the petty cash drawer was used in the morning, and one-third of the remaining money was used in the afternoon, leaving $5 in the petty cash drawer at the end of the day. How much money was in the petty cash drawer at the start of the day?

53. A formula is an equation that relates variables in a known way. Find two examples of formulas that are used in your college major. Explain what each of the variables represents.

S E C T I O N 4.2

Integer, Coin, and Stamp Problems

1 Consecutive integer problems

Recall that the integers are the numbers ..., −3, −2, −1, 0, 1, 2, 3, 4, ...

An **even integer** is an integer that is divisible by 2. Examples of even integers are −8, 0, and 22.

An **odd integer** is an integer that is not divisible by 2. Examples of odd integers are −17, 1, and 39.

Consecutive integers are integers that follow one another in order. Examples of consecutive integers are shown at the right. (Assume the variable n represents an integer.)	11, 12, 13 −8, −7, −6 $n, n + 1, n + 2$
Examples of **consecutive even integers** are shown at the right. (Assume the variable n represents an even integer.)	24, 26, 28 −10, −8, −6 $n, n + 2, n + 4$
Examples of **consecutive odd integers** are shown at the right. (Assume the variable n represents an odd integer.)	19, 21, 23 −1, 1, 3 $n, n + 2, n + 4$

Solve: The sum of three consecutive odd integers is 51. Find the integers.

STRATEGY for solving a consecutive integer problem

■ Let a variable represent one of the integers. Express each of the other integers in terms of that variable. Remember that consecutive integers differ by 1. Consecutive even or consecutive odd integers differ by 2.

First odd integer: n
Second odd integer: $n + 2$
Third odd integer: $n + 4$

■ Determine the relationship among the integers.

The sum of the three odd integers is 51.

$$n + (n + 2) + (n + 4) = 51$$
$$3n + 6 = 51$$
$$3n = 45$$
$$n = 15$$

$n + 2 = 15 + 2 = 17$ ▶ Substitute the value of n into the variable expres-
$n + 4 = 15 + 4 = 19$ sions for the second and third integers.

The three consecutive odd integers are 15, 17, and 19.

Example 1   Find three consecutive even integers such that three times the second is six more than the sum of the first and third.

Strategy ■ First even integer: n
Second even integer: $n + 2$
Third even integer: $n + 4$
■ Three times the second equals six more than the sum of the first and third.

Solution $3(n + 2) = n + (n + 4) + 6$
$3n + 6 = 2n + 10$
$n + 6 = 10$
$n = 4$ ▶ The first even integer is 4.

$n + 2 = 4 + 2 = 6$ ▶ Substitute the value of n into the
$n + 4 = 4 + 4 = 8$ variable expressions for the sec-
                            ond and third integers.

The three consecutive even integers are 4, 6, and 8.

Problem 1   Find three consecutive integers whose sum is -12.

Solution   See page A19.

2   Coin and stamp problems

In solving problems that deal with coins or stamps of different values, it is necessary to represent the value of the coins or stamps in the same unit of money. Frequently, the unit of money is cents.

The value of 3 quarters in cents is $3 \cdot 25$ or 75 cents.
The value of 4 nickels in cents is $4 \cdot 5$ or 20 cents.
The value of d dimes in cents is $d \cdot 10$ or $10d$ cents.

Solve: A coin bank contains $1.20 in dimes and quarters. In all, there are nine coins in the bank. Find the number of quarters in the bank.

STRATEGY for solving a coin problem

■ For each denomination of coin, write a numerical or variable expression for the number of coins, the value of the coin in cents, and the total value of the coins in cents. The results can be recorded in a table.

The total number of coins is 9.

Number of quarters: x
Number of dimes: $9 - x$

LOOK CLOSELY

Use the information given in the problem to fill in the number and value columns of the table. Fill in the total value column by multiplying the two expressions you wrote in each row.

Coin	Number of coins	·	Value of coin in cents	=	Total value in cents
Quarter	x	·	25	=	$25x$
Dime	$9 - x$	·	10	=	$10(9 - x)$

■ Determine the relationship between the total values of the different coins. Use the fact that the sum of the total values of the different denominations of coins is equal to the total value of all the coins.

The sum of the total values of the different denominations of coins is equal to the total value of all the coins (120 cents).

$$25x + 10(9 - x) = 120$$
$$25x + 90 - 10x = 120$$
$$15x + 90 = 120$$
$$15x = 30$$
$$x = 2$$

▶ The total value of the quarters plus the total value of the dimes equals the amount of money in the bank.

There are 2 quarters in the bank.

Example 2 A collection of stamps consists of 3¢ stamps and 8¢ stamps. The number of 8¢ stamps is five more than three times the number of 3¢ stamps. The total value of all the stamps is $1.75. Find the number of each type of stamp in the collection.

Strategy ■ Number of 3¢ stamps: x
Number of 8¢ stamps: $3x + 5$

Stamp	Number	Value	Total value
3¢	x	3	$3x$
8¢	$3x + 5$	8	$8(3x + 5)$

■ The sum of the total values of the different types of stamps equals the total value of all the stamps (175 cents).

Solution
$$3x + 8(3x + 5) = 175$$
$$3x + 24x + 40 = 175$$
$$27x + 40 = 175$$
$$27x = 135$$
$$x = 5$$

▶ The total value of the 3¢ stamps plus the total value of the 8¢ stamps equals the total value of all the stamps.

$$3x + 5 = 3(5) + 5$$
$$= 15 + 5 = 20$$

▶ Find the number of 8¢ stamps.

There are five 3¢ stamps and twenty 8¢ stamps in the collection.

Problem 2 A coin bank contains nickels, dimes, and quarters. There are five times as many nickels as dimes and six more quarters than dimes. The total value of all the coins is $6.30. Find the number of each kind of coin in the bank.

Solution See page A19.

CONCEPT REVIEW 4.2

Determine whether the statement is always true, sometimes true, or never true.

1. An even integer is a multiple of 2.

2. When $10d$ is used to represent the value of d dimes, 10 is written as the coefficient of d because it is the number of cents in one dime.

3. Given the consecutive odd integers -5 and -3, the next consecutive odd integer is -1.

4. If the first of three consecutive odd integers is n, then the second and third consecutive odd integers are represented as $n + 1$ and $n + 3$.

5. You have a total of 20 coins in nickels and dimes. If n represents the number of nickels you have, then $n - 20$ represents the number of dimes you have.

6. You have a total of 7 coins in dimes and quarters. The total value of the coins is $1.15. If d represents the number of dimes you have, then $10d + 25(7 - d) = 1.15$.

EXERCISES 4.2

1   Solve.

1. The sum of three consecutive integers is 54. Find the integers.

2. The sum of three consecutive integers is 75. Find the integers.

3. The sum of three consecutive even integers is 84. Find the integers.

4. The sum of three consecutive even integers is 48. Find the integers.

5. The sum of three consecutive odd integers is 57. Find the integers.

6. The sum of three consecutive odd integers is 81. Find the integers.

7. Find two consecutive even integers such that five times the first is equal to four times the second.

8. Find two consecutive even integers such that six times the first equals three times the second.

9. Nine times the first of two consecutive odd integers equals seven times the second. Find the integers.

10. Five times the first of two consecutive odd integers is three times the second. Find the integers.

11. Find three consecutive integers whose sum is negative twenty-four.

12. Find three consecutive even integers whose sum is negative twelve.

13. Three times the smallest of three consecutive even integers is two more than twice the largest. Find the integers.

14. Twice the smallest of three consecutive odd integers is five more than the largest. Find the integers.

15. Find three consecutive even integers such that three times the middle integer is six more than the sum of the first and third.

16. Find three consecutive odd integers such that four times the middle integer is equal to two less than the sum of the first and third.

2   Solve.

17. A bank contains 27 coins in dimes and quarters. The coins have a total value of $4.95. Find the number of dimes and quarters in the bank.

18. A coin purse contains 18 coins in nickels and dimes. The coins have a total value of $1.15. Find the number of nickels and dimes in the coin purse.

19. A business executive bought 40 stamps for $12.08. The purchase included 32¢ stamps and 23¢ stamps. How many of each type of stamp were bought?

20. A postal clerk sold some 32¢ stamps and some 23¢ stamps. Altogether, 15 stamps were sold for a total cost of $4.26. How many of each type of stamp were sold?

21. A drawer contains 29¢ stamps and 3¢ stamps. The number of 29¢ stamps is four less than three times the number of 3¢ stamps. The total value of all the stamps is $1.54. How many 29¢ stamps are in the drawer?

22. The total value of the dimes and quarters in a bank is $6.05. There are six more quarters than dimes. Find the number of each type of coin in the bank.

23. A child's piggy bank contains 44 coins in quarters and dimes. The coins have a total value of $8.60. Find the number of quarters in the bank.

24. A coin bank contains nickels and dimes. The number of dimes is 10 less than twice the number of nickels. The total value of all the coins is $2.75. Find the number of each type of coin in the bank.

25. A total of 26 bills are in a cash box. Some of the bills are one-dollar bills, and the rest are five-dollar bills. The total amount of cash in the box is $50. Find the number of each type of bill in the cash box.

26. A bank teller cashed a check for $200 using twenty-dollar bills and ten-dollar bills. In all, twelve bills were handed to the customer. Find the number of twenty-dollar bills and the number of ten-dollar bills.

27. A coin bank contains pennies, nickels, and dimes. There are six times as many nickels as pennies and four times as many dimes as pennies. The total amount of money in the bank is $7.81. Find the number of pennies in the bank.

28. A coin bank contains pennies, nickels, and quarters. There are seven times as many nickels as pennies and three times as many quarters as pennies. The total amount of money in the bank is $5.55. Find the number of pennies in the bank.

29. A collection of stamps consists of 22¢ stamps and 40¢ stamps. The number of 22¢ stamps is three more than four times the number of 40¢ stamps. The total value of the stamps is $8.34. Find the number of 22¢ stamps in the collection.

30. A collection of stamps consists of 2¢ stamps, 8¢ stamps, and 14¢ stamps. The number of 2¢ stamps is five more than twice the number of 8¢ stamps. The number of 14¢ stamps is three times the number of 8¢ stamps. The total value of the stamps is $2.26. Find the number of each type of stamp in the collection.

31. A collection of stamps consists of 3¢ stamps, 7¢ stamps, and 12¢ stamps. The number of 3¢ stamps is five less than the number of 7¢ stamps. The number of 12¢ stamps is one-half the number of 7¢ stamps. The total value of all the stamps is $2.73. Find the number of each type of stamp in the collection.

32. A collection of stamps consists of 2¢ stamps, 5¢ stamps, and 7¢ stamps. There are nine more 2¢ stamps than 5¢ stamps and twice as many 7¢ stamps as 5¢ stamps. The total value of the stamps is $1.44. Find the number of each type of stamp in the collection.

33. A collection of stamps consists of 6¢ stamps, 8¢ stamps, and 15¢ stamps. The number of 6¢ stamps is three times the number of 8¢ stamps. There are six more 15¢ stamps than there are 6¢ stamps. The total value of all the stamps is $5.16. Find the number of each type of stamp.

34. A child's piggy bank contains nickels, dimes, and quarters. There are twice as many nickels as dimes and four more quarters than nickels. The total value of all the coins is $9.40. Find the number of each type of coin.

SUPPLEMENTAL EXERCISES 4.2

Solve.

35. Find four consecutive even integers whose sum is −36.

36. Find four consecutive odd integers whose sum is −48.

37. A coin bank contains only dimes and quarters. The number of quarters in the bank is two less than twice the number of dimes. There are 34 coins in the bank. How much money is in the bank?

38. A postal clerk sold twenty stamps to a customer. The number of 32¢ stamps purchased was two more than twice the number of 23¢ stamps purchased. If the customer bought only 32¢ stamps and 23¢ stamps, how much money did the clerk collect from the customer?

39. Find three consecutive odd integers such that the sum of the first and third is twice the second.

40. Find four consecutive integers such that the sum of the first and fourth equals the sum of the second and third.

41. Explain why both consecutive even integers and consecutive odd integers are represented algebraically as $n, n + 2, n + 4, \ldots$.

SECTION 4.3

Geometry Problems

1 ■ Perimeter problems

The **perimeter** of a geometric figure is a measure of the distance around the figure. The equations for the perimeters of a rectangle and a triangle are shown below.

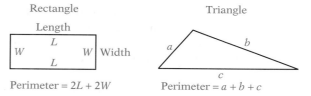

Rectangle

Length

W L W Width

L

Perimeter $= 2L + 2W$

Triangle

a b

c

Perimeter $= a + b + c$

POINT OF INTEREST

Leonardo DaVinci painted the Mona Lisa on a rectangular canvas. The height of the canvas was approximately 1.6 times its width. Rectangles with these proportions, called golden rectangles, were used extensively in Renaissance art.

Solve: The perimeter of a rectangle is 32 ft. The length of the rectangle is 1 ft more than twice the width. Find the width of the rectangle.

***STRATEGY** for solving a perimeter problem*

■ Let a variable represent the measure of one of the unknown sides of the figure. Express the measures of the remaining sides in terms of that variable.

Width: W
Length: $2W + 1$

$2W + 1$

W

■ Determine which perimeter equation to use.

Use the equation for the perimeter of a rectangle.

$$2L + 2W = P$$
$$2(2W + 1) + 2W = 32$$
$$4W + 2 + 2W = 32$$
$$6W + 2 = 32$$
$$6W = 30$$
$$W = 5$$

The width is 5 ft.

The problem that follows refers to an isosceles triangle. An **isosceles triangle** is one in which two sides are of equal measure.

Example 1   The perimeter of an isosceles triangle is 22 ft. The length of the third side is 2 ft less than the length of one of the equal sides. Find the measures of the three sides of the triangle.

Strategy   ▬ Each equal side: x
The third side: $x - 2$
▬ Use the equation for the perimeter of a triangle.

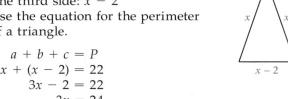

Solution

$$a + b + c = P$$
$$x + x + (x - 2) = 22$$
$$3x - 2 = 22$$
$$3x = 24$$
$$x = 8$$

$x - 2 = 8 - 2 = 6$ ▶ Substitute the value of x into the variable expression for the length of the third side.

Each of the equal sides measures 8 ft.
The third side measures 6 ft.

Problem 1   The perimeter of a rectangle is 42 m. The length of the rectangle is 3 m more than the width. Find the width.

Solution   See page A20.

2   Problems involving the angles of a triangle

In a triangle, the sum of the measures of all the angles is 180°.

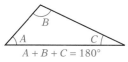

Two special types of triangles are shown at the right. A **right triangle** has one **right angle**—that is, an angle of 90°. It was stated in the previous objective that an isosceles triangle has two sides of equal measure. An **isosceles triangle** also has two equal angles.

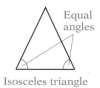

Right triangle   Isosceles triangle

The problem that follows refers to an acute angle. An **acute angle** is one that measures between 0° and 90°. In other words, it is an angle that measures more than 0° and less than 90°.

Solve: In a right triangle, the measure of one acute angle is four times the measure of the smallest angle. Find the measure of the smallest angle.

STRATEGY for solving a problem involving the angles of a triangle

■ Let a variable represent one of the unknown angles. Express the other angles in terms of that variable.

Measure of smallest angle: x
Measure of second angle: $4x$
Measure of right angle: $90°$

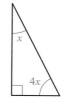

■ Use the equation $A + B + C = 180°$.

$$x + 4x + 90 = 180$$
$$5x + 90 = 180$$
$$5x = 90$$
$$x = 18$$

The measure of the smallest angle is $18°$.

Example 2 In an isosceles triangle, the measure of one angle is 40° more than twice the measure of one of the equal angles. Find the measure of one of the equal angles.

Strategy ■ Measure of one of the equal angles: x
Measure of the second equal angle: x
Measure of the third angle: $2x + 40$
■ Use the equation $A + B + C = 180°$.

Solution $x + x + (2x + 40) = 180$
$$4x + 40 = 180$$
$$4x = 140$$
$$x = 35$$

The measure of one of the equal angles is $35°$.

Problem 2 In a triangle, the measure of one angle is twice the measure of the second angle. The measure of the third angle is 8° less than the measure of the second angle. Find the measure of each angle.

Solution See page A20.

CONCEPT REVIEW 4.3

Determine whether the statement is always true, sometimes true, or never true.

1. The formula for the perimeter of a rectangle is $P = 2L + 2W$, where L represents the length and W represents the width of a rectangle.

2. In the formula for the perimeter of a triangle, $P = a + b + c$, the variables a, b, and c represent the measures of the three angles of a triangle.

3. If two angles of a triangle measure 52° and 73°, then the third angle of the triangle measures 55°.

4. An isosceles triangle has two sides of equal measure and two angles of equal measure.

5. The perimeter of a geometric figure is a measure of the area of the figure.

6. In a right triangle, the sum of the measures of the two acute angles is 90°.

EXERCISES 4.3

1   Solve.

1. The perimeter of a rectangle is 150 ft. The length of the rectangle is twice the width. Find the length and width of the rectangle.

2. The perimeter of a rectangle is 58 m. The width of the rectangle is 5 m less than the length. Find the length and width of the rectangle.

3. The width of a rectangle is 40% of the length. The perimeter of the rectangle is 364 ft. Find the length and width of the rectangle.

4. The width of a rectangle is 20% of the length. The perimeter is 240 cm. Find the length and width of the rectangle.

5. The perimeter of a triangle is 39 ft. One side of the triangle is 1 ft longer than the second side. The third side is 2 ft longer than the second side. Find the length of each side.

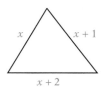

6. In an isosceles triangle, the third side is 2 m less than one of the equal sides. The perimeter is 16 m. Find the length of each side.

7. The perimeter of a triangle is 30 ft. The length of the first side is 1 ft less than twice the length of the second side. The length of the third side is 1 ft more than twice the length of the second side. Find the length of each side.

8. The perimeter of a rectangle is 36 m. The length of the rectangle is 3 m less than twice the width. Find the length and width of the rectangle.

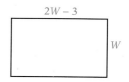

9. In an isosceles triangle, the length of the third side is 25% of the length of one of the equal sides. Find the length of each side when the perimeter is 135 ft.

10. The perimeter of a triangle is 130 cm. One side is twice as long as the second side. The third side is 30 cm longer than the second side. Find the length of each side.

11. The perimeter of a rectangle is 52.6 m. The width of the rectangle is 0.7 m less than the length. Find the length and width of the rectangle.

12. In an isosceles triangle, the length of one of the equal sides is 2.5 times the length of the third side. The perimeter is 10.8 m. Find the length of each side.

2 Solve.

13. In an isosceles triangle, one angle is three times the measure of one of the equal angles. Find the measure of each angle.

14. In an equiangular triangle, all three angles are equal. Find the measures of the equal angles.

15. One angle of a right triangle is 6° less than twice the measure of the smallest angle. Find the measure of each angle.

16. In an isosceles right triangle, two angles are equal, and the third angle is 90°. Find the measures of the equal angles.

17. In an isosceles triangle, one angle is 10° less than three times the measure of one of the equal angles. Find the measure of each angle.

18. In an isosceles triangle, one angle is 16° more than twice the measure of one of the equal angles. Find the measure of each angle.

19. In a triangle, one angle is 15° more than the measure of the second angle. The third angle is 30° more than the measure of the second angle. Find the measure of each angle.

20. In a triangle, one angle is twice the measure of the second angle. The third angle is three times the measure of the second angle. Find the measure of each angle.

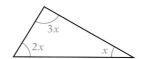

21. One angle of a triangle is three times the measure of the third angle. The second angle is 5° more than the measure of the third angle. Find the measure of each angle.

22. One angle of a triangle is five times the measure of the second angle. The third angle is six times the measure of the first angle. Find the measure of each angle.

23. The first angle of a triangle is three times the measure of the second angle. The third angle is 47° more than the measure of the first angle. Find the measure of each angle.

24. The first angle of a triangle is twice the measure of the second angle. The third angle is 5° less than the measure of the first angle. Find the measure of each angle.

SUPPLEMENTAL EXERCISES 4.3

Solve.

25. An equilateral triangle is one in which the three sides are of equal measure. Suppose a rectangle and an equilateral triangle have the same perimeter. The length of the rectangle is three times the width. Each side of the triangle is 8 cm. Find the length and width of the rectangle.

26. The length of a rectangle is 1 cm more than twice the width. If the length of the rectangle is decreased by 2 cm and the width is decreased by 1 cm, the perimeter is 20 cm. Find the length and width of the original rectangle.

27. The width of a rectangle is $8x$. The perimeter is $48x$. Find the length of the rectangle in terms of the variable x.

28. Discuss the concepts of accuracy and precision.

29. Write a report stating the arguments of the proponents of changing to the metric system in the United States or a report stating the arguments of those opposed to switching from the U.S. Customary System of measurement to the metric system.

30. Prepare a report on the use of geometric form in architecture.

S E C T I O N 4.4
Markup and Discount Problems

1 Markup problems

Cost is the price that a business pays for a product. **Selling price** is the price for which a business sells a product to a customer. The difference between selling price and cost is called **markup**. Markup is added to a retailer's cost to cover the expenses of operating a business. Markup is usually expressed as a percent of the retailer's cost. This percent is called the **markup rate**.

The basic markup equations used by a business are

$$\text{Selling price} = \text{Cost} + \text{Markup}$$
$$S \quad = \quad C \;+\; M$$

$$\text{Markup} = \text{Markup rate} \cdot \text{Cost}$$
$$M \quad = \quad r \quad\cdot\quad C$$

By substituting $r \cdot C$ for M in the first equation, we can also write selling price as

$$S = C + M$$
$$S = C + (r \cdot C)$$
$$S = C + rC$$

The equation $S = C + rC$ is the equation used to solve the markup problems in this section.

Example 1 The manager of a clothing store buys a suit for $90 and sells the suit for $126. Find the markup rate.

Strategy Given: $C = \$90$
$S = \$126$
Unknown markup rate: r
Use the equation $S = C + rC$.

Solution
$$S = C + rC$$
$$126 = 90 + 90r \qquad \blacktriangleright \text{Substitute the values of } C \text{ and } S \text{ into the equation.}$$
$$126 - 90 = 90 - 90 + 90r \qquad \blacktriangleright \text{Subtract 90 from each side of the equation.}$$
$$36 = 90r$$
$$\frac{36}{90} = \frac{90r}{90} \qquad \blacktriangleright \text{Divide each side of the equation by 90.}$$
$$0.4 = r \qquad \blacktriangleright \text{The decimal must be changed to a percent.}$$

The markup rate is 40%.

Problem 1 The cost to the manager of a sporting goods store for a tennis racket is $60. The selling price of the racket is $90. Find the markup rate.

Solution See page A20.

Example 2 The manager of a furniture store uses a markup rate of 45% on all items. The selling price of a chair is $232. Find the cost of the chair.

Strategy Given: $r = 45\% = 0.45$
$S = \$232$
Unknown cost: C
Use the equation $S = C + rC$.

Solution $S = C + rC$
$232 = C + 0.45C$
$232 = 1.45C$ ▶ $C + 0.45C = 1C + 0.45C = (1 + 0.45)C$
$160 = C$

The cost of the chair is $160.

Problem 2   A hardware store employee uses a markup rate of 40% on all items. The selling price of a lawnmower is $133. Find the cost.

Solution   See page A20.

2 Discount problems

Discount is the amount by which a retailer reduces the regular price of a product for a promotional sale. Discount is usually expressed as a percent of the regular price. This percent is called the **discount rate.**

The basic discount equations used by a business are

$$\text{Sale price} = \text{Regular price} - \text{Discount}$$
$$S = R - D$$

$$\text{Discount} = \text{Discount rate} \cdot \text{Regular price}$$
$$D = r \cdot R$$

By substituting $r \cdot R$ for D in the first equation, we can also write sale price as

$$S = R - D$$
$$S = R - (r \cdot R)$$
$$S = R - rR$$

The equation $S = R - rR$ is the equation used to solve the discount problems in this section.

Example 3   In a garden supply store, the regular price of a 100 ft garden hose is $48. During an "after-summer sale," the hose is being sold for $36. Find the discount rate.

Strategy   Given: $R = \$48$
$S = \$36$
Unknown discount rate: r
Use the equation $S = R - rR$.

Solution $S = R - rR$
$36 = 48 - 48r$ ▶ Substitute the values of R and S into the equation.

$36 - 48 = 48 - 48 - 48r$ ▶ Subtract 48 from each side of the equation.
$-12 = -48r$

$\dfrac{-12}{-48} = \dfrac{-48r}{-48}$ ▶ Divide each side of the equation by -48.

$0.25 = r$ ▶ The decimal must be changed to a percent.

The discount rate is 25%.

Problem 3 A case of motor oil that regularly sells for $29.80 is on sale for $22.35. What is the discount rate?

Solution See page A21.

Example 4 The sale price for a chemical sprayer is $27.30. This price is 35% off the regular price. Find the regular price.

Strategy Given: $S = \$27.30$
$$r = 35\% = 0.35$$
Unknown regular price: R
Use the equation $S = R - rR$.

Solution
$$S = R - rR$$
$$27.30 = R - 0.35R$$
$$27.30 = 0.65R \qquad \blacktriangleright R - 0.35R = 1R - 0.35R = (1 - 0.35)R$$
$$42 = R$$

The regular price is $42.00.

Problem 4 The sale price for a telephone is $43.50. This price is 25% off the regular price. Find the regular price.

Solution See page A21.

CONCEPT REVIEW 4.4

Determine whether the statement is always true, sometimes true, or never true.

1. In the markup equation $S = C + rC$, the variable r represents the markup rate.

2. Markup is an amount of money, whereas markup rate is a percent.

3. In the discount equation $S = R - rR$, the variable R represents the discount rate.

4. In the markup equation, S represents sale price, and in the discount equation, S represents selling price.

5. $R - 0.45R = 1R - 0.45R = (1 - 0.45)R = 0.55R$ is an example of the application of the Distributive Property.

EXERCISES 4.4

1   Solve.

1. A computer software retailer uses a markup rate of 40%. Find the selling price of a computer game that costs the retailer $25.

2. A car dealer advertises a 5% markup over cost. Find the selling price of a car that costs the dealer $16,000.

3. The pro in a golf shop purchases a one-iron for $40. The selling price of the one-iron is $75. Find the markup rate.

4. A jeweler purchases a diamond ring for $350. The selling price of the ring is $700. Find the markup rate.

5. A leather jacket costs a clothing store manager $140. Find the selling price of the leather jacket if the markup rate is 40%.

6. The cost to a landscape architect for a 25 gal tree is $65. Find the selling price of the tree if the markup rate used by the architect is 30%.

7. A compact disc costs the manager of a music store $11.40. The selling price of the disc is $15.96. Find the markup rate.

8. A grocer purchases a can of fruit juice for $.92. The selling price of the fruit juice is $1.15. Find the markup rate.

9. A tire dealer uses a markup rate of 55% on steel-belted tires. Find the selling price of a steel-belted tire that costs the dealer $72.

10. A cobbler uses a markup rate of 40% on rubber heels for shoes. Find the selling price of a rubber heel that costs the cobbler $5.50.

11. The manager of an electronics store adds $50 to the cost of every 17 in. television, regardless of the cost of the set. Find the markup rate on a television that cost the manager $215. Round to the nearest tenth of a percent.

12. A department store manager uses a markup rate of 40% on items that cost over $100 and a markup rate of 50% on items that cost less than $100. Find the selling price of a ceramic bowl that costs the department store $86.

13. The selling price for a car compact disc player is $168. The markup rate used by the seller is 40%. Find the cost of the compact disc player.

14. A manufacturer of exercise equipment uses a markup rate of 45%. One of the manufacturer's treadmills has a selling price of $580. Find the cost of the treadmill.

2 Solve.

15. A tennis racket that regularly sells for $95 is on sale for 25% off the regular price. Find the sale price.

16. A fax machine that regularly sells for $975 is on sale for $33\frac{1}{3}$% off the regular price. Find the sale price.

17. A car stereo system that regularly sells for $425 is on sale for $318.75. Find the discount rate.

18. During a year-end clearance sale, a car dealer offers $2500 off the regular price of a car. Find the discount rate for a car that regularly sells for $12,500.

19. A college bookstore sells a used book at a discount of 30% off the regular price of a new book. Find the price of a used book that would sell for $35 if it were new.

20. An airline is offering a 35% discount on round-trip air fares. Find the sale price of a round-trip ticket that normally costs $385.

21. A gold bracelet that regularly sells for $1250 is on sale for $750. Find the discount rate.

22. A pair of skis that regularly sells for $325 is on sale for $250. Find the discount rate. Round to the nearest percent.

23. A supplier of electrical equipment offers a 5% discount for a purchase that is paid for within 30 days. A transformer regularly sells for $230. Find the discount price of a transformer that is paid for 10 days after the purchase.

24. A clothing wholesaler offers a discount of 10% per shirt when 10 to 20 shirts are purchased and a discount of 15% per shirt when 21 to 50 shirts are purchased. A shirt regularly sells for $27. Find the sale price per shirt when 35 shirts are purchased.

25. A service station offers a discount of $10 per tire when two tires are purchased and a discount of $25 per tire when four tires are purchased. Find the discount rate when a customer buys four tires that regularly sell for $95 each. Round to the nearest percent.

26. A department store offers a discount of $3 per dinner plate when five or fewer plates are purchased and a discount of $5 per plate when more than five plates are purchased. Find the discount rate when a customer buys three dinner plates that regularly sell for $18 each. Round to the nearest percent.

27. The sale price of a free-weight home gym is $248, which is 20% off the regular price. Find the regular price.

28. The sale price of a 6 ft toboggan is $77, which is 30% off the regular price. Find the regular price.

SUPPLEMENTAL EXERCISES 4.4

Solve.

29. A pair of shoes that now sells for $63 has been marked up 40%. Find the markup on the pair of shoes.

30. The sale price of a typewriter is 25% off the regular price. The discount is $70. Find the sale price.

31. A refrigerator selling for $770 has a markup of $220. Find the markup rate.

32. The sale price of a word processor is $765 after a discount of $135. Find the discount rate.

33. The manager of a camera store uses a markup rate of 30%. Find the cost of a camera selling for $299.

34. The sale price of a television was $180. Find the regular price if the sale price was computed by taking $\frac{1}{3}$ off the regular price followed by an additional 25% discount on the reduced price.

35. A customer buys four tires, three at the regular price and one for 20% off the regular price. The four tires cost $304. What was the regular price of a tire?

36. A lamp, originally priced at under $100, was on sale for 25% off the regular price. When the regular price, a whole number of dollars, was discounted, the discounted price was also a whole number of dollars. Find the largest possible number of dollars in the regular price of the lamp.

37. A used car is on sale for a discount of 20% off the regular price of $5500. An additional 10% discount on the sale price was offered. Is the result a 30% discount? What is the single discount that would give the same sale price?

38. Write a report on series trade discounts. Explain how to convert a series discount to a single-discount equivalent.

S E C T I O N **4.5**

Investment Problems

1 Investment problems

The annual simple interest that an investment earns is given by the equation $I = Pr$, where I is the simple interest, P is the principal, or the amount invested, and r is the simple interest rate.

POINT OF INTEREST

You may be familiar with the simple interest formula $I = Prt$. If so, you know that t represents time. In the problems in this section, time is always 1 (one year), so the formula $I = Prt$ simplifies to

$$I = Pr(1)$$
$$I = Pr$$

The annual simple interest rate on a $4500 investment is 8%. Find the annual simple interest earned on the investment.

Given: $P = \$4500$ $\qquad I = Pr$
$\qquad r = 8\% = 0.08$ $\qquad I = 4500(0.08)$
Unknown interest: I $\qquad I = 360$

The annual simple interest is $360.

Solve: An investor has a total of $10,000 to deposit into two simple interest accounts. On one account, the annual simple interest rate is 7%. On the second account, the annual simple interest rate is 8%. How much should be invested in each account so that the total annual interest earned is $785?

STRATEGY for solving a problem involving money deposited in two simple interest accounts

■ For each amount invested, write a numerical or variable expression for the principal, the interest rate, and the interest earned. The results can be recorded in a table.

The sum of the amounts invested is $10,000.

Amount invested at 7%: x
Amount invested at 8%: $\$10,000 - x$

LOOK CLOSELY

Use the information given in the problem to fill in the principal and interest rate columns of the table. Fill in the interest earned column by multiplying the two expressions you wrote in each row.

	Principal, P	·	Interest rate, r	=	Interest earned, I
Amount at 7%	x	·	0.07	=	0.07x
Amount at 8%	$10,000 - x$	·	0.08	=	$0.08(10,000 - x)$

■ Determine how the amounts of interest earned on each amount are related. For example, the total interest earned by both accounts may be known, or it may be known that the interest earned on one account is equal to the interest earned by the other account.

The sum of the interest earned by the two investments equals the total annual interest earned ($785).

$0.07x + 0.08(10,000 - x) = 785$ ▶ The interest earned on the 7% account plus
$\qquad 0.07x + 800 - 0.08x = 785$ the interest earned on the 8% account
$\qquad\qquad -0.01x + 800 = 785$ equals the total annual interest earned.
$\qquad\qquad\qquad -0.01x = -15$
$\qquad\qquad\qquad\qquad x = 1500$

$10,000 - x = 10,000 - 1500 = 8500$ ▶ Substitute the value of x into the variable
$\qquad\qquad\qquad\qquad\qquad\qquad$ expression for the amount invested at 8%.

The amount invested at 7% is $1500.
The amount invested at 8% is $8500.

Example 1   An investment counselor invested 75% of a client's money into a 9% annual simple interest money market fund. The remainder was invested in 6% annual simple interest government securities. Find the amount invested in each if the total annual interest earned is $3300.

Strategy   ▪ Amount invested: x
Amount invested at 9%: $0.75x$
Amount invested at 6%: $0.25x$

	Principal	·	Rate	=	Interest
Amount at 9%	$0.75x$	·	0.09	=	$0.09(0.75x)$
Amount at 6%	$0.25x$	·	0.06	=	$0.06(0.25x)$

▪ The sum of the interest earned by the two investments equals the total annual interest earned ($3300).

Solution $0.09(0.75x) + 0.06(0.25x) = 3300$ ▶ The interest earned on the
$0.0675x + 0.015x = 3300$ 9% account plus the interest
$0.0825x = 3300$ earned on the 6% account
equals the total annual in-
terest earned.

$x = 40{,}000$ ▶ The amount invested is
$40,000.

$0.75x = 0.75(40{,}000) = 30{,}000$ ▶ Find the amount invested
at 9%.

$0.25x = 0.25(40{,}000) = 10{,}000$ ▶ Find the amount invested
at 6%.

The amount invested at 9% is $30,000.
The amount invested at 6% is $10,000.

Problem 1   An investment of $2500 is made at an annual simple interest rate of 7%. How much additional money must be invested at 10% so that the total interest earned will be 9% of the total investment?

Solution   See page A21.

CONCEPT REVIEW 4.5

Determine whether the statement is always true, sometimes true, or never true.

1. In the simple interest formula $I = Pr$, the variable I represents the simple interest earned.

2. For one year you have $1000 deposited in an account that pays 5% annual simple interest. You will earn exactly $5 in simple interest on this account.

3. For one year, you have x dollars deposited in an account that pays 7% annual simple interest. You will earn $0.07x$ in simple interest on this account.

4. If you have a total of $8000 deposited in two accounts and you represent the amount you have in the first account as x, then the amount in the second account is represented as $8000 - x$.

5. The amount of interest earned on one account is $0.05x$ and the amount of interest earned on a second account is $0.08(9000 - x)$. If the two accounts earn the same amount of interest, then we can write the equation $0.05x + 0.08(9000 - x)$.

6. If the amount of interest earned on one account is $0.06x$ and the amount of interest earned on a second account is $0.09(4000 - x)$, then the total interest earned on the two accounts can be represented as $0.06x + 0.09(4000 - x)$.

EXERCISES 4.5

1 Solve.

1. An engineer invested a portion of $15,000 in a 7% annual simple interest account and the remainder in a 6.5% annual simple interest government bond. The two investments earn $1020 in interest annually. How much was invested in each account?

2. An investment club invested part of $20,000 in preferred stock that earns 8% annual simple interest and the remainder in a municipal bond that earns 7% annual simple interest. The amount of interest earned each year is $1520. How much was invested in each account?

3. A grocer deposited an amount of money into a high-yield mutual fund that earns 13% annual simple interest. A second deposit, $2500 more than the first, was placed in a certificate of deposit earning 7% annual simple interest. In one year, the total interest earned on both investments was $475. How much money was invested in the mutual fund?

4. A deposit was made into a 7% annual simple interest account. Another deposit, $1500 less than the first, was placed in a certificate of deposit earning 9% annual simple interest. The total interest earned on both investments for one year was $505. How much money was deposited in the certificate of deposit?

5. A corporation gave a university $300,000 to support product safety research. The university deposited some of the money in a 10% simple interest account and the remainder in an 8.5% annual simple interest account. How much was deposited in each account if the annual interest is $28,500?

6. A financial consultant invested part of a client's $30,000 in municipal bonds that earn 6.5% annual simple interest and the remainder of the money in 8.5% corporate bonds. How much is invested in each account if the total annual interest earned is $2190?

7. To provide for retirement income, an electrician purchases a $5000 bond that earns 7.5% annual simple interest. How much money does the electrician have invested in bonds that earn 8% annual simple interest if the total annual interest earned from the two investments is $615?

8. The portfolio manager for an investment group invested $40,000 in a certificate of deposit that earns 7.25% annual simple interest. How much money is invested in certificates that earn an annual simple interest rate of 8.5% if the total annual interest earned from the two investments is $5025?

9. An investment of $2500 is made at an annual simple interest rate of 7%. How much money is invested at an annual simple interest rate of 11% if the total interest earned is 9% of the total investment?

10. A total of $6000 is invested into two simple interest accounts. The annual simple interest rate on one account is 9%. The annual simple interest rate on the second account is 6%. How much should be invested in each account so that both accounts earn the same amount of interest?

11. A charity deposited a total of $54,000 into two simple interest accounts. The annual simple interest rate on one account is 8%. The annual simple interest rate on the second account is 12%. How much was invested in each account if the total annual interest earned is 9% of the total investment?

12. A college sports foundation deposited a total of $24,000 into two simple interest accounts. The annual simple interest rate on one account is 7%. The annual simple interest rate on the second account is 11%. How much is invested in each account if the total annual interest earned is 10% of the total investment?

13. An investment banker invested 55% of the bank's available cash in an account that earns 8.25% annual simple interest. The remainder of the cash was placed in an account that earns 10% annual simple interest. The interest earned in one year was $58,743.75. Find the total amount invested.

14. A financial planner recommended that 40% of a client's cash account be invested in preferred stock earning 9% annual simple interest. The remainder of the client's cash was placed in Treasury bonds earning 7% annual simple interest. The total annual interest earned from the two investments was $2496. Find the total amount invested.

15. The manager of a mutual fund placed 30% of the fund's available cash in a 6% annual simple interest account, 25% in 8% corporate bonds, and the remainder in a money market fund earning 7.5% annual simple interest. The total annual interest earned from the investments was $35,875. Find the total amount invested.

16. The manager of a trust invested 30% of a client's cash in government bonds that earn 6.5% annual simple interest, 30% in utility stocks that earn 7% annual simple interest, and the remainder in an account that earns 8% annual simple interest. The total annual interest earned from the investments was $5437.50. Find the total amount invested.

SUPPLEMENTAL EXERCISES 4.5

Solve.

17. A sales representative invests in a stock paying 9% dividends. A research consultant invests $5000 more than the sales representative in bonds paying 8% annual simple interest. The research consultant's income from the investment is equal to the sales representative's. Find the amount of the research consultant's investment.

18. A financial manager invested 20% of a client's money in bonds paying 9% annual simple interest, 35% in an 8% simple interest account, and the remainder in 9.5% corporate bonds. Find the amount invested in each if the total annual interest earned is $5325.

19. A plant manager invested $3000 more in stocks than in bonds. The stocks paid 8% annual simple interest, and the bonds paid 9.5% annual simple interest. Both investments yielded the same income. Find the total annual interest received on both investments.

20. A bank offers a customer a 2-year certificate of deposit (CD) that earns 8% compound annual interest. This means that the interest earned each year is added to the principal before the interest for the next year is calculated. Find the value in 2 years of a nurse's investment of $2500 in this CD.

21. A bank offers a customer a 3-year certificate of deposit (CD) that earns 8.5% compound annual interest. This means that the interest earned each year is added to the principal before the interest for the next year is calculated. Find the value in 3 years of an accountant's investment of $3000 in this CD.

22. Write an essay on the topic of annual percentage rates.

23. Financial advisors may predict how much money we should have saved for retirement by the ages of 35, 45, 55, and 65. One such prediction is included in the table below.

   a. According to the estimates in the table, how much should a couple who have earnings of $75,000 have saved for retirement by age 55?

   b. Write an explanation of how interest and interest rates affect the level of savings required at any given age. What effect do inflation rates have on savings?

Minimum Levels of Savings Required for Married Couples to Be Prepared for Retirement			
Savings Accumulation by Age			
35	45	55	65
Earnings = $50,000 8,000	23,000	90,000	170,000
Earnings = $75,000 17,000	60,000	170,000	310,000
Earnings = $100,000 34,000	110,000	280,000	480,000
Earnings = $150,000 67,000	210,000	490,000	840,000

S E C T I O N **4.6**

Mixture Problems

1 Value mixture problems

A value mixture problem involves combining two ingredients that have different prices into a single blend. For example, a coffee merchant may blend two types of coffee into a single blend, or a candy manufacturer may combine two types of candy to sell as a "variety pack."

The solution of a value mixture problem is based on the equation $V = AC$, where V is the value of an ingredient, A is the amount of the ingredient, and C is the cost per unit of the ingredient.

Find the value of 5 oz of a gold alloy that costs $185 per ounce.

Given: $A = 5$ oz $V = AC$
          $C = \$185$ $V = 5(185)$
Unknown value: V $V = 925$

The value of the 5 oz of gold alloy is $925.

Solve: A coffee merchant wants to make 9 lb of a blend of coffee costing $6 per pound. The blend is made using a $7 grade and a $4 grade of coffee. How many pounds of each of these grades should be used?

STRATEGY for solving a value mixture problem

■ For each ingredient in the mixture, write a numerical or variable expression for the amount of the ingredient used, the unit cost of the ingredient, and the value of the amount used. For the blend, write a numerical or variable expression for the amount, the unit cost of the blend, and the value of the amount. The results can be recorded in a table.

The sum of the amounts is 9 lb.

Amount of $7 coffee: x
Amount of $4 coffee: $9 - x$

LOOK CLOSELY

Use the information given in the problem to fill in the amount and unit cost columns of the table. Fill in the value column by multiplying the two expressions you wrote in each row. Use the expressions in the last column to write the equation.

	Amount, A	·	Unit cost, C	=	Value, V
$7 grade	x	·	$7	=	$7x$
$4 grade	$9 - x$	·	$4	=	$4(9 - x)$
$6 blend	9	·	$6	=	$6(9)$

■ Determine how the values of the ingredients are related. Use the fact that the sum of the values of all ingredients is equal to the value of the blend.

The sum of the values of the $7 grade and the $4 grade is equal to the value of the $6 blend.

$$7x + 4(9 - x) = 6(9)$$ ▶ The value of the $7 grade plus the value of the $4 grade
$$7x + 36 - 4x = 54$$ equals the value of the blend.
$$3x + 36 = 54$$
$$3x = 18$$
$$x = 6$$

$$9 - x = 9 - 6 = 3$$ ▶ Substitute the value of x into the variable expression for the amount of the $4 grade.

The merchant must use 6 lb of the $7 coffee and 3 lb of the $4 coffee.

Example 1   How many ounces of a silver alloy that costs $6 an ounce must be mixed with 10 oz of a silver alloy that costs $8 an ounce to make a mixture that costs $6.50 an ounce?

Strategy   ■ Ounces of $6 alloy: x

	Amount	Cost	Value
$6 alloy	x	$6	$6x$
$8 alloy	10	$8	$8(10)$
$6.50 mixture	$10 + x$	$6.50	$6.50(10 + x)$

■ The sum of the values before mixing equals the value after mixing.

Solution $6x + 8(10) = 6.50(10 + x)$ ▶ The value of the $6 alloy plus the
                $6x + 80 = 65 + 6.5x$ value of the $8 alloy equals the
               $-0.5x + 80 = 65$ value of the mixture.
                     $-0.5x = -15$
                          $x = 30$

30 oz of the $6 silver alloy must be used.

Problem 1   A gardener has 20 lb of a lawn fertilizer that costs $.90 per pound. How many pounds of a fertilizer that costs $.75 per pound should be mixed with this 20 lb of lawn fertilizer to produce a mixture that costs $.85 per pound?

Solution   See page A21.

2 Percent mixture problems

The amount of a substance in a solution can be given as a percent of the total solution. For example, in a 5% saltwater solution, 5% of the total solution is salt. The remaining 95% is water.

The solution of a percent mixture problem is based on the equation $Q = Ar$, where Q is the quantity of a substance in the solution, r is the percent of concentration, and A is the amount of solution.

A 500 ml bottle contains a 3% solution of hydrogen peroxide. Find the amount of hydrogen peroxide in the solution.

Given: $A = 500$ $Q = Ar$
       $r = 3\% = 0.03$ $Q = 500(0.03)$
Unknown amount: Q $Q = 15$

The bottle contains 15 ml of hydrogen peroxide.

Solve: How many gallons of a 15% salt solution must be mixed with 4 gal of a 20% salt solution to make a 17% salt solution?

STRATEGY for solving a percent mixture problem

■ For each solution, write a numerical or variable expression for the amount of solution, the percent of concentration, and the quantity of the substance in the solution. The results can be recorded in a table.

LOOK CLOSELY
Use the information given in the problem to fill in the amount and percent columns of the table. Fill in the quantity column by multiplying the two expressions you wrote in each row. Use the expressions in the last column to write the equation.

The unknown quantity of 15% solution: x

	Amount of solution, A	·	Percent of concentration, r	=	Quantity of substance, Q
15% solution	x	·	0.15	=	$0.15x$
20% solution	4	·	0.20	=	$0.20(4)$
17% solution	$x + 4$	·	0.17	=	$0.17(x + 4)$

■ Determine how the quantities of the substance in each solution are related. Use the fact that the sum of the quantities of the substances being mixed is equal to the quantity of the substance after mixing.

The sum of the quantities of salt in the 15% solution and the 20% solution is equal to the quantity of salt in the 17% solution.

$$0.15x + 0.20(4) = 0.17(x + 4)$$
$$0.15x + 0.8 = 0.17x + 0.68$$
$$-0.02x + 0.8 = 0.68$$
$$-0.02x = -0.12$$
$$x = 6$$

▶ The amount of salt in the 15% solution plus the amount of salt in the 20% solution equals the amount of salt in the 17% solution.

6 gal of the 15% solution are required.

Example 2   A chemist wishes to make 3 L of a 7% acid solution by mixing a 9% acid solution and a 4% acid solution. How many liters of each solution should the chemist use?

Strategy   ■ Liters of 9% solution: x
Liters of 4% solution: $3 - x$

	Amount	Percent	Quantity
9%	x	0.09	$0.09x$
4%	$3 - x$	0.04	$0.04(3 - x)$
7%	3	0.07	$0.07(3)$

■ The sum of the quantities before mixing is equal to the quantity after mixing.

Solution
$$0.09x + 0.04(3 - x) = 0.07(3)$$
$$0.09x + 0.12 - 0.04x = 0.21$$
$$0.05x + 0.12 = 0.21$$
$$0.05x = 0.09$$
$$x = 1.8$$

▶ The amount of acid in the 9% solution plus the amount of acid in the 4% solution equals the amount of acid in the 7% solution.

▶ 1.8 L of the 9% solution are needed.

$$3 - x = 3 - 1.8 = 1.2$$

▶ Find the amount of the 4% solution needed.

The chemist needs 1.8 L of the 9% solution and 1.2 L of the 4% solution.

Problem 2   How many quarts of pure orange juice are added to 5 qt of fruit drink that is 10% orange juice to make an orange drink that is 25% orange juice?

Solution   See page A22.

STUDY TIPS

DETERMINING HOW MUCH TO STUDY

Instructors often advise students to spend twice the amount of time outside of class studying as they spend in the classroom. For example, if a course meets for three hours each week, customarily instructors advise students to study for six hours each week outside of class.

It is often necessary to practice a skill more than a teacher requires. For example, this textbook may provide 50 practice problems on a specific objective and the instructor may assign only 25 of them. However, some students may need to do 30, 40, or all 50 problems.

If you are an accomplished athlete, musician, or dancer, you know that long hours of practice are necessary to acquire a skill. Do not cheat yourself of the practice you need to develop the abilities taught in this course.

Study followed by reward is usually productive. Schedule something enjoyable to do following study sessions. If you know that you have only two hours to study because you have scheduled a pleasant activity for yourself, you may be inspired to make the best use of the two hours that you have set aside for studying.

CONCEPT REVIEW 4.6

Determine whether the statement is always true, sometimes true, or never true.

1. Both value mixture and percent mixture problems involve combining two or more ingredients into a single substance.

2. In the value mixture equation $V = AC$, the variable A represents the quantity of an ingredient.

3. Suppose we are mixing two salt solutions. Then the variable Q in the percent mixture equation $Q = Ar$ represents the amount of salt in a solution.

4. If we combine an alloy that costs $8 an ounce with an alloy that costs $5 an ounce, the cost of the resulting mixture will be greater than $8 an ounce.

5. If we combine a 9% acid solution with a solution that is 4% acid, the resulting solution will be less than 4% acid.

6. If 8 L of a solvent costs $75 per liter, then the value of the 8 L of solvent is $600.

7. If 100 oz of a silver alloy is 25% silver, then the alloy contains 25 oz of silver.

EXERCISES 4.6

1   Solve.

1. A high-protein diet supplement that costs $6.75 per pound is mixed with a vitamin supplement that costs $3.25 per pound. How many pounds of each should be used to make 5 lb of a mixture that costs $4.65 per pound?

2. A 20 oz alloy of platinum that costs $220 per ounce is mixed with an alloy that costs $400 per ounce. How many ounces of the $400 alloy should be used to make an alloy that costs $300 per ounce?

3. Find the cost per pound of a coffee mixture made from 8 lb of coffee that costs $9.20 per pound and 12 lb of coffee that costs $5.50 per pound.

4. How many pounds of tea that cost $4.20 per pound must be mixed with 12 lb of tea that cost $2.25 per pound to make a mixture that costs $3.40 per pound?

5. A goldsmith combined an alloy that costs $4.30 per ounce with an alloy that costs $1.80 per ounce. How many ounces of each were used to make a mixture of 200 oz costing $2.50 per ounce?

6. How many liters of a solvent that costs $80 per liter must be mixed with 6 L of a solvent that costs $25 per liter to make a solvent that costs $36 per liter?

7. Find the cost per pound of a trail mix made from 40 lb of raisins that cost $4.40 per pound and 100 lb of granola that costs $2.30 per pound.

8. Find the cost per ounce of a mixture of 200 oz of a cologne that costs $5.50 per ounce and 500 oz of a cologne that costs $2.00 per ounce.

9. How many kilograms of hard candy that cost $7.50 per kilogram must be mixed with 24 kg of jelly beans that cost $3.25 per kilogram to make a mixture that costs $4.50 per kilogram?

10. A grocery store offers a cheese and fruit sampler that combines cheddar cheese that costs $8 per kilogram with kiwis that cost $3 per kilogram. How many kilograms of each were used to make a 5 kg mixture that costs $4.50 per kilogram?

11. A ground meat mixture is formed by combining meat that costs $2.20 per pound with meat that costs $4.20 per pound. How many pounds of each were used to make a 50 lb mixture that costs $3.00 per pound?

12. A lumber company combined oak wood chips that cost $3.10 per pound with pine wood chips that cost $2.50 per pound. How many pounds of each were used to make an 80 lb mixture costing $2.65 per pound?

13. How many kilograms of soil supplement that costs $7.00 per kilogram must be mixed with 20 kg of aluminum nitrate that costs $3.50 per kilogram to make a fertilizer that costs $4.50 per kilogram?

14. A caterer made an ice cream punch by combining fruit juice that costs $2.25 per gallon with ice cream that costs $3.25 per gallon. How many gallons of each were used to make 100 gal of punch costing $2.50 per gallon?

15. The manager of a specialty food store combined almonds that cost $4.50 per pound with walnuts that cost $2.50 per pound. How many pounds of each were used to make a 100 lb mixture that costs $3.24 per pound?

16. Find the cost per gallon of a carbonated fruit drink made from 12 gal of fruit juice that costs $4.00 per gallon and 30 gal of carbonated water that costs $2.25 per gallon.

17. Find the cost per pound of a sugar-coated breakfast cereal made from 40 lb of sugar that cost $1.00 per pound and 120 lb of corn flakes that cost $.60 per pound.

18. Find the cost per ounce of a gold alloy made from 25 oz of pure gold that costs $482 per ounce and 40 oz of an alloy that costs $300 per ounce.

19. How many pounds of lima beans that cost $.90 per pound must be mixed with 16 lb of corn that costs $.50 per pound to make a mixture of vegetables that costs $.65 per pound?

20. How many liters of a blue dye that costs $1.60 per liter must be mixed with 18 L of anil that costs $2.50 per liter to make a mixture that costs $1.90 per liter?

2   Solve.

21. A chemist wants to make 50 ml of a 16% acid solution by mixing a 13% acid solution and an 18% acid solution. How many milliliters of each solution should the chemist use?

22. How many pounds of coffee that is 40% java beans must be mixed with 80 lb of coffee that is 30% java beans to make a coffee blend that is 32% java beans?

23. Thirty ounces of pure silver are added to 50 oz of a silver alloy that is 20% silver. What is the percent concentration of silver in the resulting alloy?

24. Two hundred liters of punch that contains 35% fruit juice is mixed with 300 L of a second punch. The resulting fruit punch is 20% fruit juice. Find the percent concentration of fruit juice in the second punch.

25. The manager of a garden shop mixes grass seed that is 60% rye grass with 70 lb of grass seed that is 80% rye grass to make a mixture that is 74% rye grass. How much of the 60% rye grass is used?

26. Five grams of sugar are added to a 45 g serving of a breakfast cereal that is 10% sugar. What is the percent concentration of sugar in the resulting mixture?

27. A dermatologist mixes 50 g of a cream that is 0.5% hydrocortisone with 150 g of a second hydrocortisone cream. The resulting mixture is 0.68% hydrocortisone. Find the percent concentration of hydrocortisone in the second hydrocortisone cream.

28. A carpet manufacturer blends two fibers, one 20% wool and the second 50% wool. How many pounds of each fiber should be woven together to produce 600 lb of a fabric that is 28% wool?

29. A hair dye is made by blending a 7% hydrogen peroxide solution and a 4% hydrogen peroxide solution. How many milliliters of each are used to make a 300 ml solution that is 5% hydrogen peroxide?

30. How many grams of pure salt must be added to 40 g of a 20% salt solution to make a solution that is 36% salt?

31. How many ounces of an 8% saline solution must be added to 40 oz of a 15% saline solution to make a saline solution that is 10% salt?

32. A paint that contains 21% green dye is mixed with a paint that contains 15% green dye. How many gallons of each must be used to make 60 gal of paint that is 19% green dye?

33. A goldsmith mixes 8 oz of a 30% gold alloy with 12 oz of a 25% gold alloy. What is the percent concentration of the resulting alloy?

Latex Exterior
House Paint

Satin

34. A physicist mixes 40 L of liquid oxygen with 50 L of liquid air that is 64% liquid oxygen. What is the percent concentration of liquid oxygen in the resulting mixture?

35. How many ounces of pure bran flakes must be added to 50 oz of cereal that is 40% bran flakes to produce a mixture that is 50% bran flakes?

36. How many milliliters of pure chocolate must be added to 150 ml of chocolate topping that is 50% chocolate to make a topping that is 75% chocolate?

37. A tea that is 20% jasmine is blended with a tea that is 15% jasmine. How many pounds of each tea are used to make 5 lb of tea that is 18% jasmine?

38. A clothing manufacturer has some pure silk thread and some thread that is 85% silk. How many kilograms of each must be woven together to make 75 kg of cloth that is 96% silk?

39. How many ounces of dried apricots must be added to 18 oz of a snack mix that contains 20% dried apricots to make a mixture that is 25% dried apricots?

40. A recipe for a rice dish calls for 12 oz of a rice mixture that is 20% wild rice and 8 oz of pure wild rice. What is the percent concentration of wild rice in the 20 oz mixture?

SUPPLEMENTAL EXERCISES 4.6

Solve.

41. Find the cost per ounce of a mixture of 30 oz of an alloy that costs $4.50 per ounce, 40 oz of an alloy that costs $3.50 per ounce, and 30 oz of an alloy that costs $3.00 per ounce.

42. A grocer combined walnuts that cost $2.60 per pound and cashews that cost $3.50 per pound with 20 lb of peanuts that cost $2.00 per pound. Find the amount of walnuts and the amount of cashews used to make the 50 lb mixture costing $2.72 per pound.

43. How many ounces of water evaporated from 50 oz of a 12% salt solution to produce a 15% salt solution?

44. A chemist mixed pure acid with water to make 10 L of a 30% acid solution. How much pure acid and how much water did the chemist use?

45. How many grams of pure water must be added to 50 g of pure acid to make a solution that is 40% acid?

46. A radiator contains 15 gal of a 20% antifreeze solution. How many gallons must be drained from the radiator and replaced by pure antifreeze so that the radiator will contain 15 gal of a 40% antifreeze solution?

47. Tickets to a performance by the community theater cost $5.50 for adults and $2.75 for children. A total of 120 tickets were sold for $563.75. How many adults and how many children attended the performance?

48. Explain why we look for patterns and relationships in mathematics. Include a discussion of the relationship between value mixture problems and percent mixture problems and how understanding one of these can be helpful in understanding the other. Also discuss why understanding how to solve the value mixture problems in this section can assist in solving Exercise 47.

SECTION 4.7
Uniform Motion Problems

1 Uniform motion problems

A train that travels constantly in a straight line at 50 mph is in *uniform motion*. **Uniform motion** means the speed of an object does not change.

The solution of a uniform motion problem is based on the equation $d = rt$, where d is the distance traveled, r is the rate of travel, and t is the time spent traveling.

A car travels 50 mph for 3 h.

$$d = rt$$
$$d = 50(3)$$
$$d = 150$$

The car travels a distance of 150 mi.

Solve: A car leaves a town traveling at 35 mph. Two hours later, a second car leaves the same town, on the same road, traveling at 55 mph. In how many hours after the second car leaves will the second car be passing the first car?

STRATEGY for solving a uniform motion problem

■ For each object, write a numerical or variable expression for the distance, rate, and time. The results can be recorded in a table.

The first car traveled 2 h longer than the second car.

Unknown time for the second car: t
Time for the first car: $t + 2$

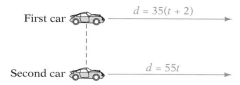

First car $d = 35(t + 2)$

Second car $d = 55t$

LOOK CLOSELY

Use the information given in the problem to fill in the rate and time columns of the table. Fill in the distance column by multiplying the two expressions you wrote in each row.

	Rate, r	·	Time, t	=	Distance, d
First car	35	·	$t + 2$	=	$35(t + 2)$
Second car	55	·	t	=	$55t$

■ Determine how the distances traveled by each object are related. For example, the total distance traveled by both objects may be known, or it may be known that the two objects traveled the same distance.

The two cars travel the same distance.

$35(t + 2) = 55t$ ▶ The distance traveled by the first car equals the distance traveled
$35t + 70 = 55t$ by the second car.
$70 = 20t$
$3.5 = t$

The second car will be passing the first car in 3.5 h.

Example 1   Two cars, one traveling 10 mph faster than the second car, start at the same time from the same point and travel in opposite directions. In 3 h, they are 288 mi apart. Find the rate of the second car.

Strategy   ▪ Rate of second car: r
Rate of first car: $r + 10$

	Rate	Time	Distance
First car	$r + 10$	3	$3(r + 10)$
Second car	r	3	$3r$

▪ The total distance traveled by the two cars is 288 mi.

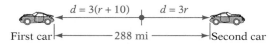

Solution $3(r + 10) + 3r = 288$ ▶ The distance traveled by the first car plus
$3r + 30 + 3r = 288$ the distance traveled by the second car
$6r + 30 = 288$ is 288 mi.
$6r = 258$
$r = 43$

The second car is traveling 43 mph.

Problem 1   Two trains, one traveling at twice the speed of the other, start at the same time from stations that are 306 mi apart and travel toward each other. In 3 h, the trains pass each other. Find the rate of each train.

Solution   See page A22.

Example 2   A bicycling club rides out into the country at a speed of 16 mph and returns over the same road at 12 mph. How far does the club ride out into the country if it travels a total of 7 h?

Strategy   ▪ Time spent riding out: t
Time spend riding back: $7 - t$

	Rate	Time	Distance
Out	16	t	$16t$
Back	12	$7 - t$	$12(7 - t)$

■ The distance out equals the distance back.

$d = 16t$

$d = 12(7 - t)$

Solution $16t = 12(7 - t)$
$16t = 84 - 12t$
$28t = 84$
$\quad t = 3$ ▶ The time is 3 h. Find the distance.

The distance out $= 16t = 16(3) = 48$.

The club rides 48 mi into the country.

Problem 2   On a survey mission, a pilot flew out to a parcel of land and back in 7 h. The rate out was 120 mph. The rate back was 90 mph. How far was the parcel of land?

Solution   See page A22.

CONCEPT REVIEW 4.7

Determine whether the statement is always true, sometimes true, or never true.

1.   The solution of a uniform motion problem is based on the equation $d = rt$, where r is a percent.

2.   If you drive at a constant speed of 60 mph for 4 h, you will travel a total distance of 240 mi.

3.   It takes a student a total of 5 h to drive to a friend's house and back. If the variable t is used to represent the time it took to drive to the friend's house, then the expression $t - 5$ is used to represent the time spent on the return trip.

4.   If the speed of one train is 20 mph slower than that of a second train, then the speeds of the two trains can be represented as r and $20 - r$.

5.   A car and a bus both travel from city A to city B. The car leaves city A one hour later than the bus, but both vehicles arrive in city B at the same time. If t represents the time the car was traveling, then $t + 1$ represents the time the bus was traveling. If t represents the time the bus was traveling, then $t - 1$ represents the time the car was traveling.

6.   Two cars travel the same distance. If the distance traveled by the first car is represented by the expression $35t$, and the distance traveled by the second car is represented by the expression $45(t + 1)$, then we can write the equation $35t = 45(t + 1)$.

EXERCISES 4.7

1   Solve.

1.   Two small planes start from the same point and fly in opposite directions. The first plane is flying 25 mph slower than the second plane. In 2 h, the planes are 470 mi apart. Find the rate of each plane.

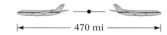
470 mi

2.   Two cyclists start from the same point and ride in opposite directions. One cyclist rides twice as fast as the other. In 3 h, they are 81 mi apart. Find the rate of each cyclist.

3.   A long-distance runner started on a course running at an average speed of 6 mph. One half-hour later, a second runner began the same course at an average speed of 7 mph. How long after the second runner started will the second runner overtake the first runner?

4.   A motorboat leaves a harbor and travels at an average speed of 9 mph toward a small island. Two hours later a cabin cruiser leaves the same harbor and travels at an average speed of 18 mph toward the same island. In how many hours after the cabin cruiser leaves will the cabin cruiser be alongside the motorboat?

5.   On a 130 mi trip, a car traveled at an average speed of 55 mph and then reduced its speed to 40 mph for the remainder of the trip. The trip took a total of 2.5 h. For how long did the car travel at 40 mph?

6.   A motorboat leaves a harbor and travels at an average speed of 18 mph to an island. The average speed on the return trip was 12 mph. How far was the island from the harbor if the total trip took 5 h?

7.   As part of flight training, a student pilot was required to fly to an airport and then return. The average speed on the way to the airport was 100 mph, and the average speed returning was 150 mph. Find the distance between the two airports if the total flying time was 5 h.

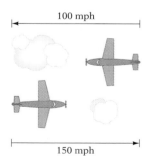
100 mph

150 mph

8.   A family drove to a resort at an average speed of 25 mph and later returned over the same road at an average speed of 40 mph. Find the distance to the resort if the total driving time was 13 h.

9.   Three campers left their campsite by canoe and paddled downstream at an average rate of 10 mph. They then turned around and paddled back upstream at an average rate of 5 mph to return to their campsite. How long did it take the campers to canoe downstream if the total trip took 1 h?

10. Running at an average rate of 8 m/s, a sprinter ran to the end of a track and then jogged back to the starting point at an average rate of 6 m/s. The sprinter took 35 s to run to the end of the track and jog back. Find the length of the track.

11. A jet plane traveling at 570 mph overtakes a propeller-driven plane that has had a 2 h head start. The propeller-driven plane is traveling at 190 mph. How far from the starting point does the jet overtake the propeller-driven plane?

12. A car traveling at 56 mph overtakes a cyclist who, riding at 14 mph, has had a 3 h head start. How far from the starting point does the car overtake the cyclist?

13. A 605 mi, 5 h plane trip was flown at two speeds. For the first part of the trip, the average speed was 115 mph. For the remainder of the trip, the average speed was 125 mph. How long did the plane fly at each speed?

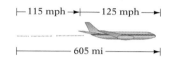

14. On a 220 mi trip, a car traveled at an average speed of 50 mph and then reduced its average speed to 35 mph for the remainder of the trip. The trip took a total of 5 h. How long did the car travel at each speed?

15. After a sailboat had been on the water for 3 h, a change in wind direction reduced the average speed of the boat by 5 mph. The entire distance sailed was 51 mi. The total time spent sailing was 6 h. How far did the sailboat travel in the first 3 h?

16. A bus traveled on a level road for 2 h at an average speed that was 20 mph faster than its average speed on a winding road. The time spent on the winding road was 3 h. Find the average speed on the winding road if the total trip was 210 mi.

17. A passenger train leaves a train depot 1 h after a freight train leaves the same depot. The freight train is traveling 15 mph slower than the passenger train. Find the rate of each train if the passenger train overtakes the freight train in 3 h.

18. A bus traveling at a rate of 60 mph overtakes a car traveling at a rate of 45 mph. If the car had a 1 h head start, how far from the starting point does the bus overtake the car?

19. Two joggers start at the same time from opposite ends of a 12 mi course. One jogger is running at a rate of 5 mph, and the other is running at a rate of 7 mph. How long after they begin will they meet?

20. Two cyclists start at the same time from opposite ends of a course that is 51 mi long. One cyclist is riding at a rate of 16 mph, and the second cyclist is riding at a rate of 18 mph. How long after they begin will they meet?

SUPPLEMENTAL EXERCISES 4.7

Solve.

21. A car and a cyclist start at 10 A.M. from the same point, headed in the same direction. The average speed of the car is 5 mph more than three times the average speed of the cyclist. In 1.5 h, the car is 46.5 mi ahead of the cyclist. Find the rate of the cyclist.

22. A cyclist and a jogger set out at 11 A.M. from the same point, headed in the same direction. The average speed of the cyclist is twice the average speed of the jogger. In 1 h, the cyclist is 7 mi ahead of the jogger. Find the rate of the cyclist.

23. A car and a bus set out at 2 P.M. from the same point, headed in the same direction. The average speed of the car is 30 mph slower than twice the average speed of the bus. In 2 h, the car is 30 mi ahead of the bus. Find the rate of the car.

24. At 10 A.M., two campers left their campsite by canoe and paddled downstream at an average speed of 12 mph. They then turned around and paddled back upstream at an average rate of 4 mph. The total trip took 1 h. At what time did the campers turn around downstream?

25. At 7 A.M., two joggers start from opposite ends of an 8 mi course. One jogger is running at a rate of 4 mph, and the other is running at a rate of 6 mph. At what time will the joggers meet?

26. A truck leaves a depot at 11 A.M. and travels at a speed of 45 mph. At noon, a van leaves the same place and travels the same route at a speed of 65 mph. At what time does the van overtake the truck?

27. A bicyclist rides for 2 h at a speed of 10 mph and then returns at a speed of 20 mph. Find the cyclist's average speed for the trip.

28. A car travels a 1 mi track at an average speed of 30 mph. At what average speed must the car travel the next mile so that the average speed for the 2 mi is 60 mph?

29. Explain why the motion problems in this section are restricted to *uniform* motion.

30. Explain why 60 mph is the same as 88 ft/s.

SECTION **4.8**

Inequalities

1   Applications of inequalities

Solving application problems requires recognition of the verbal phrases that translate into mathematical symbols. Here is a partial list of the phrases used to indicate each of the four inequality symbols.

$<$ is less than

$>$ is greater than
is more than
exceeds

$\leq$ is less than or equal to
maximum
at most
or less

$\geq$ is greater than or equal to
minimum
at least
or more

Example 1   A student must have at least 450 points out of 500 points on five tests to receive an A in a course. One student's results on the first four tests were 93, 79, 87, and 94. What scores on the last test will enable this student to receive an A in the course?

Strategy   To find the scores, write and solve an inequality using N to represent the score on the last test.

Solution

total number of points on the five tests	is greater than or equal to	450

$$93 + 79 + 87 + 94 + N \geq 450$$
$$353 + N \geq 450$$
$$353 - 353 + N \geq 450 - 353$$
$$N \geq 97$$

The student's score on the last test must be equal to or greater than 97.

Problem 1   An appliance dealer will make a profit on the sale of a television set if the cost of the new set is less than 70% of the selling price. What minimum selling price will enable the dealer to make a profit on a television set that costs the dealer $340?

Solution   See page A23.

Example 2   The base of a triangle is 8 in., and the height is $(3x - 5)$ in. Express as an integer the maximum height of the triangle when the area is less than 112 in².

Strategy   To find the maximum height:
- Replace the variables in the area formula by the given values and solve for x.
- Replace the variable in the expression $3x - 5$ with the value found for x.

Solution

one-half the base times the height	is less than	112 in²

$$\frac{1}{2}(8)(3x - 5) < 112$$
$$4(3x - 5) < 112$$
$$12x - 20 < 112$$
$$12x - 20 + 20 < 112 + 20$$
$$12x < 132$$
$$\frac{12x}{12} < \frac{132}{12}$$
$$x < 11$$

$3x - 5 < 28$ ▶ Substitute the value of x into the variable expression for the height. $3x - 5 = 3(11) - 5 = 28$. Note that the height is less than 28 because $x < 11$.

The maximum height of the triangle is 27 in.

Problem 2   Company A rents cars for $9 per day and 10¢ per mile driven. Company B rents cars for $12 per day and 8¢ per mile driven. You want to rent a car for one week. What is the maximum number of miles you can drive a Company A car if it is to cost you less than a Company B car?

Solution   See page A23.

CONCEPT REVIEW 4.8

Determine whether the statement is always true, sometimes true, or never true.

1. Both "is greater than" and "is more than" indicate the inequality symbol $\geq$.

2. A minimum refers to a lower limit, whereas a maximum refers to an upper limit.

3. Given that $x > \dfrac{32}{6}$, the minimum integer that satisfies the inequality is 6.

4. Given that $x < \dfrac{25}{4}$, the maximum integer that satisfies the inequality is 7.

5. A rental car costs $15 per day and 20¢ per mile driven. If m represents the number of miles the rental car is driven, then the expression $15 + 0.20m$ represents the cost to rent the car for one week.

6. A patient's physician recommends a maximum cholesterol level of 205 units. The patient's cholesterol level is 199 units and therefore must be lowered 6 units in order to meet the physician's recommended level.

EXERCISES 4.8

1   Solve.

1. Three-fifths of a number is greater than two-thirds. Find the smallest integer that satisfies the inequality.

2. To be eligible for a basketball tournament, a basketball team must win at least 60% of its remaining games. If the team has 17 games remaining, how many games must the team win to qualify for the tournament?

3. To avoid a tax penalty, at least 90% of a self-employed person's total annual income tax liability must be paid by April 15. What amount of income tax must be paid by April 15 by a person with an annual income tax liability of $3500?

4. A health official recommends a maximum cholesterol level of 220 units. A patient has a cholesterol level of 275. By how many units must this patient's cholesterol level be reduced to satisfy the recommended maximum level?

5. A service organization will receive a bonus of $200 for collecting more than 1850 lb of aluminum cans during its four collection drives. On the first three drives, the organization collected 505 lb, 493 lb, and 412 lb. How many pounds of cans must the organization collect on the fourth drive to receive the bonus?

6. A professor scores all tests with a maximum of 100 points. To earn an A in this course, a student must have an average of at least 92 on four tests. One student's grades on the first three tests were 89, 86, and 90. Can this student earn an A grade?

7. A student must have an average of at least 80 points on five tests to receive a B in a course. The student's grades on the first four tests were 75, 83, 86, and 78. What scores on the last test will enable this student to receive a B in the course?

8. A car sales representative receives a commission that is the greater of $250 or 8% of the selling price of a car. What dollar amounts in the sale price of a car will make the commission offer more attractive than the $250 fee?

9. A sales representative for a stereo store has the option of a monthly salary of $2000 or a 35% commission on the selling price of each item sold by the representative. What dollar amounts in sales will make the commission more attractive than the monthly salary?

10. Four times the sum of a number and five is less than six times the number. Find the smallest integer that satisfies the inequality.

11. The sales agent for a jewelry company is offered a flat monthly salary of $3200 or a salary of $1000 plus an 11% commission on the selling price of each item sold by the agent. If the agent chooses the $3200 salary, what dollar amount does the agent expect to sell in one month?

12. A baseball player is offered an annual salary of $200,000 or a base salary of $100,000 plus a bonus of $1000 for each hit over 100 hits. How many hits must the baseball player make to earn more than $200,000?

13. A computer bulletin board service charges a flat fee of $10 per month or $4 per month plus $.10 for each minute the service is used. For how many minutes must a person use this service for the cost to exceed $10?

14. A site licensing fee for a computer program is $1500. The fee allows a company to use the program at any computer terminal within the company. Alternatively, a company can choose to pay $200 for each computer terminal it has. How many computer terminals must a company have for the site licensing fee to be the more economical plan?

15. For a product to be labeled orange juice, a state agency requires that at least 80% of the drink be real orange juice. How many ounces of artificial flavors can be added to 32 oz of real orange juice if the drink is to be labeled orange juice?

16. Grade A hamburger cannot contain more than 20% fat. How much fat can a butcher mix with 300 lb of lean meat to meet the 20% requirement?

17. A shuttle service taking skiers to a ski area charges $8 per person each way. Four skiers are debating whether to take the shuttle bus or rent a car for $45 plus $.25 per mile. Assuming that the skiers will share the cost of the car and that they want the least expensive method of transportation, how far away is the ski area if they choose to take the shuttle service?

18. A residential water bill is based on a flat fee of $10 plus a charge of $.75 for each 1000 gal of water used. Find the number of gallons of water a family can use and have a monthly water bill that is less than $55.

19. Company A rents cars for $25 per day and 8¢ per mile driven. Company B rents cars for $15 per day and 14¢ per mile driven. You want to rent a car for one day. Find the maximum number of miles you can drive a Company B car if it is to cost you less than a Company A car.

20. A rectangle is 8 ft wide and $(2x + 7)$ ft long. Express as an integer the maximum length of the rectangle when the area is less than 152 ft^2.

SUPPLEMENTAL EXERCISES 4.8

Solve.

21. Find three positive consecutive odd integers such that three times the sum of the first two integers is less than four times the third integer. (*Hint:* There is more than one solution.)

22. Find three positive consecutive even integers such that four times the sum of the first two integers is less than or equal to five times the third integer. (*Hint:* There is more than one solution.)

23. A maintenance crew requires between 30 min and 45 min to prepare an aircraft for its next flight. How many aircraft can this crew prepare for flight in a 6 h period of time?

24. The charges for a long-distance telephone call are $1.21 for the first three minutes and $.42 for each additional minute or fraction of a minute. What is the largest whole number of minutes a call could last if it is to cost you less than $6?

25. Your class decides to publish a calendar to raise money. The initial cost, regardless of the number of calendars printed, is $900. After the initial cost, each calendar costs $1.50 to produce. What is the minimum number of calendars your class must sell at $6 per calendar to make a profit of at least $1200?

26. The graph below depicts U.S. voter turnout during presidential election years and during off-year elections.

   a. During which presidential election years was voter turnout greater than 60%?

   b. During which off-year elections was voter turnout less than 40%?

   c. During which presidential election years was voter turnout less than 50%?

   d. Write a paragraph describing any pattern you see in the graph.

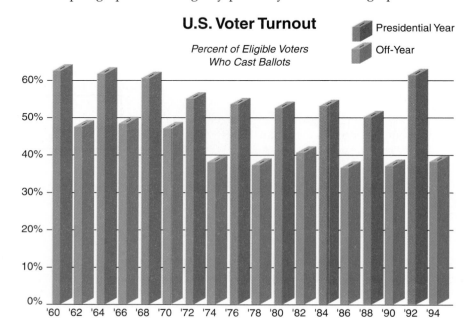

Project in Mathematics

Angles Formed by Intersecting Lines

The word *geometry* comes from the Greek words for *earth* and *measure*. The original purpose of geometry was to measure land. Today geometry is used in many fields, including physics, medicine, geology, mechanical drawing, and astronomy. Geometric forms are used in art and design.

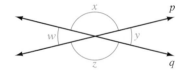

The topic of Section 4.3 is Geometry Problems. However, Section 4.3 is not the only place where geometry problems are presented in this text. Vertical angles are introduced in Section 3.2 on page 110. **Vertical angles** are two angles that are on opposite sides of the intersection of two lines. Vertical angles have the same measure. For the diagram at the left, $\angle x$ and $\angle z$ have the same measure, and $\angle w$ and $\angle y$ have the same measure. (The symbol $\angle$ is used for "angle.")

Two angles that share a common side are called **adjacent angles**. Adjacent angles of intersecting lines are supplementary angles. (Recall from page 140 in

Section 4.1 that two angles are **supplementary angles** if the sum of the measures of the angles is 180°.) For the diagram at the bottom of page 186, ∠x and ∠y are adjacent angles, as are ∠y and ∠z, ∠z and ∠w, and ∠w and ∠x.

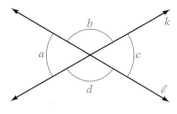

In the diagram at the left, ∠c = 65°. Find the measures of angles a, b, and d.

∠a = ∠c because ∠a and ∠c are vertical angles.　　　　∠a = 65°

∠b is supplementary to ∠c because ∠b and ∠c are adjacent angles of intersecting lines.

$$\angle b + \angle c = 180°$$
$$\angle b + 65° = 180°$$
$$\angle b = 115°$$

∠d = ∠b because ∠d and ∠b are vertical angles.　　∠d = 115°

Perpendicular lines are intersecting lines that form right angles, as shown in the diagram at the left.

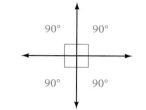

If two lines do not intersect, then they are parallel lines. **Parallel lines** never meet. The distance between them is always the same.

Parallel Lines

An **acute angle** is an angle whose measure is between 0° and 90°. ∠A at the left is an acute angle. An **obtuse angle** is an angle whose measure is between 90° and 180°. ∠B is an obtuse angle.

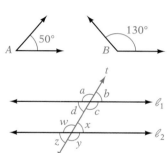

A line that intersects two other lines at different points is called a **transversal**. If the lines cut by a transversal t are parallel lines and the transversal is not perpendicular to the parallel lines, all four acute angles have the same measure and all four obtuse angles have the same measure. For the figure at the left,

$$\angle b = \angle d = \angle x = \angle z$$
$$\angle a = \angle c = \angle w = \angle y$$

Alternate interior angles are two angles that are on opposite sides of the transveral and between the parallel lines. In the figure above, ∠c and ∠w are alternate interior angles; ∠d and ∠x are alternate interior angles. Alternate interior angles have the same measure.

Alternate interior angles have the same measure.

$$\angle c = \angle w$$
$$\angle d = \angle x$$

Alternate exterior angles are two angles that are on opposite sides of the transversal and outside the parallel lines. In the figure above, ∠a and ∠y are alternate exterior angles; ∠b and ∠z are alternate exterior angles. Alternate exterior angles have the same measure.

Alternate exterior angles have the same measure.

$$\angle a = \angle y$$
$$\angle b = \angle z$$

Corresponding angles are two angles that are on the same side of the transversal and are both acute angles or are both obtuse angles. For the figure above, the following pairs of angles are corresponding angles: ∠a and ∠w, ∠d and ∠z, ∠b and ∠x, ∠c and ∠y. Corresponding angles have the same measure.

Corresponding angles have the same measure.

$$\angle a = \angle w$$
$$\angle d = \angle z$$
$$\angle b = \angle x$$
$$\angle c = \angle y$$

For the diagram at the right given that lines ℓ_1 and ℓ_2 are parallel lines and $\angle c = 58°$, find the measures of angles f, h, and g.

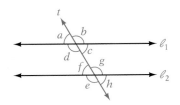

$\angle c$ and $\angle f$ are alternate interior angles.

$\angle c$ and $\angle h$ are corresponding angles.

$\angle g$ is supplementary to $\angle h$.

$\angle f = \angle c = 58°$

$\angle h = \angle c = 58°$

$\angle g + \angle h = 180°$
$\angle g + 58° = 180°$
$\angle g = 122°$

Find x.

1.

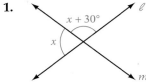

2.

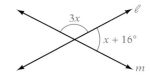

3.

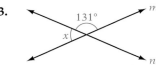

4.

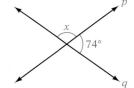

Given that lines ℓ_1 and ℓ_2 are parallel lines, find the measures of angles a and b.

5.

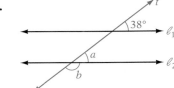

6.

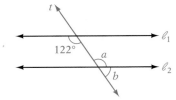

7.

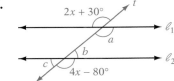

8.

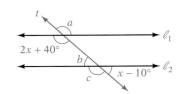

Given that lines ℓ_1 and ℓ_2 are parallel lines, find x.

9.

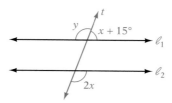

10.

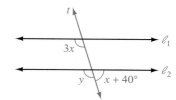

11.

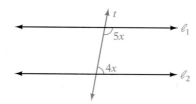

12.

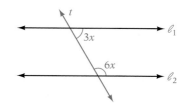

13.

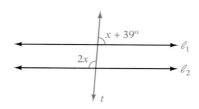

14.

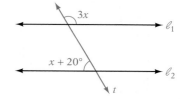

15. Cut out a triangle and then tear off two of the angles, as shown below. Position the pieces you tore off so that angle a is adjacent to angle b, and angle c is adjacent to angle b. Describe what you observe. What does this demonstrate?

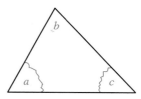

16. For the figure below, find the sum of the measures of angles x, y, and z.

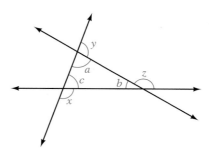

Focus on Problem Solving

Diagramming Problems

How do you best remember something? Do you remember best what you hear? The word "aural" means *pertaining to the ear*; people with a strong aural memory remember best those things that they hear. The word "visual" means *pertaining to the sense of sight*; people with a strong visual memory remember best that which they see written down. Some people claim their memory is in their writing hand—they remember something only if they write it down! The method by which you best remember something is probably also the method by which you can best learn something new.

In problem-solving situations, try to capitalize on your strengths. If you tend to understand material better when you hear it spoken, read application problems aloud or have someone else read them to you. If writing helps you to organize ideas, rewrite application problems in your own words.

No matter what your main strength, visualizing a problem can be a valuable aid in problem solving. A drawing, sketch, diagram, or chart can be a useful tool in problem solving, just as calculators and computers are tools. A diagram can be helpful in gaining an understanding of the relationships inherent in a problem-solving situation. A sketch will help you to organize the given information and can lead to your being able to focus on the method by which the solution can be determined.

A tour bus drives 5 mi south, then 4 mi west, then 3 mi north, and then 4 mi east. How far is the tour bus from the starting point?

Draw a diagram of the given information.

From the diagram, we can see that the solution can be determined by subtracting 3 from 5: $5 - 3 = 2$.

The bus is 2 mi from the starting point.

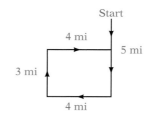

If you roll two ordinary six-sided dice and multiply the two numbers that appear on top, how many different products are there?

Make a chart of the possible products. In the chart below, repeated products are marked with an asterisk.

$1 \cdot 1 = 1$	$2 \cdot 1 = 2$ *	$3 \cdot 1 = 3$ *	$4 \cdot 1 = 4$ *	$5 \cdot 1 = 5$ *	$6 \cdot 1 = 6$ *
$1 \cdot 2 = 2$	$2 \cdot 2 = 4$ *	$3 \cdot 2 = 6$ *	$4 \cdot 2 = 8$ *	$5 \cdot 2 = 10$ *	$6 \cdot 2 = 12$ *
$1 \cdot 3 = 3$	$2 \cdot 3 = 6$ *	$3 \cdot 3 = 9$	$4 \cdot 3 = 12$ *	$5 \cdot 3 = 15$ *	$6 \cdot 3 = 18$ *
$1 \cdot 4 = 4$	$2 \cdot 4 = 8$	$3 \cdot 4 = 12$ *	$4 \cdot 4 = 16$	$5 \cdot 4 = 20$ *	$6 \cdot 4 = 24$ *
$1 \cdot 5 = 5$	$2 \cdot 5 = 10$	$3 \cdot 5 = 15$	$4 \cdot 5 = 20$	$5 \cdot 5 = 25$	$6 \cdot 5 = 30$ *
$1 \cdot 6 = 6$	$2 \cdot 6 = 12$	$3 \cdot 6 = 18$	$4 \cdot 6 = 24$	$5 \cdot 6 = 30$	$6 \cdot 6 = 36$

By counting the products that are not repeats, we can see that there are 18 different products.

In this chapter, you may have noticed that a drawing accompanies the Strategy step of application problems in Section 4.3, Geometry Problems, and Section 4.7, Uniform Motion Problems. We encourage you to draw a diagram when solving the application problems in these sections, as well as in any other situation where it can prove helpful. It will undoubtedly improve your problem-solving skills.

Chapter Summary

Key Words

Consecutive integers follow one another in order. (Objective 4.2.1)

11, 12, 13 are consecutive integers.

−9, −8, −7 are consecutive integers.

The *perimeter* of a geometric figure is a measure of the distance around the figure. (Objective 4.3.1)

The perimeter of a triangle is the sum of the lengths of the three sides.

The perimeter of a rectangle is the sum of the lengths of the four sides.

A *right angle* is an angle that measures 90°. A *right triangle* has one right angle. An *isosceles triangle* has two equal angles and two sides of the same length. (Objective 4.3.2)

Cost is the price that a business pays for a product. *Selling price* is the price for which a business sells a product to a customer. *Markup* is the difference between selling price and cost. *Markup rate* is the markup expressed as a percent of the retailer's cost. (Objective 4.4.1)

If a business pays $100 for a product and sells that product for $140, then the cost of the product is $100, the selling price is $140, and the markup is $140 − $100 = $40.

Discount is the amount by which a retailer reduces the regular price of a product. *Discount rate* is the discount expressed as a percent of the regular price. (Objective 4.4.2)

The regular price of a product is $100. The product is now on sale for $75. The discount on the product is $100 − $75 = $25.

Essential Rules and Procedures

Consecutive Integers
$n, n + 1, n + 2, \ldots$ (Objective 4.2.1)

The sum of three consecutive integers is 135.

$$n + (n + 1) + (n + 2) = 135$$

Consecutive Even or Consecutive Odd Integers
$n, n + 2, n + 4, \ldots$ (Objective 4.2.1)

The sum of three consecutive odd integers is 135.

$$n + (n + 2) + (n + 4) = 135$$

Coin and Stamp Equation

$$\frac{\text{Number}}{\text{of coins}} \cdot \frac{\text{Value of}}{\text{coin in cents}} = \frac{\text{Total value}}{\text{in cents}}$$ (Objective 4.2.2)

A coin bank contains $1.80 in dimes and quarters. In all, there are twelve coins in the bank. How many dimes are in the bank?

$$10d + 25(12 - d) = 180$$

Perimeter of a Rectangle
$P = 2L + 2W$ (Objective 4.3.1)

The perimeter of a rectangle is 108 m. The length is 6 m more than three times the width. Find the width of the rectangle.

$$108 = 2(3W + 6) + 2W$$

Perimeter of a Triangle
$P = a + b + c$ (Objective 4.3.1)

The perimeter of an isosceles triangle is 34 ft. The length of the third side is 5 ft less than the length of one of the equal sides. Find the length of one of the equal sides.

$$34 = x + x + (x - 5)$$

The Sum of the Angles of a Triangle
$A + B + C = 180°$ (Objective 4.3.2)

In a right triangle, the measure of one acute angle is 12° more than the measure of the smallest angle. Find the measure of the smallest angle.

$$x + (x + 12) + 90 = 180$$

Basic Markup Equation
$S = C + rC$ (Objective 4.4.1)

The manager of a department buys a pasta maker machine for $70 and sells the machine for $98. Find the markup rate.

$$98 = 70 + 70r$$

Basic Discount Equation
$S = R - rR$ (Objective 4.4.2)

The sale price for a leather bomber jacket is $95. This price is 24% off the regular price. Find the regular price.

$$95 = R - 0.24R$$

Annual Simple Interest Equation
$I = Pr$ (Objective 4.5.1)

You invest a portion of $10,000 in a 7% annual simple interest account and the remainder in a 6% annual simple interest bond. The two investments earn total annual interest of $680. How much is invested in the 7% account?

$$0.07x + 0.06(10,000 - x) = 680$$

Value Mixture Equation
$V = AC$ (Objective 4.6.1)

An herbalist has 30 oz of herbs costing $2 per ounce. How many ounces of herbs costing $1 per ounce should be mixed with the 30 oz to produce a mixture costing $1.60 per ounce?

$$30(2) + 1x = 1.60(30 + x)$$

Percent Mixture Equation
$Q = Ar$ (Objective 4.6.2)

Forty ounces of a 30% gold alloy are mixed with 60 oz of a 20% gold alloy. Find the percent concentration of the resulting gold alloy.

$$0.30(40) + 0.20(60) = x(100)$$

Uniform Motion Equation
$d = rt$ (Objective 4.7.1)

A boat traveled from a harbor to an island at an average speed of 20 mph. The average speed on the return trip was 15 mph. The total trip took 3.5 h. How long did it take to travel to the island?

$$20t = 15(3.5 - t)$$

Chapter Review Exercises

1. Translate "four less than the product of five and a number is sixteen" into an equation and solve.

2. A piano wire is 35 in. long. A note can be produced by dividing this wire into two parts so that three times the length of the shorter piece is twice the length of the longer piece. Find the length of the shorter piece.

3. The sum of two numbers is twenty-one. Three times the smaller number is two less than twice the larger number. Find the two numbers.

4. Translate "the product of six and three more than a number is ten less than twice the number" into an equation and solve.

5. A ceiling fan that regularly sells for $60 is on sale for $40. Find the discount rate.

6. A total of $15,000 is deposited into two simple interest accounts. The annual simple interest rate on one account is 6%. The annual simple interest rate on the second account is 7%. How much should be invested in each account so that the total interest earned is $970?

7. Find the cost per pound of a meatloaf mixture made from 3 lb of ground beef costing $1.99 per pound and 1 lb of ground turkey costing $1.39 per pound.

8. A motorcyclist and a bicyclist set out at 8 A.M. from the same point, headed in the same direction. The speed of the motorcyclist is three times the speed of the bicyclist. In 2 h, the motorcyclist is 60 mi ahead of the bicyclist. Find the rate of the motorcyclist.

9. In an isosceles triangle, the measure of one angle is 25° less than half the measure of one of the equal angles. Find the measure of each angle.

10. The manager of a sporting goods store buys indoor/outdoor basketballs for $8.50 and sells them for $14.45. Find the markup rate.

11. A dairy owner mixes 5 gal of cream that is 30% butterfat with 8 gal of milk that is 4% butterfat. Find the percent concentration of butterfat in the resulting mixture.

12. The length of a rectangle is four times the width. The perimeter is 200 ft. Find the length and width of the rectangle.

13. Company A rents cars for $6 per day and 25¢ per mile driven. Company B rents cars for $15 per day and 10¢ per mile driven. You want to rent a car for 6 days. What is the maximum number of miles you can drive a Company A car if it is to cost you less than a Company B car?

14. The sale price for a carpet sweeper is $26.56, which is 17% off the regular price. Find the regular price.

15. The owner of a health food store combined cranberry juice that cost $1.79 per quart with apple juice that cost $1.19 per quart. How many quarts of each were used to make 10 qt of a cranapple juice mixture costing $1.61 per quart?

16. The manager of a telescope supply company buys a 16 in. mirror blank for $340 and uses a 75% markup rate. Find the selling price of the mirror blank.

17. The sum of two numbers is thirty-six. The difference between the larger number and eight equals the total of four and three times the smaller number. Find the two numbers.

18. An engineering consultant invested $14,000 in an individual retirement account paying 8.15% annual simple interest. How much money does the consultant have deposited in an account paying 12% annual simple interest if the total interest earned is 9.25% of the total investment?

19. Find two consecutive integers such that five times the first integer is 15 more than three times the second integer.

20. A pharmacist has 15 L of an 80% alcohol solution. How many liters of pure water should be added to the alcohol solution to make an alcohol solution that is 75% alcohol?

21. A student's grades on five sociology exams were 68, 82, 90, 73, and 95. What is the lowest score this student can receive on the sixth exam and still have earned a total of at least 480 points?

22. Translate "fifteen is equal to the total of two-thirds of a number and three" into an equation and solve.

23. The Empire State Building is 1472 ft tall. This is 514 ft less than twice the height of the Eiffel Tower. Find the height of the Eiffel Tower.

24. One angle of a triangle is 15° more than the measure of the second angle. The third angle is 15° less than the measure of the second angle. Find the measure of each angle.

25. A ticket seller at a baseball card show had $145 in one-dollar bills and five-dollar bills. In all, there were 53 bills. Find the number of one-dollar bills held by the ticket seller.

26. A board 10 ft long is cut into two pieces. Four times the length of the shorter piece is 2 ft less than two times the length of the longer piece. Find the length of the longer piece.

27. An optical engineer's consulting fee was $600. This included $80 for supplies and $65 for each hour of consultation. Find the number of hours of consultation.

28. A restaurant diner left a tip of $2.25 in dimes and quarters. There were two more quarters than dimes. Find the number of dimes and the number of quarters left for the tip.

29. A furniture store uses a markup rate of 60%. The store sells a solid oak curio cabinet for $1074. Find the cost of the curio cabinet.

30. The largest swimming pool in the world is located in Casablanca, Morocco, and is 480 m long. Two swimmers start at the same time from opposite ends of the pool and start swimming toward each other. One swimmer's rate is 65 m/min. The other swimmer's rate is 55 m/min. In how many minutes after they begin will they meet?

31. The perimeter of a triangle is 35 in. The second side is 4 in. longer than the first side. The third side is 1 in. shorter than twice the first side. Find the measure of each side.

32. The sum of three consecutive odd integers is -45. Find the integers.

33. A rectangle is 15 ft long and $(2x - 4)$ ft wide. Express as an integer the maximum width of the rectangle when the area is less than 180 ft^2.

Chapter Test

1. Translate "the sum of six times a number and thirteen is five less than the product of three and the number" into an equation and solve.

2. Translate "the difference between three times a number and fifteen is twenty-seven" into an equation and solve.

3. The sum of two numbers is 18. The difference between four times the smaller number and seven is equal to the sum of two times the larger number and five. Find the two numbers.

4. A board 18 ft long is cut into two pieces. Two feet less than the product of five and the length of the shorter piece is equal to the difference between three times the length of the longer piece and eight. Find the length of each piece.

5. The manager of a sports shop uses a markup rate of 50%. The selling price for a set of golf clubs is $300. Find the cost of the golf clubs.

6. A portable typewriter that regularly sells for $100 is on sale for $80. Find the discount rate.

7. How many gallons of a 15% acid solution must be mixed with 5 gal of a 20% acid solution to make a 16% acid solution?

8. The perimeter of a rectangle is 38 m. The length of the rectangle is 1 m less than three times the width. Find the length and width of the rectangle.

9. Find three consecutive odd integers such that three times the first integer is one less than the sum of the second and third integers.

10. Florist A charges a $3 delivery fee plus $21 per bouquet delivered. Florist B charges a $15 delivery fee plus $18 per bouquet delivered. An organization wants to send each resident of a small nursing home a bouquet for Valentine's Day. How many residents are in the nursing home if it is more economical for the organization to use Florist B?

11. A total of $7000 is deposited into two simple interest accounts. On one account, the annual simple interest rate is 10%, and on the second account, the annual simple interest rate is 15%. How much should be invested in each account so that the total annual interest earned is $800?

12. A coffee merchant wants to make 12 lb of a blend of coffee costing $6 per pound. The blend is made using a $7 grade and a $4 grade of coffee. How many pounds of each of these grades should be used?

13. Two planes start at the same time from the same point and fly in opposite directions. The first plane is flying 100 mph faster than the second plane. In 3 h, the two planes are 1050 mi apart. Find the rate of each plane.

14. In a triangle, the first angle is 15° more than the second angle. The third angle is three times the second angle. Find the measure of each angle.

15. A coin bank contains 50 coins in nickels and quarters. The total amount of money in the bank is $9.50. Find the number of nickels and the number of quarters in the bank.

16. A club treasurer deposited $2400 in two simple interest accounts. The annual simple interest rate on one account is 6.75%. The annual simple interest rate on the other account is 9.45%. How much should be deposited in each account so that the same interest is earned on each account?

17. The width of a rectangle is 12 ft. The length is $(3x + 5)$ ft. Express as an integer the minimum length of the rectangle if the area is greater than 276 ft².

18. A file cabinet that normally sells for $99 is on sale for 20% off. Find the sale price.

Cumulative Review Exercises

1. Given $B = \{-12, -6, -3, -1\}$, which elements of set B are less than -4?

2. Simplify: $-2 + (-8) - (-16)$

3. Simplify: $\left(-\frac{2}{3}\right)^3\left(-\frac{3}{4}\right)^2$

4. Simplify: $\frac{5}{6} - \left(\frac{2}{3}\right)^2 \div \left(\frac{1}{2} - \frac{1}{3}\right)$

5. Evaluate $-|-18|$.

6. Evaluate $b^2 - (a - b)^2$ when $a = 4$ and $b = -1$.

7. Simplify: $5x - 3y - (-4x) + 7y$

8. Simplify: $-4(3 - 2x - 5x^3)$

9. Simplify: $-2[x - 3(x - 1) - 5]$

10. Simplify: $-3x^2 - (-5x^2) + 4x^2$

11. Is 2 a solution of $4 - 2x - x^2 = 2 - 4x$?

12. Solve: $9 - x = 12$

13. Solve: $-\frac{4}{5}x = 12$

14. Solve: $8 - 5x = -7$

15. Solve: $-6x - 4(3 - 2x) = 4x + 8$

16. Write 40% as a fraction.

17. Solve and graph the solution set of $4x \geq 16$.

18. Solve: $-15x \leq 45$

19. Solve: $2x - 3 > x + 15$

20. Solve: $12 - 4(x - 1) \leq 5(x - 4)$

21. Write 0.025 as a percent.

22. Write $\frac{3}{25}$ as a percent.

23. Find $16\frac{2}{3}\%$ of 18.

24. 40% of what is 18?

25. Translate "the sum of eight times a number and twelve is equal to the product of four and the number" into an equation and solve.

26. The area of the cement foundation of a house is 2000 ft². This is 200 ft² more than three times the area of the garage. Find the area of the garage.

27. An auto repair bill was $213. This includes $88 for parts and $25 for each hour of labor. Find the number of hours of labor.

28. A survey of 250 librarians showed that 50 of the libraries had a particular reference book on their shelves. What percent of the libraries had the reference book?

29. A deposit of $4000 is made into an account that earns 11% annual simple interest. How much money is also deposited in an account that pays 14% annual simple interest if the total annual interest earned is 12% of the total investment?

30. The manager of a department store buys a chain necklace for $8 and sells it for $14. Find the markup rate.

31. How many grams of a gold alloy that costs $4 a gram must be mixed with 30 g of a gold alloy that costs $7 a gram to make an alloy costing $5 a gram?

32. How many ounces of pure water must be added to 70 oz of a 10% salt solution to make a 7% salt solution?

33. In an isosceles triangle, two angles are equal. The third angle is 8° less than twice the measure of one of the equal angles. Find the measure of one of the equal angles.

34. Three times the second of three consecutive even integers is 14 more than the sum of the first and third integers. Find the middle even integer.

35. A coin bank contains dimes and quarters. The number of quarters is five less than four times the number of dimes. The total amount in the bank is $6.45. Find the number of dimes in the bank.

5

Linear Equations and Inequalities

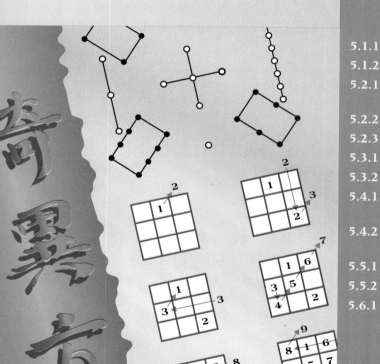

Objectives

Magic Squares

A magic square is a square array of distinct integers so arranged that the numbers along any row, column, or main diagonal have the same sum. An example of a magic square is shown at the right.

8	3	4
1	5	9
6	7	2

The oldest known example of a magic square comes from China. Estimates are that this magic square is over 4000 years old. It is shown at the left.

There is a simple way to produce a magic square with an odd number of cells. Start by writing a 1 in the top middle cell. The rule then is to proceed diagonally upward to the right with the successive integers.

When the rule takes you outside the square, write the number by shifting either across the square from right to left or down the square from top to bottom, as the case may be. For example, in Figure B, the second number (2) is outside the square above a column. Because the 2 is above a column, it should be shifted down to the bottom cell in that column. In Figure C, the 3 is outside the square to the right of a column and should therefore be shifted all the way to the left.

If the rule takes you to a square that is already filled (as shown in Figure D), then write the number in the cell directly below the last number written. Continue until the entire square is filled.

It is possible to begin a magic square with any integer and proceed by using the above rule and consecutive integers.

For an odd magic square beginning with 1, the sum of a row, column, or diagonal is $\frac{n(n^2 + 1)}{2}$, where n is the number of rows.

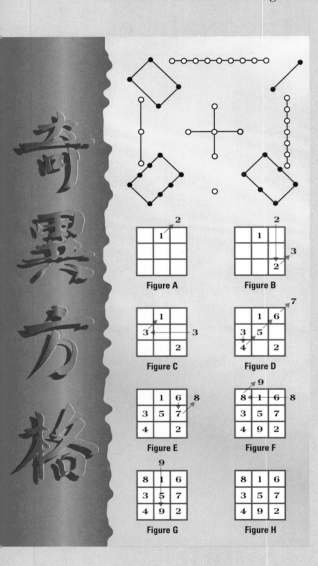

Figure A

Figure B

Figure C

Figure D

Figure E

Figure F

Figure G

Figure H

SECTION **5.1**

The Rectangular Coordinate System

1 Graph points in a rectangular coordinate system

Before the 15th century, geometry and algebra were considered separate branches of mathematics. That all changed when René Descartes, a French mathematician who lived from 1596 to 1650, founded **analytic geometry.** In this geometry, a *coordinate system* is used to study the relationships between variables.

A **rectangular coordinate system** is formed by two number lines, one horizontal and one vertical, that intersect at the zero point of each line. The point of intersection is called the **origin.** The two axes are called the **coordinate axes,** or simply the **axes.** Generally, the horizontal axis is labeled the *x*-**axis**, and the vertical axis is labeled the *y*-**axis.**

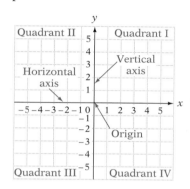

The axes determine a **plane**, which can be thought of as a large, flat sheet of paper. The two axes divide the plane into four regions called **quadrants,** which are numbered counterclockwise from I to IV starting at the upper right.

Each point in the plane can be identified by a pair of numbers called an **ordered pair.** The first number of the ordered pair measures a horizontal distance and is called the **abscissa,** or *x*-**coordinate**. The second number of the pair measures a vertical distance and is called the **ordinate,** or *y*-**coordinate**. The ordered pair (x, y) associated with a point is also called the **coordinates** of the point.

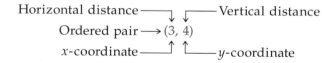

To **graph,** or **plot, a point in the plane,** place a dot at the location given by the ordered pair. For example, to graph the point $(4, 1)$, start at the origin. Move 4 units to the right and then 1 unit up. Draw a dot. To graph $(-3, -2)$, start at the origin. Move 3 units left and then 2 units down. Draw a dot.

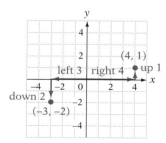

The **graph of an ordered pair** is the dot drawn at the coordinates of the point in the plane. The graphs of the ordered pairs $(4, 1)$ and $(-3, -2)$ are shown at the right.

The graphs of the points whose coordinates are (2, 3) and (3, 2) are shown at the right. Note that they are different points. The order in which the numbers in an ordered pair appear *is* important.

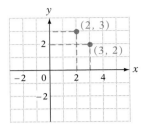

Each point in the plane is associated with an ordered pair, and each ordered pair is associated with a point in the plane. Although only integers are labeled on the coordinate grid, any ordered pair can be graphed by approximating its location. The graphs of the ordered pairs $\left(\frac{3}{2}, -\frac{4}{3}\right)$ and $(-2.4, 3.5)$ are shown at the right.

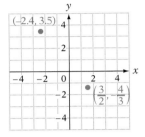

Example 1   Graph the ordered pairs $(-2, -3)$, $(3, -2)$, $(1, 3)$, and $(4, 1)$.

Solution

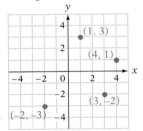

Problem 1   Graph the ordered pairs $(-1, 3)$, $(1, 4)$, $(-4, 0)$, and $(-2, -1)$.

Solution   See page A23.

Example 2   Find the coordinates of each of the points.

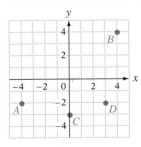

Solution $A(-4, -2)$, $B(4, 4)$, $C(0, -3)$, $D(3, -2)$

Problem 2   Find the coordinates of each of the points.

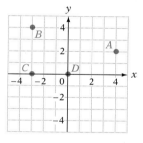

Solution   See page A23.

2   Scatter diagrams

There are many situations in which the relationship between two variables may be of interest. For example, a college admissions director wants to know the relationship between SAT scores (one variable) and success in college (the second variable). An employer is interested in the relationship between scores on a pre-employment test and ability to perform a job.

A researcher can investigate the relationship between two variables by means of regression analysis, which is a branch of statistics. The study of the relationship between two variables may begin with a **scatter diagram,** which is a graph of some of the known data.

The following table gives the record times for events of different lengths at a track meet. Record times are rounded to the nearest second. Lengths are given in meters.

Length, x	100	200	400	800	1000	1500
Time, y	10	20	50	100	130	210

The scatter diagram for these data is shown at the right. Each ordered pair represents the length of the race and the record time. For example, the ordered pair (400, 50) indicates that the record for the 400 m race is 50 s.

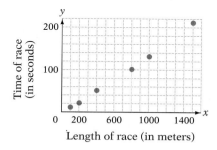

Length of race (in meters)

Example 3   To test a heart medicine, a doctor measures the heart rate, in beats per minute, of five patients before and after they take the medication. The results are recorded in the following table. Graph the scatter diagram for these data.

Before medicine, x	85	80	85	75	90
After medicine, y	75	70	70	80	80

Strategy Graph the ordered pairs on a rectangular coordinate system where the reading on the horizontal axis represents the heart rate before taking the medication and the vertical axis represents the heart rate after taking the medication.

Solution

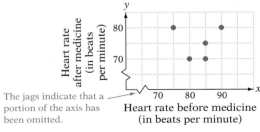

The jags indicate that a portion of the axis has been omitted.

Problem 3 The age of five cars in years and the prices paid for the cars in hundreds of dollars are recorded in the following table. Graph the scatter diagram for these data.

Age, x	4	5	3	5	2
Price, y	27	22	31	18	32

Solution See page A23.

CONCEPT REVIEW 5.1

Determine whether the statement is always true, sometimes true, or never true.

1. The point (0, 0) on a rectangular coordinate system is at the origin.

2. When a point is plotted in the rectangular coordinate system, the first number in an ordered pair indicates a movement up or down from the origin. The second number in the ordered pair indicates a movement left or right.

3. In an xy-coordinate system, the first number is the y-coordinate and the second number is the x-coordinate.

4. When a point is plotted in the rectangular coordinate system, a negative x value indicates a movement to the left. A negative y value indicates a movement down.

5. Any point plotted in quadrant III has a positive y value.

6. In the figure at the right, point C is the graph of the point whose coordinates are $(-4, -2)$.

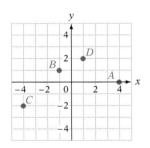

7. Any point on the x-axis has y-coordinate 0.

8. Any point on the y-axis has x-coordinate 0.

EXERCISES 5.1

1. Graph the ordered pairs $(-2, 1)$, $(3, -5)$, $(-2, 4)$, and $(0, 3)$.

2. Graph the ordered pairs $(5, -1)$, $(-3, -3)$, $(-1, 0)$, and $(1, -1)$.

3. Graph the ordered pairs $(0, 0)$, $(0, -5)$, $(-3, 0)$, and $(0, 2)$.

4. Graph the ordered pairs $(-4, 5)$, $(-3, 1)$, $(3, -4)$, and $(5, 0)$.

5. Graph the ordered pairs $(-1, 4)$, $(-2, -3)$, $(0, 2)$, and $(4, 0)$.

6. Graph the ordered pairs $(5, 2)$, $(-4, -1)$, $(0, 0)$, and $(0, 3)$.

7. Find the coordinates of each of the points.

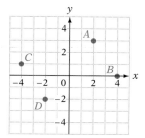

8. Find the coordinates of each of the points.

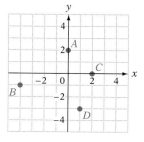

9. Find the coordinates of each of the points.

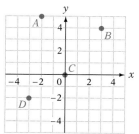

10. Find the coordinates of each of the points.

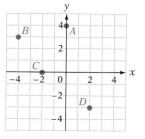

11. **a.** Name the abscissas of points A and C.
    b. Name the ordinates of points B and D.

12. **a.** Name the abscissas of points A and C.
    b. Name the ordinates of points B and D.

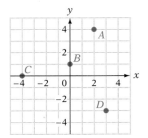
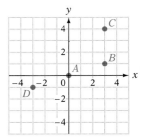

---

2 Solve.

13. The math midterm scores and the final exam scores for six students are given in the following table. Graph the scatter diagram for these data.

Midterm score, x	90	85	75	80	85	70
Final exam score, y	95	75	80	75	90	70

14. The following table shows, for each of five tennis players, the number of aces served by the player during one set and the number of games that player won in the set. Graph the scatter diagram for these data.

Aces served, x	6	8	4	10	6
Games won, y	6	5	3	6	4

15. The number of hit records in one year by each of five musical groups and the number of concerts the group performed that year are given in the following table. Graph the scatter diagram for these data.

Hit records, x	11	9	7	12	12
Concerts, y	200	150	200	175	250

16. The monthly salaries, in thousands of dollars, of six employees in a company and the number of years of college completed by each are given in the following table. Graph the scatter diagram for these data.

Years of college, x	4	4	5	4	6	8
Monthly salary (in thousands), y	3	3.5	4	2.5	5	5.5

17. The number of lift tickets sold during one day and the number of skiing accidents reported that day are recorded in the following table. Graph the scatter diagram for these data.

Tickets sold, x	1500	2500	4000	6000
Accidents reported, y	25	35	50	55

18. A study of six algebra classes measured the number of hours the class met per week and the average score of the class on the final exam. The results are recorded in the following table. Graph the scatter diagram for these data.

Average score, x	72	72	68	75	76	70
Hours in class, y	3	4	3	3	5	4

SUPPLEMENTAL EXERCISES 5.1

In which quadrant is the given point located?

19. $(2, -4)$ **20.** $(-3, 2)$ **21.** $(-1, -6)$

What is the distance from the given point to the horizontal axis?

22. $(-5, 1)$ **23.** $(3, -4)$ **24.** $(-6, 0)$

What is the distance from the given point to the vertical axis?

25. $(-2, 4)$ **26.** $(1, -3)$ **27.** $(5, 0)$

28. Name the coordinates of a point plotted at the origin of the rectangular coordinate system.

29. Describe the signs of the coordinates of a point plotted in:
   a. quadrant I   **b.** quadrant II
   c. quadrant III   **d.** quadrant IV

30. A computer screen has a coordinate system that is different from the xy-coordinate system we have discussed. In one mode, the origin of the coordinate system is the top left point of the screen, as shown at the right. Plot the points whose coordinates are (200, 400), (0, 100), and (100, 300).

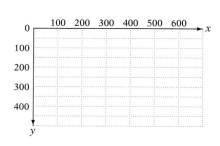

31. Write a paragraph explaining how to plot points in a rectangular coordinate system.

32. Decide on two quantities that may be related, and collect at least ten pairs of values. Here are some examples: height and weight, time studying for a test and the test grade, age of a car and its cost. Draw a scatter diagram for the data. Is there any trend? That is, as the values on the horizontal axis increase, do the values on the vertical axis increase or decrease?

33. There is a coordinate system on Earth that consists of *longitude* and *latitude*. Write a report on how location is determined on the surface of Earth.

SECTION 5.2
Graphs of Straight Lines

1 Determine solutions of linear equations in two variables

An equation of the form $y = mx + b$, where m is the coefficient of x and b is a constant, is a **linear equation in two variables.** Examples of linear equations are shown below.

$$y = 3x + 4 \qquad (m = 3, b = 4)$$
$$y = 2x - 3 \qquad (m = 2, b = -3)$$
$$y = -\frac{2}{3}x + 1 \qquad \left(m = -\frac{2}{3}, b = 1\right)$$
$$y = -2x \qquad (m = -2, b = 0)$$
$$y = x + 2 \qquad (m = 1, b = 2)$$

In a linear equation, the exponent of each variable is 1. The equations $y = 2x^2 - 1$ and $y = \frac{1}{x}$ are not linear equations.

A **solution of a linear equation in two variables** is an ordered pair of numbers (x, y) that makes the equation a true statement.

LOOK CLOSELY

An ordered pair is of the form (x, y). For the ordered pair $(1, -2)$, 1 is the x value and -2 is the y value. Substitute 1 for x and -2 for y.

Is $(1, -2)$ a solution of $y = 3x - 5$?

Replace x with 1. Replace y with -2.

Compare the results. If the results are equal, the given ordered pair is a solution. If the results are not equal, the given ordered pair is not a solution.

$$
\begin{array}{c|c}
y & = 3x - 5 \\
\hline
-2 & 3(1) - 5 \\
 & 3 - 5 \\
\end{array}
$$
$$-2 = -2$$

Yes, $(1, -2)$ is a solution of the equation $y = 3x - 5$.

Besides the ordered pair $(1, -2)$, there are many other ordered-pair solutions of the equation $y = 3x - 5$. For example, the method used above can be used to show that $(2, 1)$, $(-1, -8)$, $\left(\frac{2}{3}, -3\right)$, and $(0, -5)$ are also solutions.

Example 1   Is $(-3, 2)$ a solution of $y = 2x + 2$?

Solution
$$y = 2x + 2$$

$$\begin{array}{c|l} 2 & 2(-3) + 2 \\ & -6 + 2 \\ & -4 \end{array}$$ ▶ Replace x with -3 and y with 2.

$$2 \neq -4$$

No, $(-3, 2)$ is not a solution of $y = 2x + 2$.

Problem 1   Is $(2, -4)$ a solution of $y = -\dfrac{1}{2}x - 3$?

Solution   See page A24.

In general, a linear equation in two variables has an infinite number of ordered-pair solutions. By choosing any value for x and substituting that value into the linear equation, we can find a corresponding value of y.

Find the ordered-pair solution of $y = 2x - 5$ that corresponds to $x = 1$.

Substitute 1 for x.
Solve for y.

$$y = 2x - 5$$
$$y = 2 \cdot 1 - 5$$
$$y = 2 - 5$$
$$y = -3$$

The ordered-pair solution is $(1, -3)$.

Example 2   Find the ordered-pair solution of $y = \dfrac{2}{3}x - 1$ that corresponds to $x = 3$.

Solution $y = \dfrac{2}{3}x - 1$

$y = \dfrac{2}{3}(3) - 1$ ▶ Substitute 3 for x.

$y = 2 - 1$ ▶ Solve for y.
$y = 1$ ▶ When $x = 3$, $y = 1$.

The ordered-pair solution is $(3, 1)$.

Problem 2   Find the ordered-pair solution of $y = -\dfrac{1}{4}x + 1$ that corresponds to $x = 4$.

Solution   See page A24.

2   Graph equations of the form $y = mx + b$

The **graph of an equation in two variables** is a drawing of the ordered-pair solutions of the equation. For a linear equation in two variables, the graph is a straight line.

To graph a linear equation, find ordered-pair solutions of the equation. Do this by choosing any value of x and finding the corresponding value of y. Repeat this procedure, choosing different values for x, until you have found the number of solutions desired.

Because the graph of a linear equation in two variables is a straight line, and a straight line is determined by two points, it is necessary to find only two solutions. However, finding at least three points will ensure accuracy.

LOOK CLOSELY

If the three points you graph do not lie on a straight line, you have made an arithmetic error in calculating a point or you have plotted a point incorrectly.

To graph $y = 2x + 1$, choose any values of x, and then find the corresponding values of y. The numbers 0, 2, and -1 were chosen arbitrarily for x. It is convenient to record these solutions in a table.

x	$y = 2x + 1$	y
0	$2 \cdot 0 + 1$	1
2	$2 \cdot 2 + 1$	5
-1	$2(-1) + 1$	-1

Graph the ordered-pair solutions $(0, 1)$, $(2, 5)$, and $(-1, -1)$. Draw a line through the ordered-pair solutions.

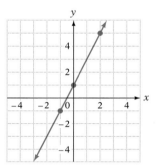

Note that the points whose coordinates are $(-2, -3)$ and $(1, 3)$ are on the graph and that these ordered pairs are solutions of the equation $y = 2x + 1$.

Remember that a graph is a drawing of the ordered-pair solutions of the equation. Therefore, every point on the graph is a solution of the equation, and every solution of the equation is a point on the graph.

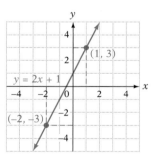

Example 3   Graph: $y = 3x - 2$

Solution

x	y
0	-2
2	4
-1	-5

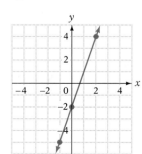

Problem 3   Graph: $y = 3x + 1$

Solution   See page A24.

Graphing utilities create graphs by plotting points and then connecting the points to form a curve. Using a graphing utility, enter the equation $y = 3x - 2$ and verify the graph drawn in Example 3. (Refer to the Appendix on page A1 for instructions on using a graphing calculator to graph a linear equation.) Trace along the graph and verify that $(0, -2)$, $(2, 4)$, and $(-1, -5)$ are coordinates of points on the graph. Now enter the equation $y = 3x + 1$ given in Problem 3. Verify that the ordered pairs you found for this equation are the coordinates of points on the graph.

When m is a fraction in the equation $y = mx + b$, choose values of x that will simplify the evaluation. For example, to graph $y = \frac{1}{3}x - 1$, we might choose the numbers 0, 3, and -3 for x. Note that these numbers are multiples of the denominator.

x	y
0	-1
3	0
-3	-2

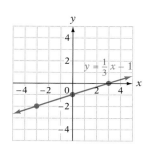

Example 4   Graph: $y = \frac{1}{2}x - 1$

Solution

x	y
0	-1
2	0
-2	-2

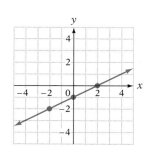

Problem 4   Graph: $y = \frac{1}{3}x - 3$

Solution   See page A24.

Using a graphing utility, enter the equation $y = \frac{1}{2}x - 1$ and verify the graph drawn in Example 4. Trace along the graph and verify that $(0, -1)$, $(2, 0)$, and $(-2, -2)$ are the coordinates of points on the graph. Follow the same procedure for Problem 4. (See the Appendix on page A1 for instructions on entering a fractional coefficient of x.)

3   Graph equations of the form $Ax + By = C$

An equation in the form $Ax + By = C$, where A and B are coefficients and C is a constant, is also a linear equation. Examples of these equations are shown below.

$$2x + 3y = 6 \qquad (A = 2, B = 3, C = 6)$$
$$x - 2y = -4 \qquad (A = 1, B = -2, C = -4)$$
$$2x + y = 0 \qquad (A = 2, B = 1, C = 0)$$
$$4x - 5y = 2 \qquad (A = 4, B = -5, C = 2)$$

One method of graphing an equation of the form $Ax + By = C$ involves first solving the equation for y and then following the same procedure used for graphing an equation of the form $y = mx + b$. To solve the equation for y means to rewrite the equation so that y is alone on one side of the equation and the term containing x and the constant are on the other side of the equation. The Addition and Multiplication Properties of Equations are used to rewrite an equation of the form $Ax + By = C$ in the form $y = mx + b$.

Solve the equation $3x + 2y = 4$ for y.

The equation is in the form $Ax + By = C$.
$$3x + 2y = 4$$

Use the Addition Property of Equations to subtract the term $3x$ from each side of the equation.
$$3x - 3x + 2y = -3x + 4$$

Simplify. Note that on the right side of the equation, the term containing x is first, followed by the constant.
$$2y = -3x + 4$$

Use the Multiplication Property of Equations to multiply each side of the equation by the reciprocal of the coefficient of y. (The coefficient of y is 2; the reciprocal of 2 is $\frac{1}{2}$.)
$$\frac{1}{2} \cdot 2y = \frac{1}{2}(-3x + 4)$$

Simplify. Use the Distributive Property on the right side of the equation.
$$y = \frac{1}{2}(-3x) + \frac{1}{2}(4)$$

The equation is now in the form $y = mx + b$, with $m = -\frac{3}{2}$ and $b = 2$.
$$y = -\frac{3}{2}x + 2$$

In solving the equation $3x + 2y = 4$ for y, where we multiplied both sides of the equation by $\frac{1}{2}$, we could have divided both sides of the equation by 2, as shown at the right. In simplifying the right side after dividing both sides by 2, be sure to divide *each term* by 2.

$$2y = -3x + 4$$
$$\frac{2y}{2} = \frac{-3x + 4}{2}$$
$$y = \frac{-3x}{2} + \frac{4}{2}$$
$$y = -\frac{3}{2}x + 2$$

Being able to solve an equation of the form $Ax + By = C$ for y is important because most graphing utilities require that an equation be in the form $y = mx + b$ when the equation of the line is entered for graphing.

Example 5   Solve $3x - 4y = 12$ for y. Write the equation in the form $y = mx + b$.

Solution

$$3x - 4y = 12$$
$$3x - 3x - 4y = -3x + 12 \quad \blacktriangleright \text{Subtract } 3x \text{ from each side of the equation.}$$
$$-4y = -3x + 12 \quad \blacktriangleright \text{Simplify.}$$
$$\frac{-4y}{-4} = \frac{-3x + 12}{-4} \quad \blacktriangleright \text{Divide each side of the equation by } -4.$$
$$y = \frac{-3x}{-4} + \frac{12}{-4}$$
$$y = \frac{3}{4}x - 3$$

Problem 5   Solve $5x - 2y = 10$ for y. Write the equation in the form $y = mx + b$.

Solution   See page A24.

To graph an equation of the form $Ax + By = C$, we can first solve the equation for y and then follow the same procedure used for graphing an equation of the form $y = mx + b$. An example is shown below.

Graph: $3x + 4y = 12$

Solve the equation for y.

$$3x + 4y = 12$$
$$4y = -3x + 12$$
$$y = -\frac{3}{4}x + 3$$

Find at least three solutions.
Display the ordered pairs in a table.

x	y
0	3
4	0
-4	6

Graph the ordered pairs on a rectangular coordinate system, and draw a straight line through the points.

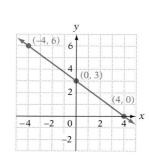

Example 6 Graph.   A. $2x - 5y = 10$ B. $x + 2y = 6$

Solution A. $2x - 5y = 10$ B. $x + 2y = 6$
$$-5y = -2x + 10$$ $$2y = -x + 6$$
$$y = \frac{2}{5}x - 2$$ $$y = -\frac{1}{2}x + 3$$

x	y
0	−2
5	0
−5	−4

x	y
0	3
−2	4
4	1

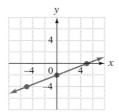   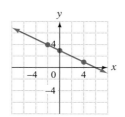

Problem 6 Graph.   A. $5x - 2y = 10$ B. $x - 3y = 9$

Solution See page A24.

The graph of the equation $2x + 3y = 6$ is shown at the right. The graph crosses the x-axis at $(3, 0)$. This point is called the **x-intercept**. The graph also crosses the y-axis at $(0, 2)$. This point is called the **y-intercept**.

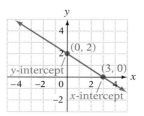

Find the x-intercept and the y-intercept of the graph of the equation $2x + 3y = 6$ algebraically.

To find the x-intercept, let $y = 0$.
(Any point on the x-axis has y-coordinate 0.)

$$2x + 3y = 6$$
$$2x + 3(0) = 6$$
$$2x = 6$$
$$x = 3$$

LOOK CLOSELY

To find the x-intercept, let $y = 0$. To find the y-intercept, let $x = 0$.

The x-intercept is $(3, 0)$.

To find the y-intercept, let $x = 0$.
(Any point on the y-axis has x-coordinate 0.)

$$2x + 3y = 6$$
$$2(0) + 3y = 6$$
$$3y = 6$$
$$y = 2$$

The y-intercept is $(0, 2)$.

Another method of graphing some equations of the form $Ax + By = C$ is to find the x- and y-intercepts, plot both intercepts, and then draw a line through the two points.

Example 7   Find the x- and y-intercepts for $x - 2y = 4$. Graph the line.

Solution x-intercept: $x - 2y = 4$
$x - 2(0) = 4$ ▶ To find the x-intercept, let $y = 0$.
$x = 4$

The x-intercept is $(4, 0)$.

y-intercept: $x - 2y = 4$
$0 - 2y = 4$ ▶ To find the y-intercept, let $x = 0$.
$-2y = 4$
$y = -2$

The y-intercept is $(0, -2)$.

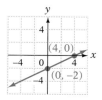

▶ Graph the ordered pairs $(4, 0)$ and $(0, -2)$. Draw a straight line through the points.

Problem 7   Find the x- and y-intercepts for $4x - y = 4$. Graph the line.

Solution   See page A24.

The graph of an equation in which one of the variables is missing is either a horizontal line or a vertical line.

The equation $y = 2$ could be written $0 \cdot x + y = 2$. No matter what value of x is chosen, y is always 2. Some solutions to the equation are $(3, 2)$, $(-1, 2)$, $(0, 2)$ and $(-4, 2)$. The graph is shown at the right.

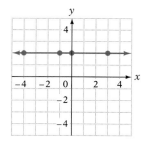

The **graph of** $y = b$ is a horizontal line passing through the point whose coordinates are $(0, b)$.

Note that $(0, b)$ is the y-intercept of the graph of $y = b$. An equation of the form $y = b$ does not have an x-intercept.

The equation $x = -2$ could be written $x + 0 \cdot y = -2$. No matter what value of y is chosen, x is always -2. Some solutions to the equation are $(-2, 3)$, $(-2, -2)$, $(-2, 0)$, and $(-2, 2)$. The graph is shown at the right.

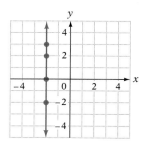

The **graph of** $x = a$ is a vertical line passing through the point whose coordinates are $(a, 0)$.

Note that $(a, 0)$ is the x-intercept of the graph of $x = a$. An equation of the form $x = a$ does not have a y-intercept.

Example 8 Graph. A. $y = -2$ B. $x = 3$

Solution A. The graph of an equation of the form $y = b$ is a horizontal line with y-intercept $(0, b)$.

B. The graph of an equation of the form $x = a$ is a vertical line with x-intercept $(a, 0)$.

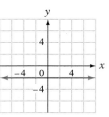

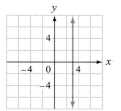

Problem 8 Graph. A. $y = 3$ B. $x = -4$

Solution See page A25.

CONCEPT REVIEW 5.2

Determine whether the statement is always true, sometimes true, or never true.

1. The equation $y = 3x^2 - 6$ is an example of a linear equation in two variables.

2. The value of m in the equation $y = 4x + 9$ is 9.

3. $(3, 5)$ is a solution of the equation $y = x + 2$.

4. The graph of a linear equation in two variables is a straight line.

5. To find the x-intercept of the graph of a linear equation, let $x = 0$.

6. The graph of an equation of the form $y = b$ is a horizontal line.

7. The graph of the equation $x = -4$ has an x-intercept of $(0, -4)$.

8. The graph shown at the right has a y-intercept of $(0, 3)$.

9. The graph of the equation $x = 2$ is a line that goes through every point that has an x-coordinate of 2.

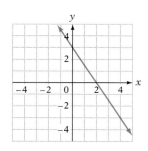

EXERCISES 5.2

1 For the given linear equation, find the value of m and the value of b.

1. $y = 4x + 1$

2. $y = -3x - 2$

3. $y = \dfrac{5}{6}x + \dfrac{1}{6}$

4. $y = -x$

5. Is (3, 4) a solution of $y = -x + 7$?

6. Is (2, −3) a solution of $y = x + 5$?

7. Is (−1, 2) a solution of $y = \frac{1}{2}x - 1$?

8. Is (1, −3) a solution of $y = -2x - 1$?

9. Is (4, 1) a solution of $y = \frac{1}{4}x + 1$?

10. Is (−5, 3) a solution of $y = -\frac{2}{5}x + 1$?

11. Is (0, 4) a solution of $y = \frac{3}{4}x + 4$?

12. Is (−2, 0) a solution of $y = -\frac{1}{2}x - 1$?

13. Is (0, 0) a solution of $y = 3x + 2$?

14. Is (0, 0) a solution of $y = -\frac{3}{4}x$?

15. Find the ordered-pair solution of $y = 3x - 2$ that corresponds to $x = 3$.

16. Find the ordered-pair solution of $y = 4x + 1$ that corresponds to $x = -1$.

17. Find the ordered-pair solution of $y = \frac{2}{3}x - 1$ that corresponds to $x = 6$.

18. Find the ordered-pair solution of $y = \frac{3}{4}x - 2$ that corresponds to $x = 4$.

19. Find the ordered-pair solution of $y = -3x + 1$ that corresponds to $x = 0$.

20. Find the ordered-pair solution of $y = \frac{2}{5}x - 5$ that corresponds to $x = 0$.

21. Find the ordered-pair solution of $y = \frac{2}{5}x + 2$ that corresponds to $x = -5$.

22. Find the ordered-pair solution of $y = -\frac{1}{6}x - 2$ that corresponds to $x = 12$.

2 Graph.

23. $y = 2x - 3$

24. $y = -2x + 2$

25. $y = \frac{1}{3}x$

26. $y = -3x$

27. $y = \frac{2}{3}x - 1$

28. $y = \frac{3}{4}x + 2$

29. $y = -\frac{1}{4}x + 2$

30. $y = -\frac{1}{3}x + 1$

31. $y = -\frac{2}{5}x + 1$

32. $y = -\frac{1}{2}x + 3$

33. $y = 2x - 4$

34. $y = 3x - 4$

35. $y = -x + 2$

36. $y = -x - 1$

37. $y = -\frac{2}{3}x + 1$

38. $y = 5x - 4$ **39.** $y = -3x + 2$ **40.** $y = -x + 3$

Graph using a graphing utility.

41. $y = 3x - 4$ **42.** $y = -2x + 3$ **43.** $y = 2x - 3$ **44.** $y = -2x - 3$

45. $y = -\dfrac{2}{3}x$ **46.** $y = \dfrac{3}{2}x$ **47.** $y = \dfrac{3}{4}x + 2$ **48.** $y = -\dfrac{1}{3}x + 2$

49. $y = -\dfrac{3}{2}x - 3$ **50.** $y = -\dfrac{2}{5}x + 2$ **51.** $y = \dfrac{1}{4}x - 2$ **52.** $y = \dfrac{2}{3}x - 4$

3   Solve for y. Write the equation in the form $y = mx + b$.

53. $3x + y = 10$ **54.** $2x + y = 5$ **55.** $4x - y = 3$ **56.** $5x - y = 7$

57. $3x + 2y = 6$ **58.** $2x + 3y = 9$ **59.** $2x - 5y = 10$ **60.** $5x - 2y = 4$

61. $2x + 7y = 14$ **62.** $6x - 5y = 10$ **63.** $x + 3y = 6$ **64.** $x - 4y = 12$

Graph.

65. $3x + y = 3$ **66.** $2x + y = 4$ **67.** $2x + 3y = 6$ **68.** $3x + 2y = 4$

69. $x - 2y = 4$ **70.** $x - 3y = 6$ **71.** $2x - 3y = 6$ **72.** $3x - 2y = 8$

73. $y = 4$ **74.** $y = -4$ **75.** $x = -2$ **76.** $x = 3$

Find the x- and y-intercepts.

77. $x - y = 3$ **78.** $3x + 4y = 12$ **79.** $y = 2x - 6$ **80.** $y = 2x + 10$

81. $x - 5y = 10$ **82.** $3x + 2y = 12$ **83.** $y = 3x + 12$ **84.** $y = 5x + 10$

85. $2x - 3y = 0$ **86.** $3x + 4y = 0$ **87.** $y = \dfrac{1}{2}x + 3$ **88.** $y = \dfrac{2}{3}x - 4$

Graph by using x- and y-intercepts.

89. $5x + 2y = 10$ **90.** $x - 3y = 6$ **91.** $3x - 4y = 12$

92. $2x - 3y = -6$ **93.** $2x + 3y = 6$ **94.** $x + 2y = 4$

95. $2x + 5y = 10$ **96.** $3x + 4y = 12$ **97.** $x - 3y = 6$

98. $3x - y = 6$ **99.** $x - 4y = 8$ **100.** $4x + 3y = 12$

101. $2x + y = 3$ **102.** $3x + y = -5$ **103.** $4x - 3y = 6$

Graph using a graphing utility. Verify that the graph has the correct x- and y-intercepts.

104. $x - 2y = -4$ **105.** $3x + 4y = -12$ **106.** $2x - 3y = -6$

107. $2x - y = 4$ **108.** $3x - 4y = 4$ **109.** $2x - 3y = 9$

SUPPLEMENTAL EXERCISES 5.2

Solve.

110. **a.** Show that the equation $y + 3 = 2(x + 4)$ is a linear equation by writing it in the form $y = mx + b$.

   b. Find the ordered-pair solution that corresponds to $x = -4$.

111. **a.** Show that the equation $y + 4 = -\frac{1}{2}(x + 2)$ is a linear equation by writing it in the form $y = mx + b$.

   b. Find the ordered-pair solution that corresponds to $x = -2$.

112. For the linear equation $y = 2x - 3$, what is the increase in y that results when x is increased by 1?

113. For the linear equation $y = -x - 4$, what is the decrease in y that results when x is increased by 1?

114. Write the equation of a line that has $(0, 0)$ as both the x-intercept and the y-intercept.

115. Explain **a.** why the y-coordinate of any point on the x-axis is 0 and **b.** why the x-coordinate of any point on the y-axis is 0.

S E C T I O N 5.3
Slopes of Straight Lines

1 Find the slope of a straight line

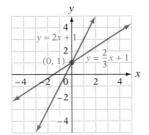

The graphs of $y = \frac{2}{3}x + 1$ and $y = 2x + 1$ are shown at the left. Each graph crosses the y-axis at $(0, 1)$, but the graphs have different slants. The **slope** of a line is a measure of the slant of the line. The symbol for slope is m.

The slope of a line is the ratio of the change in the y-coordinates between any two points to the change in the x-coordinates.

The line containing the points whose coordinates are $(-2, -3)$ and $(6, 1)$ is graphed at the right. The change in y is the difference between the two y-coordinates.

Change in $y = 1 - (-3) = 4$

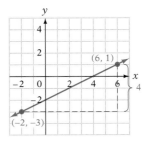

The change in x is the difference between the two x-coordinates.

Change in $x = 6 - (-2) = 8$

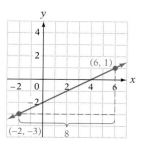

For the line containing the points whose coordinates are $(-2, -3)$ and $(6, 1)$,

$$\text{slope} = m = \frac{\text{change in } y}{\text{change in } x} = \frac{4}{8} = \frac{1}{2}$$

The slope of the line can also be described as the ratio of the vertical change (4 units) to the horizontal change (8 units) from the point whose coordinates are $(-2, -3)$ to the point whose coordinates are $(6, 1)$.

Slope Formula

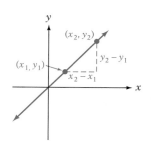

> The slope of a line containing two points P_1 and P_2, whose coordinates are (x_1, y_1) and (x_2, y_2), is given by
>
> $$\text{slope} = m = \frac{y_2 - y_1}{x_2 - x_1}, \; x_1 \neq x_2$$

In the slope formula, the points P_1 and P_2 are any two points on the line. The slope of a line is constant; therefore, the slope calculated using any two points on the line will be the same.

Find the slope of the line containing the points whose coordinates are $(-1, 1)$ and $(2, 3)$.

Let $P_1 = (-1, 1)$ and $P_2 = (2, 3)$. Then $x_1 = -1$, $y_1 = 1$, $x_2 = 2$, and $y_2 = 3$.

$$m = \frac{y_2 - y_1}{x_2 - x_1} = \frac{3 - 1}{2 - (-1)} = \frac{2}{3}$$

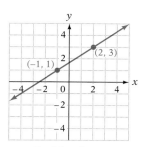

Positive slope

It does not matter which point is named P_1 and which P_2; the slope of the line will be the same. If the points are reversed, then $P_1 = (2, 3)$ and $P_2 = (-1, 1)$.

$$m = \frac{y_2 - y_1}{x_2 - x_1} = \frac{1 - 3}{-1 - 2} = \frac{-2}{-3} = \frac{2}{3}$$

This is the same result. Here the slope is a positive number. A line that slants upward to the right has a **positive slope.**

LOOK CLOSELY

Positive slope means that the value of y increases as the value of x increases.

Find the slope of the line containing the points whose coordinates are $(-3, 4)$ and $(2, -2)$.

Let $P_1 = (-3, 4)$ and $P_2 = (2, -2)$.

$$m = \frac{y_2 - y_1}{x_2 - x_1} = \frac{-2 - 4}{2 - (-3)} = \frac{-6}{5} = -\frac{6}{5}$$

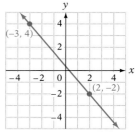

Here the slope is a negative number. A line that slants downward to the right has a **negative slope.**

Negative slope

Find the slope of the line containing the points whose coordinates are $(-1, 3)$ and $(4, 3)$.

Let $P_1 = (-1, 3)$ and $P_2 = (4, 3)$.

$$m = \frac{y_2 - y_1}{x_2 - x_1} = \frac{3 - 3}{4 - (-1)} = \frac{0}{5} = 0$$

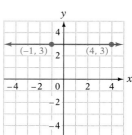

When $y_1 = y_2$, the graph is a horizontal line.

A horizontal line has **zero slope.**

Zero slope

Find the slope of the line containing the points whose coordinates are $(2, -2)$ and $(2, 4)$.

Let $P_1 = (2, -2)$ and $P_2 = (2, 4)$.

$$m = \frac{y_2 - y_1}{x_2 - x_1} = \frac{4 - (-2)}{2 - 2} = \frac{6}{0} \leftarrow \text{Not a real number}$$

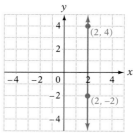

When $x_1 = x_2$, the denominator of $\frac{y_2 - y_1}{x_2 - x_1}$ is 0 and the graph is a vertical line. Because division by zero is not defined, the slope of the line is not defined.

The slope of a vertical line is **undefined.**

Undefined Slope

Remember that zero slope and undefined slope are different. The graph of a line with zero slope is horizontal. The graph of a line with undefined slope is vertical.

Example 1   Find the slope of the line containing the points P_1 and P_2.
   A. $P_1(-2, -1)$, $P_2(3, 4)$ B. $P_1(-3, 1)$, $P_2(2, -2)$
   C. $P_1(-1, 4)$, $P_2(-1, 0)$ D. $P_1(-1, 2)$, $P_2(4, 2)$

Solution   A. $m = \frac{y_2 - y_1}{x_2 - x_1} = \frac{4 - (-1)}{3 - (-2)} = \frac{5}{5} = 1$

The slope is 1.

B. $m = \dfrac{y_2 - y_1}{x_2 - x_1} = \dfrac{-2 - 1}{2 - (-3)} = \dfrac{-3}{5}$

The slope is $-\dfrac{3}{5}$.

C. $m = \dfrac{y_2 - y_1}{x_2 - x_1} = \dfrac{0 - 4}{-1 - (-1)} = \dfrac{-4}{0}$

The slope is undefined.

D. $m = \dfrac{y_2 - y_1}{x_2 - x_1} = \dfrac{2 - 2}{4 - (-1)} = \dfrac{0}{5} = 0$

The slope is 0.

Problem 1   Find the slope of the line containing the points P_1 and P_2.

A. $P_1(-1, 2)$, $P_2(1, 3)$ B. $P_1(1, 2)$, $P_2(4, -5)$
C. $P_1(2, 3)$, $P_2(2, 7)$ D. $P_1(1, -3)$, $P_2(-5, -3)$

Solution   See page A25.

There are many applications of the concept of slope. Here are two possibilities.

In 1988, when Florence Griffith-Joyner set the world record for the 100 m dash, her average rate of speed was approximately 9.5 m/s. The graph at the right shows the distance she ran during her record-setting run. From the graph, note that after 4 s she had traveled 38 m and that after 6 s she had traveled 57 m. The slope of the line between these two points is

$$m = \dfrac{57 - 38}{6 - 4} = \dfrac{19}{2} = 9.5$$

Note that the slope of the line is the same as the rate she was running, 9.5 m/s. The average speed of an object is related to slope.

Here is an example of slope taken from economics. According to the Department of Commerce, from 1987 to 1994, the value of U.S. goods exported to other countries increased at a rate of approximately $2.5 billion per year. The graph at the right shows the value of exports for each year. From the graph, we learn that in 1988 exports were $27 billion and that in 1992 exports were $37 billion. The slope of the line between these two points is

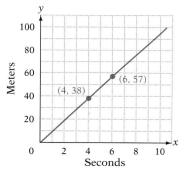

$$m = \dfrac{37 - 27}{1992 - 1988} = \dfrac{10}{4} = 2.5$$

Observe that the slope of the line is the same as the rate at which exports are increasing, $2.5 billion per year.

In general, any quantity that is expressed by using the word *per* is represented mathematically as slope. In the first example on page 223, slope was 9.5 meters *per* second; in the second example, slope was $2.5 billion *per* year.

Example 2   The graph at the right shows the prices, in dollars, of a Macintosh IIci computer from January to June (shown as the numbers 1 to 6). Find the slope of the line. Write a sentence that states the meaning of the slope.

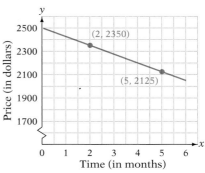

Solution $m = \dfrac{2125 - 2350}{5 - 2} = \dfrac{-225}{3} = -75$ ► Use any two points to find the slope. The points (5, 2125) and (2, 2350) are used here.

A slope of -75 means that during this period, the price of a Macintosh IIci was *decreasing* at a rate of $75 per month.

Problem 2   The graph at the right shows the approximate decline in the value of a used car over a five-year period. Find the slope of the line. Write a sentence that states the meaning of the slope.

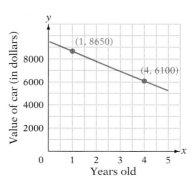

Solution   See page A25.

---

2   Graph a line using the slope and *y*-intercept

Recall that we can find the *y*-intercept of a linear equation by letting $x = 0$.

To find the *y*-intercept of $y = 3x + 4$, let $x = 0$. $y = 3x + 4$
$y = 3(0) + 4$
$y = 4$

The *y*-intercept is (0, 4).

The constant term of $y = 3x + 4$ is the *y*-coordinate of the *y*-intercept.

In general, for any equation of the form $y = mx + b$, the *y*-intercept is (0, b).

The graph of the equation $y = \frac{2}{3}x + 1$ is shown at the right. The points whose coordinates are $(-3, -1)$ and $(3, 3)$ are on the graph. The slope of the line is

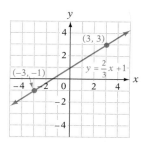

$$m = \frac{3 - (-1)}{3 - (-3)} = \frac{4}{6} = \frac{2}{3}$$

Note that the slope of the line has the same value as the coefficient of x.

Slope-Intercept Form of a Straight Line

For any equation of the form $y = mx + b$, the slope of the line is m, the coefficient of x. The y-intercept is $(0, b)$. The equation

$$y = mx + b$$

is called the **slope-intercept form of a straight line.**

The slope of the line $y = -\frac{3}{4}x + 1$ is $-\frac{3}{4}$. The y-intercept is $(0, 1)$.

$$y = \boxed{m}\; x + \boxed{b}$$
$$y = \boxed{-\frac{3}{4}}\; x + \boxed{1}$$

$$\text{Slope} = m = -\frac{3}{4} \qquad\qquad y\text{-intercept} = (0, b) = (0, 1)$$

Find the slope and y-intercept of the line $y = \frac{5}{2}x - 6$.

$$\text{Slope} = m = \frac{5}{2} \qquad b = -6$$

The slope is $\frac{5}{2}$.   The y-intercept is $(0, -6)$.

When the equation of a straight line is in the form $y = mx + b$, the graph can be drawn using the slope and y-intercept. First locate the y-intercept. Use the slope to find a second point on the line. Then draw a line through the two points.

Graph $y = 2x - 3$ by using the slope and y-intercept.

y-intercept $= (0, b) = (0, -3)$

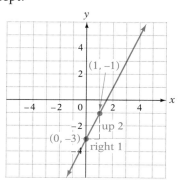

$$m = 2 = \frac{2}{1} = \frac{\text{change in } y}{\text{change in } x}$$

Beginning at the y-intercept, move right 1 unit (change in x) and then up 2 units (change in y).

$(1, -1)$ are the coordinates of the second point on the graph.

Draw a line through $(0, -3)$ and $(1, -1)$.

▦ Using a graphing utility, enter the equation $y = 2x - 3$ and verify the graph shown on page 225. Trace along the graph and verify that $(0, -3)$ is the y-intercept and that the point whose coordinates are $(1, -1)$ is on the graph.

Example 3   Graph $y = -\dfrac{2}{3}x + 1$ by using the slope and y-intercept.

Solution y-intercept $= (0, b) = (0, 1)$

$$m = -\frac{2}{3} = \frac{-2}{3}$$

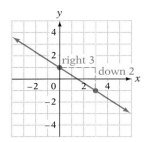

▶ A slope of $\dfrac{-2}{3}$ means to move right 3 units and then down 2 units.

Problem 3   Graph $y = -\dfrac{1}{4}x - 1$ by using the slope and y-intercept.

Solution   See page A25.

Example 4   Graph $2x - 3y = 6$ by using the slope and y-intercept.

Solution $2x - 3y = 6$
$-3y = -2x + 6$ ▶ Solve the equation for y.
$y = \dfrac{2}{3}x - 2$

y-intercept $= (0, -2)$

$$m = \frac{2}{3}$$

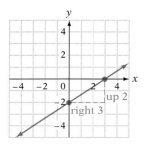

Problem 4   Graph $x - 2y = 4$ by using the slope and y-intercept.

Solution   See page A25.

CONCEPT REVIEW 5.3

Determine whether the statement is always true, sometimes true, or never true.

1. In the equation $y = mx + b$, m is the symbol for slope.

2. The formula for slope is $m = \frac{y_2 - y_1}{x_2 - x_1}$.

3. A vertical line has zero slope, and the slope of a horizontal line is undefined.

4. Any two points on a line can be used to find the slope of the line.

5. Shown at the right is the graph of a line with negative slope.

6. The y-intercept of the graph of the equation $y = -2x + 5$ is $(5, 0)$.

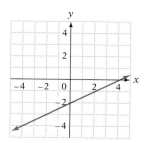

EXERCISES 5.3

1 Find the slope of the line containing the points P_1 and P_2.

1. $P_1(4, 2)$, $P_2(3, 4)$

2. $P_1(2, 1)$, $P_2(3, 4)$

3. $P_1(-1, 3)$, $P_2(2, 4)$

4. $P_1(-2, 1)$, $P_2(2, 2)$

5. $P_1(2, 4)$, $P_2(4, -1)$

6. $P_1(1, 3)$, $P_2(5, -3)$

7. $P_1(-2, 3)$, $P_2(2, 1)$

8. $P_1(5, -2)$, $P_2(1, 0)$

9. $P_1(8, -3)$, $P_2(4, 1)$

10. $P_1(0, 3)$, $P_2(2, -1)$

11. $P_1(3, -4)$, $P_2(3, 5)$

12. $P_1(-1, 2)$, $P_2(-1, 3)$

13. $P_1(4, -2)$, $P_2(3, -2)$

14. $P_1(5, 1)$, $P_2(-2, 1)$

15. $P_1(0, -1)$, $P_2(3, -2)$

16. $P_1(3, 0)$, $P_2(2, -1)$

17. $P_1(-2, 3)$, $P_2(1, 3)$

18. $P_1(4, -1)$, $P_2(-3, -1)$

19. $P_1(-2, 4)$, $P_2(-1, -1)$

20. $P_1(6, -4)$, $P_2(4, -2)$

21. $P_1(-2, -3)$, $P_2(-2, 1)$

22. $P_1(5, 1)$, $P_2(5, -2)$

23. $P_1(-1, 5)$, $P_2(5, 1)$

24. $P_1(-1, 5)$, $P_2(7, 1)$

25. The graph below shows the cost, in dollars, to make a transatlantic telephone call. Find the slope of the line. Write a sentence that states the meaning of the slope.

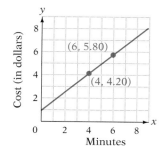

26. The graph below shows the pressure, in pounds per square inch, on a diver. Find the slope of the line. Write a sentence that states the meaning of the slope.

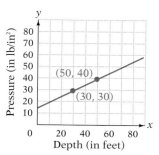

27. The graph below shows the approximate decline in the median family income, in thousands of dollars, in the United States from 1990 to 1994. Find the slope of the line. Write a sentence that states the meaning of the slope.

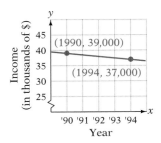

28. The graph below shows the decrease in the median price, in dollars, of a home in a city over a six-month period. Find the slope of the line. Write a sentence that states the meaning of the slope.

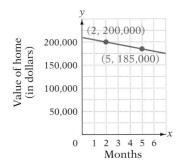

2   Graph by using the slope and y-intercept.

29. $y = 3x + 1$

30. $y = -2x - 1$

31. $y = \frac{2}{5}x - 2$

32. $y = \frac{3}{4}x + 1$

33. $2x + y = 3$

34. $3x - y = 1$

35. $x - 2y = 4$

36. $x + 3y = 6$

37. $y = \frac{2}{3}x$

38. $y = \frac{1}{2}x$

39. $y = -x + 1$

40. $y = -x - 3$

41. $3x - 4y = 12$

42. $5x - 2y = 10$

43. $y = -4x + 2$

44. $y = 5x - 2$

45. $4x - 5y = 20$

46. $x - 3y = 6$

SUPPLEMENTAL EXERCISES 5.3

47. What effect does increasing the coefficient of x have on the graph of $y = mx + b$?

48. What effect does decreasing the coefficient of x have on the graph of $y = mx + b$?

49. What effect does increasing the constant term have on the graph of $y = mx + b$?

50. What effect does decreasing the constant term have on the graph of $y = mx + b$?

51. Do the graphs of all straight lines have a y-intercept? If not, give an example of one that does not.

52. If two lines have the same slope and the same y-intercept, must the graphs of the lines be the same? If not, give an example.

53. What does the highway sign shown at the right have to do with slope?

S E C T I O N **5.4**

Equations of Straight Lines

1 Find the equation of a line using the equation $y = mx + b$

When the slope of a line and a point on the line are known, the equation of the line can be written using the slope-intercept form, $y = mx + b$. In the first example shown below, the known point is the y-intercept. In the second example, the known point is a point other than the y-intercept.

Find the equation of the line that has slope 3 and y-intercept $(0, 2)$.

The given slope, 3, is m. $y = mx + b$

Replace m with 3. $y = 3x + b$

The given point, $(0, 2)$, is the y-intercept. Replace b with 2. $y = 3x + 2$

The equation of the line that has slope 3 and y-intercept 2 is $y = 3x + 2$.

Find the equation of the line that has slope $\frac{1}{2}$ and contains the point whose coordinates are $(-2, 4)$.

The given slope, $\frac{1}{2}$, is m.	$y = mx + b$
Replace m with $\frac{1}{2}$.	$y = \frac{1}{2}x + b$

LOOK CLOSELY

Every ordered pair is of the form (x, y). For the point $(-2, 4)$, -2 is the x value and 4 is the y value. Substitute -2 for x and 4 for y.

The given point, $(-2, 4)$, is a solution of the equation of the line. Replace x and y in the equation with the coordinates of the point.

$$4 = \frac{1}{2}(-2) + b$$

Solve for b, the y-intercept.

$$4 = -1 + b$$
$$5 = b$$

Write the equation of the line by replacing m and b in the equation by their values.

$$y = mx + b$$
$$y = \frac{1}{2}x + 5$$

The equation of the line that has slope $\frac{1}{2}$ and contains the point whose coordinates are $(-2, 4)$ is $y = \frac{1}{2}x + 5$.

Example 1   Find the equation of the line that contains the point whose coordinates are $(3, -3)$ and has slope $\frac{2}{3}$.

Solution $y = mx + b$

$y = \frac{2}{3}x + b$ ▶ Replace m with the given slope.

$-3 = \frac{2}{3}(3) + b$ ▶ Replace x and y in the equation with the coordinates of the given point.

$-3 = 2 + b$ ▶ Solve for b.
$-5 = b$

$y = \frac{2}{3}x - 5$ ▶ Write the equation of the line by replacing m and b in $y = mx + b$ by their values.

Problem 1   Find the equation of the line that contains the point whose coordinates are $(4, -2)$ and has slope $\frac{3}{2}$.

Solution   See page A25.

---

2   Find the equation of a line using the point-slope formula

An alternative method for finding the equation of a line, given the slope and the coordinates of a point on the line, involves use of the point-slope formula. The point-slope formula is derived from the formula for slope.

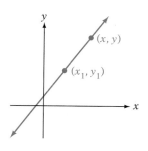

Let (x_1, y_1) be the coordinates of the given point on the line, and let (x, y) be the coordinates of any other point on the line.

Use the formula for slope. $\dfrac{y - y_1}{x - x_1} = m$

Multiply both sides of the equation by $(x - x_1)$. $\dfrac{y - y_1}{x - x_1}(x - x_1) = m(x - x_1)$

Simplify. $y - y_1 = m(x - x_1)$

Point-Slope Formula

> The equation of the line that has slope m and contains the point whose coordinates are (x_1, y_1) can be found by the point-slope formula:
>
> $$y - y_1 = m(x - x_1)$$

Example 2   Use the point-slope formula to find the equation of a line that passes through the point whose coordinates are $(-2, -1)$ and has slope $\dfrac{3}{2}$.

Solution $(x_1, y_1) = (-2, -1)$ ▶ Let (x_1, y_1) be the given point.

$m = \dfrac{3}{2}$ ▶ m is the given slope.

$y - y_1 = m(x - x_1)$ ▶ This is the point-slope formula.

$y - (-1) = \dfrac{3}{2}[x - (-2)]$ ▶ Substitute -2 for x_1, -1 for y_1, and $\dfrac{3}{2}$ for m.

$y + 1 = \dfrac{3}{2}(x + 2)$ ▶ Rewrite the equation in the form $y = mx + b$.

$y + 1 = \dfrac{3}{2}x + 3$

$y = \dfrac{3}{2}x + 2$

Problem 2   Use the point-slope formula to find the equation of a line that passes through the point whose coordinates are $(5, 4)$ and has slope $\dfrac{2}{5}$.

Solution   See page A25.

CONCEPT REVIEW 5.4

Determine whether the statement is always true, sometimes true, or never true.

1.   The graph of the equation $y = 5x + 7$ has slope 5 and y-intercept $(0, 7)$.

2.   If the equation of a line has y-intercept $(0, 4)$, then 4 can be substituted for b in the equation $y = mx + b$.

3.   If the equation of a line contains the point $(-3, 1)$, then when y is -3, x is 1.

4. The point-slope formula is $y - y_1 = mx - x_1$.

5. The point-slope formula can be used to find the equation of a line with zero slope.

6. If it is stated that the y-intercept is 2, then the y-intercept is the point (2, 0).

EXERCISES 5.4

1   Use the slope-intercept form.

1. Find the equation of the line that contains the point whose coordinates are (0, 2) and has slope 2.

2. Find the equation of the line that contains the point whose coordinates are (0, −1) and has slope −2.

3. Find the equation of the line that contains the point whose coordinates are (−1, 2) and has slope −3.

4. Find the equation of the line that contains the point whose coordinates are (2, −3) and has slope 3.

5. Find the equation of the line that contains the point whose coordinates are (3, 1) and has slope $\frac{1}{3}$.

6. Find the equation of the line that contains the point whose coordinates are (−2, 3) and has slope $\frac{1}{2}$.

7. Find the equation of the line that contains the point whose coordinates are (4, −2) and has slope $\frac{3}{4}$.

8. Find the equation of the line that contains the point whose coordinates are (2, 3) and has slope $-\frac{1}{2}$.

9. Find the equation of the line that contains the point whose coordinates are (5, −3) and has slope $-\frac{3}{5}$.

10. Find the equation of the line that contains the point whose coordinates are (5, −1) and has slope $\frac{1}{5}$.

11. Find the equation of the line that contains the point whose coordinates are (2, 3) and has slope $\frac{1}{4}$.

12. Find the equation of the line that contains the point whose coordinates are (−1, 2) and has slope $-\frac{1}{2}$.

13. Find the equation of the line that contains the point whose coordinates are $(-3, -5)$ and has slope $-\dfrac{2}{3}$.

14. Find the equation of the line that contains the point whose coordinates are $(-4, 0)$ and has slope $\dfrac{5}{2}$.

2 Use the point-slope formula.

15. Find the equation of the line that passes through the point whose coordinates are $(1, -1)$ and has slope 2.

16. Find the equation of the line that passes through the point whose coordinates are $(2, 3)$ and has slope -1.

17. Find the equation of the line that passes through the point whose coordinates are $(-2, 1)$ and has slope -2.

18. Find the equation of the line that passes through the point whose coordinates are $(-1, -3)$ and has slope -3.

19. Find the equation of the line that passes through the point whose coordinates are $(0, 0)$ and has slope $\dfrac{2}{3}$.

20. Find the equation of the line that passes through the point whose coordinates are $(0, 0)$ and has slope $-\dfrac{1}{5}$.

21. Find the equation of the line that passes through the point whose coordinates are $(2, 3)$ and has slope $\dfrac{1}{2}$.

22. Find the equation of the line that passes through the point whose coordinates are $(3, -1)$ and has slope $\dfrac{2}{3}$.

23. Find the equation of the line that passes through the point whose coordinates are $(-4, 1)$ and has slope $-\dfrac{3}{4}$.

24. Find the equation of the line that passes through the point whose coordinates are $(-5, 0)$ and has slope $-\dfrac{1}{5}$.

25. Find the equation of the line that passes through the point whose coordinates are $(-2, 1)$ and has slope $\dfrac{3}{4}$.

26. Find the equation of the line that passes through the point whose coordinates are $(3, -2)$ and has slope $\dfrac{1}{6}$.

27. Find the equation of the line that passes through the point whose coordinates are $(-3, -5)$ and has slope $-\frac{4}{3}$.

28. Find the equation of the line that passes through the point whose coordinates are $(3, -1)$ and has slope $\frac{3}{5}$.

SUPPLEMENTAL EXERCISES 5.4

Is there a linear equation that contains all the given ordered pairs? If there is, find the equation.

29. $(5, 1), (4, 2), (0, 6)$

30. $(-2, -4), (0, -3), (4, -1)$

31. $(-1, -5), (2, 4), (0, 2)$

32. $(3, -1), (12, -4), (-6, 2)$

The given ordered pairs are solutions to the same linear equation. Find n.

33. $(0, 1), (4, 9), (3, n)$

34. $(2, 2), (-1, 5), (3, n)$

35. $(2, -2), (-2, -4), (4, n)$

36. $(1, -2), (-2, 4), (4, n)$

The relationship between Celsius and Fahrenheit temperature can be given by a linear equation. Water freezes at 0°C or at 32°F. Water boils at 100°C or at 212°F.

37. Write a linear equation expressing Fahrenheit temperature in terms of Celsius temperature.

38. Graph the equation found in Exercise 37.

The formula $y - y_1 = \frac{y_2 - y_1}{x_2 - x_1}(x - x_1)$, where $x_1 \neq x_2$, is called the **two-point formula** for a straight line. This formula can be used to find the equation of a line given two points. Use this formula for Exercises 39 and 40.

39. Find the equation of the line that passes through $(-2, 3)$ and $(4, -1)$.

40. Find the equation of the line that passes through the points $(3, -1)$ and $(4, -3)$.

41. Explain why the condition $x_1 \neq x_2$ is placed on the two-point formula given above.

42. Explain how the two-point formula given above can be derived from the point-slope formula.

SECTION 5.5

Functions

1   Introduction to functions

The definition of set given in Chapter 1 was that a set is a collection of objects. Recall that the objects in a set are called the elements of the set. The elements in a set can be anything.

The set of planets in our solar system is

{Mercury, Venus, Earth, Mars, Jupiter, Saturn, Uranus, Neptune, Pluto}

The set of colors in a rainbow is

{red, orange, yellow, green, blue, indigo, violet}

The objects in a set can be ordered pairs. When the elements in a set are ordered pairs, the set is called a relation. A **relation** is any set of ordered pairs.

The set {(1, 1), (2, 4), (3, 9), (4, 16), (5, 25)} is a relation. There are five elements in the set. The elements are the ordered pairs (1, 1), (2, 4), (3, 9), (4, 16), and (5, 25).

The following table shows the number of hours that each of eight students spent in the math lab during the week of the math midterm exam and the score that each of these students received on the math midterm.

Hours	2	3	4	4	5	6	6	7
Score	60	70	70	80	85	85	95	90

This information can be written as the relation

{(2, 60), (3, 70), (4, 70), (4, 80), (5, 85), (6, 85), (6, 95), (7, 90)}

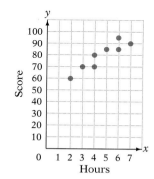

where the first coordinate of each ordered pair is the hours spent in the math lab and the second coordinate is the score on the midterm exam.

The **domain** of a relation is the set of first coordinates of the ordered pairs. The **range** is the set of second coordinates of the ordered pairs. For the relation above,

domain = {2, 3, 4, 5, 6, 7}   range = {60, 70, 80, 85, 90, 95}

The **graph of a relation** is the graph of the ordered pairs that belong to the relation. The graph of the relation given above is shown at the left. The horizontal axis represents the domain (the hours spent in the math lab) and the vertical axis represents the range (the exam score).

A **function** is a special type of relation in which no two ordered pairs have the same first coordinate and different second coordinates. The relation above is not a function because the ordered pairs (4, 70) and (4, 80) have the same first coordinate and different second coordinates. The ordered pairs (6, 85) and (6, 95) also have the same first coordinate and different second coordinates.

The table at the right describes a grading scale that defines a relationship between a test score and a letter grade. Some of the ordered pairs in this relation are (38, F), (73, C), and (94, A).

Score	Letter Grade
90–100	A
80–89	B
70–79	C
60–69	D
0–59	F

This relation defines a function because no two ordered pairs can have the *same* first coordinate and *different* second coordinates. For instance, it is not possible to have an average of 73 paired with any grade other than C. Both $(73, C)$ and $(73, A)$ cannot be ordered pairs belonging to the function, or two students with the same score would receive different grades. Note that $(81, B)$ and $(88, B)$ are ordered pairs of this function. Ordered pairs of a function may have the same *second* coordinate paired with different *first* coordinates.

The domain of this function is {0, 1, 2, 3, ..., 98, 99, 100}.

The range of this function is {A, B, C, D, F}.

Example 1   Find the domain and range of the relation {(−5, 1), (−3, 3), (−1, 5)}. Is the relation a function?

Solution   The domain is {−5, −3, −1}.   ▶ The domain of the relation is the set of the first components of the ordered pairs.

The range is {1, 3, 5}.   ▶ The range of the relation is the set of the second components of the ordered pairs.

No two ordered pairs have the same first coordinate. The relation is a function.

Problem 1   Find the domain and range of the relation {(1, 0), (1, 1), (1, 2), (1, 3), (1, 4)}. Is the relation a function?

Solution   See page A26.

Although a function can be described in terms of ordered pairs or in a table, functions are often described by an equation. The letter f is commonly used to represent a function, but any letter can be used.

The "square" function assigns to each real number its square. The square function is described by the equation

$$f(x) = x^2 \qquad \text{read } f(x) \text{ as "}f\text{ of }x\text{" or "the value of }f\text{ at }x\text{."}$$

$f(x)$ is the symbol for the number that is paired with x. In terms of ordered pairs, this is written $(x, f(x))$. $f(x)$ is the **value of the function** at x because it is the result of evaluating the variable expression. For example, $f(4)$ means to replace x by 4 and then simplify the resulting numerical expression. This process is called **evaluating the function.**

The notation $f(4)$ is used to indicate the number that is paired with 4. To evaluate $f(x) = x^2$ at 4, replace x with 4 and simplify.

$$f(x) = x^2$$
$$f(4) = 4^2$$
$$f(4) = 16$$

The square function squares a number, and when 4 is squared, the result is 16. For the square function, the number 4 is paired with 16. In other words, when x is 4, $f(x)$ is 16. The ordered pair $(4, 16)$ is an element of the function.

It is important to remember that $f(x)$ does not mean f times x. The letter f stands for the function, and $f(x)$ is the number that is paired with x.

Example 2 Evaluate $f(x) = 2x - 4$ at $x = 3$. Write an ordered pair that is an element of the function.

Solution $f(x) = 2x - 4$
$f(3) = 2(3) - 4$ ▶ $f(3)$ is the number that is paired with 3. Replace
$f(3) = 6 - 4$ x by 3 and evaluate.
$f(3) = 2$

The ordered pair $(3, 2)$ is an element of the function.

Problem 2 Evaluate $f(x) = -5x + 1$ at $x = 2$. Write an ordered pair that is an element of the function.

Solution See page A26.

When a function is described by an equation and the domain is specified, the range of the function can be found by evaluating the function at each point of the domain.

Example 3 Find the range of the function given by the equation $f(x) = -3x + 2$ if the domain is $\{-4, -2, 0, 2, 4\}$. Write five ordered pairs that belong to the function.

Solution $f(x) = -3x + 2$
$f(-4) = -3(-4) + 2 = 12 + 2 = 14$ ▶ Replace x by each mem-
$f(-2) = -3(-2) + 2 = 6 + 2 = 8$ ber of the domain.
$f(0) = -3(0) + 2 = 0 + 2 = 2$
$f(2) = -3(2) + 2 = -6 + 2 = -4$
$f(4) = -3(4) + 2 = -12 + 2 = -10$

The range is $\{-10, -4, 2, 8, 14\}$.

The ordered pairs $(-4, 14)$, $(-2, 8)$, $(0, 2)$, $(2, -4)$, and $(4, -10)$ belong to the function.

Problem 3 Find the range of the function given by the equation $f(x) = 4x - 3$ if the domain is $\{-5, -3, -1, 1\}$. Write four ordered pairs that belong to the function.

Solution See page A26.

2   Graphs of linear functions

The solutions of the equation

$$y = 7x - 3$$

are ordered pairs (x, y). For example, the ordered pairs $(-1, -10)$, $(0, -3)$, and $(1, 4)$ are solutions of the equation. Therefore, this equation defines a relation.

It is not possible to substitute one value of x into the equation $y = 7x - 3$ and get two different values of y. For example, the number 1 in the domain cannot be paired with any number other than 4 in the range. (*Remember:* A function cannot have ordered pairs in which the same first coordinate is paired with different second coordinates.) Therefore, the equation defines a function.

The equation $y = 7x - 3$ is an equation of the form $y = mx + b$. In general, any equation of the form $y = mx + b$ is a function.

In the equation $y = 7x - 3$, the variable y is called the **dependent variable** because its value *depends* on the value of x. The variable x is called the **independent variable.** We choose a value for x and substitute that value into the equation to determine the value of y. We say that y is a function of x.

When an equation defines y as a function of x, functional notation is frequently used to emphasize that the relation is a function. In this case, it is common to use the notation $f(x)$. Therefore, we can write the equation

$$y = 7x - 3$$

LOOK CLOSELY

When y is a function of x, y and $f(x)$ are interchangeable.

in functional notation as

$$f(x) = 7x - 3$$

The **graph of a function** is a graph of the ordered pairs (x, y) of the function. Because the graph of the equation $y = mx + b$ is a straight line, a function of the form $f(x) = mx + b$ is a **linear function.**

Graph: $f(x) = \dfrac{2}{3}x - 1$

Think of the function as the equation $y = \dfrac{2}{3}x - 1$.

This is the equation of a straight line.

The y-intercept is $(0, -1)$. The slope is $\dfrac{2}{3}$.

Graph the point $(0, -1)$. From the y-intercept, go right 3 units and then up 2 units. Graph the point $(3, 1)$.

Draw a line through the two points.

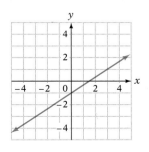

Example 4   Graph: $f(x) = \dfrac{3}{4}x + 2$

Solution $f(x) = \dfrac{3}{4}x + 2$

$$y = \dfrac{3}{4}x + 2$$

▶ Think of the function as the equation $y = \dfrac{3}{4}x + 2$.

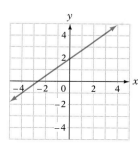

▶ The graph is a straight line with y-intercept $(0, 2)$ and slope $\dfrac{3}{4}$.

Problem 4   Graph: $f(x) = -\dfrac{1}{2}x - 3$

Solution   See page A26.

Graphing utilities are used to graph functions. Using a graphing utility, enter the equation $y = \dfrac{3}{4}x + 2$ and verify the graph drawn in Example 4. Trace along the graph and verify that $(-4, -1)$, $(0, 2)$, and $(4, 5)$ are coordinates of points on the graph. Now enter the equation given in Problem 4 and verify the graph you drew.

There are a variety of applications of linear functions. For example, suppose an installer of marble kitchen countertops charges $250 plus $180 per foot of countertop. The equation that describes the total cost, C (in dollars), for having x ft of countertop installed is $C = 180x + 250$. In this situation, C is a function of x; the total cost depends on how many feet of countertop are installed. Therefore, we could rewrite the equation

$$C = 180x + 250$$

as the function

$$f(x) = 180x + 250$$

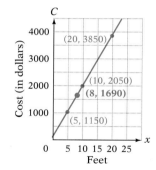

To graph this equation, we first choose a reasonable domain—for example, values of x between 0 and 25. It would not be reasonable to have $x \leq 0$, because no one would order the installation of 0 ft of countertop, and any amount less than 0 would be a negative amount of countertop. The upper limit of 25 is chosen because most kitchens have less than 25 ft of countertop.

Choosing $x = 5$, 10, and 20 results in the ordered pairs $(5, 1150)$, $(10, 2050)$, and $(20, 3850)$. Graph these points and draw a line through them. The graph is shown at the left.

The point whose coordinates are $(8, 1690)$ is on the graph. This ordered pair can be interpreted to mean that 8 ft of countertop costs $1690 to install.

Example 5   The value, V, of an investment of $2500 at an annual simple interest rate of 6% is given by the equation $V = 150t + 2500$, where t is the amount of time, in years, that the money is invested.

a. Write the equation in functional notation.

b. Graph the equation for values of t between 0 and 10.

c. The point whose coordinates are (5, 3250) is on the graph. Write a sentence that explains the meaning of this ordered pair.

Solution   a. $V = 150t + 2500$

$f(t) = 150t + 2500$

▶ The value V of the investment depends on the amount of time t it is invested. The value V is a function of the time t.

b.

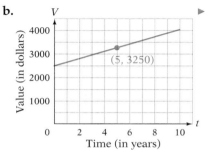

▶ Some ordered pairs of the function are (2, 2800), (4, 3100), and (6, 3400).

c. The ordered pair (5, 3250) means that in 5 years the value of the investment will be $3250.

Problem 5   A car is traveling at a uniform speed of 40 mph. The distance, d (in miles), the car travels in t hours is given by the equation $d = 40t$.

a. Write the equation in functional notation.

b. Use the coordinate axes at the right to graph this equation for values of t between 0 and 5.

c. The point whose coordinates are (3, 120) is on the graph. Write a sentence that explains the meaning of this ordered pair.

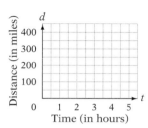

Solution   See page A26.

CONCEPT REVIEW 5.5

Determine whether the statement is always true, sometimes true, or never true.

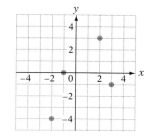

1. The graph shown at the right is the graph of a function.

2. For the relation graphed at the right, the domain is $\{-2, -1, 2, 3\}$.

3. The graph of the function $f(x) = -\frac{2}{3}x + 5$ has a positive slope.

4. The graphs of $y = \frac{1}{4}x - 6$ and $f(x) = \frac{1}{4}x - 6$ are identical.

5. The value of the function $\{(-3, 3), (-2, 2), (-1, 1), (0, 0)\}$ at -2 is 2.

6. A function can be represented in a table, by an equation, by a graph, or in a chart.

7. The graph of a function of the form $f(x) = mx + b$ is a straight line.

8. The equation $y = 7x - 1$ defines a function.

EXERCISES 5.5

1. To test a heart medicine, a physician measures the heart rate, in beats per minute, of five patients before and after they take the medicine. The results are given in the table below.

Before medication	85	80	85	75	90
After medication	75	70	70	80	80

    Write a relation in which the first coordinate is the heart rate before taking the medicine and the second coordinate is the heart rate after taking the medicine. Is the relation a function?

2. A professor records the midterm and final exam scores of five graduate students. The results are given in the table below.

Midterm exam	82	78	81	87	81
Final exam	91	86	96	79	87

    Write a relation in which the first coordinate is the score on the midterm and the second coordinate is the score on the final exam. Is the relation a function?

3. The runs scored by a baseball team and whether the team won, W, or lost, L, its game are recorded in the table below.

Runs scored	4	6	4	2	1	6
Win or loss	L	W	W	W	L	W

    Write a relation in which the first coordinate is the runs scored and the second coordinate is a win or loss. Is the relation a function?

4. The number of aces served by a tennis player during a set and the number of games that player won in the set are given in the table below.

Aces served	6	8	4	10	6
Games won	6	5	3	6	4

    Write a relation in which the first coordinate is the aces served and the second coordinate is the number of games won. Is the relation a function?

5. The number of concerts given in one year by five musical groups and the number of hit records the group had that year are given in the table below.

Hit records	11	9	7	12	12
Number of concerts	200	150	200	175	250

    Write a relation in which the first coordinate is the number of hit records and the second coordinate is the number of concerts. Is the relation a function?

6. The monthly salary, in thousands of dollars, for each employee of a small firm and the number of years of college completed by that employee are given in the table.

Years of college	4	4	5	4	6	8
Monthly salary	3	3.5	4	2.5	5	5.5

    Write a relation in which the first coordinate is the years of college and the second coordinate is the monthly salary. Is the relation a function?

Find the domain and range of the relation. State whether or not the relation is a function.

7. {(0, 0), (2, 0), (4, 0), (6, 0)}

8. {(−2, 2), (0, 2), (1, 2), (2, 2)}

9. {(2, 2), (2, 4), (2, 6), (2, 8)}

10. {(−4, 4), (−2, 2), (0, 0), (−2, −2)}

11. {(0, 0), (1, 1), (2, 2), (3, 3)}

12. {(0, 5), (1, 4), (2, 3), (3, 2), (4, 1), (5, 0)}

13. {(−2, −3), (2, 3), (−1, 2), (1, 2), (−3, 4), (3, 4)}

14. {(−1, 0), (0, −1), (1, 0), (2, 3), (3, 5)}

Evaluate the function at the given value of x. Write an ordered pair that is an element of the function.

15. $f(x) = 4x$; $x = 10$

16. $f(x) = 8x$; $x = 11$

17. $f(x) = x - 5$; $x = -6$

18. $f(x) = x + 7$; $x = -9$

19. $f(x) = 3x^2$; $x = -2$

20. $f(x) = x^2 - 1$; $x = -8$

21. $f(x) = 5x + 1$; $x = \dfrac{1}{2}$

22. $f(x) = 2x - 6$; $x = \dfrac{3}{4}$

23. $f(x) = \dfrac{2}{5}x + 4$; $x = -5$

24. $f(x) = \dfrac{3}{2}x - 5$; $x = 2$

25. $f(x) = 2x^2$; $x = -4$

26. $f(x) = 4x^2 + 2$; $x = -3$

Find the range of the function defined by the given equation. Write five ordered pairs that belong to the function.

27. $f(x) = 3x - 4$; domain $= \{-5, -3, -1, 1, 3\}$

28. $f(x) = 2x + 5$; domain $= \{-10, -5, 0, 5, 10\}$

29. $f(x) = \dfrac{1}{2}x + 3$; domain $= \{-4, -2, 0, 2, 4\}$

30. $f(x) = \dfrac{3}{4}x - 1$; domain $= \{-8, -4, 0, 4, 8\}$

31. $f(x) = x^2 + 6$; domain $= \{-3, -1, 0, 1, 3\}$

32. $f(x) = 3x^2 + 6$; domain $= \{-2, -1, 0, 1, 2\}$

2   Graph.

33. $f(x) = 5x$

34. $f(x) = -4x$

35. $f(x) = x + 2$

36. $f(x) = x - 3$

37. $f(x) = 6x - 1$

38. $f(x) = 3x + 4$

39. $f(x) = -2x + 3$

40. $f(x) = -5x - 2$

41. $f(x) = \dfrac{1}{3}x - 4$

42. $f(x) = \dfrac{3}{5}x + 1$

43. $f(x) = 4$

44. $f(x) = -3$

Graph using a graphing utility.

45. $f(x) = 2x - 1$

46. $f(x) = -3x - 1$

47. $f(x) = -\dfrac{1}{2}x + 1$

48. $f(x) = \dfrac{2}{3}x + 4$

49. $f(x) = \dfrac{3}{4}x + 1$

50. $f(x) = -\dfrac{1}{2}x - 2$

51. $f(x) = -\dfrac{4}{3}x + 5$

52. $f(x) = \dfrac{5}{2}x$

53. $f(x) = -1$

Solve.

54. Depreciation is the declining value of an asset. For instance, a company that purchases a truck for \$20,000 has an asset worth \$20,000. In 5 years, however, the value of the truck will have declined and it may be worth only \$4000. An equation that represents this decline is $V = 20{,}000 - 3200x$, where V is the value, in dollars, of the truck after x years.

  a. Write the equation in functional notation.

  b. Use the coordinate axes at the right to graph the equation for values of x between 0 and 5.

  c. The point (4, 7200) is on the graph. Write a sentence that explains the meaning of this ordered pair.

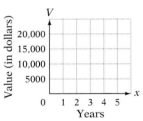

55. A company uses the equation $V = 30{,}000 - 5000x$ to estimate the depreciated value, in dollars, of a computer. (See Exercise 54.)

  a. Write the equation in functional notation.

  b. Use the coordinate axes at the right to graph the equation for values of x between 0 and 5.

  c. The point (1, 25,000) is on the graph. Write a sentence that explains the meaning of this ordered pair.

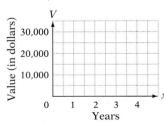

56. An architect charges a fee of $500 plus $2.65 per square foot to design a house. The equation that represents the architect's fee is given by $F = 2.65s + 500$, where F is the fee, in dollars, and s is the number of square feet in the house.

   a. Write the equation in functional notation.

   b. Use the coordinate axes at the right to graph the equation for values of s between 0 and 5000.

   c. The point (3500, 9775) is on the graph. Write a sentence that explains the meaning of this ordered pair.

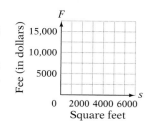

57. A rental car company charges a "drop-off" fee of $50 to return a car to a location different from that from which it was rented. In addition, it charges a fee of $.18 per mile the car is driven. An equation that represents the total cost to rent a car from this company is $C = 0.18m + 50$, where C is the total cost, in dollars, and m is the number of miles the car is driven.

   a. Write the equation in functional notation.

   b. Use the coordinate axes at the right to graph the equation for values of m between 0 and 1000.

   c. The point (500, 140) is on the graph. Write a sentence that explains the meaning of this ordered pair.

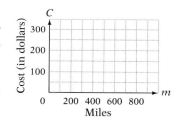

SUPPLEMENTAL EXERCISES 5.5

58. **a.** A function f consists of the ordered pairs {(−4, −6), (−2, −2), (0, 2), (2, 6), (4, 10)}. Find $f(2)$.

   b. A function f consists of the ordered pairs {(0, 9), (1, 8), (2, 7), (3, 6), (4, 5)}. Find $f(1)$.

59. According to the National Computer Security Association, the cost in lost time and productivity because of "computer virus infections" in the United States is as shown in the table at the right.

Year	Cost in Millions
1990	$100
1991	$300
1992	$700
1993	$1400
1994	$2700

   a. Do these data represent a function? Why or why not?

   b. If the data represent a function, do they represent a linear function? Why or why not?

   c. If the data represent a function, is the year a function of the cost, or is the cost a function of the year?

   d. If you had to predict the cost in 1995 on the basis of the information in the table, would you predict that it would be more or less than $2.7 billion? Explain your answer.

60. Write a few sentences describing the similarities and differences between relations and functions.

61. Investigating a relationship between two variables is an important task in the application of mathematics. For example, botanists study the relationship between the number of bushels of wheat yielded per acre and the amount of watering per acre. Environmental scientists study the relationship between the incidence of skin cancer and the amount of ozone in the atmosphere. Business analysts study the relationship between the price of a product and the number of products that are sold at that price. Describe a relationship that is important to your major field of study.

62. Functions are a part of our everyday lives. For example, the cost to mail a package via first-class mail is a function of the weight of the package. The tuition paid by a part-time student is a function of the number of credit hours the student registers for. Provide other examples of functions.

S E C T I O N **5.6**

Graphing Linear Inequalities

1 Graph inequalities in two variables

POINT OF INTEREST

Linear inequalities play an important role in applied mathematics. They are used in a branch of mathematics called *linear programming,* which was developed during World War II to solve problems in supplying the Air Force with the machine parts necessary to keep planes flying. Today its applications have been broadened to many other disciplines.

The graph of the linear equation $y = x - 2$ separates a plane into three sets:

the set of points on the line

the set of points above the line

the set of points below the line

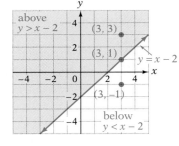

The point whose coordinates are (3, 1) is a solution of $y = x - 2$.

$$y = x - 2$$
$$1 \mid 3 - 2$$
$$1 = 1$$

The point whose coordinates are (3, 3) is a solution of $y > x - 2$.

$$y > x - 2$$
$$3 \mid 3 - 2$$
$$3 > 1$$

Any point above the line is a solution of $y > x - 2$.

The point whose coordinates are (3, −1) is a solution of $y < x - 2$.

$$y < x - 2$$
$$-1 \mid 3 - 2$$
$$-1 < 1$$

Any point below the line is a solution of $y < x - 2$.

The solution set of $y = x - 2$ is all points on the line. The solution set of $y > x - 2$ is all points above the line. The solution set of $y < x - 2$ is all points below the line. The solution set of an inequality in two variables is a **half-plane.**

The following illustrates the procedure for graphing a linear inequality.

Graph the solution set of $2x + 3y \leq 6$.

Solve the inequality for y.

$$2x + 3y \leq 6$$
$$2x - 2x + 3y \leq -2x + 6$$
$$3y \leq -2x + 6$$
$$\frac{3y}{3} \leq \frac{-2x + 6}{3}$$
$$y \leq -\frac{2}{3}x + 2$$

Change the inequality to an equality and graph the line. If the inequality is $\geq$ or $\leq$, the line is in the solution set and is shown by a **solid line.** If the inequality is $>$ or $<$, the line is not a part of the solution set and is shown by a **dashed line.**

$$y = -\frac{2}{3}x + 2$$

If the inequality is in the form $y > mx + b$ or $y \geq mx + b$, shade the **upper half-plane.** If the inequality is in the form $y < mx + b$ or $y \leq mx + b$, shade the **lower half-plane.**

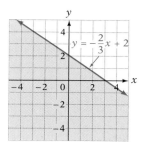

The equation $y \leq -\frac{2}{3}x + 2$ is in the form $y \leq mx + b$. Draw a solid line and shade the lower half-plane.

The inequality $2x + 3y \leq 6$ can also be graphed as shown below.

Change the inequality to an equality.

$$2x + 3y = 6$$

Find the x- and y-intercepts of the equation. To find the x-intercept, let $y = 0$.

$$2x + 3(0) = 6$$
$$2x = 6$$
$$x = 3$$

The x-intercept is $(3, 0)$.

To find the y-intercept, let $x = 0$.

$$2(0) + 3y = 6$$
$$3y = 6$$
$$y = 2$$

The y-intercept is $(0, 2)$.

Graph the ordered pairs $(3, 0)$ and $(0, 2)$. Draw a solid line through the points because the inequality is $\leq$. (See the graph on the next page.)

The point $(0, 0)$ can be used to determine which region to shade. If $(0, 0)$ is a solution of the inequality, then shade the region that includes the point $(0, 0)$. If $(0, 0)$ is not a solution of the inequality, then shade the region that does not include the point $(0, 0)$. For this example, $(0, 0)$ is a solution of the inequality. The region that contains the point $(0, 0)$ is shaded.

If the line passes through the point $(0, 0)$, another point must be used to determine which region to shade. For example, use the point $(1, 0)$.

$$2x + 3y \le 6$$
$$2(0) + 3(0) \le 6$$
$$0 \le 6 \quad \text{True}$$

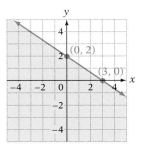

It is important to note that every point in the shaded region is a solution of the inequality and that every solution of the inequality is a point in the shaded region. No point outside of the shaded region is a solution of the inequality.

Example 1   Graph the solution set of $3x + y > -2$.

Solution
$$3x + y > -2$$
$$3x - 3x + y > -3x - 2 \quad \blacktriangleright \text{ Solve the inequality for } y.$$
$$y > -3x - 2$$

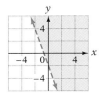

▸ Graph $y = -3x - 2$ as a dashed line. Shade the upper half-plane.

Problem 1   Graph the solution set of $x - 3y < 2$.

Solution   See page A26.

Example 2   Graph the solution set of $y > 3$.

Solution

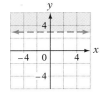

▸ The inequality is solved for y. Graph $y = 3$ as a dashed line. Shade the upper half-plane.

Problem 2   Graph the solution set of $x < 3$.

Solution   See page A26.

CONCEPT REVIEW 5.6

Determine whether the statement is always true, sometimes true, or never true.

1. $y \geq \frac{1}{2}x - 5$ is an example of a linear equation in two variables.

2. $(-2, 4)$ is a solution of $y > x + 2$.

3. $(3, 0)$ is a solution of $y \leq x - 1$.

4. Every point above the line $y = x - 3$ is a solution of $y < x - 3$.

5. $(0, 0)$ is a not a solution of $y > x + 4$. Therefore, the point $(0, 0)$ should not be in the shaded region that indicates the solution set of the inequality.

6. In the graph of an inequality in two variables, any point in the shaded region is a solution of the inequality, and any point not in the shaded region is not a solution of the inequality.

EXERCISES 5.6

1   Graph the solution set.

1. $y > 2x + 3$

2. $y > 3x - 9$

3. $y > \frac{3}{2}x - 4$

4. $y > -\frac{5}{4}x + 1$

5. $y \leq -\frac{3}{4}x - 1$

6. $y \leq -\frac{5}{2}x - 4$

7. $y \leq -\frac{6}{5}x - 2$

8. $y < \frac{4}{5}x + 3$

9. $x + y > 4$

10. $x - y > -3$

11. $2x + y \geq 4$

12. $3x + y \geq 6$

13. $y \leq -2$

14. $y > 3$

15. $2x + 2y \leq -4$

16. $-4x + 3y < -12$

SUPPLEMENTAL EXERCISES 5.6

Write the inequality given its graph.

17.

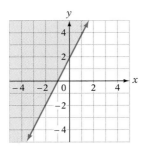

18.

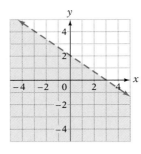

19.

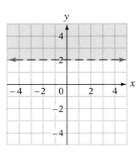

20.

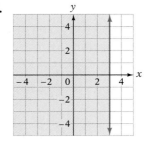

Graph the solution set.

21. $y - 5 < 4(x - 2)$

22. $y + 3 < 6(x + 1)$

23. $3x - 2(y + 1) \leq y - (5 - x)$

24. $2x - 3(y + 1) \geq y - (4 - x)$

25. Does an inequality in two variables define a relation? Why or why not? Does an inequality in two variables define a function? Why or why not?

Project in Mathematics

Graphs of Motion

A graph can be useful in analyzing the motion of a body. For example, consider an airplane in uniform motion traveling at 100 m/s. The table at the right shows the distance, in meters, traveled by the plane at the end of each of five 1-second intervals.

Time (in seconds)	Distance (in meters)
0	0
1	100
2	200
3	300
4	400
5	500

These data can be graphed on a rectangular coordinate system and a straight line drawn through the points plotted. The travel time is shown along the horizontal axis, and the distance traveled by the plane is shown along the vertical axis. (Note that the units along the two axes are not the same length.)

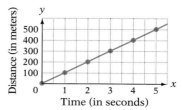

To write the equation for the line just graphed, use the coordinates of any two points on the line to find the slope. The y-intercept is $(0, 0)$.

Let $(x_1, y_1) = (1, 100)$ and $(x_2, y_2) = (2, 200)$.

Then $m = \dfrac{y_2 - y_1}{x_2 - x_1} = \dfrac{200 - 100}{2 - 1} = \dfrac{100}{1} = 100$; $b = 0$.

$y = mx + b$
$y = 100x + 0$
$y = 100x$

Note that the slope of the line, 100, is equal to the speed, 100 m/s.

The slope of a distance-time graph represents the speed of the object. The greater the slope of a distance-time graph, the greater the speed of the object it represents.

The distance-time graphs for two planes are shown at the right. One plane is traveling at 100 m/s, and the other is traveling at 200 m/s. The slope of the line representing the faster plane is greater than the slope of the line representing the slower plane.

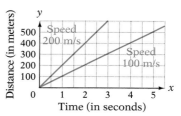

In the graph at the right, the travel time of a plane flying at 100 m/s is shown along the horizontal axis, and the speed of the plane is shown along the vertical axis. Because the speed is constant, the graph is a horizontal line.

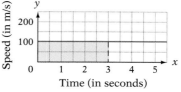

In the last graph on page 251, the area between the horizontal line graphed and the horizontal axis is equal to the distance traveled by the plane up to that time. For example, the area of the shaded region on the graph is

$$\text{Length} \cdot \text{Width} = (3 \text{ s})(100 \text{ m/s}) = \frac{3 \text{ s}}{1} \cdot \frac{100 \text{ m}}{\text{s}} = 300 \text{ m}$$

The distance traveled by the plane in 3 s is equal to 300 m.

Solve.

1. A car in uniform motion is traveling at 20 m/s.
    a. Prepare a distance-time graph for the car for 0 s to 5 s.
    b. Find the slope of the line.
    c. Find the equation of the line.
    d. Prepare a speed-time graph for the car for 0 s to 5 s.
    e. Find the distance traveled by the car after 3 s.

2. One car in uniform motion is traveling at 10 m/s. A second car in uniform motion is traveling at 15 m/s.
    a. Prepare one distance-time graph for both cars for 0 s to 5 s.
    b. Find the slope of each line.
    c. Find the equation of each line graphed.
    d. Assuming that the cars started at the same point at 0 s, find the distance between the cars at the end of 5 s.

3. a. In a distance-time graph, is it possible for the graph to be a horizontal line?
    b. What does a horizontal line reveal about the motion of the object during that time period?

4. a. In a distance-time graph, is it possible for the graph to be a vertical line?
    b. What does a vertical line reveal about the motion of the object?

Focus on Problem Solving

Deductive Reasoning

Suppose that during the last week of your math class, your instructor tells you that if you receive an A on the final exam, you will earn an A in the course. When the final exam grades are posted, you learn that you received an A on the final exam. You can then assume that you will earn an A in the course.

The process used to determine your grade in the math course is deductive reasoning. **Deductive reasoning** involves drawing a conclusion that is based on given facts; in other words, we go from general statements to a specific conclusion.

Some of the exercises in this text require deductive reasoning. For example, Exercise 58a in Section 5.5, requires that you understand the general statement on page 236 that explains the symbol $f(x)$ and that you then make a specific conclusion concerning $f(2)$ in that exercise.

Here are two more examples of using deductive reasoning.

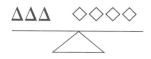

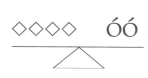

Given that $\triangle\triangle\triangle = \diamondsuit\diamondsuit\diamondsuit\diamondsuit$ and $\diamondsuit\diamondsuit\diamondsuit\diamondsuit = \acute{O}\acute{O}$, then $\triangle\triangle\triangle\triangle\triangle\triangle$ is equivalent to how many $\acute{O}$s?

Because $3 \triangle$s $= 4 \diamondsuit$s and $4 \diamondsuit$s $= 2 \acute{O}$s, $3 \triangle$s $= 2 \acute{O}$s.

$6 \triangle$s is twice $3 \triangle$s. We need to find twice $2 \acute{O}$s, which is $4 \acute{O}$s.

Therefore, $\triangle\triangle\triangle\triangle\triangle\triangle = \acute{O}\acute{O}\acute{O}\acute{O}$.

Lomax, Parish, Thorpe, and Wong are neighbors. Each drives a different type of vehicle: a compact car, a sedan, a sports car, or a station wagon. From the following statements, determine which type of vehicle each of the neighbors drives.

(1) Although the vehicle owned by Lomax has more mileage on it than does either the sedan or the sports car, it does not have the highest mileage of all four cars.

(2) Wong and the owner of the sports car live on one side of the street, and Thorpe and the owner of the compact car live on the other side of the street.

(3) Thorpe owns the vehicle with the most mileage on it.

To determine which type of vehicle each neighbor drives, complete the chart on page 254. Here X is used to indicate that a possibility has been eliminated, and $\sqrt{}$ is used to show that a match has been found. Note that when a row or column has three Xs, a $\sqrt{}$ is written in the fourth box. Also, whenever a $\sqrt{}$ is written in a box, an X is written in the remaining open boxes in that row and column of the chart.

From statement 1, Lomax does not drive a sedan or a sports car. X1 is written in each of these two boxes. (X1 refers to an elimination based on statement 1.)

From statement 2, Wong does not drive a sports car or a compact car. The same can be said of Thorpe. X2 is written in each of these four boxes. (X2 refers to statement 2.) Now there are three Xs in the sports car column; put a $\sqrt{}$ in the empty box in that column. Parish drives the sports car. Therefore, Parish does not drive the compact car, sedan, or wagon; write X2 in each of these three boxes. Now there are three Xs in the compact car column; put a $\sqrt{}$ in the empty box in that column. Lomax drives the compact car. Therefore, Lomax does not drive the station wagon; write X2 in that box.

From statement 3, Thorpe drives the vehicle with the most mileage on it. From statement 1, neither the sedan nor the sports car has the most mileage on it. The box for Thorpe owning the sports car already has an X in it; write X3 in the box for Thorpe owning the sedan. There are now three Xs in the Thorpe row; put a √ in the empty box in that row. Thorpe drives the station wagon. By elimination, Wong drives the sedan.

	Compact	Sedan	Sports Car	Wagon
Lomax	√	X1	X1	X2
Parish	X2	X2	√	X2
Thorpe	X2	X3	X2	√
Wong	X2		X2	

Lomax drives the compact car, Parish drives the sports car, Thorpe drives the station wagon, and Wong drives the sedan.

Try each of the following exercises. Each exercise requires deductive reasoning.

1. Given that ‡ ‡ = ● ● ● and ● ● ● = ⌐, then ‡ ‡ ‡ ‡ is equivalent to how many ⌐s?

2. Given that Ó Ó Ó = Ω Ω and ✷ = Ω Ω, then ✷ ✷ is equivalent to how many Ós?

3. Four neighbors, Chris, Dana, Leslie, and Pat, have different occupations (accountant, banker, chef, and manager). From the following statements, determine the occupation of each neighbor.
    (1) Dana usually gets home from work after the banker but before the manager.
    (2) Leslie, who is usually the last to get home from work, is not the accountant.
    (3) The manager and Leslie usually leave for work about the same time.
    (4) The banker lives next door to Pat.

4. The Ontkeans, Kedrovas, McIvers, and Levinsons are neighbors. Each of the four families specializes in a different national cuisine (Chinese, French, Italian, or Mexican). From the following statements, determine which cuisine each family specializes in.
    (1) The Ontkeans invited the family that specializes in Chinese cuisine and the family that specializes in Mexican cuisine for dinner last night.
    (2) The McIvers live between the family that specializes in Italian cuisine and the Ontkeans. The Levinsons live between the Kedrovas and the family that specializes in Chinese cuisine.
    (3) The Kedrovas and the family that specializes in Italian cuisine both subscribe to the same culinary magazine.

5. Anna, Kay, Megan, and Nicole decide to travel together during spring break, but they need to find a destination where each of them will be able to

participate in her favorite sport (golf, horseback riding, sailing, or tennis). From the following statements, determine the favorite sport of each student.

(1) Anna and the student whose favorite sport is sailing both like to swim, whereas Nicole and the student whose favorite sport is tennis would prefer to scuba dive.

(2) Megan and the student whose favorite sport is sailing are roommates. Nicole and the student whose favorite sport is golf each live in a single.

6. Chang, Nick, Pablo, and Saul each take a different form of transportation (bus, car, subway, or taxi) from the office to the airport. From the following statements, determine which form of transportation each takes.

(1) Chang spent more on transportation than did the fellow who took the bus, but less than the fellow who took the taxi.

(2) Pablo, who did not travel by bus and who spent the least on transportation, arrived at the airport after Nick but before the fellow who took the subway.

(3) Saul spent less on transportation than either Chang or Nick.

Chapter Summary

Key Words

A *rectangular coordinate system* is formed by two number lines, one horizontal and one vertical, that intersect at the zero point of each line. The number lines that make up a rectangular coordinate system are called the *coordinate axes,* or simply the *axes*. The *origin* is the point of intersection of the two coordinate axes. Generally, the horizontal axis is labeled the *x-axis*, and the vertical axis is labeled the *y-axis*. A rectangular coordinate system divides the plane determined by the axes into four regions called *quadrants*.   (Objective 5.1.1)

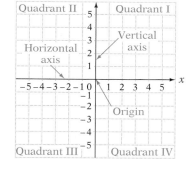

The coordinate axes determine a *plane*. Every point in the plane can be identified by an *ordered pair (x, y)*. The first number in an ordered pair is called the *x-coordinate* or the *abscissa*. The second number is called the *y-coordinate* or the *ordinate*. The *coordinates* of a point are the numbers in the ordered pair associated with the point.   (Objective 5.1.1)

The ordered pair $(4, -9)$ has x-coordinate 4 and y-coordinate -9.

An equation of the form $y = mx + b$, where m is the coefficient of x and b is a constant, is a *linear equation in two variables*. A *solution of a linear equation in two variables* is an ordered pair (x, y) that makes the equation a true statement. (Objective 5.2.1)

$y = 2x + 3$ is an example of a linear equation in two variables. The ordered pair $(1, 5)$ is a solution of this equation because when 1 is substituted for x and 5 is substituted for y, the result is a true statement.

The point at which a graph crosses the x-axis is called the *x-intercept*. At the x-intercept, the y-coordinate is 0. The point at which a graph crosses the y-axis is called the *y-intercept*. At the y-intercept, the x-coordinate is 0.   (Objective 5.2.3)

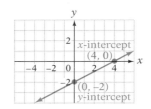

An equation of the form $Ax + By = C$ is also a *linear equation in two variables*.   (Objective 5.2.3)

$5x - 2y = 10$ is an example of an equation in $Ax + By = C$ form.

The *graph of $y = b$* is a horizontal line with y-intercept $(0, b)$. The *graph of $x = a$* is a vertical line with x-intercept $(a, 0)$. (Objective 5.2.3)

The graph of $y = -3$ is a horizontal line with y-intercept $(0, -3)$. The graph of $x = 2$ is a vertical line with x-intercept $(2, 0)$.

The *slope* of a line is a measure of the slant of the line. The symbol for slope is m. A line that slants upward to the right has a *positive slope*. A line that slants downward to the right has a *negative slope*. A horizontal line has *zero slope*. The slope of a vertical line is *undefined*.   (Objective 5.3.1)

For the line $y = 4x - 3$, $m = 4$; the slope is positive.

For the line $y = -6x + 1$, $m = -6$; the slope is negative.

For the line $y = -7$, $m = 0$; the slope is 0.

For the line $x = 8$, the slope is undefined.

A *relation* is any set of ordered pairs. The *domain* of a relation is the set of first coordinates of the ordered pairs. The *range* is the set of second coordinates of the ordered pairs. (Objective 5.5.1)

For the relation $\{(-2, -5), (0, -3), (4, -1)\}$, the domain is $\{-2, 0, 4\}$ and the range is $\{-5, -3, -1\}$.

A *function* is a relation in which no two ordered pairs have the same first coordinate and different second coordinates. (Objective 5.5.1)

$\{(-2, -5), (0, -3), (4, -1)\}$ is a function. $\{(-2, -5), (-2, -3), (4, -1)\}$ is not a function because two ordered pairs have x-coordinate -2 and different y-coordinates.

A function of the form $f(x) = mx + b$ is a *linear function*. Its graph is a straight line.   (Objective 5.5.2)

$f(x) = -3x + 8$ is an example of a linear function.

The solution set of an inequality in two variables is a *half-plane*.   (Objective 5.6.1)

The solution set of $y > x - 2$ is all points above the line $y = x - 2$. The solution set of $y < x - 2$ is all points below the line $y = x - 2$.

Essential Rules and Procedures

To find the x-intercept, let $y = 0$.
To find the y-intercept, let $x = 0$.   (Objective 5.2.3)

To find the x-intercept of $4x - 3y = 12$, let $y = 0$. To find the y-intercept, let $x = 0$.

$$4x - 3y = 12$$
$$4x - 3(0) = 12$$
$$4x - 0 = 12$$
$$4x = 12$$
$$x = 3$$

The x-intercept is $(3, 0)$.

$$4x - 3y = 12$$
$$4(0) - 3y = 12$$
$$0 - 3y = 12$$
$$-3y = 12$$
$$y = -4$$

The y-intercept is $(0, -4)$.

Slope of a Linear Equation

Slope $= m = \frac{y_2 - y_1}{x_2 - x_1}$, $x_2 \neq x_1$ (Objective 5.3.1)

To find the slope of the line between the points $(2, -3)$ and $(-1, 6)$, let $P_1 = (2, -3)$ and $P_2 = (-1, 6)$. Then $(x_1, y_1) = (2, -3)$ and $(x_2, y_2) = (-1, 6)$.

$$m = \frac{y_2 - y_1}{x_2 - x_1} = \frac{6 - (-3)}{-1 - 2} = \frac{9}{-3} = -3$$

Slope-Intercept Form of a Linear Equation

$y = mx + b$ (Objective 5.3.2)

For the equation $y = -4x + 7$, the slope $= m = -4$ and the y-intercept $= (0, b) = (0, 7)$.

Point-Slope Formula

$y - y_1 = m(x - x_1)$ (Objective 5.4.2)

To find the equation of the line that contains the point $(6, -1)$ and has slope 2, let $(x_1, y_1) = (6, -1)$ and $m = 2$.

$$y - y_1 = m(x - x_1)$$
$$y - (-1) = 2(x - 6)$$
$$y + 1 = 2x - 12$$
$$y = 2x - 13$$

Chapter Review Exercises

1. Find the ordered-pair solution of $y = -\frac{2}{3}x + 2$ that corresponds to $x = 3$.

2. Find the equation of the line that contains the point whose coordinates are $(0, -1)$ and has slope 3.

3. Graph: $x = -3$

4. Graph the ordered pairs $(-3, 1)$ and $(0, 2)$.

5. Evaluate $f(x) = 3x^2 + 4$ at $x = -5$.

6. Find the slope of the line that contains the points whose coordinates are $(3, -4)$ and $(1, -4)$.

7. Graph: $y = 3x + 1$

8. Graph: $3x - 2y = 6$

9. Find the equation of the line that contains the point whose coordinates are $(-1, 2)$ and has slope $-\dfrac{2}{3}$.

10. Find the x- and y-intercepts for $6x - 4y = 12$.

11. Graph the line that has slope $\dfrac{1}{2}$ and y-intercept $(0, -1)$.

12. Graph: $f(x) = -\dfrac{2}{3}x + 4$

13. Find the equation of the line that contains the point whose coordinates are $(-3, 1)$ and has slope $\dfrac{2}{3}$.

14. Find the slope of the line that contains the points whose coordinates are $(2, -3)$ and $(4, 1)$.

15. Evaluate $f(x) = \dfrac{3}{5}x + 2$ at $x = -10$.

16. Find the domain and range of the relation $\{(-20, -10), (-10, -5), (0, 0), (10, 5)\}$. Is the relation a function?

17. Graph: $y = -\dfrac{3}{4}x + 3$

18. Graph the line that has slope 2 and y-intercept -2.

19. Graph the ordered pairs $(-2, -3)$ and $(2, 4)$.

20. Graph: $f(x) = 5x + 1$

21. Find the x- and y-intercepts for $2x - 3y = 12$.

22. Find the equation of the line that contains the point whose coordinates are $(-1, 0)$ and has slope 2.

23. Graph: $y = 3$

24. Graph the solution set of $3x + 2y \le 12$.

25. Find the slope of the line that contains the points whose coordinates are $(2, -3)$ and $(-3, 4)$.

26. Find the equation of the line that contains the point whose coordinates are $(2, 3)$ and has slope $\frac{1}{2}$.

27. Graph the line that has slope -1 and y-intercept $(0, 2)$.

28. Graph: $2x - 3y = 6$

29. Find the ordered-pair solution of $y = 2x - 1$ that corresponds to $x = -2$.

30. Find the equation of the line that contains the point $(0, 2)$ and has slope -3.

31. Graph: $f(x) = 2x$

32. Evaluate $f(x) = 3x - 5$ at $x = \frac{5}{3}$.

33. Find the slope of the line that contains the points whose coordinates are $(3, -2)$ and $(3, 5)$.

34. Find the equation of the line that contains the point whose coordinates are $(2, -1)$ and has slope $\frac{1}{2}$.

35. Is $(-10, 0)$ a solution of $y = \frac{1}{5}x + 2$?

36. Find the ordered-pair solution of $y = 4x - 9$ that corresponds to $x = 2$.

37. Find the x- and y-intercepts for $4x - 3y = 0$.

38. Find the equation of the line that contains the point whose coordinates are $(-2, 3)$ and has zero slope.

39. Graph the solution set of $6x - y > 6$.

40. Graph the line that has slope -3 and y-intercept $(0, 1)$.

41. Find the equation of the line that contains the point whose coordinates are $(0, -4)$ and has slope 3.

42. Graph: $x + 2y = -4$

43. Find the domain and range of the relation {(−10, −5), (−5, 0), (5, 0), (−10, 0)}. Is the relation a function?

44. Find the range of the function given by the equation $f(x) = 3x + 7$ if the domain is {−20, −10, 0, 10, 20}.

45. Find the range of the function given by the equation $f(x) = \frac{1}{3}x + 4$ if the domain is {−6, −3, 0, 3, 6}.

46. The following table gives the number of people, in thousands, collecting unemployment benefits in a northeastern state during a four-month period and the number of housing starts, in thousands, in that state during the same period. Graph the scatter diagram of these data.

People collecting unemployment, x	40	45	65	90
Housing starts, y	40	35	25	20

47. A study of six algebra students measured the average number of hours per week a student spent in the math lab and the student's score on the final exam. The results are recorded in the following table. Graph a scatter diagram for these data.

Hours in the math lab, x	3	2	4	5	3	1
Final exam score, y	78	66	75	82	74	63

48. The table below shows the tread depth, in millimeters, of a tire and the number of miles, in thousands, that have been driven on that tire.

Miles driven	25	35	40	20	45
Tread depth	4.8	3.5	2.1	5.5	1.0

Write a relation in which the first coordinate is the number of miles driven and the second coordinate is the tread depth. Is the relation a function?

49. A contractor uses the equation $C = 70s + 40,000$ to estimate the cost of building a new house. In this equation, C is the total cost, in dollars, and s is the number of square feet in the home.

  a. Write the equation in functional notation.

  b. Use the coordinate axes at the right to graph the equation for values of s between 0 and 5000.

  c. The point (1500, 145,000) is on the graph. Write a sentence that explains the meaning of this ordered pair.

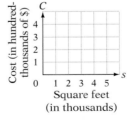

50. The graph at the right shows the number of heart surgeries performed over a ten-year period. Find the slope of the line. Write a sentence that states the meaning of the slope.

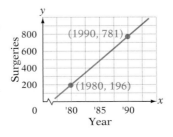

Chapter Test

1. Find the equation of the line that contains the point whose coordinates are $(9, -3)$ and has slope $-\dfrac{1}{3}$.

2. Find the slope of the line that contains the points whose coordinates are $(9, 8)$ and $(-2, 1)$.

3. Find the x- and y-intercepts for $3x - 2y = 24$.

4. Find the ordered-pair solution of $y = -\dfrac{4}{3}x - 1$ that corresponds to $x = 9$.

5. Graph: $5x + 3y = 15$

6. Graph: $y = \dfrac{1}{4}x + 3$

7. Evaluate $f(x) = 4x + 7$ at $x = \dfrac{3}{4}$.

8. Is $(6, 3)$ a solution of $y = \dfrac{2}{3}x + 1$?

9. Graph the line that has slope $-\dfrac{2}{3}$ and y-intercept $(0, 4)$.

10. Graph the solution set of $2x - y \geq 2$.

11. Graph the ordered pairs $(3, -2)$ and $(0, 4)$.

12. Graph the line that has slope 2 and y-intercept $(0, -4)$.

13. Find the slope of the line that contains the points whose coordinates are $(4, -3)$ and $(-2, -3)$.

14. Find the equation of the line that contains the point whose coordinates are $(0, 7)$ and has slope $-\dfrac{2}{5}$.

15. Find the equation of the line that contains the point whose coordinates are $(2, 1)$ and has slope 4.

16. Evaluate $f(x) = 4x^2 - 3$ at $x = -2$.

17. Graph: $y = -2x - 1$

18. Graph the solution set of $y > 2$.

19. Graph: $x = -3$

20. Graph the line that has slope $-\frac{2}{3}$ and y-intercept $(0, 2)$.

21. Graph: $f(x) = \frac{1}{2}x - 3$

22. Graph: $f(x) = -5x$

23. Evaluate $f(x) = -3x + 7$ at $x = -6$.

24. Find the domain and range of the relation $\{(8, 8), (8, 4), (8, 2), (8, 0)\}$. Is the relation a function?

25. Find the equation of the line that contains the point whose coordinates are $(5, 0)$ and has slope $\frac{3}{5}$.

26. Find the range of the function given by the equation $f(x) = -3x + 5$ if the domain is $\{-7, -2, 0, 4, 9\}$.

27. The distance, in miles, a house is from a fire station and the amount, in thousands of dollars, of fire damage that house sustained in a fire are given in the following table. Graph the scatter diagram for these data.

Distance (in miles), x	3.5	4.0	5.5	6.0
Damage (in thousands of dollars), y	25	30	40	35

28. A company that manufactures toasters has fixed costs of $1000 each month. The manufacturing cost per toaster is $8. An equation that represents the total cost to manufacture the toasters is $C = 8t + 1000$, where C is the total cost, in dollars, and t is the number of toasters manufactured each month.

a. Write the equation in functional notation.

b. Use the coordinate axes at the right to graph the equation for values of t between 0 and 500.

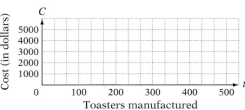

c. The point $(340, 3720)$ is on the graph. Write a sentence that explains the meaning of this ordered pair.

29. The data in the following table show a reading test grade and the final exam grade in a history class.

Reading test grade	8.5	9.4	10.1	11.4	12.0
History exam	64	68	76	87	92

Write a relation in which the first coordinate is the reading test grade and the second coordinate is the score on the history final exam. Is the relation a function?

30. The graph at the right shows the cost, in dollars, per 1000 board feet of lumber over a six-month period. Find the slope of the line. Write a sentence that states the meaning of the slope.

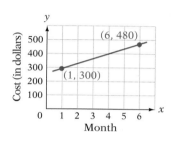

Cumulative Review Exercises

1. Simplify: $12 - 18 \div 3 \cdot (-2)^2$

2. Evaluate $\dfrac{a - b}{a^2 - c}$ when $a = -2$, $b = 3$, and $c = -4$.

3. Simplify: $4(2 - 3x) - 5(x - 4)$

4. Solve: $2x - \dfrac{2}{3} = \dfrac{7}{3}$

5. Solve: $3x - 2[x - 3(2 - 3x)] = x - 6$

6. Write $6\dfrac{2}{3}\%$ as a fraction.

7. Use the roster method to write the set of natural numbers less than 9.

8. Given $D = \{-23, -18, -4, 0, 5\}$, which elements of set D are greater than -16?

9. Solve: $8a - 3 \geq 5a - 6$

10. Solve $4x - 5y = 15$ for y.

11. Find the ordered-pair solution of $y = 3x - 1$ that corresponds to $x = -2$.

12. Find the slope of the line that contains the points whose coordinates are $(2, 3)$ and $(-2, 3)$.

13. Find the x- and y-intercepts for $5x + 2y = 20$.

14. Find the equation of the line that contains the point whose coordinates are $(3, 2)$ and has slope -1.

15. Graph: $y = \dfrac{1}{2}x + 2$

16. Graph: $3x + y = 2$

17. Graph: $f(x) = -4x - 1$

18. Graph the solution set of $x - y \le 5$.

19. Find the domain and range of the relation $\{(0, 4), (1, 3), (2, 2), (3, 1), (4, 0)\}$. Is the relation a function?

20. Evaluate $f(x) = -4x + 9$ at $x = 5$.

21. Find the range of the function given by the equation $f(x) = -\dfrac{5}{3}x + 3$ if the domain is $\{-9, -6, -3, 0, 3, 6\}$.

22. The sum of two numbers is 24. Twice the smaller number is three less than the larger number. Find the two numbers.

23. A lever is 8 ft long. A force of 80 lb is applied to one end of the lever, and a force of 560 lb is applied to the other end. Where is the fulcrum located when the system balances?

24. The perimeter of a triangle is 49 ft. The length of the first side is twice the length of the third side, and the length of the second side is 5 ft more than the length of the third side. Find the length of the first side.

25. A suit that regularly sells for $89 is on sale for 30% off the regular price. Find the sale price.

6

Systems of Linear Equations

Objectives

Input-Output Analysis

The economies of the industrial nations are very complex; they comprise hundreds of different industries, and each industry supplies other industries with goods and services needed in the production process. For example, the steel industry requires coal to produce steel, and the coal industry requires steel (in the form of machinery) to mine and transport coal.

Wassily Leontief, a Russian-born economist, developed a method of describing mathematically the interactions of an economic system. His technique was to examine various sectors of an economy (steel industry, oil, farms, autos, and so on) and determine how each sector interacted with the others. More than five hundred sectors of the economy were studied.

The interaction of each sector with the others was written as a series of equations. This series of equations is called a *system of equations*. Using a computer, economists searched for a solution to the system of equations that would determine the output levels various sectors would have to meet to satisfy the requests from other sectors. The method is called input-output analysis.

Input-output analysis has many applications. For example, it is used today to predict the production needs of large corporations and to determine the effect of price changes on the economy. In recognition of the importance of his ideas, Leontief was awarded the Nobel Prize in Economics in 1973.

This chapter begins the study of systems of equations.

SECTION **6.1**

Solving Systems of Linear Equations by Graphing

1 Solve systems of linear equations by graphing

Equations considered together are called a **system of equations.** A system of equations is shown at the right.

$$2x + y = 3$$
$$x + y = 1$$

A **solution of a system of equations in two variables** is an ordered pair that is a solution of each equation of the system.

For example, $(2, -1)$ is a solution of the system of equations given above because it is a solution of each equation in the system.

$2x + y = 3$		$x + y = 1$	
$2(2) + (-1)$	3	$2 + (-1)$	1
$4 + (-1)$	3	$1 = 1$	
$3 = 3$			

However, $(3, -3)$ is not a solution of this system because it is not a solution of each equation in the system.

$2x + y = 3$		$x + y = 1$	
$2(3) + (-3)$	3	$3 + (-3)$	1
$6 + (-3)$	3	$0 \neq 1$	
$3 = 3$			

Example 1 Is $(1, -3)$ a solution of the system $3x + 2y = -3$
$$x - 3y = 6?$$

Solution

$3x + 2y = -3$		$x - 3y = 6$	
$3 \cdot 1 + 2(-3)$	-3	$1 - 3(-3)$	6
$3 + (-6)$	-3	$1 - (-9)$	6
$-3 = -3$		$10 \neq 6$	

No, $(1, -3)$ is not a solution of the system of equations.

Problem 1 Is $(-1, -2)$ a solution of the system $2x - 5y = 8$
$$-x + 3y = -5?$$

Solution See page A27.

The solution of a system of linear equations in two variables can be found by graphing the two equations on the same coordinate system. Three possible conditions result.

1. The lines graphed can intersect at one point. The point of intersection of the lines is the ordered pair that is a solution of each equation of the system. It is the solution of the system of equations. The system of equations is **independent**.

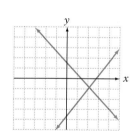

2. The lines graphed can be parallel and not intersect at all. The system of equations is **inconsistent** and has no solution.

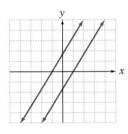

3. The lines graphed can represent the same line. The lines intersect at infinitely many points; therefore, there are infinitely many solutions. The system of equations is **dependent**. The solutions are the ordered pairs that are solutions of either one of the two equations in the system.

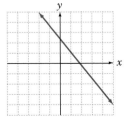

Solve by graphing: $2x + 3y = 6$
$2x + y = -2$

Graph each line.

The equations intersect at one point.
The system of equations is independent.

Find the point of intersection.

The solution is $(-3, 4)$.

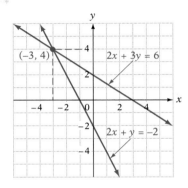

If a graphing utility is used to estimate the solution of the system of equations given above, the estimated solution might be $(-3.03, 4.05)$ or some other ordered pair that contains decimals. When these values are rounded to the nearest integer, the ordered pair becomes $(-3, 4)$. This ordered pair is the solution, which can be verified by replacing x by -3 and y by 4 in the system of equations.

Solve by graphing: $2x - y = 1$
$6x - 3y = 12$

Graph each line.

The lines are parallel and therefore do not intersect.

The system of equations is inconsistent and has no solution.

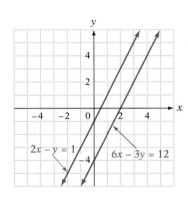

Solve by graphing: $2x + 3y = 6$
$6x + 9y = 18$

Graph each line.

The two equations represent the same line. The system of equations is dependent and, therefore, has an infinite number of solutions.

The solutions are the ordered pairs that are solutions of the equation $2x + 3y = 6$.

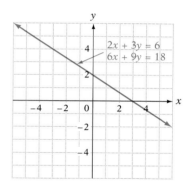

By choosing values for x, substituting these values into the equation $2x + 3y = 6$, and finding the corresponding values for y, we can find some specific ordered-pair solutions. For example, $(3, 0)$, $(0, 2)$, and $(6, -2)$ are solutions of this system of equations.

Example 2 Solve by graphing. A. $x - 2y = 2$ B. $4x - 2y = 6$
$x + y = 5$ $y = 2x - 3$

Solution A.

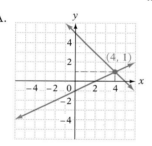

B.

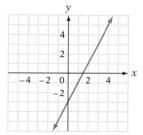

The solution is $(4, 1)$.

The system of equations is dependent. The solutions are the ordered pairs that satisfy the equation $y = 2x - 3$.

Problem 2 Solve by graphing. A. $x + 3y = 3$ B. $y = 3x - 1$
$-x + y = 5$ $6x - 2y = -6$

Solution See page A27.

CONCEPT REVIEW 6.1

Determine whether the statement is always true, sometimes true, or never true.

1. A solution of a system of linear equations in two variables is an ordered pair (x, y).

2. Graphically, the solution of an independent system of linear equations in two variables is the point of intersection of the graphs of the two equations.

3. If an ordered pair is a solution of one equation in a system of linear equations and not of the other equation, then the system has no solution.

4. Either a system of linear equations has one solution, represented graphically by two lines intersecting at exactly one point, or no solutions, represented graphically by two parallel lines.

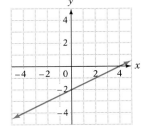

5. The graph at the right is the graph of a system of two linear equations.

6. An independent system of equations has no solution.

EXERCISES 6.1

1

1. Is $(2, 3)$ a solution of the system $3x + 4y = 18$
    $2x - y = 1$?

2. Is $(2, -1)$ a solution of the system $x - 2y = 4$
    $2x + y = 3$?

3. Is $(1, -2)$ a solution of the system
    $3x - y = 5$
    $2x + 5y = -8$?

4. Is $(-1, -1)$ a solution of the system
    $x - 4y = 3$
    $3x + y = 2$?

5. Is $(4, 3)$ a solution of the system $5x - 2y = 14$
    $x + y = 8$?

6. Is $(2, 5)$ a solution of the system $3x + 2y = 16$
    $2x - 3y = 4$?

7. Is $(-1, 3)$ a solution of the system $4x - y = -5$
    $2x + 5y = 13$?

8. Is $(4, -1)$ a solution of the system
    $x - 4y = 9$
    $2x - 3y = 11$?

9. Is $(0, 0)$ a solution of the system $4x + 3y = 0$
    $2x - y = 1$?

10. Is $(2, 0)$ a solution of the system $3x - y = 6$
    $x + 3y = 2$?

11. Is $(2, -3)$ a solution of the system
    $y = 2x - 7$
    $3x - y = 9$?

12. Is $(-1, -2)$ a solution of the system
    $3x - 4y = 5$
    $y = x - 1$?

13. Is $(5, 2)$ a solution of the system $y = 2x - 8$
    $y = 3x - 13$?

14. Is $(-4, 3)$ a solution of the system
    $y = 2x + 11$
    $y = 5x - 19$?

Solve by graphing.

15. $x - y = 3$
    $x + y = 5$

16. $2x - y = 4$
    $x + y = 5$

17. $x + 2y = 6$
    $x - y = 3$

18. $3x - y = 3$
    $2x + y = 2$

19. $3x - 2y = 6$
    $y = 3$

20. $x = 2$
    $3x + 2y = 4$

21. $x = 3$
    $y = -2$

22. $x + 1 = 0$
    $y - 3 = 0$

23. $y = 2x - 6$
    $x + y = 0$

24. $5x - 2y = 11$
        $y = 2x - 5$

25. $2x + y = -2$
    $6x + 3y = 6$

26. $x + y = 5$
    $3x + 3y = 6$

27. $4x - 2y = 4$
        $y = 2x - 2$

28. $2x + 6y = 6$
        $y = -\dfrac{1}{3}x + 1$

29. $x - y = 5$
    $2x - y = 6$

30. $5x - 2y = 10$
    $3x + 2y = 6$

31. $3x + 4y = 0$
    $2x - 5y = 0$

32. $2x - 3y = 0$
        $y = -\dfrac{1}{3}x$

33. $x - 3y = 3$
    $2x - 6y = 12$

34. $4x + 6y = 12$
    $6x + 9y = 18$

35. $3x + 2y = -4$
        $x = 2y + 4$

36. $5x + 2y = -14$
    $3x - 4y = 2$

37. $4x - y = 5$
    $3x - 2y = 5$

38. $2x - 3y = 9$
    $4x + 3y = -9$

Solve by graphing. Then use a graphing utility to verify your solution.

39. $5x - 2y = 10$
    $3x + 2y = 6$

40. $x - y = 5$
    $2x + y = 4$

41. $2x - 5y = 4$
        $x - y = -1$

42. $x - 2y = -5$
    $3x + 4y = -15$

43. $2x + 3y = 6$
        $y = -\dfrac{2}{3}x + 1$

44. $2x - 5y = 10$
        $y = \dfrac{2}{5}x - 2$

SUPPLEMENTAL EXERCISES 6.1

Write a system of equations given the graph.

45.

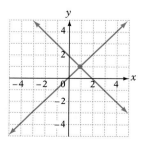

46.

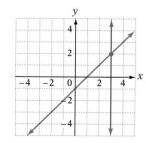

47.

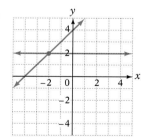

48.

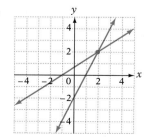

49. Determine whether the statement is always true, sometimes true, or never true.

   a. Two parallel lines have the same slope.

   b. Two different lines with the same y-intercept are parallel.

   c. Two different lines with the same slope are parallel.

50. Write three different systems of equations: **a.** one that has $(-3, 5)$ as its only solution, **b.** one for which there is no solution, and **c.** one that is a dependent system of equations.

51. Explain how you can determine from the graph of a system of two equations in two variables whether it is an independent system of equations. Explain how you can determine whether it is an inconsistent system of equations.

52. The following graph shows the life expectancy at birth for males and females. Write an essay describing your interpretation of the data presented. Be sure to include in your discussion an interpretation of the point at which the two lines intersect.

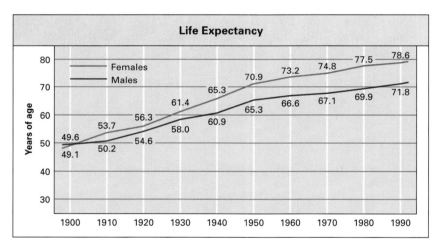

S E C T I O N   **6.2**

Solving Systems of Linear Equations by the Substitution Method

1   Solve systems of linear equations by the substitution method

A graphical solution of a system of equations may give only an approximate solution of the system. For example, the point whose coordinates are $\left(\dfrac{1}{4}, \dfrac{1}{2}\right)$ would be difficult to read from the graph. An algebraic method called the **substitution method** can be used to find an exact solution of a system.

In the system of equations at the right, equation (2) states that $y = 3x - 9$.

(1) $2x + 5y = -11$
(2) $y = 3x - 9$

Substitute $3x - 9$ for y in equation (1).

$2x + 5(3x - 9) = -11$

Solve for x.

$2x + 15x - 45 = -11$
$17x - 45 = -11$
$17x = 34$
$x = 2$

Substitute the value of x into equation (2) and solve for y.

(2)

$y = 3x - 9$
$y = 3 \cdot 2 - 9$
$y = 6 - 9$
$y = -3$

The solution is $(2, -3)$.

The graph of the equations in this system of equations is shown at the right. Note that the lines intersect at the point whose coordinates are $(2, -3)$, which is the algebraic solution we determined by the substitution method.

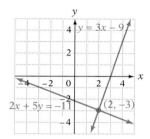

In the system of equations at the right, equation (1) is the easier equation to solve for one variable in terms of the other.

(1) $5x + y = 4$
(2) $2x - 3y = 5$

Solve equation (1) for y.

$5x + y = 4$
$y = -5x + 4$

Substitute $-5x + 4$ for y in equation (2). Solve for x.

$2x - 3(-5x + 4) = 5$
$2x + 15x - 12 = 5$
$17x - 12 = 5$
$17x = 17$
$x = 1$

LOOK CLOSELY

You can *always* check the solution to an independent system of equations. Use the skill you developed in Objective 6.1.1 to check that the ordered pair is a solution of each equation in the system.

Substitute the value of x in equation (1) and solve for y.

(1)

$5x + y = 4$
$5(1) + y = 4$
$5 + y = 4$
$y = -1$

The solution is $(1, -1)$.

Solve by substitution: $y = 3x - 1$ (1)
$y = -2x - 6$ (2)

Substitute $-2x - 6$ for y in equation (1).

$$y = 3x - 1$$
$$-2x - 6 = 3x - 1$$

Solve for x.

$$-5x - 6 = -1$$
$$-5x = 5$$
$$x = -1$$

Substitute the value of x into either equation and solve for y. Equation (1) is used here.

$$y = 3x - 1$$
$$y = 3(-1) - 1$$
$$y = -3 - 1$$
$$y = -4$$

The solution is $(-1, -4)$.

Example 1   Solve by substitution: $3x + 4y = -2$ (1)
$-x + 2y = 4$ (2)

Solution

$$-x + 2y = 4$$
$$-x = -2y + 4$$
$$x = 2y - 4$$

▶ Solve equation (2) for x.

$$3x + 4y = -2$$
$$3(2y - 4) + 4y = -2$$
$$6y - 12 + 4y = -2$$
$$10y - 12 = -2$$
$$10y = 10$$
$$y = 1$$

▶ Substitute $2y - 4$ for x in equation (1).
▶ Solve for y.

$$-x + 2y = 4$$
$$-x + 2(1) = 4$$
$$-x + 2 = 4$$
$$-x = 2$$
$$x = -2$$

▶ Substitute the value of y into equation (2).
▶ Solve for x.

The solution is $(-2, 1)$.

Problem 1   Solve by substitution: $7x - y = 4$
$3x + 2y = 9$

Solution   See page A27.

Example 2   Solve by substitution: $4x + 2y = 5$ (1)
$y = -2x + 1$ (2)

Solution

$$4x + 2y = 5$$
$$4x + 2(-2x + 1) = 5$$
$$4x - 4x + 2 = 5$$
$$2 = 5$$

▶ Substitute $-2x + 1$ for y in equation (1).

$2 = 5$ is not a true equation. The system of equations is inconsistent. The system does not have a solution.

LOOK CLOSELY

Solve equation (1) for y. The resulting equation is $y = -2x + \dfrac{5}{2}$. Thus both lines have the same slope, -2, but different y-intercepts. The lines are parallel.

Problem 2 Solve by substitution: $3x - y = 4$
$$y = 3x + 2$$

Solution See page A27.

Example 3 Solve by substitution: $6x - 2y = 4$ (1)
$$y = 3x - 2 \quad (2)$$

LOOK CLOSELY

Solve equation (1) for y. The resulting equation is $y = 3x - 2$, which is the same as equation (2).

Solution

$$6x - 2y = 4 \quad \blacktriangleright \text{Substitute } 3x - 2 \text{ for } y \text{ in equation (1).}$$
$$6x - 2(3x - 2) = 4$$
$$6x - 6x + 4 = 4$$
$$4 = 4$$

$4 = 4$ is a true equation. The system of equations is dependent. The solutions are the ordered pairs that satisfy the equation $y = 3x - 2$.

Problem 3 Solve by substitution: $y = -2x + 1$
$$6x + 3y = 3$$

Solution See page A27.

Note from Examples 2 and 3 above: If, when you are solving a system of equations, the variable is eliminated and the result is a false equation, such as $2 = 5$, the system is inconsistent and does not have a solution. If the result is a true equation, such as $4 = 4$, the system is dependent and has an infinite number of solutions.

CONCEPT REVIEW 6.2

Determine whether the statement is always true, sometimes true, or never true.

1. If one of the equations in a system of two linear equations is $y = x + 2$, then $x + 2$ can be substituted for y in the other equation in the system.

2. If a system of equations contains the equations $y = 2x + 1$ and $x + y = 5$, then $x + 2x + 1 = 5$.

3. If the equation $x = 4$ results from solving a system of equations by the substitution method, then the solution of the system of equations is 4.

4. If the true equation $6 = 6$ results from solving a system of equations by the substitution method, then the system of equations has an infinite number of solutions.

5. If the false equation $0 = 7$ results from solving a system of equations by the substitution method, then the system of equations is independent.

6. The ordered pair $(0, 0)$ is a solution of a system of linear equations.

EXERCISES 6.2

1 Solve by substitution.

1. $2x + 3y = 7$
$x = 2$

2. $y = 3$
$3x - 2y = 6$

3. $y = x - 3$
$x + y = 5$

4. $y = x + 2$
$x + y = 6$

5. $x = y - 2$
$x + 3y = 2$

6. $x = y + 1$
$x + 2y = 7$

7. $2x + 3y = 9$
$y = x - 2$

8. $3x + 2y = 11$
$y = x + 3$

9. $3x - y = 2$
$y = 2x - 1$

10. $2x - y = -5$
$y = x + 4$

11. $x = 2y - 3$
$2x - 3y = -5$

12. $x = 3y - 1$
$3x + 4y = 10$

13. $y = 4 - 3x$
$3x + y = 5$

14. $y = 2 - 3x$
$6x + 2y = 7$

15. $x = 3y + 3$
$2x - 6y = 12$

16. $x = 2 - y$
$3x + 3y = 6$

17. $3x + 5y = -6$
$x = 5y + 3$

18. $y = 2x + 3$
$4x - 3y = 1$

19. $x = 4y - 3$
$2x - 3y = 0$

20. $x = 2y$
$-2x + 4y = 6$

21. $y = 2x - 9$
$3x - y = 2$

22. $y = x + 4$
$2x - y = 6$

23. $3x + 4y = 7$
$x = 3 - 2y$

24. $3x - 5y = 1$
$y = 4x - 2$

25. $2x - y = 4$
$3x + 2y = 6$

26. $x + y = 12$
$3x - 2y = 6$

27. $4x - 3y = 5$
$x + 2y = 4$

28. $3x - 5y = 2$
$2x - y = 4$

29. $7x - y = 4$
$5x + 2y = 1$

30. $x - 7y = 4$
$-3x + 2y = 6$

31. $4x - 3y = -1$
$y = 2x - 3$

32. $3x - 7y = 28$
$x = 3 - 4y$

33. $7x + y = 14$
$2x - 5y = -33$

34. $3x + y = 4$
$4x - 3y = 1$

35. $x - 4y = 9$
$2x - 3y = 11$

36. $3x - y = 6$
$x + 3y = 2$

37. $4x - y = -5$
$2x + 5y = 13$

38. $3x - y = 5$
$2x + 5y = -8$

39. $3x + 4y = 18$
$2x - y = 1$

40. $4x + 3y = 0$
$2x - y = 0$

41. $5x + 2y = 0$
$x - 3y = 0$

42. $6x - 3y = 6$
$2x - y = 2$

43. $3x + y = 4$
$9x + 3y = 12$

44. $x - 5y = 6$
$2x - 7y = 9$

45. $x + 7y = -5$
$2x - 3y = 5$

46. $y = 2x + 11$
$y = 5x - 19$

47. $y = 2x - 8$
$y = 3x - 13$

48. $y = -4x + 2$
$y = -3x - 1$

49. $x = 3y + 7$
$x = 2y - 1$

50. $x = 4y - 2$
$x = 6y + 8$

51. $x = 3 - 2y$
$x = 5y - 10$

52. $y = 2x - 7$
$y = 4x + 5$

53. $3x - y = 11$
$2x + 5y = -4$

54. $-x + 6y = 8$
$2x + 5y = 1$

SUPPLEMENTAL EXERCISES 6.2

Rewrite each equation so that the coefficients and constant are integers. Then solve the system of equations.

55. $0.1x - 0.6y = -0.4$
$-0.7x + 0.2y = 0.5$

56. $0.8x - 0.1y = 0.3$
$0.5x - 0.2y = -0.5$

57. $0.4x + 0.5y = 0.2$
$0.3x - 0.1y = 1.1$

58. $-0.1x + 0.3y = 1.1$
$0.4x - 0.1y = -2.2$

59. $1.2x + 0.1y = 1.9$
$0.1x + 0.3y = 2.2$

60. $1.25x - 0.01y = 1.5$
$0.24x - 0.02y = -1.52$

For what value of k does the system of equations have no solution?

61. $2x - 3y = 7$
$kx - 3y = 4$

62. $8x - 4y = 1$
$2x - ky = 3$

63. $x = 4y + 4$
$kx - 8y = 4$

64. The following was offered as a solution to the system of equations shown at the right.

(1) $\quad y = \dfrac{1}{2}x + 2$

(2) $\quad 2x + 5y = 10$

$2x + 5y = 10 \quad$ ▶ Equation (2)

$2x + 5\left(\dfrac{1}{2}x + 2\right) = 10 \quad$ ▶ Substitute $\dfrac{1}{2}x + 2$ for y.

$2x + \dfrac{5}{2}x + 10 = 10 \quad$ ▶ Solve for x.

$\dfrac{9}{2}x = 0$

$x = 0$

At this point the student stated that because $x = 0$, the system of equations has no solution. If this assertion is correct, is the system of equations independent, dependent, or inconsistent? If the assertion is not correct, what is the correct solution?

65. Describe in your own words the process of solving a system of equations by the substitution method.

66. When you solve a system of equations by the substitution method, how do you determine whether the system of equations is dependent? How do you determine whether the system of equations is inconsistent?

Solving Systems of Linear Equations by the Addition Method

■ **1** Solve systems of linear equations by the addition method

LOOK CLOSELY

Equation (1) states $3x + 2y$ equals 4, and equation (2) states $4x - 2y$ equals 10. Thus adding equations (1) and (2) is like adding

$$
\begin{array}{r}
4 = 4 \\
+ \ 10 = 10 \\
\hline
14 = 14 \quad \text{A true equation}
\end{array}
$$

The addition method of solving a system of equations is based on adding the same number to each side of the equation.

Another algebraic method for solving a system of equations is called the **addition method.** It is based on the Addition Property of Equations.

In the system of equations at the right, note the effect of adding equation (2) to equation (1). Because $2y$ and $-2y$ are opposites, adding the equations results in an equation with only one variable.

$$
\begin{array}{ll}
(1) & 3x + 2y = 4 \\
(2) & 4x - 2y = 10 \\
& 7x + 0y = 14 \\
& 7x = 14
\end{array}
$$

The solution of the resulting equation is the first coordinate of the ordered-pair solution of the system.

$$
\begin{array}{l}
7x = 14 \\
x = 2
\end{array}
$$

The second coordinate is found by substituting the value of x into equation (1) or (2) and then solving for y. Equation (1) is used here.

$$
\begin{array}{ll}
(1) & 3x + 2y = 4 \\
& 3 \cdot 2 + 2y = 4 \\
& 6 + 2y = 4 \\
& 2y = -2 \\
& y = -1
\end{array}
$$

The solution is $(2, -1)$.

Sometimes adding the two equations does not eliminate one of the variables. In this case, use the Multiplication Property of Equations to rewrite one or both of the equations so that when the equations are added, one of the variables is eliminated.

To do this, first choose which variable to eliminate. The coefficients of that variable must be opposites. Multiply each equation by a constant that will produce coefficients that are opposites.

To eliminate y in the system of equations at the right, multiply each side of equation (1) by 2.

$$
\begin{array}{ll}
(1) & 3x + 2y = 7 \\
(2) & 5x - 4y = 19
\end{array}
$$

$$
\begin{array}{l}
2(3x + 2y) = 2 \cdot 7 \\
5x - 4y = 19
\end{array}
$$

Now the coefficients of the y terms are opposites.

$$
\begin{array}{l}
6x + 4y = 14 \\
5x - 4y = 19
\end{array}
$$

Add the equations.
Solve for x.

$$
\begin{array}{l}
11x + 0y = 33 \\
11x = 33 \\
x = 3
\end{array}
$$

Substitute the value of x into one of the equations and solve for y. Equation (2) is used here.

(2) $\quad 5x - 4y = 19$
$5 \cdot 3 - 4y = 19$
$15 - 4y = 19$
$-4y = 4$
$y = -1$

The solution is $(3, -1)$.

To eliminate x in the system of equations at the right, multiply each side of equation (1) by 2 and each side of equation (2) by -5.

(1) $\quad 5x + 6y = 3$
(2) $\quad 2x - 5y = 16$

Note how the constants are selected. The negative sign is used so that the coefficients will be opposites.

$2 \diagdown (5x + 6y) = 2 \cdot 3$
$-5 \diagdown (2x - 5y) = -5 \cdot 16$

Now the coefficients of the x terms are opposites.

$10x + 12y = 6$
$-10x + 25y = -80$

Add the equations.
Solve for y.

$0x + 37y = -74$
$37y = -74$
$y = -2$

Substitute the value of y into one of the equations and solve for x. Equation (1) is used here.

(1) $\quad 5x + 6y = 3$
$5x + 6(-2) = 3$
$5x - 12 = 3$
$5x = 15$
$x = 3$

The solution is $(3, -2)$.

To solve the system of equations at the right, first write equation (1) in the form $Ax + By = C$.

(1) $\quad 5x = 2y - 7$
(2) $\quad 3x + 4y = 1$

$5x - 2y = -7$
$3x + 4y = 1$

Eliminate y. Multiply each side of equation (1) by 2.

$2(5x - 2y) = 2(-7)$
$3x + 4y = 1$

Now the coefficients of the y terms are opposites.

$10x - 4y = -14$
$3x + 4y = 1$

Add the equations.
Solve for x.

$13x + 0y = -13$
$13x = -13$
$x = -1$

Substitute the value of x into one of the equations and solve for y. Equation (1) is used here.

$5x = 2y - 7$
$5(-1) = 2y - 7$
$-5 = 2y - 7$
$2 = 2y$
$1 = y$

The solution is $(-1, 1)$.

To eliminate y in the system of equations at the right, multiply each side of equation (1) by -2.

(1) $2x + y = 2$
(2) $4x + 2y = -5$

$$-4x - 2y = -4$$
$$4x + 2y = -5$$

Add the equations.
This is not a true equation.

$$0x + 0y = -9$$
$$0 = -9$$

The system of equations is inconsistent. The system does not have a solution.

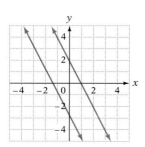

The graphs of the two equations in the system of equations above are shown at the left. Note that the graphs are parallel and therefore do not intersect. Thus the system of equations has no solutions.

Example 1   Solve by the addition method: $2x + 4y = 7$ (1)
                                                   $5x - 3y = -2$ (2)

Solution

$$5(2x + 4y) = 5 \cdot 7$$
$$-2(5x - 3y) = -2 \cdot (-2)$$

▶ Eliminate x. Multiply each side of equation (1) by 5 and each side of equation (2) by -2.

$$10x + 20y = 35$$
$$-10x + 6y = 4$$
$$26y = 39$$

▶ Add the equations.

$$y = \frac{39}{26} = \frac{3}{2}$$

▶ Solve for y.

$$2x + 4\left(\frac{3}{2}\right) = 7$$

▶ Substitute the value of y in equation (1).

$$2x + 6 = 7$$
$$2x = 1$$

▶ Solve for x.

$$x = \frac{1}{2}$$

The solution is $\left(\frac{1}{2}, \frac{3}{2}\right)$.

Problem 1   Solve by the addition method: $x - 2y = 1$
                                                  $2x + 4y = 0$

Solution   See page A28.

Example 2   Solve by the addition method: $6x + 9y = 15$ (1)
                                                $4x + 6y = 10$ (2)

Solution

$$4(6x + 9y) = 4 \cdot 15$$
$$-6(4x + 6y) = -6 \cdot 10$$

▶ Eliminate x. Multiply each side of equation (1) by 4 and each side of equation (2) by -6.

$$24x + 36y = 60$$
$$-24x - 36y = -60$$
$$0 = 0$$

▶ Add the equations.

$0 = 0$ is a true equation. The system of equations is dependent. The solutions are ordered pairs that satisfy the equation $6x + 9y = 15$.

> **Problem 2**   Solve by the addition method: $2x - 3y = 4$
> $-4x + 6y = -8$
>
> **Solution**   See page A28.

CONCEPT REVIEW 6.3

Determine whether the statement is always true, sometimes true, or never true.

1. When using the addition method of solving a system of linear equations, if you multiply one side of an equation by a number, then you must multiply the other side of the equation by the same number.

2. When a system of linear equations is being solved by the addition method, in order for one of the variables to be eliminated, the coefficients of one of the variables must be opposites.

3. If the two equations in a system of equations have the same slope and different y-intercepts, then the system of equations has an infinite number of solutions.

4. You are using the addition method to solve the system of $3x + 2y = 6$
    equations shown at the right. The first step you will per- $2x + 3y = -6$
    form is to add the two equations.

5. If the equation $y = -2$ results from solving a system of equations by the addition method, then the solution of the system of equations is -2.

6. If the true equation $-3 = -3$ results from solving a system of equations by the addition method, then the system of equations is dependent.

EXERCISES 6.3

1   Solve by the addition method.

1. $x + y = 4$
    $x - y = 6$

2. $2x + y = 3$
    $x - y = 3$

3. $x + y = 4$
    $2x + y = 5$

4. $x - 3y = 2$
    $x + 2y = -3$

5. $2x - y = 1$
    $x + 3y = 4$

6. $x - 2y = 4$
    $3x + 4y = 2$

7. $4x - 5y = 22$
    $x + 2y = -1$

8. $3x - y = 11$
    $2x + 5y = 13$

9. $2x - y = 1$
    $4x - 2y = 2$

10. $x + 3y = 2$
    $3x + 9y = 6$

11. $4x + 3y = 15$
    $2x - 5y = 1$

12. $3x - 7y = 13$
    $6x + 5y = 7$

13. $2x - 3y = 5$
     $4x - 6y = 3$

14. $2x + 4y = 3$
     $3x + 6y = 8$

15. $5x - 2y = -1$
     $x + 3y = -5$

16. $4x - 3y = 1$
     $8x + 5y = 13$

17. $5x + 7y = 10$
     $3x - 14y = 6$

18. $7x + 10y = 13$
     $4x + 5y = 6$

19. $3x - 2y = 0$
     $6x + 5y = 0$

20. $5x + 2y = 0$
     $3x + 5y = 0$

21. $2x - 3y = 16$
     $3x + 4y = 7$

22. $3x + 4y = 10$
     $4x + 3y = 11$

23. $5x + 3y = 7$
     $2x + 5y = 1$

24. $-2x + 7y = 9$
     $3x + 2y = -1$

25. $7x - 2y = 13$
     $5x + 3y = 27$

26. $12x + 5y = 23$
     $2x - 7y = 39$

27. $8x - 3y = 11$
     $6x - 5y = 11$

28. $4x - 8y = 36$
     $3x - 6y = 27$

29. $5x + 15y = 20$
     $2x + 6y = 8$

30. $y = 2x - 3$
     $4x + 4y = -1$

31. $3x = 2y + 7$
     $5x - 2y = 13$

32. $2y = 4 - 9x$
     $9x - y = 25$

33. $2x + 9y = 16$
     $5x = 1 - 3y$

34. $3x - 4 = y + 18$
     $4x + 5y = -21$

35. $2x + 3y = 7 - 2x$
     $7x + 2y = 9$

36. $5x - 3y = 3y + 4$
     $4x + 3y = 11$

37. $3x + y = 1$
     $5x + y = 2$

38. $2x - y = 1$
     $2x - 5y = -1$

39. $4x + 3y = 3$
     $x + 3y = 1$

40. $2x - 5y = 4$
     $x + 5y = 1$

41. $3x - 4y = 1$
     $4x + 3y = 1$

42. $2x - 7y = -17$
     $3x + 5y = 17$

43. $2x - 3y = 4$
     $-x + 4y = 3$

44. $4x - 2y = 5$
     $2x + 3y = 4$

SUPPLEMENTAL EXERCISES 6.3

Solve.

45. $x - 0.2y = 0.2$
     $0.2x + 0.5y = 2.2$

46. $0.5x - 1.2y = 0.3$
      $0.2x + y = 1.6$

47. $1.25x - 1.5y = -1.75$
     $2.5x - 1.75y = -1$

48. The point of intersection of the graphs of the equations $Ax + 2y = 2$ and $2x + By = 10$ is $(2, -2)$. Find A and B.

49. The point of intersection of the graphs of the equations $Ax - 4y = 9$ and $4x + By = -1$ is $(-1, -3)$. Find A and B.

50. Given that the graphs of the equations $2x - y = 6$, $3x - 4y = 4$, and $Ax - 2y = 0$ all intersect at the same point, find A.

51. Given that the graphs of the equations $3x - 2y = -2$, $2x - y = 0$, and $Ax + y = 8$ all intersect at the same point, find A.

52. For what value of k is the system of equations dependent?

a. $2x + 3y = 7$
     $4x + 6y = k$

b. $y = \dfrac{2}{3}x - 3$
     $y = kx - 3$

c. $x = ky - 1$
     $y = 2x + 2$

53. For what values of k is the system of equations independent?

a. $x + y = 7$
     $kx + y = 3$

b. $x + 2y = 4$
     $kx + 3y = 2$

c. $2x + ky = 1$
     $x + 2y = 2$

54. Describe in your own words the process of solving a system of equations by the addition method.

S E C T I O N **6.4**

Application Problems in Two Variables

1 Rate-of-wind and water-current problems

Solving motion problems that involve an object moving with or against a wind or current normally requires two variables. One variable represents the speed of the moving object in calm air or still water, and a second variable represents the rate of the wind or current.

A plane flying with the wind will travel a greater distance per hour than it would travel without the wind. The resulting rate of the plane is represented by the sum of the plane's speed and the rate of the wind.

A plane traveling against the wind, on the other hand, will travel a shorter distance per hour than it would travel without the wind. The resulting rate of the plane is represented by the difference between the plane's speed and the rate of the wind.

The same principle is used to describe the rate of a boat traveling with or against a water current.

Solve: Flying with the wind, a small plane can fly 750 mi in 3 h. Against the wind, the plane can fly the same distance in 5 h. Find the rate of the plane in calm air and the rate of the wind.

STRATEGY for solving rate-of-wind and water-current problems

■ Choose one variable to represent the rate of the object in calm conditions and a second variable to represent the rate of the wind or current. Using these variables, express the rate of the object with and against the wind or current. Use the equation $d = rt$ to write expressions for the distance traveled by the object. The results can be recorded in a table.

Rate of plane in calm air: p
Rate of wind: w

With wind: 750 mi in 3 h

Against wind: 750 mi in 5 h

	Rate	·	Time	=	Distance
With the wind	$p + w$	·	3	=	$3(p + w)$
Against the wind	$p - w$	·	5	=	$5(p - w)$

■ Determine how the expressions for distance are related.

The distance traveled with the wind is 750 mi. $3(p + w) = 750$
The distance traveled against the wind is 750 mi. $5(p - w) = 750$

Solve the system of equations.

$3(p + w) = 750$ ➡ $\dfrac{3(p + w)}{3} = \dfrac{750}{3}$ ➡ $p + w = 250$

$5(p - w) = 750$ ➡ $\dfrac{5(p - w)}{5} = \dfrac{750}{5}$ ➡ $p - w = 150$

$$2p = 400$$
$$p = 200$$

Substitute the value of p in the equation $p + w = 250$. $p + w = 250$
Solve for w. $200 + w = 250$
                                                               $w = 50$

The rate of the plane in calm air is 200 mph.
The rate of the wind is 50 mph.

Example 1   A 600 mi trip from one city to another takes 4 h when a plane is flying with the wind. The return trip against the wind takes 5 h. Find the rate of the plane in still air and the rate of the wind.

Strategy   With wind: 600 mi in 4 h

Against wind: 600 mi in 5 h

■ Rate of the plane in still air: p
Rate of the wind: w

	Rate	Time	Distance
With wind	$p + w$	4	$4(p + w)$
Against wind	$p - w$	5	$5(p - w)$

■ The distance traveled with the wind is 600 mi.
The distance traveled against the wind is 600 mi.

Solution $4(p + w) = 600$ (1)
$5(p - w) = 600$ (2)

$\dfrac{4(p + w)}{4} = \dfrac{600}{4}$ ▶ Simplify equation (1) by dividing each side of the equation by 4.

$\dfrac{5(p - w)}{5} = \dfrac{600}{5}$ ▶ Simplify equation (2) by dividing each side of the equation by 5.

$p + w = 150$
$p - w = 120$
$2p = 270$ ▶ Add the two equations.
$p = 135$ ▶ Solve for p, the rate of the plane in still air.

$p + w = 150$ ▶ Substitute the value of p into one of the equations.
$135 + w = 150$
$w = 15$ ▶ Solve for w, the rate of the wind.

The rate of the plane in still air is 135 mph.
The rate of the wind is 15 mph.

Problem 1   A canoeist paddling with the current can travel 24 mi in 3 h. Against the current, it takes 4 h to travel the same distance. Find the rate of the current and the rate of the canoeist in calm water.

Solution   See page A28.

2   Application problems

The application problems in this section are varieties of those problems solved earlier in the text. Each of the strategies for the problems in this section will result in a system of equations.

Solve: A jeweler purchased 5 oz of a gold alloy and 20 oz of a silver alloy for a total cost of $700. The next day, at the same prices per ounce, the jeweler purchased 4 oz of the gold alloy and 30 oz of the silver alloy for a total cost of $630. Find the cost per ounce of the silver alloy.

STRATEGY *for solving an application problem in two variables*

■ Choose one variable to represent one of the unknown quantities and a second variable to represent the other unknown quantity. Write numerical or variable expressions for all the remaining quantities. These results can be recorded in two tables, one for each of the conditions.

Cost per ounce of gold: g
Cost per ounce of silver: s

First day

	Amount	·	Unit cost	=	Value
Gold	5	·	g	=	$5g$
Silver	20	·	s	=	$20s$

Second day

	Amount	·	Unit cost	=	Value
Gold	4	·	g	=	$4g$
Silver	30	·	s	=	$30s$

■ Determine a system of equations. The strategies presented in Chapter 4 can be used to determine the relationships between the expressions in the tables. Each table will give one equation of the system.

The total value of the purchase on the first day was $700. $5g + 20s = 700$
The total value of the purchase on the second day was $630. $4g + 30s = 630$

Solve the system of equations.

$5g + 20s = 700$ $4(5g + 20s) = 4 \cdot 700$ $20g + 80s = 2800$
$4g + 30s = 630$ ➡ $-5(4g + 30s) = -5 \cdot 630$ ➡ $-20g - 150s = -3150$
                                                                        $-70s = -350$
                                                                            $s = 5$

The cost per ounce of the silver alloy was $5.

Example 2   A store owner purchased 20 incandescent light bulbs and 30 fluorescent bulbs for a total cost of $40. A second purchase, at the same prices, included 30 incandescent bulbs and 10 fluorescent bulbs for a total cost of $25. Find the cost of an incandescent bulb and of a fluorescent bulb.

Strategy   ■ Cost of an incandescent bulb: I
Cost of a fluorescent bulb: F

First purchase

	Amount	Unit cost	Value
Incandescent	20	I	$20I$
Fluorescent	30	F	$30F$

Second purchase

	Amount	Unit cost	Value
Incandescent	30	I	$30I$
Fluorescent	10	F	$10F$

■ The total of the first purchase was $40.
The total of the second purchase was $25.

Solution

$20I + 30F = 40$ (1)
$30I + 10F = 25$ (2)

$3(20I + 30F) = 3(40)$
$-2(30I + 10F) = -2(25)$

▸ Eliminate I. Multiply each side of equation (1) by 3 and each side of equation (2) by -2.

$60I + 90F = 120$
$-60I - 20F = -50$
$70F = 70$
$F = 1$

▸ Add the two equations.
▸ Solve for F, the cost of a fluorescent bulb.

$20I + 30F = 40$
$20I + 30(1) = 40$
$20I = 10$

$I = \dfrac{1}{2}$

▸ Substitute the value of F into one of the equations.
▸ Solve for I.

The cost of an incandescent bulb was $.50.
The cost of a fluorescent bulb was $1.00.

Problem 2 Two coin banks contain only dimes and quarters. In the first bank, the total value of the coins is $3.90. In the second bank, there are twice as many dimes as in the first bank and one-half the number of quarters. The total value of the coins in the second bank is $3.30. Find the number of dimes and the number of quarters in the first bank.

Solution See page A29.

CONCEPT REVIEW 6.4

Determine whether the statement is always true, sometimes true, or never true.

1. A plane flying with the wind is traveling faster than it would be traveling without the wind.

2. The uniform motion equation $r = dt$ is used to solve rate-of-wind and water-current problems.

3. If b represents the rate of a boat in calm water and c represents the rate of the water current, then $b + c$ represents the rate of the boat while it is traveling against the current.

4. Both sides of an equation can be divided by the same number without changing the solution of the equation.

5. If, in a system of equations, p represents the rate of a plane in calm air and w represents the rate of the wind, and the solution of the system is $p = 100$, this means that the rate of the wind is 100.

6. The system of equations at the right represents the following problem:

$$2(p + w) = 600$$
$$3(p - w) = 600$$

A plane flying with the wind flew 600 mi in 2 h. Flying against the wind, the plane could fly the same distance in 3 h. Find the rate of the plane in calm air.

EXERCISES 6.4

1   Solve.

1. A plane flying with the jet stream flew from Los Angeles to Chicago, a distance of 2250 mi, in 5 h. Flying against the jet stream, the plane could fly only 1750 mi in the same amount of time. Find the rate of the plane in calm air and the rate of the wind.

2. A rowing team rowing with the current traveled 40 km in 2 h. Rowing against the current, the team could travel only 16 km in 2 h. Find the team's rowing rate in calm water and the rate of the current.

3. A motorboat traveling with the current went 35 mi in 3.5 h. Traveling against the current, the boat went 12 mi in 3 h. Find the rate of the boat in calm water and the rate of the current.

4. A small plane, flying into a headwind, flew 270 mi in 3 h. Flying with the wind, the plane traveled 260 mi in 2 h. Find the rate of the plane in calm air and the rate of the wind.

5. A plane flying with a tailwind flew 300 mi in 2 h. Against the wind, it took 3 h to travel the same distance. Find the rate of the plane in calm air and the rate of the wind.

6. A rowing team rowing with the current traveled 18 mi in 2 h. Against the current, the team rowed a distance of 8 mi in the same amount of time. Find the rate of the rowing team in calm water and the rate of the current.

7. A seaplane flying with the wind flew from an ocean port to a lake, a distance of 240 mi, in 2 h. Flying against the wind, it made the trip from the lake to the ocean port in 3 h. Find the rate of the plane in calm air and the rate of the wind.

8. Rowing with the current, a canoeist paddled 14 mi in 2 h. Against the current, the canoeist could paddle only 10 mi in the same amount of time. Find the rate of the canoeist in calm water and the rate of the current.

9. Flying with the wind, a small plane flew 280 mi in 2 h. Flying against the wind, the plane flew 160 mi in 2 h. Find the rate of the plane in calm air and the rate of the wind.

10. With the wind, a quarterback passes a football 140 ft in 2 s. Against the wind, the same pass would have traveled 80 ft in 2 s. Find the rate of the pass and the rate of the wind.

11. Flying with the wind, a plane flew 1000 km in 5 h. Against the wind, the plane could fly only 800 km in the same amount of time. Find the rate of the plane in calm air and the rate of the wind.

12. Traveling with the current, a cruise ship sailed between two islands, a distance of 90 mi, in 3 h. The return trip against the current required 4 h and 30 min. Find the rate of the cruise ship in calm water and the rate of the current.

2 Solve.

13. The manager of a computer software store received two shipments of software. The cost of the first shipment, which contained 12 identical word processing programs and 10 identical spreadsheet programs, was $6190. The second shipment, at the same prices, contained 5 copies of the word processing program and 8 copies of the spreadsheet program. The cost of the second shipment was $3825. Find the cost for one copy of the word processing program.

14. A baker purchased 12 lb of wheat flour and 15 lb of rye flour for a total cost of $18.30. A second purchase, at the same prices, included 15 lb of wheat flour and 10 lb of rye flour. The cost of the second purchase was $16.75. Find the cost per pound of the wheat and rye flours.

15. An investor owned 300 shares of an oil company and 200 shares of a movie company. The quarterly dividend from the two stocks was $165. After the investor sold 100 shares of the oil company and bought an additional 100 shares of the movie company, the quarterly dividend was $185. Find the dividend per share for each stock.

16. For using a computerized financial news network for 25 min during prime time and 35 min during non-prime time, a customer was charged $10.75. A second customer was charged $13.35 for using the network for 30 min of prime time and 45 min of non-prime time. Find the cost per minute for using the financial news network during prime time.

17. A basketball team scored 87 points in two-point baskets and three-point baskets. If the two-point baskets had been three-point baskets and the three-point baskets had been two-point baskets, the team would have scored 93 points. Find how many two-point baskets and how many three-point baskets the team scored.

18. A football team scored 30 points in one game with only touchdowns and field goals. If the touchdowns had been field goals and the field goals had been touchdowns, the score would have been 33 points. Find the number of touchdowns and the number of field goals scored. Use 6 points for a touchdown and 3 points for a field goal.

19. Two coin banks contain only nickels and quarters. The total value of the coins in the first bank is $2.90. In the second bank, there are two more quarters than in the first bank and twice as many nickels. The total value of the coins in the second bank is $3.80. Find the number of nickels and the number of quarters in the first bank.

20. Two coin banks contain only nickels and dimes. The total value of the coins in the first bank is $3. In the second bank, there are 4 more nickels than in the first bank and one-half as many dimes. The total value of the coins in the second bank is $2. Find the number of nickels and the number of dimes in the first bank.

21. The total value of the dimes and quarters in a coin bank is $3.30. If the quarters were dimes and the dimes were quarters, the total value of the coins would be $3. Find the number of dimes and the number of quarters in the bank.

22. The total value of the nickels and dimes in a coin bank is $3. If the nickels were dimes and the dimes were nickels, the total value of the coins would be $3.75. Find the number of nickels and the number of dimes in the bank.

SUPPLEMENTAL EXERCISES 6.4

Solve.

23. Two angles are supplementary. The larger angle is 15° more than twice the measure of the smaller angle. Find the measure of the two angles. (Supplementary angles are two angles whose sum is 180°.)

24. Two angles are complementary. The larger angle is four times the measure of the smaller angle. Find the measure of the two angles. (Complementary angles are two angles whose sum is 90°.)

25. An investment club placed a portion of its funds in a 9% annual simple interest account and the remainder in an 8% annual simple interest account. The amount of interest earned for one year was $860. If the amounts placed in each account had been reversed, the interest earned would have been $840. How much was invested in each account?

26. An investor has $5000 to invest in two accounts. The first account earns 8% annual simple interest, and the second account earns 10% annual simple interest. How much money should be invested in each account so that the annual interest earned is $600?

27. The value of the nickels and dimes in a coin bank is $.25. If the number of nickels and the number of dimes were doubled, the value of the coins would be $.50. How many nickels and dimes are in the bank?

28. A coin bank contains nickels and dimes, but there are no more than 27 coins. The value of the coins is $2.10. How many different combinations of nickels and/or dimes could be in the bank?

29. The coin and stamp problems in Chapter 4 can be solved using a system of equations. For example, look at the problem on page 143. Let q represent the number of quarters in the bank, and let d represent the number of dimes. There is a total of 9 coins, so $q + d = 9$. The total value of the money in the bank is $1.20, so we can write the equation $25q + 10d = 120$. The solution of the system of equations

$$q + d = 9$$
$$25q + 10d = 120$$

is (2, 7), so there are 2 quarters and 7 dimes in the bank. Explain how to use a system of equations to solve the investment problems in Chapter 4. You might use the problem on page 161 as a basis for your explanation.

Project in Mathematics

Break-Even Analysis

Break-even analysis is a method used to determine the sales volume required for a company to break even, or experience neither a profit nor a loss, on the sale of its product. The break-even point represents the number of units that must be made and sold in order for income from sales to equal the cost of the product.

The break-even point can be determined by graphing two equations on the same coordinate grid. The first equation is $R = SN$, where R is the revenue earned, S is the selling price per unit, and N is the number of units sold. The second equation is $T = VN + F$, where T is the total cost, F is the fixed costs, V is the variable costs per unit, and N is the number of units sold. The break-even point is the point where the graphs of the two equations intersect, which is the point where revenue is equal to cost.

Solve.

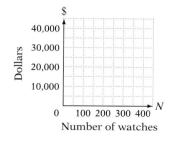

1. **a.** A company manufactures and sells digital watches. The fixed costs are $20,000, the variable costs per unit are $25, and the selling price per watch is $125. Write two equations for this information.

    b. Graph the two equations on the same coordinate grid, using only quadrant I. The horizontal axis is the number of units sold. Use the model shown at the left.

    c. How many watches must the company sell in order to break even?

2. **a.** A company manufactures and sells basketballs. The fixed costs are $12,000, the variable costs per unit are $10, and the selling price per basketball is $50. Write two equations for this information.

    b. Graph the two equations on the same coordinate grid, using only quadrant I. The horizontal axis is the number of units sold. Use the model graph shown in Exercise 1, changing the label on the horizontal axis to "number of basketballs."

    c. How many basketballs must the company sell in order to break even?

3. **a.** A company manufactures and sells calculators. The fixed costs are $10,000, the variable costs per unit are $5, and the selling price per calculator is $25. How many calculators must the company sell in order to experience neither a profit nor a loss?

    b. If the company changes the selling price per calculator to $30, what is the break-even point?

Focus on Problem Solving

Calculators

A calculator is an important tool of problem solving. It can be used as an aid to guessing or estimating a solution to a problem. Here are a few problems to solve with a calculator.

1. Choose any positive integer less than 9. Multiply the number by 1507. Now multiply the result by 7519. What is the answer? Choose another positive single-digit number and again multiply by 1507 and 7519. What is the answer? What pattern do you see? Why does this work?

2. Are there enough people in the United States so that if they held hands in a line, they would stretch around the world at the equator? To answer this question, begin by determining what information you need. What assumptions must you make?

3. Which of the reciprocals of the first 16 natural numbers have a terminating decimal representation and which have a repeating decimal representation?

4. What is the largest natural number n for which $4^n > 1 \cdot 2 \cdot 3 \cdots \cdot n$?

5. Calculate 15^2, 35^2, 65^2, and 85^2. Study the results. Make a conjecture about a relationship between a number ending in 5 and its square. Use your conjecture to find 75^2 and 95^2. Does your conjecture work for 125^2?

6. Find the sum of the first 1000 natural numbers. (*Hint:* You could just start adding $1 + 2 + 3 + 4 + \cdots$, but even if you performed one operation each second, it would take over 15 minutes to find the sum. Instead, try pairing the numbers and then adding the numbers in each pair. Pair 1 and 1000, 2 and 999, 3 and 998, and so on. What is the sum of each pair? How many pairs are there? Use this information to answer the original question.)

7. For a borrower to qualify for a home loan, a bank requires that the monthly mortgage payment be less than 25% of the borrower's monthly take-home income. A laboratory technician has deductions for taxes, insurance, and retirement that amount to 25% of the technician's monthly gross income. What minimum gross monthly income must this technician earn to receive a bank loan that has a mortgage payment of $1200 per month?

A calculator can also be used to solve a system of equations. By using the addition method, it is possible to solve the system of equations

$$ax + by = c$$
$$dx + ey = f$$

The solution is

$$x = \frac{ce - bf}{ae - bd} \quad \text{and} \quad y = \frac{af - cd}{ae - bd}$$

where $ae - bd \neq 0$. The denominators of these expressions are identical. Perform the calculation of the denominator first, and store this number in the calculator's memory. Here is an example.

Solve: $2x - 5y = 9$
           $4x + 3y = 2$

Make a list of the values of $a, b, c, d, e,$ and f.	$a = 2 \qquad b = -5 \qquad c = 9$ $d = 4 \qquad e = 3 \qquad f = 2$
Calculate the denominator, D, and store the result in memory.	$D = ae - bd = 2(3) - (-5)(4) = 26$
To find x, calculate the numerator. Divide the result by the number in memory. To the nearest hundredth, the display should read 1.42.	$x = \frac{ce - bf}{D} = \frac{9(3) - (-5)(2)}{26}$
To find y, calculate the numerator. Divide the result by the number in memory. To the nearest hundredth, the display should read -1.23.	$y = \frac{af - cd}{D} = \frac{2(2) - 9(4)}{26}$

The solution of the system is approximately $(1.42, -1.23)$.

If the value of the denominator D is zero, the system of equations is dependent or inconsistent, and this calculator method cannot be used.

Use the calculator method to solve the system of equations. Round to the nearest hundredth.

8. $2x - 5y = 1$
    $7x + 3y = 2$

9. $6x + 5y = 3$
    $8x - 3y = 7$

Chapter Summary

Key Words

Equations considered together are called a *system of equations.* (Objective 6.1.1)

An example of a system of equations is
$$4x + y = 6$$
$$3x + 2y = 7$$

A *solution of a system of equations in two variables* is an ordered pair that is a solution of each equation of the system. (Objective 6.1.1)

The solution of the system of equations shown above is the ordered pair (1, 2) because it is a solution of each equation in the system.

An *independent system of equations* has one solution. The graphs of the equations in an independent system of linear equations intersect at one point. (Objective 6.1.1)

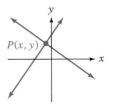

An *inconsistent system of equations* has no solution. The graphs of the equations in an inconsistent system of linear equations are parallel lines. If, when you are solving a system of equations algebraically, the variable is eliminated and the result is a false equation, such as $-3 = 8$, the system is inconsistent. (Objective 6.1.1/6.2.1/6.3.1)

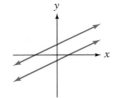

A *dependent system of equations* has an infinite number of solutions. The graphs of the equations in a dependent system of linear equations represent the same line. If, when you are solving a system of equations algebraically, the variable is eliminated and the result is a true equation, such as $1 = 1$, the system is dependent. (Objective 6.1.1/6.2.1/6.3.1)

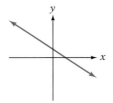

Essential Rules and Procedures

To solve a system of linear equations in two variables by graphing, graph each equation on the same coordinate system. If the lines graphed intersect at one point, the point of intersection is the ordered pair that is a solution of the system. If the lines graphed are parallel, the system is inconsistent. If the lines represent the same line, the system is dependent. (Objective 6.1.1)

Solve by graphing: $x + 2y = 4$
$$2x + y = -1$$

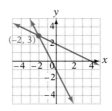

The solution is $(-2, 3)$.

To solve a system of linear equations by the substitution method, one variable must be written in terms of the other variable.   (Objective 6.2.1)

Solve by substitution: $2x - y = 5$ (1)
$3x + y = 5$ (2)

$3x + y = 5$ ▶ Solve equation (2)
$\quad\quad y = -3x + 5$ for y.

$\quad\quad\quad\quad 2x - y = 5$ ▶ Substitute for y in
$2x - (-3x + 5) = 5$ equation (1).
$\quad\quad 2x + 3x - 5 = 5$
$\quad\quad\quad\quad\quad\quad 5x = 10$
$\quad\quad\quad\quad\quad\quad\ x = 2$

$y = -3x + 5$ ▶ Substitute the value
$y = -3(2) + 5$ of x to find y.
$y = -1$

The solution is $(2, -1)$.

To solve a system of linear equations by the addition method, use the Multiplication Property of Equations to rewrite one or both of the equations so that the coefficients of one variable are opposites. Then add the two equations and solve for the variables.   (Objective 6.3.1)

Solve by the addition method:
$\ 3x + y = 4$ (1)
$\quad x + y = 6$ (2)

$\ 3x + y = 4$
$-x - y = -6$ ▶ Multiply both sides of
$\quad\quad\quad\quad\quad\quad\quad\quad$ equation (2) by -1.
$\quad\quad\quad 2x = -2$ ▶ Add the two equations.
$\quad\quad\quad\ x = -1$ ▶ Solve for x.

$\quad x + y = 6$ ▶ Substitute the value of x
$-1 + y = 6$ to find y.
$\quad\quad\quad y = 7$

The solution is $(-1, 7)$.

Chapter Review Exercises

1. Solve by substitution: $4x + 7y = 3$
$x = y - 2$

2. Solve by graphing: $3x + y = 3$
$x = 2$

3. Solve by the addition method: $3x + 8y = -1$
$x - 2y = -5$

4. Solve by substitution: $8x - y = 2$
$y = 5x + 1$

5. Solve by graphing: $x + y = 2$
$x - y = 0$

6. Solve by the addition method: $4x - y = 9$
$2x + 3y = -13$

7. Solve by substitution: $-2x + y = -4$
$x = y + 1$

8. Solve by the addition method:
$8x - y = 25$
$32x - 4y = 100$

9. Solve by the addition method: $5x - 15y = 30$
$2x + 6y = 0$

10. Solve by graphing: $3x - y = 6$
$y = -3$

11. Solve by the addition method: $7x - 2y = 0$
$2x + y = -11$

12. Solve by substitution: $x - 5y = 4$
$y = x - 4$

13. Is $(-1, -3)$ a solution of the system
$5x + 4y = -17$
$2x - y = 1$?

14. Solve by the addition method:
$6x + 4y = -3$
$12x - 10y = -15$

15. Solve by the addition method: $5x + 2y = -9$
$12x - 7y = 2$

16. Solve by graphing: $x - 3y = 12$
$y = x - 6$

17. Solve by the addition method:
$5x + 7y = 21$
$20x + 28y = 63$

18. Solve by substitution: $9x + 12y = -1$
$x - 4y = -1$

19. Solve by graphing: $4x - 2y = 8$
$y = 2x - 4$

20. Solve by the addition method:
$3x + y = -2$
$-9x - 3y = 6$

21. Solve by the addition method: $11x - 2y = 4$
$25x - 4y = 2$

22. Solve by substitution: $4x + 3y = 12$
$y = -\dfrac{4}{3}x + 4$

23. Solve by graphing: $y = -\dfrac{1}{4}x + 3$
$2x - y = 6$

24. Solve by the addition method: $2x - y = 5$
$10x - 5y = 20$

25. Solve by substitution: $6x + 5y = -2$
$y = 2x - 2$

26. Is $(-2, 0)$ a solution of the system
$-x + 9y = 2$
$6x - 4y = 12$?

27. Solve by the addition method: $6x - 18y = 7$
$9x + 24y = 2$

28. Solve by substitution: $12x - 9y = 18$
$y = \dfrac{4}{3}x - 3$

29. Solve by substitution: $9x - y = -3$
$18x - y = 0$

30. Solve by graphing: $x + 2y = 3$
$y = -\dfrac{1}{2}x + 1$

31. Solve by the addition method: $7x - 9y = 9$
$3x - y = 1$

32. Solve by substitution: $7x + 3y = -16$
$x - 2y = 5$

33. Solve by substitution: $5x - 3y = 6$
$x - y = 2$

34. Solve by the addition method: $6x + y = 12$
$9x + 2y = 18$

35. Solve by the addition method: $5x + 12y = 4$
$x + 6y = 8$

36. Solve by substitution: $6x - y = 0$
$7x - y = 1$

37. Flying with the wind, a plane can travel 800 mi in 4 h. Against the wind, the plane requires 5 h to fly the same distance. Find the rate of the plane in calm air and the rate of the wind.

38. Admission to a movie theater is $7 for adults and $5 for children. If the receipts from 200 tickets were $1120, how many adults' tickets and how many children's tickets were sold?

39. A canoeist traveling with the current traveled the 30 mi between two riverside campsites in 3 h. The return trip took 5 h. Find the rate of the canoeist in still water and the rate of the current.

40. A small wood carving company mailed 190 advertisements, some requiring 32¢ in postage and others requiring 55¢ in postage. If the total cost for the mailings was $74.60, how many mailings that required 32¢ were sent?

41. A boat traveling with the current went 48 km in 3 h. Against the current, the boat traveled 24 km in 2 h. Find the rate of the boat in calm water and the rate of the current.

42. A music shop sells some compact discs for $15 and some for $10. A customer spent $120 on 10 compact discs. How many at each price did the customer purchase?

43. With a tailwind, a flight crew flew 420 km in 3 h. Flying against the tailwind, the crew flew 440 km in 4 h. Find the rate of the plane in calm air and the rate of the wind.

44. A paddle boat can travel 4 mi downstream in 1 h. Paddling upstream, against the current, the paddle boat traveled 2 mi in 1 h. Find the rate of the boat in calm water and the rate of the current.

45. A silo contains a mixture of lentils and corn. If 50 bushels of lentils were added to this mixture, there would be twice as many bushels of lentils as corn. If 150 bushels of corn were added to the original mixture, there would be the same amount of corn as lentils. How many bushels of each are in the silo?

46. Flying with the wind, a small plane flew 360 mi in 3 h. Against the wind, the plane took 4 h to fly the same distance. Find the rate of the plane in calm air and the rate of the wind.

47. A coin purse contains 80¢ in nickels and dimes. If there are 12 coins in all, how many dimes are in the coin purse?

48. A pilot flying with the wind flew 2100 mi from one city to another in 6 h. The return trip against the wind took 7 h. Find the rate of the plane in calm air and the rate of the wind.

49. An investor buys 1500 shares of stock, some costing $6 per share and the rest costing $25 per share. If the total cost of the stock is $12,800, how many shares of each did the investor buy?

50. Rowing with the wind, a sculling team went 24 mi in 2 h. Rowing against the current, the team went 18 mi in 3 h. Find the rate of the sculling team in calm water and the rate of the current.

Chapter Test

1. Solve by substitution: $4x - y = 11$
$$y = 2x - 5$$

2. Solve by the addition method: $4x + 3y = 11$
$$5x - 3y = 7$$

3. Is $(-2, 3)$ a solution of the system
$2x + 5y = 11$
$x + 3y = 7$?

4. Solve by substitution: $x = 2y + 3$
$$3x - 2y = 5$$

5. Solve by the addition method: $2x - 5y = 6$
$$4x + 3y = -1$$

6. Solve by graphing: $3x + 2y = 6$
$$5x + 2y = 2$$

7. Solve by substitution: $4x + 2y = 3$
$$y = -2x + 1$$

8. Solve by substitution: $3x + 5y = 1$
$$2x - y = 5$$

9. Solve by the addition method: $7x + 3y = 11$
$$2x - 5y = 9$$

10. Solve by substitution: $3x - 5y = 13$
$$x + 3y = 1$$

11. Solve by the addition method: $5x + 6y = -7$
$$3x + 4y = -5$$

12. Is $(2, 1)$ a solution of the system $3x - 2y = 8$
$$4x + 5y = 3?$$

13. Solve by substitution: $3x - y = 5$
$$y = 2x - 3$$

14. Solve by the addition method: $3x + 2y = 2$
$$5x - 2y = 14$$

15. Solve by graphing: $3x + 2y = 6$
$$3x - 2y = 6$$

16. Solve by substitution: $x = 3y + 1$
$$2x + 5y = 13$$

17. Solve by the addition method: $5x + 4y = 7$
$$3x - 2y = 13$$

18. Solve by graphing: $3x + 6y = 2$
$$y = -\frac{1}{2}x + \frac{1}{3}$$

19. Solve by substitution: $4x - 3y = 1$
$$2x + y = 3$$

20. Solve by the addition method: $5x - 3y = 29$
$$4x + 7y = -5$$

21. Solve by substitution: $3x - 5y = -23$
$$x + 2y = -4$$

22. Solve by the addition method: $9x - 2y = 17$
$$5x + 3y = -7$$

23. With the wind, a plane flies 240 mi in 2 h. Against the wind, the plane requires 3 h to fly the same distance. Find the rate of the plane in calm air and the rate of the wind.

24. With the current, a motorboat can travel 48 mi in 3 h. Against the current, the boat requires 4 h to travel the same distance. Find the rate of the boat in calm water and the rate of the current.

25. Two coin banks contain only dimes and nickels. In the first bank, the total value of the coins is $5.50. In the second bank, there are one-half as many dimes as in the first bank and 10 fewer nickels. The total value of the coins in the second bank is $3. Find the number of dimes and the number of nickels in the first bank.

Cumulative Review Exercises

1. Given $A = \{-8, -4, 0\}$, which elements of set A are less than or equal to -4?

2. Use the roster method to write the set of positive integers less than or equal to 10.

3. Simplify: $12 - 2(7 - 5)^2 \div 4$

4. Simplify: $2[5a - 3(2 - 5a) - 8]$

5. Evaluate $\dfrac{a^2 - b^2}{2a}$ when $a = 4$ and $b = -2$.

6. Solve: $-\dfrac{3}{4}x = \dfrac{9}{8}$

7. Solve: $4 - 3(2 - 3x) = 7x - 9$

8. Solve: $3[2 - 4(x + 1)] = 6x - 2$

9. Solve: $-7x - 5 > 4x + 50$

10. Solve: $5 + 2(x + 1) \le 13$

11. What percent of 50 is 12?

12. Find the x- and y-intercepts of $3x - 6y = 12$.

13. Find the slope of the line that contains the points whose coordinates are $(2, -3)$ and $(-3, 4)$.

14. Find the equation of the line that contains the point whose coordinates are $(-2, 3)$ and has slope $-\dfrac{3}{2}$.

15. Graph: $3x - 2y = 6$

16. Graph: $y = -\dfrac{1}{3}x + 3$

17. Graph the solution set of $y > -3x + 4$.

18. Graph: $f(x) = \dfrac{3}{4}x + 2$

19. Find the domain and range of the relation $\{(-5, 5), (0, 5), (1, 5), (5, 5)\}$. Is the relation a function?

20. Evaluate $f(x) = -2x - 5$ at $x = -4$.

21. Is $(2, 0)$ a solution of the system $5x - 3y = 10$
$4x + 7y = 8$?

22. Solve by substitution: $2x - 3y = -7$
$x + 4y = 2$

23. Solve by graphing: $2x + 3y = 6$
$3x + y = 2$

24. Solve by the addition method: $5x - 2y = 8$
$4x + 3y = 11$

25. Find the range of the function given by the equation $f(x) = 5x - 2$ if the domain is $\{-4, -2, 0, 2, 4\}$.

26. A business manager has determined that the cost per unit for a camera is $90 and that the fixed costs per month are $3500. Find the number of cameras produced during a month in which the total cost was $21,500. Use the equation $T = U \cdot N + F$, where T is the total cost, U is the cost per unit, N is the number of units produced, and F is the fixed cost.

27. A total of $8750 is invested in two simple interest accounts. On one account, the annual simple interest rate is 9.6%. On the second account, the annual simple interest rate is 7.2%. How much should be invested in each account so that both accounts earn the same amount of interest?

28. A plane can travel 160 mph in calm air. Flying with the wind, the plane can fly 570 mi in the same amount of time that it takes to fly 390 mi against the wind. Find the rate of the wind.

29. With the current, a motorboat can travel 24 mi in 2 h. Against the current, the boat requires 3 h to travel the same distance. Find the rate of the boat in calm water.

30. Two coin banks contain only dimes and nickels. In the first bank, the total value of the coins is $5.25. In the second bank, there are one-half as many dimes as in the first bank and 15 fewer nickels. The total value of the coins in the second bank is $2.50. Find the number of dimes in the first bank.

7

Polynomials

Objectives

$$433$$
$$-225$$
$$4\overset{2\ 13}{\cancel{3}}3$$

$$-225$$
$$208$$

$$8 \quad + \quad 200 \quad = \quad 208$$

Early Egyptian Arithmetic Operations

The early Egyptian arithmetic processes are recorded on the Rhind Papyrus, but the underlying principles are not included. Scholars of today can only guess how these early developments were discovered.

Egyptian hieroglyphics used a base-ten system of numbers in which a vertical line represented 1; a heel bone, ∩, represented 10; and a scroll, ꝰ, represented 100.

The symbols shown below represent the number 237. There are 7 vertical lines, 3 heel bones, and 2 scrolls. Thus the symbols represent 7 + 30 + 200, or 237.

Addition in hieroglyphic notation does not require memorization of addition facts. Addition is done just by counting symbols.

The top example shown at the left shows that addition is a simple grouping operation. Write down the total of each kind of symbol. Then group 10 straight lines into one heel bone.

Subtraction in the hieroglyphic system is similar to making change. For example, what change do you get from a $1.00 bill when buying a $.55 item?

In the bottom example shown at the left, 5 cannot be subtracted from 3, so a 10 is "borrowed," and 10 ones are added.

Note that no zero is provided in this number system. That place value symbol is just not used. As shown at the left, the heel bone is not used when writing 208 because there are no 10's necessary in 208.

HIEROGLYPHIC NOTATION	MODERN NOTATION
‖‖ ∩∩ ꝰꝰꝰ	324
+ ‖‖ ∩∩∩ ꝰ	+138
‖‖ ‖‖ ‖‖ ∩∩∩∩ ꝰꝰꝰꝰ ∩	
12 + 50 + 400 = 462	
‖ ∩∩∩∩ ꝰꝰꝰꝰ ∩∩	

‖‖ ∩∩∩ ꝰꝰꝰꝰ	433
– ‖‖ ∩∩ ꝰꝰ	–225
‖‖ ‖‖ ∩∩ ꝰꝰꝰꝰ	4 2̸ 13̸ 3
– ‖‖ ∩∩ ꝰꝰ	–225
‖‖ ‖‖ ꝰꝰ	208
8 + 200 = 208	

302

S E C T I O N **7.1**

Addition and Subtraction of Polynomials

1   Add polynomials

A **monomial** is a number, a variable, or a product of numbers and variables. For instance,

7	b	$\dfrac{2}{3}a$	$12xy^2$
A number	A variable	A product of a number and a variable	A product of a number and variables

The expression $3\sqrt{x}$ is not a monomial because $\sqrt{x}$ cannot be written as a product of variables. The expression $\dfrac{2x}{y^2}$ is not a monomial because it is a *quotient* of variables.

A **polynomial** is a variable expression in which the terms are monomials.

A polynomial of *one* term is a **monomial**. $-7x^2$ is a monomial.

A polynomial of *two* terms is a **binomial**. $4x + 2$ is a binomial.

A polynomial of *three* terms is a **trinomial**. $7x^2 + 5x - 7$ is a trinomial.

The terms of a polynomial in one variable are usually arranged so that the exponents of the variable decrease from left to right. This is called **descending order.**

$$4x^3 - 3x^2 + 6x - 1$$

$$5y^4 - 2y^3 + y^2 - 7y + 8$$

The **degree of a polynomial in one variable** is its largest exponent. The degree of $4x^3 - 3x^2 + 6x - 1$ is 3. The degree of $5y^4 - 2y^3 + y^2 - 7y + 8$ is 4.

Polynomials can be added, using either a vertical or a horizontal format, by combining like terms.

Example 1   Add: $(2x^2 + x - 1) + (3x^3 + 4x^2 - 5)$
Use a vertical format.

Solution
$$\begin{array}{r} 2x^2 + x - 1 \\ 3x^3 + 4x^2 \quad\; - 5 \\ \hline 3x^3 + 6x^2 + x - 6 \end{array}$$

▶ Arrange the terms of each polynomial in descending order with like terms in the same column.
▶ Combine the terms in each column.

Problem 1   Add: $(2x^2 + 4x - 3) + (5x^2 - 6x)$
Use a vertical format.

Solution   See page A29.

Example 2   Add: $(3x^3 - 7x + 2) + (7x^2 + 2x - 7)$
Use a horizontal format.

Solution $(3x^3 - 7x + 2) + (7x^2 + 2x - 7)$
$= 3x^3 + 7x^2 + (-7x + 2x) + (2 - 7)$ ► Use the Commutative and Associative Properties of Addition to rearrange and group like terms.

$= 3x^3 + 7x^2 - 5x - 5$ ► Combine like terms, and write the polynomial in descending order.

Problem 2   Add: $(-4x^2 - 3xy + 2y^2) + (3x^2 - 4y^2)$
Use a horizontal format.

Solution   See page A29.

---

[2]   ## Subtract polynomials

The **opposite** of the polynomial $x^2 - 2x + 3$ is $-(x^2 - 2x + 3)$.

To simplify the opposite of a polynomial, remove the parentheses and change the sign of every term inside the parentheses.

$$-(x^2 - 2x + 3) = -x^2 + 2x - 3$$

Polynomials can be subtracted using either a vertical or horizontal format. To subtract, add the opposite of the second polynomial to the first.

Example 3   Subtract: $(-3x^2 - 7) - (-8x^2 + 3x - 4)$
Use a vertical format.

Solution   The opposite of $-8x^2 + 3x - 4$ is $8x^2 - 3x + 4$.

$$
\begin{array}{r}
-3x^2 \qquad\;\; - 7 \\
8x^2 - 3x + 4 \\
\hline
5x^2 - 3x - 3
\end{array}
$$

► Write the terms of each polynomial in descending order with like terms in the same column.

► Combine the terms in each column.

Problem 3   Subtract: $(8y^2 - 4xy + x^2) - (2y^2 - xy + 5x^2)$
Use a vertical format.

Solution   See page A30.

Example 4   Subtract: $(5x^2 - 3x + 4) - (-3x^3 - 2x + 8)$
Use a horizontal format.

Solution $(5x^2 - 3x + 4) - (-3x^3 - 2x + 8)$
$= (5x^2 - 3x + 4) + (3x^3 + 2x - 8)$ ► Rewrite subtraction as addition of the opposite.
$= 3x^3 + 5x^2 + (-3x + 2x) + (4 - 8)$

$= 3x^3 + 5x^2 - x - 4$ ► Combine like terms. Write the polynomial in descending order.

Problem 4   Subtract: $(-3a^2 - 4a + 2) - (5a^3 + 2a - 6)$
Use a horizontal format.

Solution   See page A30.

CONCEPT REVIEW 7.1

Determine whether the statement is always true, sometimes true, or never true.

1. The terms of a polynomial are monomials.

2. The polynomial $5x^3 + 4x^6 + 3x + 2$ is written in descending order.

3. The degree of the polynomial $7x^2 - 8x + 1$ is 3.

4. Like terms have the same coefficient and the same variable part.

5. The opposite of the polynomial $ax^3 - bx^2 + cx + d$ is $-ax^3 + bx^2 - cx - d$.

6. Subtraction is addition of the opposite.

EXERCISES 7.1

1   Add. Use a vertical format.

1. $(x^2 + 7x) + (-3x^2 - 4x)$

2. $(3y^2 - 2y) + (5y^2 + 6y)$

3. $(y^2 + 4y) + (-4y - 8)$

4. $(3x^2 + 9x) + (6x - 24)$

5. $(2x^2 + 6x + 12) + (3x^2 + x + 8)$

6. $(x^2 + x + 5) + (3x^2 - 10x + 4)$

7. $(x^3 - 7x + 4) + (2x^2 + x - 10)$

8. $(3y^3 + y^2 + 1) + (-4y^3 - 6y - 3)$

9. $(2a^3 - 7a + 1) + (-3a^2 - 4a + 1)$

10. $(5r^3 - 6r^2 + 3r) + (r^2 - 2r - 3)$

Add. Use a horizontal format.

11. $(4x^2 + 2x) + (x^2 + 6x)$

12. $(-3y^2 + y) + (4y^2 + 6y)$

13. $(4x^2 - 5xy) + (3x^2 + 6xy - 4y^2)$

14. $(2x^2 - 4y^2) + (6x^2 - 2xy + 4y^2)$

15. $(2a^2 - 7a + 10) + (a^2 + 4a + 7)$

16. $(-6x^2 + 7x + 3) + (3x^2 + x + 3)$

17. $(5x^3 + 7x - 7) + (10x^2 - 8x + 3)$

18. $(3y^3 + 4y + 9) + (2y^2 + 4y - 21)$

19. $(2r^2 - 5r + 7) + (3r^3 - 6r)$

20. $(3y^3 + 4y + 14) + (-4y^2 + 21)$

21. $(3x^2 + 7x + 10) + (-2x^3 + 3x + 1)$

22. $(7x^3 + 4x - 1) + (2x^2 - 6x + 2)$

2 Subtract. Use a vertical format.

23. $(x^2 - 6x) - (x^2 - 10x)$

24. $(y^2 + 4y) - (y^2 + 10y)$

25. $(2y^2 - 4y) - (-y^2 + 2)$

26. $(-3a^2 - 2a) - (4a^2 - 4)$

27. $(x^2 - 2x + 1) - (x^2 + 5x + 8)$

28. $(3x^2 + 2x - 2) - (5x^2 - 5x + 6)$

29. $(4x^3 + 5x + 2) - (-3x^2 + 2x + 1)$

30. $(5y^2 - y + 2) - (-2y^3 + 3y - 3)$

31. $(2y^3 + 6y - 2) - (y^3 + y^2 + 4)$

32. $(-2x^2 - x + 4) - (-x^3 + 3x - 2)$

Subtract. Use a horizontal format.

33. $(y^2 - 10xy) - (2y^2 + 3xy)$

34. $(x^2 - 3xy) - (-2x^2 + xy)$

35. $(3x^2 + x - 3) - (x^2 + 4x - 2)$

36. $(5y^2 - 2y + 1) - (-3y^2 - y - 2)$

37. $(-2x^3 + x - 1) - (-x^2 + x - 3)$

38. $(2x^2 + 5x - 3) - (3x^3 + 2x - 5)$

39. $(4a^3 - 2a + 1) - (a^3 - 2a + 3)$

40. $(b^2 - 8b + 7) - (4b^3 - 7b - 8)$

41. $(4y^3 - y - 1) - (2y^2 - 3y + 3)$

42. $(3x^2 - 2x - 3) - (2x^3 - 2x^2 + 4)$

SUPPLEMENTAL EXERCISES 7.1

State whether the polynomial is a monomial, a binomial, or a trinomial.

43. $8x^4 - 6x^2$

44. $4a^2b^2 + 9ab + 10$

45. $7x^3y^4$

State whether or not the expression is a monomial.

46. $3\sqrt{x}$

47. $\dfrac{4}{x}$

48. x^2y^2

State whether or not the expression is a polynomial.

49. $\dfrac{1}{5}x^3 + \dfrac{1}{2}x$

50. $\dfrac{1}{5x^2} + \dfrac{1}{2x}$

51. $x + \sqrt{5}$

Simplify.

52. $\left(\dfrac{2}{3}a^2 + \dfrac{1}{2}a - \dfrac{3}{4}\right) - \left(\dfrac{5}{3}a^2 + \dfrac{1}{2}a + \dfrac{1}{4}\right)$

53. $\left(\dfrac{3}{5}x^2 + \dfrac{1}{6}x - \dfrac{5}{8}\right) + \left(\dfrac{2}{5}x^2 + \dfrac{5}{6}x - \dfrac{3}{8}\right)$

Solve.

54. What polynomial must be added to $3x^2 - 4x - 2$ so that the sum is $-x^2 + 2x + 1$?

55. What polynomial must be added to $-2x^3 + 4x - 7$ so that the sum is $x^2 - x - 1$?

56. What polynomial must be subtracted from $6x^2 - 4x - 2$ so that the difference is $2x^2 + 2x - 5$?

57. What polynomial must be subtracted from $2x^3 - x^2 + 4x - 2$ so that the difference is $x^3 + 2x - 8$?

58. Is it possible to subtract two polynomials, each of degree 3, and have the difference be a polynomial of degree 2? If so, give an example. If not, explain why not.

59. Is it possible to add two polynomials, each of degree 3, and have the sum be a polynomial of degree 2? If so, give an example. If not, explain why not.

60. In your own words, explain the terms monomial, binomial, trinomial, and polynomial. Give an example of each.

S E C T I O N **7.2**

Multiplication of Monomials

1   Multiply monomials

Recall that in the exponential expression x^5, x is the base and 5 is the exponent. The exponent indicates the number of times the base occurs as a factor.

The product of exponential expressions with the *same* base can be simplified by writing each expression in factored form and writing the result with an exponent.

$$x^3 \cdot x^2 = \overbrace{(x \cdot x \cdot x)}^{3 \text{ factors}} \cdot \overbrace{(x \cdot x)}^{2 \text{ factors}}$$
$$\underbrace{}_{5 \text{ factors}}$$
$$= x \cdot x \cdot x \cdot x \cdot x$$
$$= x^5$$

Adding the exponents results in the same product.

$$x^3 \cdot x^2 = x^{3+2} = x^5$$

Rule for Multiplying Exponential Expressions

If m and n are integers, then $x^m \cdot x^n = x^{m+n}$.

For example, in the expression $a^2 \cdot a^6 \cdot a$, the bases are the same. The expression can be simplified by adding the exponents. Recall that $a = a^1$.

$$a^2 \cdot a^6 \cdot a = a^{2+6+1} = a^9$$

LOOK CLOSELY

The Rule for Multiplying Exponential Expressions requires that the bases be the same. The expression x^3y^2 cannot be simplified.

Example 1   Multiply: $(2xy)(3x^2y)$

Solution $(2xy)(3x^2y)$
$= (2 \cdot 3)(x \cdot x^2)(y \cdot y)$ ▶ Use the Commutative and Associative Properties of Multiplication to rearrange and group factors.

$= 6x^{1+2}y^{1+1}$ ▶ Multiply variables with the same base by adding the exponents.

$= 6x^3y^2$

Problem 1   Multiply: $(3x^2)(6x^3)$

Solution   See page A30.

Example 2   Multiply: $(2x^2y)(-5xy^4)$

Solution $(2x^2y)(-5xy^4)$
$= [2(-5)](x^2 \cdot x)(y \cdot y^4)$ ▶ Use the Properties of Multiplication to rearrange and group factors.

$= -10x^3y^5$ ▶ Multiply variables with the same base by adding the exponents.

Problem 2   Multiply: $(-3xy^2)(-4x^2y^3)$

Solution   See page A30.

2   Simplify powers of monomials

POINT OF INTEREST

One of the first symbolic representations of powers was given by Diophantus (c. A.D. 250) in his book *Arithmetica*. He used Δ^γ for x^2 and κ^γ for x^3. The symbol Δ^γ was the first two letters of the Greek word *dunamis* meaning power; κ^γ was from the Greek word *kubos* meaning cube. He also combined these symbols to denote higher powers. For instance, $\Delta\kappa^\gamma$ was the symbol for x^5.

A power of a monomial can be simplified by rewriting the expression in factored form and then using the Rule for Multiplying Exponential Expressions.

$(x^2)^3 = x^2 \cdot x^2 \cdot x^2$
$= x^{2+2+2}$
$= x^6$

$(x^4y^3)^2 = (x^4y^3)(x^4y^3)$
$= x^4 \cdot y^3 \cdot x^4 \cdot y^3$
$= (x^4 \cdot x^4)(y^3 \cdot y^3)$
$= x^{4+4}y^{3+3}$
$= x^8y^6$

Note that multiplying each exponent inside the parentheses by the exponent outside the parentheses gives the same result.

$(x^2)^3 = x^{2 \cdot 3} = x^6$

$(x^4y^3)^2 = x^{4 \cdot 2}y^{3 \cdot 2} = x^8y^6$

Rule for Simplifying Powers of Exponential Expressions

If m and n are integers, then $(x^m)^n = x^{mn}$.

Rule for Simplifying Powers of Products

If m, n, and p are integers, then $(x^my^n)^p = x^{mp}y^{np}$.

To simplify $(x^5)^2$, multiply the exponents.

$$(x^5)^2 = x^{5 \cdot 2} = x^{10}$$

To simplify $(3a^2b)^3$, multiply each exponent inside the parentheses by the exponent outside the parentheses.

$$(3a^2b)^3 = 3^{1 \cdot 3}a^{2 \cdot 3}b^{1 \cdot 3}$$
$$= 3^3a^6b^3$$
$$= 27a^6b^3$$

Example 3   Simplify: $(-2x)(-3xy^2)^3$

Solution $(-2x)(-3xy^2)^3$
$= (-2x)(-3)^3x^3y^6$ ▶ Multiply each exponent in $-3xy^2$ by the exponent outside the parentheses.
$= (-2x)(-27)x^3y^6$ ▶ Simplify $(-3)^3$.
$= [-2(-27)](x \cdot x^3)y^6$ ▶ Use the Properties of Multiplication to rearrange and group factors.
$= 54x^4y^6$ ▶ Multiply variables with the same base by adding the exponents.

Problem 3   Simplify: $(3x)(2x^2y)^3$

Solution   See page A30.

CONCEPT REVIEW 7.2

Determine whether the statement is always true, sometimes true, or never true.

1.   In the expression $(6x)^4$, x is the base and 4 is the exponent.

2.   To multiply $x^m \cdot x^n$, multiply the exponents.

3.   The expression "a power of a monomial" means the monomial is the base of an exponential expression.

4. $x^6 \cdot x \cdot x^8 = x^{14}$

5. $(4x^3y^5)^2 = 4x^6y^{10}$

6. $a^2 \cdot b^7 = ab^9$

EXERCISES 7.2

1   Multiply.

1. $(x)(2x)$

2. $(-3y)(y)$

3. $(3x)(4x)$

4. $(7y^3)(7y^2)$

5. $(-2a^3)(-3a^4)$

6. $(5a^6)(-2a^5)$

7. $(x^2y)(xy^4)$

8. $(x^2y^4)(xy^7)$

9. $(-2x^4)(5x^5y)$

10. $(-3a^3)(2a^2b^4)$

11. $(x^2y^4)(x^5y^4)$

12. $(a^2b^4)(ab^3)$

13. $(2xy)(-3x^2y^4)$ **14.** $(-3a^2b)(-2ab^3)$ **15.** $(x^2yz)(x^2y^4)$

16. $(-ab^2c)(a^2b^5)$ **17.** $(a^2b^3)(ab^2c^4)$ **18.** $(x^2y^3z)(x^3y^4)$

19. $(-a^2b^2)(a^3b^6)$ **20.** $(xy^4)(-xy^3)$ **21.** $(-6a^3)(a^2b)$

22. $(2a^2b^3)(-4ab^2)$ **23.** $(-5y^4z)(-8y^6z^5)$ **24.** $(3x^2y)(-4xy^2)$

25. $(10ab^2)(-2ab)$ **26.** $(x^2y)(yz)(xyz)$ **27.** $(xy^2z)(x^2y)(z^2y^2)$

28. $(-2x^2y^3)(3xy)(-5x^3y^4)$ **29.** $(4a^2b)(-3a^3b^4)(a^5b^2)$ **30.** $(3ab^2)(-2abc)(4ac^2)$

2   Simplify.

31. $(2^2)^3$ **32.** $(3^2)^2$ **33.** $(-2)^2$ **34.** $(-3)^3$

35. $(-2^2)^3$ **36.** $(-2^3)^3$ **37.** $(x^3)^3$ **38.** $(y^4)^2$

39. $(x^7)^2$ **40.** $(y^5)^3$ **41.** $(-x^2)^2$ **42.** $(-x^2)^3$

43. $(2x)^2$ **44.** $(3y)^3$ **45.** $(-2x^2)^3$ **46.** $(-3y^3)^2$

47. $(x^2y^3)^2$ **48.** $(x^3y^4)^5$ **49.** $(3x^2y)^2$ **50.** $(-2ab^3)^4$

51. $(a^2)(3a^2)^3$ **52.** $(b^2)(2a^3)^4$ **53.** $(-2x)(2x^3)^2$

54. $(2y)(-3y^4)^3$ **55.** $(x^2y)(x^2y)^3$ **56.** $(a^3b)(ab)^3$

57. $(ab^2)^2(ab)^2$ **58.** $(x^2y)^2(x^3y)^3$ **59.** $(-2x)(-2x^3y)^3$

60. $(-3y)(-4x^2y^3)^3$ **61.** $(-2x)(-3xy^2)^2$ **62.** $(-3y)(-2x^2y)^3$

63. $(ab^2)(-2a^2b)^3$ **64.** $(a^2b^2)(-3ab^4)^2$ **65.** $(-2a^3)(3a^2b)^3$

66. $(-3b^2)(2ab^2)^3$ **67.** $(-3ab)^2(-2ab)^3$ **68.** $(-3a^2b)^3(-3ab)^3$

SUPPLEMENTAL EXERCISES 7.2

Simplify.

69. $(6x)(2x^2) + (4x^2)(5x)$ **70.** $(2a^7)(7a^2) - (6a^3)(5a^6)$

71. $(3a^2b^2)(2ab) - (9ab^2)(a^2b)$ **72.** $(3x^2y^2)^2 - (2xy)^4$

73. $(5xy^3)(3x^4y^2) - (2x^3y)(x^2y^4)$ **74.** $a^2(ab^2)^3 - a^3(ab^3)^2$

75. $4a^2(2ab)^3 - 5b^2(a^5b)$ **76.** $9x^3(3x^2y)^2 - x(x^3y)^2$

77. $-2xy(x^2y)^3 - 3x^5(xy^2)^2$ **78.** $5a^2b(ab^2)^2 + b^3(2a^2b)^2$

79. $a^n \cdot a^n$ **80.** $(a^n)^2$ **81.** $(a^2)^n$ **82.** $a^2 \cdot a^n$

For Exercises 83–86, answer true or false. If the answer is false, correct the right side of the equation.

83. $(-a)^5 = -a^5$

84. $(-b)^8 = b^8$

85. $(x^2)^5 = x^{2+5} = x^7$

86. $x^3 + x^3 = 2x^{3+3} = 2x^6$

87. Evaluate $(2^3)^2$ and $2^{(3^2)}$. Are the results the same? If not, which expression has the larger value?

88. What is the Order of Operations for the expression x^{m^n}?

89. If n is a positive integer and $x^n = y^n$, when is $x = y$?

90. The length of a rectangle is $4ab$. The width is $2ab$. Find the perimeter of the rectangle in terms of ab.

4ab

2ab

91. Explain in your own words how to multiply monomials.

92. The distance a rock will fall in t seconds is $16t^2$ ft (neglecting air resistance). Find other examples of quantities that can be expressed in terms of an exponential expression, and explain where the expression is used.

SECTION **7.3**

Multiplication of Polynomials

1 Multiply a polynomial by a monomial

To multiply a polynomial by a monomial, use the Distributive Property and the Rule for Multiplying Exponential Expressions.

To simplify $-2x(x^2 - 4x - 3)$, use the Distributive Property and the Rule for Multiplying Exponential Expressions.

$$-2x(x^2 - 4x - 3) = -2x(x^2) - (-2x)(4x) - (-2x)(3)$$
$$= -2x^3 + 8x^2 + 6x$$

Example 1   Multiply.   A. $(5x + 4)(-2x)$ B. $x^3(2x^2 - 3x + 2)$

Solution   A. $(5x + 4)(-2x)$
$= 5x(-2x) + 4(-2x)$ ▶ Use the Distributive Property.
$= -10x^2 - 8x$ ▶ Use the Rule for Multiplying Exponential Expressions.

B. $x^3(2x^2 - 3x + 2)$
$= 2x^5 - 3x^4 + 2x^3$

Problem 1   Multiply.   A. $(-2y + 3)(-4y)$ B. $-a^2(3a^2 + 2a - 7)$

Solution   See page A30.

2   Multiply two polynomials

Multiplication of two polynomials requires the repeated application of the Distributive Property.

$$(y - 2)(y^2 + 3y + 1) = (y - 2)(y^2) + (y - 2)(3y) + (y - 2)(1)$$
$$= y^3 - 2y^2 + 3y^2 - 6y + y - 2$$
$$= y^3 + y^2 - 5y - 2$$

A convenient method of multiplying two polynomials is to use a vertical format similar to that used for multiplication of whole numbers.

Multiply each term in the trinomial by -2.
Multiply each term in the trinomial by y.
Like terms must be in the same column.
Add the terms in each column.

$$
\begin{array}{r}
y^2 + 3y + 1 \\
y - 2 \\
\hline
-2y^2 - 6y - 2 \\
y^3 + 3y^2 + y \\
\hline
y^3 + y^2 - 5y - 2
\end{array}
$$

Example 2   Multiply: $(2b^3 - b + 1)(2b + 3)$

    Solution

$$
\begin{array}{r}
2b^3 - b + 1 \\
2b + 3 \\
\hline
6b^3 - 3b + 3 \\
4b^4 - 2b^2 + 2b \\
\hline
4b^4 + 6b^3 - 2b^2 - b + 3
\end{array}
$$

     ► Multiply $2b^3 - b + 1$ by 3.
     ► Multiply $2b^3 - b + 1$ by $2b$. Arrange the terms in descending order.
     ► Add the terms in each column.

Problem 2   Multiply: $(2y^3 + 2y^2 - 3)(3y - 1)$

    Solution   See page A30.

Example 3   Multiply: $(4a^3 - 5a - 2)(3a - 2)$

    Solution

$$
\begin{array}{r}
4a^3 - 5a - 2 \\
3a - 2 \\
\hline
-8a^3 + 10a + 4 \\
12a^4 - 15a^2 - 6a \\
\hline
12a^4 - 8a^3 - 15a^2 + 4a + 4
\end{array}
$$

     ► Multiply $4a^3 - 5a - 2$ by -2.
     ► Multiply $4a^3 - 5a - 2$ by $3a$.
     ► Add the terms in each column.

Problem 3   Multiply: $(3x^3 - 2x^2 + x - 3)(2x + 5)$

    Solution   See page A30.

3   Multiply two binomials

It is often necessary to find the product of two binomials. The product can be found using a method called **FOIL**, which is based on the Distributive Property. The letters of FOIL stand for **First**, **Outer**, **Inner**, and **Last**.

LOOK CLOSELY

FOIL is not really a different way of multiplying. It is based on the Distributive Property.

$(2x + 3)(x + 5)$
$= 2x(x + 5) + 3(x + 5)$
       F   O   I   L
$= 2x^2 + 10x + 3x + 15$
$= 2x^2 + 13x + 15$

Multiply: $(2x + 3)(x + 5)$

Multiply the First terms.	$(2x + 3)(x + 5)$	$2x \cdot x = 2x^2$
Multiply the Outer terms.	$(2x + 3)(x + 5)$	$2x \cdot 5 = 10x$
Multiply the Inner terms.	$(2x + 3)(x + 5)$	$3 \cdot x = 3x$
Multiply the Last terms.	$(2x + 3)(x + 5)$	$3 \cdot 5 = 15$

Add the products.
Combine like terms.

$(2x + 3)(x + 5)$

$\quad$ **F** $\quad$ **O** $\quad$ **I** $\quad$ **L**
$= 2x^2 + 10x + 3x + 15$
$= 2x^2 + 13x + 15$

Example 4 Multiply: $(4x - 3)(3x - 2)$

Solution $(4x - 3)(3x - 2)$
$= 4x(3x) + 4x(-2) + (-3)(3x) + (-3)(-2)$ ▶ Use the FOIL
$= 12x^2 - 8x - 9x + 6$ method.
$= 12x^2 - 17x + 6$ ▶ Combine like terms.

Problem 4 Multiply: $(4y - 5)(3y - 3)$

Solution See page A30.

Example 5 Multiply: $(3x - 2y)(x + 4y)$

Solution $(3x - 2y)(x + 4y)$
$= 3x(x) + 3x(4y) + (-2y)(x) + (-2y)(4y)$ ▶ Use the FOIL
$= 3x^2 + 12xy - 2xy - 8y^2$ method.
$= 3x^2 + 10xy - 8y^2$ ▶ Combine like terms.

Problem 5 Multiply: $(3a + 2b)(3a - 5b)$

Solution See page A30.

---

4   # Multiply binomials that have special products

The expression $(a + b)(a - b)$ is the product of the sum and difference of two terms. The first binomial in the expression is a sum; the second is a difference. The two terms are a and b. The first term in each binomial is a. The second term in each binomial is b.

The expression $(a + b)^2$ is the square of a binomial. The first term in the binomial is a. The second term in the binomial is b.

Using FOIL, it is possible to find a pattern for the product of the sum and difference of two terms and for the square of a binomial.

The Sum and Difference of Two Terms

$$(a + b)(a - b) = a^2 - ab + ab - b^2$$
$$= a^2 - b^2$$

Square of first term ——————⟶

Square of second term ——————⟶

The Square of a Binomial

$$(a + b)^2 = (a + b)(a + b) = a^2 + ab + ab + b^2$$
$$= a^2 + 2ab + b^2$$

Square of first term ——————⟶

Twice the product of the two terms ——————⟶

Square of last term ——————⟶

Example 6   Multiply: $(2x + 3)(2x - 3)$

Solution $(2x + 3)(2x - 3)$ ▶ $(2x + 3)(2x - 3)$ is the product of the sum and difference of two terms.

$= (2x)^2 - 3^2$ ▶ Square the first term. Square the second term.
$= 4x^2 - 9$ ▶ Simplify.

Problem 6   Multiply: $(2a + 5c)(2a - 5c)$

Solution   See page A30.

Example 7   Multiply: $(3x - 2)^2$

Solution $(3x - 2)^2$ ▶ $(3x - 2)^2$ is the square of a binomial.

$= (3x)^2 + 2(3x)(-2) + (-2)^2$ ▶ Square the first term. Find twice the product of the two terms. Square the last term.

$= 9x^2 - 12x + 4$ ▶ Simplify.

Problem 7   Multiply: $(3x + 2y)^2$

Solution   See page A30.

5   Application problems

Example 8   The radius of a circle is $(x - 4)$ ft. Find the area of the circle in terms of the variable x. Leave the answer in terms of π.

Strategy To find the area, replace the variable r in the formula $A = \pi r^2$ with the given value, and solve for A.

Solution $A = \pi r^2$
$A = \pi(x - 4)^2$ ▶ This is the square of a binomial.
$A = \pi(x^2 - 8x + 16)$
$A = \pi x^2 - 8\pi x + 16\pi$

The area is $(\pi x^2 - 8\pi x + 16\pi)$ ft².

Problem 8 The length of a rectangle is $(x + 7)$ m. The width is $(x - 4)$ m. Find the area of the rectangle in terms of the variable x.

Solution See page A30.

Example 9 The length of a side of a square is $(3x + 5)$ in. Find the area of the square in terms of the variable x.

Strategy To find the area of the square, replace the variable s in the formula $A = s^2$ with the given value and simplify.

Solution $A = s^2$
$A = (3x + 5)^2$ ▶ This is the square of a binomial.
$A = 9x^2 + 30x + 25$

The area is $(9x^2 + 30x + 25)$ in².

Problem 9 The base of a triangle is $(x + 3)$ cm and the height is $(4x - 6)$ cm. Find the area of the triangle in terms of the variable x.

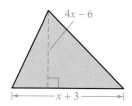

Solution See page A30.

CONCEPT REVIEW 7.3

Determine whether the statement is always true, sometimes true, or never true.

1. To multiply a monomial times a polynomial, use the Distributive Property to multiply each term of the polynomial by the monomial.

2. To multiply two polynomials, multiply each term of one polynomial by the other polynomial.

3. A binomial is a polynomial of degree 2.

4. $(x + 7)(x - 7)$ is the product of the sum and difference of the same two terms.

5. To square a binomial means to multiply it times itself.

6. The square of a binomial is a trinomial.

7. The FOIL method is used to multiply two polynomials.

8. Using the FOIL method, the terms $3x$ and 5 are the "First" terms in $(3x + 5)(2x + 7)$.

EXERCISES 7.3

1 Multiply.

1. $x(x - 2)$

2. $y(3 - y)$

3. $-x(x + 7)$

4. $-y(7 - y)$

5. $3a^2(a - 2)$

6. $4b^2(b + 8)$

7. $-5x^2(x^2 - x)$

8. $-6y^2(y + 2y^2)$

9. $-x^3(3x^2 - 7)$

10. $-y^4(2y^2 - y^6)$

11. $2x(6x^2 - 3x)$

12. $3y(4y - y^2)$

13. $(2x - 4)3x$

14. $(2x + 1)2x$

15. $-xy(x^2 - y^2)$

16. $-x^2y(2xy - y^2)$

17. $x(2x^3 - 3x + 2)$

18. $y(-3y^2 - 2y + 6)$

19. $-a(-2a^2 - 3a - 2)$

20. $-b(5b^2 + 7b - 35)$

21. $x^2(3x^4 - 3x^2 - 2)$

22. $y^3(-4y^3 - 6y + 7)$

23. $2y^2(-3y^2 - 6y + 7)$

24. $4x^2(3x^2 - 2x + 6)$

25. $(a^2 + 3a - 4)(-2a)$

26. $(b^3 - 2b + 2)(-5b)$

27. $-3y^2(-2y^2 + y - 2)$

28. $-5x^2(3x^2 - 3x - 7)$

29. $xy(x^2 - 3xy + y^2)$

30. $ab(2a^2 - 4ab - 6b^2)$

2 Multiply.

31. $(x^2 + 3x + 2)(x + 1)$

32. $(x^2 - 2x + 7)(x - 2)$

33. $(a - 3)(a^2 - 3a + 4)$

34. $(2x - 3)(x^2 - 3x + 5)$

35. $(-2b^2 - 3b + 4)(b - 5)$

36. $(-a^2 + 3a - 2)(2a - 1)$

37. $(3x - 5)(-2x^2 + 7x - 2)$

38. $(2a - 1)(-a^2 - 2a + 3)$

39. $(x^3 - 3x + 2)(x - 4)$

40. $(y^3 + 4y^2 - 8)(2y - 1)$

41. $(3y - 8)(5y^2 + 8y - 2)$

42. $(4y - 3)(3y^2 + 3y - 5)$

43. $(5a^3 - 15a + 2)(a - 4)$

44. $(3b^3 - 5b^2 + 7)(6b - 1)$

45. $(y + 2)(y^3 + 2y^2 - 3y + 1)$

46. $(2a - 3)(2a^3 - 3a^2 + 2a - 1)$

3 Multiply.

47. $(x + 1)(x + 3)$

48. $(y + 2)(y + 5)$

49. $(a - 3)(a + 4)$

50. $(b - 6)(b + 3)$

51. $(y + 3)(y - 8)$

52. $(x + 10)(x - 5)$

53. $(y - 7)(y - 3)$

54. $(a - 8)(a - 9)$

55. $(2x + 1)(x + 7)$

56. $(y + 2)(5y + 1)$

57. $(3x - 1)(x + 4)$

58. $(7x - 2)(x + 4)$

59. $(4x - 3)(x - 7)$

60. $(2x - 3)(4x - 7)$

61. $(3y - 8)(y + 2)$

62. $(5y - 9)(y + 5)$

63. $(3x + 7)(3x + 11)$

64. $(5a + 6)(6a + 5)$

65. $(7a - 16)(3a - 5)$

66. $(5a - 12)(3a - 7)$

67. $(3b + 13)(5b - 6)$

68. $(x + y)(2x + y)$

69. $(2a + b)(a + 3b)$

70. $(3x - 4y)(x - 2y)$

71. $(2a - b)(3a + 2b)$

72. $(5a - 3b)(2a + 4b)$

73. $(2x + y)(x - 2y)$

74. $(3x - 7y)(3x + 5y)$

75. $(2x + 3y)(5x + 7y)$

76. $(5x + 3y)(7x + 2y)$

77. $(3a - 2b)(2a - 7b)$

78. $(5a - b)(7a - b)$

79. $(a - 9b)(2a + 7b)$

80. $(2a + 5b)(7a - 2b)$

81. $(10a - 3b)(10a - 7b)$

82. $(12a - 5b)(3a - 4b)$

83. $(5x + 12y)(3x + 4y)$

84. $(11x + 2y)(3x + 7y)$

85. $(2x - 15y)(7x + 4y)$

86. $(5x + 2y)(2x - 5y)$

87. $(8x - 3y)(7x - 5y)$

88. $(2x - 9y)(8x - 3y)$

4 Multiply.

89. $(y - 5)(y + 5)$

90. $(y + 6)(y - 6)$

91. $(2x + 3)(2x - 3)$

92. $(4x - 7)(4x + 7)$

93. $(x + 1)^2$

94. $(y - 3)^2$

95. $(3a - 5)^2$

96. $(6x - 5)^2$

97. $(3x - 7)(3x + 7)$

98. $(9x - 2)(9x + 2)$

99. $(2a + b)^2$

100. $(x + 3y)^2$

101. $(x - 2y)^2$

102. $(2x - 3y)^2$

103. $(4 - 3y)(4 + 3y)$

104. $(4x - 9y)(4x + 9y)$

105. $(5x + 2y)^2$

106. $(2a - 9b)^2$

5   Solve.

107. The length of a rectangle is $5x$ ft. The width is $(2x - 7)$ ft. Find the area of the rectangle in terms of the variable x.

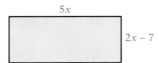

108. The width of a rectangle is $(x - 6)$ m. The length is $(2x + 3)$ m. Find the area of the rectangle in terms of the variable x.

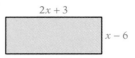

109. The width of a rectangle is $(3x + 1)$ in. The length of the rectangle is twice the width. Find the area of the rectangle in terms of the variable x.

110. The width of a rectangle is $(4x - 3)$ cm. The length of the rectangle is twice the width. Find the area of the rectangle in terms of the variable x.

111. The length of a side of a square is $(2x + 1)$ km. Find the area of the square in terms of the variable x.

112. The length of a side of a square is $(2x - 3)$ yd. Find the area of the square in terms of the variable x.

113. The base of a triangle is $4x$ m and the height is $(2x + 5)$ m. Find the area of the triangle in terms of the variable x.

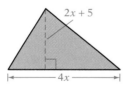

114. The base of a triangle is $(2x + 6)$ in. and the height is $(x - 8)$ in. Find the area of the triangle in terms of the variable x.

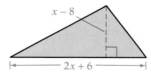

115. The radius of a circle is $(x + 4)$ cm. Find the area of the circle in terms of the variable x. Leave the answer in terms of π.

116. The radius of a circle is $(x - 3)$ ft. Find the area of the circle in terms of the variable x. Leave the answer in terms of π.

117. A softball diamond has dimensions 45 ft by 45 ft. A base path border x ft wide lies on both the first-base side and the third-base side of the diamond. Express the total area of the softball diamond and the base path in terms of the variable x.

118. An athletic field has dimensions 30 yd by 100 yd. An end zone that is w yd wide borders each end of the field. Express the total area of the field and the endzones in terms of the variable w.

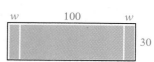

SUPPLEMENTAL EXERCISES 7.3

Simplify.

119. $(a + b)^2 - (a - b)^2$

120. $(x + 3y)^2 + (x + 3y)(x - 3y)$

121. $(3a^2 - 4a + 2)^2$

122. $(x + 4)^3$

123. $3x^2(2x^3 + 4x - 1) - 6x^3(x^2 - 2)$

124. $(3b + 2)(b - 6) + (4 + 2b)(3 - b)$

125. $x^n(x^n + 1)$

126. $(x^n + 1)(x^n - 1)$

127. $(x^n + 1)(x^n + 1)$

128. $(x^n - 1)^2$

129. $(x + 1)(x - 1)$

130. $(x + 1)(-x^2 + x - 1)$

131. $(x + 1)(x^3 - x^2 + x - 1)$

132. $(x + 1)(-x^4 + x^3 - x^2 + x - 1)$

Use the pattern of the answers to Exercises 129–132 to write the product.

133. $(x + 1)(x^5 - x^4 + x^3 - x^2 + x - 1)$

134. $(x + 1)(-x^6 + x^5 - x^4 + x^3 - x^2 + x - 1)$

Solve.

135. Find $(4n^3)^2$ if $2n - 3 = 4n - 7$.

136. What polynomial has quotient $x^2 + 2x - 1$ when divided by $x + 3$?

137. What polynomial has quotient $3x - 4$ when divided by $4x + 5$?

138. Subtract $4x^2 - x - 5$ from the product of $x^2 + x + 3$ and $x - 4$.

139. Add $x^2 + 2x - 3$ to the product of $2x - 5$ and $3x + 1$.

140. If a polynomial of degree 3 is multiplied by a polynomial of degree 2, what is the degree of the resulting polynomial?

141. Is it possible to multiply a polynomial of degree 2 by a polynomial of degree 2 and have the product be a polynomial of degree 3? If so, give an example. If not, explain why not.

SECTION 7.4

Division of Polynomials

CT

1   Integer exponents

The quotient of two exponential expressions with the *same* base can be simplified by writing each expression in factored form, dividing by the common factors, and then writing the result with an exponent.

$$\frac{x^5}{x^2} = \frac{\overset{1}{\cancel{x}} \cdot \overset{1}{\cancel{x}} \cdot x \cdot x \cdot x}{\underset{1}{\cancel{x}} \cdot \underset{1}{\cancel{x}}} = x^3$$

Note that subtracting the exponents results in the same quotient.

$$\frac{x^5}{x^2} = x^{5-2} = x^3$$

To divide two monomials with the same base, subtract the exponents of the like bases.

Simplify: $\dfrac{a^7}{a^3}$

The bases are the same. Subtract the exponent in the denominator from the exponent in the numerator.

$$\frac{a^7}{a^3} = a^{7-3} = a^4$$

Simplify: $\dfrac{r^8 s^6}{r^7 s}$

Subtract the exponents of the like bases.

$$\frac{r^8 s^6}{r^7 s} = r^{8-7} s^{6-1} = r s^5$$

Recall that for any number a, $a \neq 0$, $\dfrac{a}{a} = 1$. This property is true for exponential expressions as well. For example, for $x \neq 0$, $\dfrac{x^4}{x^4} = 1$.

This expression also can be simplified using the rule for dividing exponential expressions with the same base.

$$\frac{x^4}{x^4} = x^{4-4} = x^0$$

Because $\dfrac{x^4}{x^4} = 1$ and $\dfrac{x^4}{x^4} = x^{4-4} = x^0$, the following definition of zero as an exponent is used.

Zero as an Exponent

If $x \neq 0$, then $x^0 = 1$. The expression 0^0 is not defined.

Simplify: $(12a^3)^0$, $a \neq 0$

Any nonzero expression to the zero power is 1. $(12a^3)^0 = 1$

Simplify: $-(xy^4)^0$, $x \neq 0$, $y \neq 0$

Any nonzero expression to the zero power is 1. Because the negative sign is outside the parentheses, the answer is -1. $-(xy^4)^0 = -(1) = -1$

The meaning of a negative exponent can be developed by examining the quotient $\dfrac{x^4}{x^6}$.

The expression can be simplified by writing the numerator and denominator in factored form, dividing by the common factors, and then writing the result with an exponent.

$$\frac{x^4}{x^6} = \frac{\overset{1}{\cancel{x}} \cdot \overset{1}{\cancel{x}} \cdot \overset{1}{\cancel{x}} \cdot \overset{1}{\cancel{x}}}{\underset{1}{\cancel{x}} \cdot \underset{1}{\cancel{x}} \cdot \underset{1}{\cancel{x}} \cdot \underset{1}{\cancel{x}} \cdot x \cdot x} = \frac{1}{x^2}$$

Now simplify the same expression by subtracting the exponents of the like bases.

$$\frac{x^4}{x^6} = x^{4-6} = x^{-2}$$

Because $\dfrac{x^4}{x^6} = \dfrac{1}{x^2}$ and $\dfrac{x^4}{x^6} = x^{-2}$, the expressions $\dfrac{1}{x^2}$ and x^{-2} must be equal. The following definition of a negative exponent is used.

Definition of Negative Exponents

If n is a positive integer and $x \neq 0$, then $x^{-n} = \dfrac{1}{x^n}$ and $\dfrac{1}{x^{-n}} = x^n$.

To evaluate 2^{-4}, write the expression with a positive exponent. Then simplify. $2^{-4} = \dfrac{1}{2^4} = \dfrac{1}{16}$

Now that negative exponents have been defined, the Rule for Dividing Exponential Expressions can be stated.

Rule for Dividing Exponential Expressions

If m and n are integers and $x \neq 0$, then $\dfrac{x^m}{x^n} = x^{m-n}$.

POINT OF INTEREST

In the 15th century, the expression $12^{2\overline{m}}$ was used to mean $12x^{-2}$. The use of $\overline{m}$ reflected an Italian influence. In Italy, m was used for minus and p was used for plus. It was understood that $2\overline{m}$ referred to an unnamed variable. Isaac Newton, in the 17th century, advocated the current use of a negative exponent.

LOOK CLOSELY

Note from the example at the right that 2^{-4} is a *positive* number. A negative exponent does not indicate a negative number.

Example 1   Write $\dfrac{3^{-3}}{3^2}$ with a positive exponent. Then evaluate.

Solution $\dfrac{3^{-3}}{3^2} = 3^{-3-2}$ ▶ 3^{-3} and 3^2 have the same base. Subtract the exponents.

$= 3^{-5}$

$= \dfrac{1}{3^5}$ ▶ Use the Definition of Negative Exponents to write the expression with a positive exponent.

$= \dfrac{1}{243}$ ▶ Evaluate.

Problem 1   Write $\dfrac{2^{-2}}{2^3}$ with a positive exponent. Then evaluate.

Solution   See page A30.

The rules for simplifying exponential expressions and powers of exponential expressions are true for all integers. These rules are restated here.

Rules of Exponents

If m, n, and p are integers, then

$x^m \cdot x^n = x^{m+n}$ $(x^m)^n = x^{mn}$ $(x^m y^n)^p = x^{mp} y^{np}$

$\dfrac{x^m}{x^n} = x^{m-n}, \ x \neq 0$ $x^{-n} = \dfrac{1}{x^n}, \ x \neq 0$ $x^0 = 1, \ x \neq 0$

An exponential expression is in simplest form when it is written with only positive exponents.

To write the expression $a^{-7}b^3$ in simplest form, rewrite a^{-7} with a positive exponent. $a^{-7}b^3 = \dfrac{b^3}{a^7}$

Example 2   Simplify: $\dfrac{x^{-4}y^6}{xy^2}$

Solution $\dfrac{x^{-4}y^6}{xy^2} = x^{-5}y^4$ ▶ Divide variables with the same base by subtracting the exponents.

$= \dfrac{y^4}{x^5}$ ▶ Write the expression with only positive exponents.

Problem 2   Simplify: $\dfrac{b^8}{a^{-5}b^6}$

Solution   See page A30.

Example 3 Simplify.   A. $\dfrac{-35a^6b^{-2}}{25a^{-2}b^5}$ B. $(-2x)(3x^{-2})^{-3}$

Solution A. $\dfrac{-35a^6b^{-2}}{25a^{-2}b^5} = -\dfrac{35a^6b^{-2}}{25a^{-2}b^5}$

▶ A negative sign is placed in front of a fraction.

$$= -\dfrac{\overset{7}{\cancel{5}} \cdot 7a^{6-(-2)}b^{-2-5}}{\underset{1}{\cancel{5}} \cdot 5}$$

▶ Factor the coefficients. Divide by the common factors. Divide variables with the same base by subtracting the exponents.

$$= -\dfrac{7a^8b^{-7}}{5}$$

$$= -\dfrac{7a^8}{5b^7}$$

▶ Write the expression with only positive exponents.

B. $(-2x)(3x^{-2})^{-3} = (-2x)(3^{-3}x^6)$

▶ Use the Rule for Simplifying Powers of Products.

$$= \dfrac{-2x \cdot x^6}{3^3}$$

▶ Write the expression with positive exponents.

$$= -\dfrac{2x^7}{27}$$

▶ Use the Rule for Multiplying Exponential Expressions, and simplify the numerical exponential expression.

Problem 3 Simplify.   A. $\dfrac{12x^{-8}y^4}{-16xy^{-3}}$ B. $(-3ab)(2a^3b^{-2})^{-3}$

Solution See page A31.

---

2 Scientific notation

Very large and very small numbers are encountered in the fields of science and engineering. For example, the charge of an electron is 0.00000000000000000000160 coulomb. These numbers can be written more easily in scientific notation. In **scientific notation,** a number is expressed as a product of two factors, one a number between 1 and 10 and the other a power of 10.

To change a number written in decimal notation to scientific notation, write it in the form $a \cdot 10^n$, where a is a number between 1 and 10 and n is an integer.

For numbers greater than 10, move the decimal point to the right of the first digit. The exponent n is positive and equal to the number of places the decimal point has been moved.

$240,000 = 2.4 \cdot 10^5$

$93,000,000 = 9.3 \cdot 10^7$

For numbers less than 1, move the decimal point to the right of the first nonzero digit. The exponent n is negative. The absolute value of the exponent is equal to the number of places the decimal point has been moved.

$0.00030 = 3.0 \cdot 10^{-4}$

$0.0000832 = 8.32 \cdot 10^{-5}$

Look at the last example on page 323: $0.0000832 = 8.32 \cdot 10^{-5}$. Using the Definition of Negative Exponents,

$$10^{-5} = \frac{1}{10^5} = \frac{1}{100,000} = 0.00001$$

Because $10^{-5} = 0.00001$, we can write

$$8.32 \cdot 10^{-5} = 8.32 \cdot 0.00001 = 0.0000832$$

which is the number we started with. We have not changed the value of the number; we have just written it in another form.

Example 4   Write the number in scientific notation.
A. 824,300,000,000   B. 0.000000961

Solution   A. $824,300,000,000 = 8.243 \cdot 10^{11}$ ▸ Move the decimal point 11 places to the left. The exponent on 10 is 11.

B. $0.000000961 = 9.61 \cdot 10^{-7}$ ▸ Move the decimal point 7 places to the right. The exponent on 10 is −7.

Problem 4   Write the number in scientific notation.
A. 57,000,000,000   B. 0.000000017

Solution   See page A31.

Changing a number written in scientific notation to decimal notation also requires moving the decimal point.

When the exponent on 10 is positive, move the decimal point to the right the same number of places as the exponent.

$3.45 \cdot 10^9 = 3,450,000,000$

$2.3 \cdot 10^8 = 230,000,000$

When the exponent on 10 is negative, move the decimal point to the left the same number of places as the absolute value of the exponent.

$8.1 \cdot 10^{-3} = 0.0081$

$6.34 \cdot 10^{-6} = 0.00000634$

Example 5   Write the number in decimal notation.
A. $7.329 \cdot 10^6$ B. $6.8 \cdot 10^{-10}$

Solution   A. $7.329 \cdot 10^6 = 7,329,000$ ▸ The exponent on 10 is positive. Move the decimal point 6 places to the right.

B. $6.8 \cdot 10^{-10} = 0.00000000068$ ▸ The exponent on 10 is negative. Move the decimal point 10 places to the left.

Problem 5   Write the number in scientific notation.
A. $5 \cdot 10^{12}$ B. $4.0162 \cdot 10^{-9}$

Solution   See page A31.

The rules for multiplying and dividing with numbers in scientific notation are the same as those for calculating with algebraic expressions. The power of 10 corresponds to the variable, and the number between 1 and 10 corresponds to the coefficient of the variable.

	Algebraic Expressions	**Scientific Notation**
Multiplication	$(4x^{-3})(2x^5) = 8x^2$	$(4 \cdot 10^{-3})(2 \cdot 10^5) = 8 \cdot 10^2$
Division	$\dfrac{6x^5}{3x^{-2}} = 2x^{5-(-2)} = 2x^7$	$\dfrac{6 \cdot 10^5}{3 \cdot 10^{-2}} = 2 \cdot 10^{5-(-2)} = 2 \cdot 10^7$

Example 6   Multiply or divide.   A. $(3.0 \cdot 10^5)(1.1 \cdot 10^{-8})$ B. $\dfrac{7.2 \cdot 10^{13}}{2.4 \cdot 10^{-3}}$

Solution   A. $(3.0 \cdot 10^5)(1.1 \cdot 10^{-8}) = 3.3 \cdot 10^{-3}$ ▶ Multiply 3.0 and 1.1. Add the exponents on 10.

B. $\dfrac{7.2 \cdot 10^{13}}{2.4 \cdot 10^{-3}} = 3 \cdot 10^{16}$ ▶ Divide 7.2 by 2.4. Subtract the exponents on 10.

Problem 6   Multiply or divide.   A. $(2.4 \cdot 10^{-9})(1.6 \cdot 10^3)$ B. $\dfrac{5.4 \cdot 10^{-2}}{1.8 \cdot 10^{-4}}$

Solution   See page A31.

3   Divide a polynomial by a monomial

Note that $\dfrac{8 + 4}{2}$ can be simplified by first adding the terms in the numerator and then dividing the result. It can also be simplified by first dividing each term in the numerator by the denominator and then adding the result.

$$\frac{8 + 4}{2} = \frac{12}{2} = 6$$

$$\frac{8 + 4}{2} = \frac{8}{2} + \frac{4}{2} = 4 + 2 = 6$$

To divide a polynomial by a monomial, divide each term in the numerator by the denominator, and write the sum of the quotients.

$$\frac{a + b}{c} = \frac{a}{c} + \frac{b}{c}$$

Divide: $\dfrac{6x^2 + 4x}{2x}$

Divide each term of the polynomial $6x^2 + 4x$ by the monomial $2x$. Simplify.

$$\frac{6x^2 + 4x}{2x} = \frac{6x^2}{2x} + \frac{4x}{2x}$$

$$= 3x + 2$$

Example 7   Divide: $\dfrac{6x^3 - 3x^2 + 9x}{3x}$

Solution $\dfrac{6x^3 - 3x^2 + 9x}{3x} = \dfrac{6x^3}{3x} - \dfrac{3x^2}{3x} + \dfrac{9x}{3x}$ ▶ Divide each term of the polynomial by the monomial $3x$.

$= 2x^2 - x + 3$ ▶ Simplify each expression.

Problem 7   Divide: $\dfrac{4x^3y + 8x^2y^2 - 4xy^3}{2xy}$

Solution   See page A31.

Example 8   Divide: $\dfrac{12x^2y - 6xy + 4x^2}{2xy}$

Solution $\dfrac{12x^2y - 6xy + 4x^2}{2xy}$

$= \dfrac{12x^2y}{2xy} - \dfrac{6xy}{2xy} + \dfrac{4x^2}{2xy}$ ▶ Divide each term of the polynomial by the monomial $2xy$.

$= 6x - 3 + \dfrac{2x}{y}$ ▶ Simplify each expression.

Problem 8   Divide: $\dfrac{24x^2y^2 - 18xy + 6y}{6xy}$

Solution   See page A31.

4   Divide polynomials

To divide polynomials, use a method similar to that used for division of whole numbers. The same equation used to check division of whole numbers is used to check polynomial division.

$$\textbf{Dividend = (Quotient} \times \textbf{Divisor) + Remainder}$$

Divide: $(x^2 - 5x + 8) \div (x - 3)$

Step 1

$$
\begin{array}{r}
x \\
x - 3 \overline{\smash{)}\, x^2 - 5x + 8} \\
\underline{x^2 - 3x} \\
-2x + 8
\end{array}
$$

Think: $x \overline{\smash{)}\, x^2} = \dfrac{x^2}{x} = x$

Multiply: $x(x - 3) = x^2 - 3x$

Subtract: $(x^2 - 5x) - (x^2 - 3x) = -2x$
Bring down $+ 8$.

Step 2

$$
\begin{array}{r}
x - 2 \\
x - 3 \overline{\smash{)}\, x^2 - 5x + 8} \\
\underline{x^2 - 3x} \\
-2x + 8 \\
\underline{-2x + 6} \\
2
\end{array}
$$

Think: $x \overline{\smash{)}\, {-2x}} = \dfrac{-2x}{x} = -2$

Multiply: $-2(x - 3) = -2x + 6$

Subtract: $(-2x + 8) - (-2x + 6) = 2$
The remainder is 2.

Check: $(x - 2)(x - 3) + 2 = x^2 - 3x - 2x + 6 + 2 = x^2 - 5x + 8$

$(x^2 - 5x + 8) \div (x - 3) = x - 2 + \dfrac{2}{x - 3}$

Example 9   Divide: $(6x + 2x^3 + 26) \div (x + 2)$

Solution

$$
\begin{array}{r}
2x^2 - 4x + 14 \\
x + 2\overline{)2x^3 + 0x^2 + 6x + 26} \\
\underline{2x^3 + 4x^2} \\
-4x^2 + 6x \\
\underline{-4x^2 - 8x} \\
14x + 26 \\
\underline{14x + 28} \\
-2
\end{array}
$$

▶ Arrange the terms in descending order. There is no term of x^2 in $2x^3 + 6x + 26$. Insert $0x^2$ for the missing term so that like terms will be in columns.

$$(6x + 2x^3 + 26) \div (x + 2) = 2x^2 - 4x + 14 - \frac{2}{x + 2}$$

Problem 9   Divide: $(x^3 - 2x - 4) \div (x - 2)$

Solution   See page A31.

CONCEPT REVIEW 7.4

Determine whether the statement is always true, sometimes true, or never true.

1. The expression $\dfrac{x^5}{y^3}$ can be simplified by subtracting the exponents.

2. The rules of exponents can be applied to expressions that contain an exponent of zero or contain negative exponents.

3. The expression 3^{-2} represents the reciprocal of 3^2.

4. $5x^0 = 0$

5. The expression 4^{-3} represents a negative number.

6. To be in simplest form, an exponential expression cannot contain any negative exponents.

7. $2x^{-5} = \dfrac{1}{2x^5}$

8. $x^4y^{-6} = \dfrac{1}{x^4y^6}$

EXERCISES 7.4

1   Write with a positive or zero exponent. Then evaluate.

1. 5^{-2}

2. 3^{-3}

3. $\dfrac{1}{8^{-2}}$

4. $\dfrac{1}{12^{-1}}$

5. $\dfrac{3^{-2}}{3}$

6. $\dfrac{5^{-3}}{5}$

7. $\dfrac{2^3}{2^3}$

8. $\dfrac{3^{-2}}{3^{-2}}$

Simplify.

9. $\dfrac{y^7}{y^3}$ **10.** $\dfrac{z^9}{z^2}$ **11.** $\dfrac{a^8}{a^5}$ **12.** $\dfrac{c^{12}}{c^5}$

13. $\dfrac{p^5}{p}$ **14.** $\dfrac{w^9}{w}$ **15.** $\dfrac{4x^8}{2x^5}$ **16.** $\dfrac{12z^7}{4z^3}$

17. $\dfrac{22k^5}{11k^4}$ **18.** $\dfrac{14m^{11}}{7m^{10}}$ **19.** $\dfrac{m^9n^7}{m^4n^5}$ **20.** $\dfrac{y^5z^6}{yz^3}$

21. $\dfrac{6r^4}{4r^2}$ **22.** $\dfrac{8x^9}{12x^6}$ **23.** $\dfrac{-16a^7}{24a^6}$ **24.** $\dfrac{-18b^5}{27b^4}$

25. x^{-2} **26.** y^{-10} **27.** $\dfrac{1}{a^{-6}}$ **28.** $\dfrac{1}{b^{-4}}$

29. $4x^{-7}$ **30.** $-6y^{-1}$ **31.** $\dfrac{5}{b^{-8}}$ **32.** $\dfrac{-3}{v^{-3}}$

33. $\dfrac{1}{3x^{-2}}$ **34.** $\dfrac{2}{5c^{-6}}$ **35.** $(ab^5)^0$ **36.** $(32x^3y^4)^0$

37. $\dfrac{y^3}{y^8}$ **38.** $\dfrac{z^4}{z^6}$ **39.** $\dfrac{a^5}{a^{11}}$ **40.** $\dfrac{m}{m^7}$

41. $\dfrac{4x^2}{12x^5}$ **42.** $\dfrac{6y^8}{8y^9}$ **43.** $\dfrac{-12x}{-18x^6}$ **44.** $\dfrac{-24c^2}{-36c^{11}}$

45. $\dfrac{x^6y^5}{x^8y}$ **46.** $\dfrac{a^3b^2}{a^2b^3}$ **47.** $\dfrac{2m^6n^2}{5m^9n^{10}}$ **48.** $\dfrac{5r^3t^7}{6r^5t^7}$

49. $\dfrac{pq^3}{p^4q^4}$ **50.** $\dfrac{a^4b^5}{a^5b^6}$ **51.** $\dfrac{3x^4y^5}{6x^4y^8}$ **52.** $\dfrac{14a^3b^6}{21a^5b^6}$

53. $\dfrac{14x^4y^6z^2}{16x^3y^9z}$ **54.** $\dfrac{24a^2b^7c^9}{36a^7b^5c}$ **55.** $\dfrac{15mn^9p^3}{30m^4n^9p}$

56. $\dfrac{25x^4y^7z^2}{20x^5y^9z^{11}}$ **57.** $(-2xy^{-2})^3$ **58.** $(-3x^{-1}y^2)^2$

59. $(3x^{-1}y^{-2})^2$ **60.** $(5xy^{-3})^{-2}$ **61.** $(2x^{-1})(x^{-3})$

62. $(-2x^{-5})x^7$ **63.** $(-5a^2)(a^{-5})^2$ **64.** $(2a^{-3})(a^7b^{-1})^3$

65. $(-2ab^{-2})(4a^{-2}b)^{-2}$ **66.** $(3ab^{-2})(2a^{-1}b)^{-3}$ **67.** $(-5x^{-2}y)(-2x^{-2}y^2)$

68. $\dfrac{a^{-3}b^{-4}}{a^2b^2}$ **69.** $\dfrac{3x^{-2}y^2}{6xy^2}$ **70.** $\dfrac{2x^{-2}y}{8xy}$

71. $\dfrac{3x^{-2}y}{xy}$ **72.** $\dfrac{2x^{-1}y^4}{x^2y^3}$ **73.** $\dfrac{2x^{-1}y^{-4}}{4xy^2}$

74. $\dfrac{12a^2b^3}{-27a^2b^2}$ **75.** $\dfrac{-16xy^4}{96x^4y^4}$ **76.** $\dfrac{-8x^2y^4}{44y^2z^5}$

2 Write the number in scientific notation.

77. 2,370,000 **78.** 75,000 **79.** 0.00045 **80.** 0.000076

81. 309,000 **82.** 819,000,000 **83.** 0.000000601 **84.** 0.00000000096

Write the number in decimal notation.

85. $7.1 \cdot 10^5$

86. $2.3 \cdot 10^7$

87. $4.3 \cdot 10^{-5}$

88. $9.21 \cdot 10^{-7}$

89. $6.71 \cdot 10^8$

90. $5.75 \cdot 10^9$

91. $7.13 \cdot 10^{-6}$

92. $3.54 \cdot 10^{-8}$

93. Light travels approximately 16,000,000,000 mi in one day. Write this number in scientific notation.

94. Avogadro's number is used in chemistry. Its value is approximately 602,300,000,000,000,000,000,000. Write this number in scientific notation.

95. The approximate mass of Earth is 5,980,000,000,000,000,000,000,000 kg. Write this number in scientific notation.

96. A parsec is a distance measurement that is used by astronomers. One parsec is 3,086,000,000,000,000,000 cm. Write this number in scientific notation.

97. The electric charge on an electron is 0.00000000000000000016 coulomb. Write this number in scientific notation.

98. The length of an ultraviolet light wave is approximately 0.0000037 m. Write this number in scientific notation.

99. One unit used to measure the speed of a computer is the picosecond. One picosecond is 0.000000000001 second. Write this number in scientific notation.

100. One light-year is the distance traveled by light in one year. One light-year is 5,880,000,000,000 mi. Write this number in scientific notation.

Simplify.

101. $(1.9 \cdot 10^{12})(3.5 \cdot 10^7)$

102. $(4.2 \cdot 10^7)(1.8 \cdot 10^{-5})$

103. $(2.3 \cdot 10^{-8})(1.4 \cdot 10^{-6})$

104. $(3 \cdot 10^{-20})(2.4 \cdot 10^9)$

105. $\dfrac{6.12 \cdot 10^{14}}{1.7 \cdot 10^9}$

106. $\dfrac{6 \cdot 10^{-8}}{2.5 \cdot 10^{-2}}$

107. $\dfrac{5.58 \cdot 10^{-7}}{3.1 \cdot 10^{11}}$

108. $\dfrac{9.03 \cdot 10^6}{4.3 \cdot 10^{-5}}$

3   Divide.

109. $\dfrac{2x + 2}{2}$

110. $\dfrac{5y + 5}{5}$

111. $\dfrac{10a - 25}{5}$

112. $\dfrac{16b - 40}{8}$

113. $\dfrac{3a^2 + 2a}{a}$

114. $\dfrac{6y^2 + 4y}{y}$

115. $\dfrac{4b^3 - 3b}{b}$

116. $\dfrac{12x^2 - 7x}{x}$

117. $\dfrac{3x^2 - 6x}{3x}$

118. $\dfrac{10y^2 - 6y}{2y}$

119. $\dfrac{5x^2 - 10x}{-5x}$

120. $\dfrac{3y^2 - 27y}{-3y}$

121. $\dfrac{x^3 + 3x^2 - 5x}{x}$

122. $\dfrac{a^3 - 5a^2 + 7a}{a}$

123. $\dfrac{x^6 - 3x^4 - x^2}{x^2}$

124. $\dfrac{a^8 - 5a^5 - 3a^3}{a^2}$

125. $\dfrac{5x^2y^2 + 10xy}{5xy}$

126. $\dfrac{8x^2y^2 - 24xy}{8xy}$

127. $\dfrac{9y^6 - 15y^3}{-3y^3}$

128. $\dfrac{4x^4 - 6x^2}{-2x^2}$

129. $\dfrac{3x^2 - 2x + 1}{x}$

130. $\dfrac{8y^2 + 2y - 3}{y}$

131. $\dfrac{-3x^2 + 7x - 6}{x}$

132. $\dfrac{2y^2 - 6y + 9}{y}$

133. $\dfrac{16a^2b - 20ab + 24ab^2}{4ab}$

134. $\dfrac{22a^2b + 11ab - 33ab^2}{11ab}$

135. $\dfrac{9x^2y + 6xy - 3xy^2}{xy}$

4   Divide.

136. $(b^2 - 14b + 49) \div (b - 7)$

137. $(x^2 - x - 6) \div (x - 3)$

138. $(y^2 + 2y - 35) \div (y + 7)$

139. $(2x^2 + 5x + 2) \div (x + 2)$

140. $(2y^2 - 13y + 21) \div (y - 3)$

141. $(4x^2 - 16) \div (2x + 4)$

142. $(2y^2 + 7) \div (y - 3)$

143. $(x^2 + 1) \div (x - 1)$

144. $(x^2 + 4) \div (x + 2)$

145. $(6x^2 - 7x) \div (3x - 2)$

146. $(6y^2 + 2y) \div (2y + 4)$

147. $(5x^2 + 7x) \div (x - 1)$

148. $(6x^2 - 5) \div (x + 2)$

149. $(a^2 + 5a + 10) \div (a + 2)$

150. $(b^2 - 8b - 9) \div (b - 3)$

151. $(2y^2 - 9y + 8) \div (2y + 3)$

152. $(3x^2 + 5x - 4) \div (x - 4)$

153. $(8x + 3 + 4x^2) \div (2x - 1)$

154. $(10 + 21y + 10y^2) \div (2y + 3)$

155. $(15a^2 - 8a - 8) \div (3a + 2)$

156. $(12a^2 - 25a - 7) \div (3a - 7)$

157. $(5 - 23x + 12x^2) \div (4x - 1)$

158. $(24 + 6a^2 + 25a) \div (3a - 1)$

159. $(x^3 + 3x^2 + 5x + 3) \div (x + 1)$

160. $(x^3 - 6x^2 + 7x - 2) \div (x - 1)$

161. $(x^4 - x^2 - 6) \div (x^2 + 2)$

162. $(x^4 + 3x^2 - 10) \div (x^2 - 2)$

SUPPLEMENTAL EXERCISES 7.4

Evaluate.

163. $8^{-2} + 2^{-5}$

164. $9^{-2} + 3^{-3}$

165. Evaluate 2^x and 2^{-x} when $x = -2, -1, 0, 1,$ and 2.

166. Evaluate 3^x and 3^{-x} when $x = -2, -1, 0, 1,$ and 2.

Write in decimal notation.

167. 2^{-4}

168. 25^{-2}

Simplify.

169. $\left(\dfrac{9x^2y^4}{3xy^2}\right) - \left(\dfrac{12x^5y^6}{6x^4y^4}\right)$

170. $\left(\dfrac{6x^2 + 9x}{3x}\right) + \left(\dfrac{8xy^2 + 4y^2}{4y^2}\right)$

171. $\left(\dfrac{6x^4yz^3}{2x^2y^3}\right)\left(\dfrac{2x^2z^3}{4y^2z}\right) \div \left(\dfrac{6x^2y^3}{x^4y^2z}\right)$

172. $\left(\dfrac{5x^2yz^3}{3x^4yz^3}\right) \div \left(\dfrac{10x^2y^5z^4}{2y^3z}\right) \div \left(\dfrac{5y^4z^2}{x^2y^6z}\right)$

Complete.

173. If $m = n$ and $a \neq 0$, then $\dfrac{a^m}{a^n} = $ _____ .

174. If $m = n + 1$ and $a \neq 0$, then $\dfrac{a^m}{a^n} = $ _____ .

Solve.

175. $(-4.8)^x = 1$

Determine whether each equation is true or false. If the equation is false, change the right side of the equation to make a true equation.

176. $(2a)^{-3} = \dfrac{2}{a^3}$

177. $((a^{-1})^{-1})^{-1} = \dfrac{1}{a}$

178. $(2 + 3)^{-1} = 2^{-1} + 3^{-1}$

179. If $x \neq \dfrac{1}{3}$, then $(3x - 1)^0 = (1 - 3x)^0$.

Solve.

180. The product of a monomial and $4b$ is $12a^2b$. Find the monomial.

181. The product of a monomial and $6x$ is $24xy^2$. Find the monomial.

182. The quotient of a polynomial and $2x + 1$ is $2x - 4 + \dfrac{7}{2x + 1}$. Find the polynomial.

183. The quotient of a polynomial and $x - 3$ is $x^2 - x + 8 + \dfrac{22}{x - 3}$. Find the polynomial.

184. Why is the condition $x \neq \dfrac{1}{3}$ given in Exercise 179?

185. In your own words, explain how to divide exponential expressions.

186. If x is a nonzero real number, is x^{-2} always positive, always negative, or positive or negative depending on whether x is positive or negative? Explain your answer.

187. If x is a nonzero real number, is x^{-3} always positive, always negative, or positive or negative depending on whether x is positive or negative? Explain your answer.

Project in Mathematics

Pascal's Triangle

Simplifying the power of a binomial is called *expanding the binomial*. The expansion of the first three powers of a binomial is shown below.

$$(a + b)^1 = a + b$$
$$(a + b)^2 = (a + b)(a + b) = a^2 + 2ab + b^2$$
$$(a + b)^3 = (a + b)^2(a + b) = (a^2 + 2ab + b^2)(a + b) = a^3 + 3a^2b + 3ab^2 + b^3$$

Find $(a + b)^4$. [*Hint:* $(a + b)^4 = (a + b)^3(a + b)$]

Find $(a + b)^5$. [*Hint:* $(a + b)^5 = (a + b)^4(a + b)$]

If we continue in this way, the result for $(a + b)^6$ is

$$(a + b)^6 = a^6 + 6a^5b + 15a^4b^2 + 20a^3b^3 + 15a^2b^4 + 6ab^5 + b^6$$

Now expand $(a + b)^8$. Before you begin, see if you can find a pattern that will help you write the expansion of $(a + b)^8$ without having to multiply it out. Here are some hints.

1. Write out the variable terms of each binomial expansion from $(a + b)^1$ through $(a + b)^6$. Observe how the exponents on the variables change.

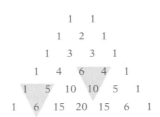

2. Write out the coefficients of all the terms without the variable parts. It will be helpful to make a triangular arrangement as shown at the left. Note that each row begins and ends with a 1. Also note in the two shaded regions that any number in a row is the sum of the two closest numbers above it. For instance, $1 + 5 = 6$ and $6 + 4 = 10$.

The triangle of numbers shown at the left is called Pascal's Triangle. To find the expansion of $(a + b)^8$, you will need to find the eighth row of Pascal's Triangle. First find row seven. Then find row eight.

Use the patterns you have observed to write the expansion of $(a + b)^8$.

Pascal's Triangle has been the subject of extensive analysis, and many patterns have been found. See if you can find some of them.

Focus on Problem Solving

Dimensional Analysis

In solving application problems, it may be useful to include the units in order to organize the problem so that the answer is in the proper units. Using units to organize and check the correctness of an application is called **dimensional analysis**. We use the operations of multiplying units and dividing units in applying dimensional analysis to application problems.

The Rule for Multiplying Exponential Expressions states that we multiply two expressions with the same base by adding the exponents.

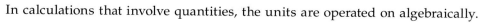

$$x^4 \cdot x^6 = x^{4+6} = x^{10}$$

In calculations that involve quantities, the units are operated on algebraically.

A rectangle measures 3 m by 5 m. Find the area of the rectangle.

$$A = LW = (3 \text{ m})(5 \text{ m}) = (3 \cdot 5)(\text{m} \cdot \text{m}) = 15 \text{ m}^2$$

The area of the rectangle is 15 m².

A box measures 10 cm by 5 cm by 3 cm. Find the volume of the box.

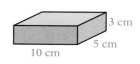

$$V = LWH = (10 \text{ cm})(5 \text{ cm})(3 \text{ cm}) = (10 \cdot 5 \cdot 3)(\text{cm} \cdot \text{cm} \cdot \text{cm}) = 150 \text{ cm}^3$$

The volume of the box is 150 cm³.

Example 9 on page 315 asks for the area of a square whose side measures $(3x + 5)$ in.

$$A = s^2 = [(3x + 5) \text{ in.}]^2 = (3x + 5)^2 \text{ in}^2 = (9x^2 + 30x + 25) \text{ in}^2$$

The area of the square is $(9x^2 + 30x + 25)$ in².

Dimensional analysis is used in the conversion of units.

The following example converts the unit miles to feet. The equivalent measures 1 mi = 5280 ft are used to form the following rates, which are called conversion

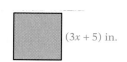

factors: $\dfrac{1 \text{ mi}}{5280 \text{ ft}}$ and $\dfrac{5280 \text{ ft}}{1 \text{ mi}}$. Because 1 mi = 5280 ft, both of the conversion factors

$\dfrac{1 \text{ mi}}{5280 \text{ ft}}$ and $\dfrac{5280 \text{ ft}}{1 \text{ mi}}$ are equal to 1.

To convert 3 mi to feet, multiply 3 mi by the conversion factor $\dfrac{5280 \text{ ft}}{1 \text{ mi}}$.

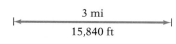

$$3 \text{ mi} = 3 \text{ mi} \cdot 1 = \frac{3 \text{ mi}}{1} \cdot \frac{5280 \text{ ft}}{1 \text{ mi}} = \frac{3 \text{ mi} \cdot 5280 \text{ ft}}{1 \text{ mi}} = 3 \cdot 5280 \text{ ft} = 15{,}840 \text{ ft}$$

There are two important points in the above illustration. First, you can think of dividing the numerator and denominator by the common unit "mile" just as you would divide the numerator and denominator of a fraction by a common factor. Second, the conversion factor $\dfrac{5280 \text{ ft}}{1 \text{ mi}}$ is equal to 1, and multiplying an expression by 1 does not change the value of the expression.

In the application problem that follows, the units are kept in the problem while the problem is worked.

In 1980, a horse named Fiddle Isle ran a 1.5 mi race in 2 min 23 s. Find Fiddle Isle's average speed in miles per hour for that race. Round to the nearest tenth.

To find the average speed, first convert 2 min 23 s to minutes. Then use the formula $r = \dfrac{d}{t}$, where r is the speed, d is the distance, and t is the time.

Convert 2 min 23 s to decimal minutes.

$$2 \text{ min } 23 \text{ s} = 2\frac{23}{60} \text{ min} \approx 2.383 \text{ min}$$

Divide the distance by the time.

$$r = \frac{d}{t} = \frac{1.5 \text{ mi}}{2.383 \text{ min}}$$

Use the conversion factor $\frac{60 \text{ min}}{1 \text{ h}}$, which equals 1.

$$= \frac{1.5 \text{ mi}}{2.383 \text{ min}} \cdot \frac{60 \text{ min}}{1 \text{ h}}$$

$$= \frac{90 \text{ mi}}{2.383 \text{ h}} \approx 37.8 \text{ mph}$$

The horse's average speed was 37.8 mph.

Solve.

1. A carpet is to be placed in a meeting hall that is 36 ft wide and 80 ft long. At $17.25 per square yard, how much will it cost to carpet the meeting hall? Use the formula $A = LW$, where A is the area, L is the length, and W is the width.

2. In the 1992 Winter Olympics, Bonnie Blair skated 1000 m in 81.9 s. Find her speed in kilometers per hour. Round to the nearest hundredth.

3. A $\frac{1}{4}$-acre commercial lot is on sale for $2.15 per square foot. Find the sale price of the commercial lot. (1 acre = 43,560 ft^2)

4. A piston-engined dragster traveled 440 yd in 4.936 s at Ennis, Texas, on October 9, 1988. Find the average speed of the dragster in miles per hour. Use the formula $r = \frac{d}{t}$, where r is the average speed, d is the distance, and t is the time. Round to the nearest hundredth.

5. A 0.75-acre industrial parcel was sold for $98,010. Find the price per square foot for the parcel. (1 acre = 43,560 ft^2)

6. A new driveway will require 800 ft^3 of concrete. Concrete is ordered by the cubic yard. How much concrete should be ordered? Round to the nearest tenth.

Chapter Summary

Key Words

A *monomial* is a number, a variable, or a product of numbers and variables.   (Objective 7.1.1)

5 is a number. y is a variable. $8a^2b^3$ is a product of numbers and variables. 5, y, and $8a^2b^3$ are monomials.

A *polynomial* is a variable expression in which the terms are monomials.   (Objective 7.1.1)

As shown above, 5, y, and $8a^2b^3$ are monomials. Therefore, $5 + y + 8a^2b^3$ is a polynomial.

A polynomial of one term is a *monomial*.   (Objective 7.1.1)

5, y, and $8a^2b^3$ are monomials.

A polynomial of two terms is a *binomial*.   (Objective 7.1.1)

$x + 9$, $y^2 - 3$, and $6a + 7b$ are binomials.

A polynomial of three terms is a *trinomial*.   (Objective 7.1.1)

$x^2 + 2x - 1$ is a trinomial.

The *degree of a polynomial in one variable* is the largest exponent on a variable.   (Objective 7.1.1)

The degree of $8x^3 - 5x^2 + 4x - 12$ is 3.

Essential Rules and Procedures

Rule for Multiplying Exponential Expressions
If m and n are integers, then $x^m \cdot x^n = x^{m+n}$.   (Objective 7.2.1)

$b^5 \cdot b^4 = b^{5+4} = b^9$

Rule for Simplifying Powers of Exponential Expressions
If m and n are integers, then $(x^m)^n = x^{mn}$.   (Objective 7.2.2)

$(y^3)^7 = y^{3(7)} = y^{21}$

Rule for Simplifying Powers of Products
If m, n, and p are integers, then $(x^m y^n)^p = x^{mp} y^{np}$.
(Objective 7.2.2)

$(a^6 b^2)^3 = a^{6(3)} b^{2(3)} = a^{18} b^6$

FOIL Method
To find the product of two binomials, add the products of the First terms, the Outer terms, the Inner terms, and the Last terms.   (Objective 7.3.3)

$(4x + 3)(2x - 1)$
$= (4x)(2x) + (4x)(-1) + (3)(2x) + (3)(-1)$
$= 8x^2 - 4x + 6x - 3 = 8x^2 + 2x - 3$

The Sum and Difference of Two Terms
$(a + b)(a - b) = a^2 - b^2$ (Objective 7.3.4)

$(3x + 4)(3x - 4) = (3x)^2 - 4^2 = 9x^2 - 16$

The Square of a Binomial
$(a + b)^2 = a^2 + 2ab + b^2$

$(a - b)^2 = a^2 - 2ab + b^2$
(Objective 7.3.4)

$(2x + 5)^2$
$= (2x)^2 + 2(2x)(5) + 5^2 = 4x^2 + 20x + 25$
$(2x - 5)^2$
$= (2x)^2 - 2(2x)(5) + 5^2 = 4x^2 - 20x + 25$

Zero as an Exponent
Any nonzero expression to the zero power equals 1.
(Objective 7.4.1)

$17^0 = 1$ $(5y)^0 = 1$

Definition of Negative Exponents
If n is a positive integer and $x \neq 0$, then $x^{-n} = \dfrac{1}{x^n}$ and $\dfrac{1}{x^{-n}} = x^n$.
(Objective 7.4.1)

$x^{-6} = \dfrac{1}{x^6}$ and $\dfrac{1}{x^{-6}} = x^6$

Rule for Dividing Exponential Expressions
If m and n are integers and $x \neq 0$, then $\dfrac{x^m}{x^n} = x^{m-n}$.
(Objective 7.4.1)

$\dfrac{y^8}{y^3} = y^{8-3} = y^5$

Scientific Notation

To express a number in scientific notation, write it in the form $a \cdot 10^n$, where a is a number between 1 and 10 and n is an integer. If the number is greater than 1, the exponent on 10 will be positive. If the number is less than 1, the exponent on 10 will be negative.

To change a number written in scientific notation to decimal notation, move the decimal point to the right if the exponent on 10 is positive and to the left if the exponent on 10 is negative. Move the decimal point the same number of places as the absolute value of the exponent on 10.   (Objective 7.4.2)

$367{,}000{,}000 = 3.67 \cdot 10^8$

$0.0000059 = 5.9 \cdot 10^{-6}$

$2.418 \cdot 10^7 = 24{,}180{,}000$

$9.06 \cdot 10^{-5} = 0.0000906$

Chapter Review Exercises

1. Add: $(12y^2 + 17y - 4) + (9y^2 - 13y + 3)$

2. Multiply: $(5xy^2)(-4x^2y^3)$

3. Multiply: $-2x(4x^2 + 7x - 9)$

4. Multiply: $(5a - 7)(2a + 9)$

5. Divide: $\dfrac{36x^2 - 42x + 60}{6}$

6. Subtract: $(5x^2 - 2x - 1) - (3x^2 - 5x + 7)$

7. Simplify: $(-3^2)^3$

8. Multiply: $(x^2 - 5x + 2)(x - 1)$

9. Multiply: $(a + 7)(a - 7)$

10. Evaluate: $\dfrac{6^2}{6^{-2}}$

11. Divide: $(x^2 + x - 42) \div (x + 7)$

12. Add: $(2x^3 + 7x^2 + x) + (2x^2 - 4x - 12)$

13. Multiply: $(6a^2b^5)(3a^6b)$

14. Multiply: $x^2y(3x^2 - 2x + 12)$

15. Multiply: $(2b - 3)(4b + 5)$

16. Divide: $\dfrac{16y^2 - 32y}{-4y}$

17. Subtract: $(13y^3 - 7y - 2) - (12y^2 - 2y - 1)$

18. Simplify: $(2^3)^2$

19. Multiply: $(3y^2 + 4y - 7)(2y + 3)$

20. Multiply: $(2b - 9)(2b + 9)$

21. Simplify: $(a^{-2}b^3c)^2$

22. Divide: $(6y^2 - 35y + 36) \div (3y - 4)$

23. Write 0.00000397 in scientific notation.

24. Multiply: $(xy^5z^3)(x^3y^3z)$

25. Multiply: $(6y^2 - 2y + 9)(-2y^3)$

26. Multiply: $(6x - 12)(3x - 2)$

27. Write $6.23 \cdot 10^{-5}$ in decimal notation.

28. Subtract: $(8a^2 - a) - (15a^2 - 4)$

29. Simplify: $(-3x^2y^3)^2$

30. Multiply: $(4a^2 - 3)(3a - 2)$

31. Simplify: $(5y - 7)^2$

32. Simplify: $(-3x^{-2}y^{-3})^{-2}$

33. Divide: $(x^2 + 17x + 64) \div (x + 12)$

34. Write $2.4 \cdot 10^5$ in decimal notation.

35. Multiply: $(a^2b^7c^6)(ab^3c)(a^3bc^2)$

36. Multiply: $2ab^3(4a^2 - 2ab + 3b^2)$

37. Multiply: $(3x + 4y)(2x - 5y)$

38. Divide: $\dfrac{12b^7 + 36b^5 - 3b^3}{3b^3}$

39. Subtract: $(b^2 - 11b + 19) - (5b^2 + 2b - 9)$

40. Simplify: $(5a^7b^6)^2(4ab)$

41. Multiply: $(6b^3 - 2b^2 - 5)(2b^2 - 1)$

42. Multiply: $(6 - 5x)(6 + 5x)$

43. Simplify: $\dfrac{6x^{-2}y^4}{3xy}$

44. Divide: $(a^3 + a^2 + 18) \div (a + 3)$

45. Add: $(4b^3 - 7b^2 + 10) + (2b^2 - 9b - 3)$

46. Multiply: $(2a^{12}b^3)(-9b^2c^6)(3ac)$

47. Multiply: $-9x^2(2x^2 + 3x - 7)$

48. Multiply: $(10y - 3)(3y - 10)$

49. Write 9,176,000,000,000 in scientific notation.

50. Subtract: $(6y^2 + 2y + 7) - (8y^2 + y + 12)$

51. Simplify: $(6x^4y^7z^2)^2(-2x^3y^2z^6)^2$

52. Multiply: $(-3x^3 - 2x^2 + x - 9)(4x + 3)$

53. Simplify: $(8a + 1)^2$

54. Simplify: $\dfrac{4a^{-2}b^{-8}}{2a^{-1}b^{-2}}$

55. Divide: $(b^3 - 2b^2 - 33b - 7) \div (b - 7)$

56. The length of a rectangle is $5x$ m. The width is $(4x - 7)$ m. Find the area of the rectangle in terms of the variable x.

57. The length of a side of a square is $(5x + 4)$ in. Find the area of the square in terms of the variable x.

58. The base of a triangle is $(3x - 2)$ ft and the height is $(6x + 4)$ ft. Find the area of the triangle in terms of the variable x.

59. The radius of a circle is $(x - 6)$ cm. Find the area of the circle in terms of the variable x. Leave the answer in terms of π.

60. The width of a rectangle is $(3x - 8)$ mi. The length is $(5x + 4)$ mi. Find the area of the rectangle in terms of the variable x.

Chapter Test

1. Add: $(3x^3 - 2x^2 - 4) + (8x^2 - 8x + 7)$

2. Multiply: $(-2x^3 + x^2 - 7)(2x - 3)$

3. Multiply: $2x(2x^2 - 3x)$

4. Simplify: $(-2a^2b)^3$

5. Simplify: $\dfrac{12x^2}{-3x^{-4}}$

6. Simplify: $(2ab^{-3})(3a^{-2}b^4)$

7. Subtract: $(3a^2 - 2a - 7) - (5a^3 + 2a - 10)$

8. Multiply: $(a - 2b)(a + 5b)$

9. Divide: $\dfrac{16x^5 - 8x^3 + 20x}{4x}$

10. Divide: $(4x^2 - 7) \div (2x - 3)$

11. Multiply: $(-2xy^2)(3x^2y^4)$

12. Multiply: $-3y^2(-2y^2 + 3y - 6)$

13. Simplify: $\dfrac{27xy^3}{3x^4y^3}$

14. Simplify: $(2x - 5)^2$

15. Multiply: $(2x - 7y)(5x - 4y)$

16. Multiply: $(x - 3)(x^2 - 4x + 5)$

17. Simplify: $(a^2b^{-3})^2$

18. Write 0.000029 in scientific notation.

19. Multiply: $(4y - 3)(4y + 3)$

20. Subtract: $(3y^3 - 5y + 8) - (-2y^2 + 5y + 8)$

21. Simplify: $(-3a^3b^2)^2$

22. Multiply: $(2a - 7)(5a^2 - 2a + 3)$

23. Simplify: $(3b + 2)^2$

24. Simplify: $\dfrac{-2a^2b^3}{8a^4b^8}$

25. Divide: $(8x^2 + 4x - 3) \div (2x - 3)$

26. Multiply: $(a^2b^5)(ab^2)$

27. Multiply: $(a - 3b)(a + 4b)$

28. Write $3.5 \cdot 10^{-8}$ in decimal notation.

29. The length of a side of a square is $(2x + 3)$ m. Find the area of the square in terms of the variable x.

30. The radius of a circle is $(x - 5)$ in. Find the area of the circle in terms of the variable x. Leave the answer in terms of π.

Cumulative Review Exercises

1. Simplify: $\dfrac{3}{16} - \left(-\dfrac{3}{8}\right) - \dfrac{5}{9}$

2. Simplify: $-5^2 \cdot \left(\dfrac{2}{3}\right)^3 \cdot \left(-\dfrac{3}{8}\right)$

3. Simplify: $\left(-\dfrac{1}{2}\right)^2 \div \left(\dfrac{5}{8} - \dfrac{5}{6}\right) + 2$

4. Find the opposite of -87.

5. Write $\dfrac{31}{40}$ as a decimal.

6. Evaluate $\dfrac{b - (a - b)^2}{b^2}$ when $a = 3$ and $b = -2$.

7. Simplify: $-3x - (-xy) + 2x - 5xy$

8. Simplify: $(16x)\left(-\dfrac{3}{4}\right)$

9. Simplify: $-2[3x - 4(3 - 2x) + 2]$

10. Complete the statement by using the Inverse Property of Addition.
$$-8 + ? = 0$$

11. Solve: $12 = -\dfrac{2}{3}x$

12. Solve: $3x - 7 = 2x + 9$

13. Solve: $3 - 4(2 - x) = 3x + 7$

14. Solve: $-\dfrac{4}{5}x = 16 - x$

15. 38.4 is what percent of 160?

16. Solve: $7x - 8 \geq -29$

17. Find the slope of the line that contains the points whose coordinates are $(3, -4)$ and $(-2, 5)$.

18. Find the equation of the line that contains the point whose coordinates are $(1, -3)$ and has slope $-\dfrac{3}{2}$.

19. Graph: $3x - 2y = -6$

20. Graph the solution set of $y \leq \dfrac{4}{5}x - 3$.

21. Find the domain and range of the relation $\{(-8, -7), (-6, -5), (-4, -2), (-2, 0)\}$. Is the relation a function?

22. Evaluate $f(x) = -2x + 10$ at $x = 6$.

23. Solve by substitution:
$$x = 3y + 1$$
$$2x + 5y = 13$$

24. Solve by the addition method: $9x - 2y = 17$
$$5x + 3y = -7$$

25. Subtract: $(5b^3 - 4b^2 - 7) - (3b^2 - 8b + 3)$

26. Multiply: $(3x - 4)(5x^2 - 2x + 1)$

27. Multiply: $(4b - 3)(5b - 8)$

28. Simplify: $(5b + 3)^2$

29. Simplify: $\dfrac{-3a^3b^2}{12a^4b^{-2}}$

30. Divide: $\dfrac{-15y^2 + 12y - 3}{-3y}$

31. Divide: $(a^2 - 3a - 28) \div (a + 4)$

32. Simplify: $(-3x^{-4}y)(-3x^{-2}y)$

33. Find the range of the function given by the equation $f(x) = -\dfrac{4}{3}x + 9$ if the domain is $\{-12, -9, -6, 0, 6\}$.

34. Translate and simplify "the product of five and the difference between a number and twelve."

35. Translate "the difference between eight times a number and twice the number is eighteen" into an equation and solve.

36. The width of a rectangle is 40% of the length. The perimeter of the rectangle is 42 m. Find the length and width of the rectangle.

37. A calculator costs a retailer $24. Find the selling price when the markup rate is 80%.

38. Fifty ounces of pure orange juice are added to 200 oz of a fruit punch that is 10% orange juice. What is the percent concentration of orange juice in the resulting mixture?

39. A car traveling at 50 mph overtakes a cyclist who, riding at 10 mph, has had a 2 h head start. How far from the starting point does the car overtake the cyclist?

40. The length of a side of a square is $(3x + 2)$ ft. Find the area of the square in terms of the variable x.

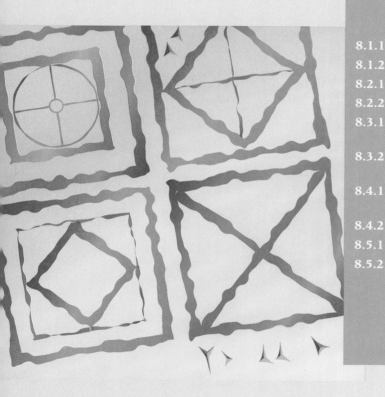

8

Factoring

Objectives

Algebra from Geometry

The early Babylonians made substantial progress in both algebra and geometry. Often the progress they made in algebra was based on geometric concepts.

Some geometric proofs of algebraic identities are shown at the left below.

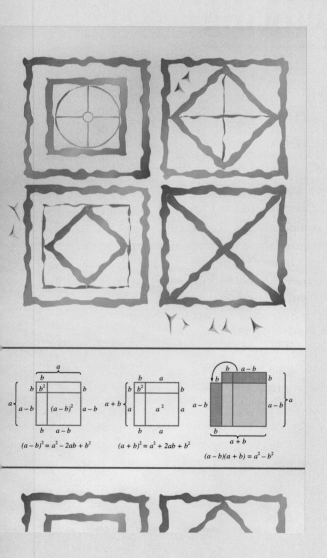

$$(a-b)^2 = a^2 - 2ab + b^2$$

$$(a+b)^2 = a^2 + 2ab + b^2$$

$$(a-b)(a+b) = a^2 - b^2$$

S E C T I O N **8.1**

Common Factors

1   Factor a monomial from a polynomial

The **greatest common factor (GCF)** of two or more integers is the greatest integer that is a factor of all the integers.

$$24 = 2 \cdot 2 \cdot 2 \cdot 3$$
$$60 = 2 \cdot 2 \cdot 3 \cdot 5$$
$$\text{GCF} = 2 \cdot 2 \cdot 3 = 12$$

The GCF of two or more monomials is the product of the GCF of the coefficients and the common variable factors.

$$6x^3y = 2 \cdot 3 \cdot x \cdot x \cdot x \cdot y$$
$$8x^2y^2 = 2 \cdot 2 \cdot 2 \cdot x \cdot x \cdot y \cdot y$$
$$\text{GCF} = 2 \cdot x \cdot x \cdot y = 2x^2y$$

Note that the exponent of each variable in the GCF is the same as the *smallest* exponent of that variable in either of the monomials.

The GCF of $6x^3y$ and $8x^2y^2$ is $2x^2y$.

Example 1   Find the GCF of $12a^4b$ and $18a^2b^2c$.

    Solution $12a^4b = 2 \cdot 2 \cdot 3 \cdot a^4 \cdot b$ ▶ Factor each monomial.
                 $18a^2b^2c = 2 \cdot 3 \cdot 3 \cdot a^2 \cdot b^2 \cdot c$

                 $\text{GCF} = 2 \cdot 3 \cdot a^2 \cdot b = 6a^2b$ ▶ The common variable factors are a^2 and b. c is not a common variable factor.

Problem 1   Find the GCF of $4x^6y$ and $18x^2y^6$.

    Solution   See page A31.

The Distributive Property is used to multiply factors of a polynomial. To **factor a polynomial** means to write the polynomial as a product of other polynomials.

     ┌──────── Multiply ────────┐
     Factors   **Polynomial**
     $2x(x + 5)$ $=$ $2x^2 + 10x$
     └──────── Factor ────────┘

In the example above, $2x$ is the GCF of the terms $2x^2$ and $10x$. It is a **common monomial factor** of the terms. $x + 5$ is a **binomial factor** of $2x^2 + 10x$.

Example 2   Factor.   A. $5x^3 - 35x^2 + 10x$ B. $16x^2y + 8x^4y^2 - 12x^4y^5$

    Solution   A. $5x^3 = 5 \cdot x^3$ ▶ Find the GCF of the terms
                $35x^2 = 5 \cdot 7 \cdot x^2$ of the polynomial.
                $10x = 2 \cdot 5 \cdot x$
                The GCF is $5x$.

                $\dfrac{5x^3}{5x} = x^2, \dfrac{-35x^2}{5x} = -7x, \dfrac{10x}{5x} = 2$ ▶ Divide each term of the polynomial by the GCF.

$$5x^3 - 35x^2 + 10x$$
$$= 5x(x^2) + 5x(-7x) + 5x(2)$$

▶ Use the quotients to rewrite the polynomial, expressing each term as a product with the GCF as one of the factors.

$$= 5x(x^2 - 7x + 2)$$

▶ Use the Distributive Property to write the polynomial as a product of factors.

B. $16x^2y = 2 \cdot 2 \cdot 2 \cdot 2 \cdot x^2 \cdot y$
$8x^4y^2 = 2 \cdot 2 \cdot 2 \cdot x^4 \cdot y^2$
$12x^4y^5 = 2 \cdot 2 \cdot 3 \cdot x^4 \cdot y^5$
The GCF is $4x^2y$.

$$16x^2y + 8x^4y^2 - 12x^4y^5$$
$$= 4x^2y(4) + 4x^2y(2x^2y) + 4x^2y(-3x^2y^4)$$
$$= 4x^2y(4 + 2x^2y - 3x^2y^4)$$

Problem 2   Factor.   A. $14a^2 - 21a^4b$ B. $6x^4y^2 - 9x^3y^2 + 12x^2y^4$

Solution   See page A31.

2   Factor by grouping

In the examples below, the binomials in parentheses are called **binomial factors.**

$$2a(a + b) \text{ and } 3xy(x - y)$$

The Distributive Property is used to factor a common binomial factor from an expression.

In the expression at the right, the common binomial factor is $y - 3$. The Distributive Property is used to write the expression as a product of factors.

$$x(y - 3) + 4(y - 3)$$
$$= (y - 3)(x + 4)$$

Example 3   Factor: $y(x + 2) + 3(x + 2)$

Solution $y(x + 2) + 3(x + 2)$ ▶ The common binomial factor is $x + 2$.
$$= (x + 2)(y + 3)$$

Problem 3   Factor: $a(b - 7) + b(b - 7)$

Solution   See page A31.

Sometimes a binomial factor must be rewritten before a common binomial factor can be found.

Factor: $a(a - b) + 5(b - a)$

$a - b$ and $b - a$ are different binomials.

Note that $(b - a) = (-a + b) = -(a - b)$.

Rewrite $(b - a)$ as $-(a - b)$ so that there is a common factor.

$$a(a - b) + 5(b - a) = a(a - b) + 5[-(a - b)]$$
$$= a(a - b) - 5(a - b)$$
$$= (a - b)(a - 5)$$

Example 4 Factor: $2x(x - 5) + y(5 - x)$

Solution $2x(x - 5) + y(5 - x)$
$= 2x(x - 5) - y(x - 5)$ ▶ Rewrite $5 - x$ as $-(x - 5)$ so that there is a common factor.
$= (x - 5)(2x - y)$ ▶ Write the expression as a product of factors.

Problem 4 Factor: $3y(5x - 2) - 4(2 - 5x)$

Solution See page A31.

Some polynomials can be factored by grouping the terms so that a common binomial factor is found.

Factor: $2x^3 - 3x^2 + 4x - 6$

Group the first two terms and the last two terms.

$$2x^3 - 3x^2 + 4x - 6$$
$$= (2x^3 - 3x^2) + (4x - 6)$$

Factor out the GCF from each group.

$$= x^2(2x - 3) + 2(2x - 3)$$

Write the expression as a product of factors.

$$= (2x - 3)(x^2 + 2)$$

Example 5 Factor: $3y^3 - 4y^2 - 6y + 8$

Solution $3y^3 - 4y^2 - 6y + 8$
$= (3y^3 - 4y^2) - (6y - 8)$ ▶ Group the first two terms and the last two terms. Note that $-6y + 8 = -(6y - 8)$.
$= y^2(3y - 4) - 2(3y - 4)$ ▶ Factor out the GCF from each group.
$= (3y - 4)(y^2 - 2)$ ▶ Write the expression as a product of factors.

Problem 5 Factor: $y^5 - 5y^3 + 4y^2 - 20$

Solution See page A31.

CONCEPT REVIEW 8.1

Determine whether the statement is always true, sometimes true, or never true.

1. To factor a polynomial means to rewrite it as multiplication.

2. The expression $3x(x + 5)$ is a sum, and the expression $3x^2 + 15x$ is a product.

3. The greatest common factor of two numbers is the largest number that divides evenly into both numbers.

4. x is a monomial factor of $x(2x - 1)$. $2x - 1$ is a binomial factor of $x(2x - 1)$.

5. $y(y + 4) + 6$ is in factored form.

6. A common monomial factor is a factor of every term of a polynomial.

7. $a^2 - 8a - ab + 8b = a(a - 8) - b(a + 8)$

8. The binomial $d + 9$ is a common binomial factor in $2c(d + 9) - 7(d + 9)$.

EXERCISES 8.1

1 Find the greatest common factor.

1. x^7, x^3
2. y^6, y^{12}
3. x^2y^4, xy^6
4. a^5b^3, a^3b^8

5. $x^2y^4z^6, xy^8z^2$
6. ab^2c^3, a^3b^2c
7. $14a^3, 49a^7$
8. $12y^2, 27y^4$

9. $3x^2y^2, 5ab^2$
10. $8x^2y^3, 7ab^4$
11. $9a^2b^4, 24a^4b^2$
12. $15a^4b^2, 9ab^5$

13. $ab^3, 4a^2b, 12a^2b^3$
14. $12x^2y, x^4y, 16x$
15. $2x^2y, 4xy, 8x$

16. $16x^2, 8x^4y^2, 12xy$
17. $3x^2y^2, 6x, 9x^3y^3$
18. $4a^2b^3, 8a^3, 12ab^4$

Factor.

19. $5a + 5$
20. $7b - 7$
21. $16 - 8a^2$
22. $12 + 12y^2$

23. $8x + 12$
24. $16a - 24$
25. $30a - 6$
26. $20b + 5$

27. $7x^2 - 3x$
28. $12y^2 - 5y$
29. $3a^2 + 5a^5$
30. $9x - 5x^2$

31. $14y^2 + 11y$
32. $6b^3 - 5b^2$
33. $2x^4 - 4x$
34. $3y^4 - 9y$

35. $10x^4 - 12x^2$
36. $12a^5 - 32a^2$
37. $x^2y - xy^3$
38. $a^2b + a^4b^2$

39. $2a^5b + 3xy^3$
40. $5x^2y - 7ab^3$
41. $6a^2b^3 - 12b^2$
42. $8x^2y^3 - 4x^2$

43. $6a^2bc + 4ab^2c$
44. $10x^2yz^2 + 15xy^3z$
45. $18x^2y^2 - 9a^2b^2$
46. $9a^2x - 27a^3x^3$

47. $6x^3y^3 - 12x^6y^6$
48. $3a^2b^2 - 12a^5b^5$
49. $x^3 - 3x^2 - x$
50. $a^3 + 4a^2 + 8a$

51. $2x^2 + 8x - 12$
52. $a^3 - 3a^2 + 5a$
53. $b^3 - 5b^2 - 7b$
54. $5x^2 - 15x + 35$

55. $8y^2 - 12y + 32$
56. $3x^3 + 6x^2 + 9x$
57. $5y^3 - 20y^2 + 10y$
58. $2x^4 - 4x^3 + 6x^2$

59. $3y^4 - 9y^3 - 6y^2$
60. $2x^3 + 6x^2 - 14x$
61. $3y^3 - 9y^2 + 24y$
62. $2y^5 - 3y^4 + 7y^3$

63. $6a^5 - 3a^3 - 2a^2$

64. $x^3y - 3x^2y^2 + 7xy^3$

65. $8x^2y^2 - 4x^2y + x^2$

66. $x^4y^4 - 3x^3y^3 + 6x^2y^2$

67. $4x^5y^5 - 8x^4y^4 + x^3y^3$

68. $16x^2y - 8x^3y^4 - 48x^2y^2$

2 Factor.

69. $x(a + b) + 2(a + b)$

70. $a(x + y) + 4(x + y)$

71. $x(b + 2) - y(b + 2)$

72. $a(y - 4) - b(y - 4)$

73. $a(x - 2) + 5(2 - x)$

74. $a(x - 7) + b(7 - x)$

75. $b(y - 3) + 3(3 - y)$

76. $c(a - 2) - b(2 - a)$

77. $a(x - y) - 2(y - x)$

78. $3(a - b) - x(b - a)$

79. $x^3 + 4x^2 + 3x + 12$

80. $x^3 - 4x^2 - 3x + 12$

81. $2y^3 + 4y^2 + 3y + 6$

82. $3y^3 - 12y^2 + y - 4$

83. $ab + 3b - 2a - 6$

84. $yz + 6z - 3y - 18$

85. $x^2a - 2x^2 - 3a + 6$

86. $x^2y + 4x^2 + 3y + 12$

87. $3ax - 3bx - 2ay + 2by$

88. $8 + 2c + 4a^2 + a^2c$

89. $x^2 - 3x + 4ax - 12a$

90. $t^2 + 4t - st - 4s$

91. $xy - 5y - 2x + 10$

92. $2y^2 - 10y + 7xy - 35x$

93. $21x^2 + 6xy - 49x - 14y$

94. $4a^2 + 5ab - 10b - 8a$

95. $2ra + a^2 - 2r - a$

96. $2ab - 3b^2 - 3b + 2a$

97. $4x^2 + 3xy - 12y - 16x$

98. $8s + 12r - 6s^2 - 9rs$

99. $10xy^2 - 15xy + 6y - 9$

100. $10a^2b - 15ab - 4a + 6$

SUPPLEMENTAL EXERCISES 8.1

Factor by grouping.

101. **a.** $2x^2 + 6x + 5x + 15$ **b.** $2x^2 + 5x + 6x + 15$

102. **a.** $3x^2 + 3xy - xy - y^2$ **b.** $3x^2 - xy + 3xy - y^2$

103. **a.** $2a^2 - 2ab - 3ab + 3b^2$ **b.** $2a^2 - 3ab - 2ab + 3b^2$

Compare your answers to parts (a) and (b) of Exercises 101–103 in order to answer Exercise 104.

104. Do different groupings of the terms in a polynomial affect the binomial factoring?

A whole number is a perfect number if it is the sum of all of its factors that are less than itself. For example, 6 is a perfect number because all the factors of 6 that are less than 6 are 1, 2, and 3, and $1 + 2 + 3 = 6$.

105. Find a perfect number between 20 and 30.   **106.** Find a perfect number between 490 and 500.

Solve.

107. In the equation $P = 2L + 2W$, what is the effect on P when the quantity $L + W$ doubles?

108. Write an expression in factored form for the shaded portion in the diagram.

a.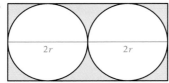

b.

S E C T I O N   **8.2**

Factoring Polynomials of the Form $x^2 + bx + c$

1 Factor trinomials of the form $x^2 + bx + c$

Trinomials of the form $x^2 + bx + c$, where b and c are integers, are shown at the right.

$x^2 + 9x + 14,$ $b = 9,$ $c = 14$
$x^2 - x - 12,$ $b = -1,$ $c = -12$
$x^2 - 2x - 15,$ $b = -2,$ $c = -15$

To factor a trinomial of this form means to express the trinomial as the product of two binomials. Some trinomials expressed as the product of binomials (factored form) are shown at the right.

Trinomial		Factored Form
$x^2 + 9x + 14$	$=$	$(x + 2)(x + 7)$
$x^2 - x - 12$	$=$	$(x + 3)(x - 4)$
$x^2 - 2x - 15$	$=$	$(x + 3)(x - 5)$

The method by which the factors of a trinomial are found is based on FOIL. Consider the following binomial products, noting the relationship between the constant terms of the binomials and the terms of the trinomials.

$(x + 6)(x + 2) = x^2 + 2x + 6x + (6)(2) = x^2 + 8x + 12$

Sum of 6 and 2

Product of 6 and 2

Signs in the binomials are the same

$(x - 3)(x - 4) = x^2 - 4x - 3x + (-3)(-4) = x^2 - 7x + 12$

Sum of -3 and -4

Product of -3 and -4

Signs in the binomials are opposite
$$\left\{ \begin{array}{l} (x + 3)(x - 5) = x^2 - 5x + 3x + (3)(-5) = x^2 - 2x - 15 \\ \qquad\qquad \text{Sum of } 3 \text{ and } -5 \\ \qquad\qquad \text{Product of } 3 \text{ and } -5 \\ (x - 4)(x + 6) = x^2 + 6x - 4x + (-4)(6) = x^2 + 2x - 24 \\ \qquad\qquad \text{Sum of } -4 \text{ and } 6 \\ \qquad\qquad \text{Product of } -4 \text{ and } 6 \end{array} \right.$$

Points to Remember in Factoring $x^2 + bx + c$

1. In the trinomial, the coefficient of x is the sum of the constant terms of the binomials.

2. In the trinomial, the constant term is the product of the constant terms of the binomials.

3. When the constant term of the trinomial is positive, the constant terms of the binomials have the same sign as the coefficient of x in the trinomial.

4. When the constant term of the trinomial is negative, the constant terms of the binomials have opposite signs.

Success at factoring a trinomial depends on remembering these four points. For example, to factor

$$x^2 - 2x - 24$$

find two numbers whose sum is -2 and whose product is -24 [Points 1 and 2]. Because the constant term of the trinomial is negative (-24), the numbers will have opposite signs [Point 4].

A systematic method of finding these numbers involves listing the factors of the constant term of the trinomial and the sum of those factors.

Factors of -24	Sum of the Factors
1, -24	$1 + (-24) = -23$
-1, 24	$-1 + 24 = 23$
2, -12	$2 + (-12) = -10$
-2, 12	$-2 + 12 = 10$
3, -8	$3 + (-8) = -5$
-3, 8	$-3 + 8 = 5$
4, -6	**$4 + (-6) = -2$**
-4, 6	$-4 + 6 = 2$

4 and -6 are two numbers whose sum is -2 and whose product is -24. Write the binomial factors of the trinomial.

$$x^2 - 2x - 24 = (x + 4)(x - 6)$$

LOOK CLOSELY

Always check your proposed factorization to ensure accuracy.

Check: $(x + 4)(x - 6) = x^2 - 6x + 4x - 24 = x^2 - 2x - 24$

By the Commutative Property of Multiplication, the binomial factors can also be written as

$$x^2 - 2x - 24 = (x - 6)(x + 4)$$

Example 1   Factor: $x^2 + 18x + 32$

Solution

Factors of 32	Sum
1, 32	33
2, 16	**18**
4, 8	12

▶ Try only positive factors of 32 [Point 3].

▶ Once the correct pair is found, the other factors need not be tried.

$x^2 + 18x + 32 = (x + 2)(x + 16)$ ▶ Write the factors of the trinomial.

Check $(x + 2)(x + 16) = x^2 + 16x + 2x + 32$
$$= x^2 + 18x + 32$$

Problem 1   Factor: $x^2 - 8x + 15$

Solution   See page A32.

Example 2   Factor: $x^2 - 6x - 16$

Solution

Factors of −16	Sum
1, −16	−15
−1, 16	15
2, −8	**−6**
−2, 8	6
4, −4	0

▶ The factors must be of opposite signs [Point 4].

$x^2 - 6x - 16 = (x + 2)(x - 8)$ ▶ Write the factors of the trinomial.

Check $(x + 2)(x - 8) = x^2 - 8x + 2x - 16$
$$= x^2 - 6x - 16$$

Problem 2   Factor: $x^2 + 3x - 18$

Solution   See page A32.

Not all trinomials can be factored when using only integers. Consider the trinomial $x^2 - 6x - 8$.

Factors of −8	Sum
1, −8	−7
−1, 8	7
2, −4	−2
−2, 4	2

Because none of the pairs of factors of −8 have a sum of −6, the trinomial is not factorable. The trinomial is said to be **nonfactorable over the integers.**

2 Factor completely

A polynomial is factored completely when it is written as a product of factors that are nonfactorable over the integers.

Example 3   Factor: $3x^3 + 15x^2 + 18x$

Solution   The GCF of $3x^3$, $15x^2$, and $18x$ is $3x$.   ▶ Find the GCF of the terms of the polynomial.

$3x^3 + 15x^2 + 18x$
$= 3x(x^2) + 3x(5x) + 3x(6)$ ▶ Factor out the GCF.
$= 3x(x^2 + 5x + 6)$ ▶ Write the polynomial as a product of factors.

Factors of 6	Sum
1, 6	7
2, 3	5

▶ Factor the trinomial $x^2 + 5x + 6$. Try only positive factors of 6.

$3x^3 + 15x^2 + 18x = 3x(x + 2)(x + 3)$

Check $3x(x + 2)(x + 3) = 3x(x^2 + 3x + 2x + 6)$
$= 3x(x^2 + 5x + 6)$
$= 3x^3 + 15x^2 + 18x$

Problem 3   Factor: $3a^2b - 18ab - 81b$

Solution   See page A32.

Example 4   Factor: $x^2 + 9xy + 20y^2$

Solution

Factors of 20	Sum
1, 20	21
2, 10	12
4, 5	9

▶ Try only positive factors of 20.

$x^2 + 9xy + 20y^2 = (x + 4y)(x + 5y)$

Check $(x + 4y)(x + 5y) = x^2 + 5xy + 4xy + 20y^2$
$= x^2 + 9xy + 20y^2$

Problem 4   Factor: $4x^2 - 40xy + 84y^2$

Solution   See page A32.

CONCEPT REVIEW 8.2

Determine whether the statement is always true, sometimes true, or never true.

1.   The value of b in the trinomial $x^2 - 8x + 7$ is 8.

2.   The factored form of $x^2 - 5x - 14$ is $(x - 7)(x + 2)$.

3. To factor a trinomial of the form $x^2 + bx + c$ means to rewrite the polynomial as a product of two binomials.

4. In the factoring of a trinomial, if the constant term is positive, then the signs in both binomial factors will be the same.

5. In the factoring of a trinomial, if the constant term is negative, then the signs in both binomial factors will be negative.

6. The first step in factoring a trinomial is to determine whether the terms of the trinomial have a common factor.

EXERCISES 8.2

1   Factor.

1. $x^2 + 3x + 2$
2. $x^2 + 5x + 6$
3. $x^2 - x - 2$
4. $x^2 + x - 6$

5. $a^2 + a - 12$
6. $a^2 - 2a - 35$
7. $a^2 - 3a + 2$
8. $a^2 - 5a + 4$

9. $a^2 + a - 2$
10. $a^2 - 2a - 3$
11. $b^2 - 6b + 9$
12. $b^2 + 8b + 16$

13. $b^2 + 7b - 8$
14. $y^2 - y - 6$
15. $y^2 + 6y - 55$
16. $z^2 - 4z - 45$

17. $y^2 - 5y + 6$
18. $y^2 - 8y + 15$
19. $z^2 - 14z + 45$
20. $z^2 - 14z + 49$

21. $z^2 - 12z - 160$
22. $p^2 + 2p - 35$
23. $p^2 + 12p + 27$
24. $p^2 - 6p + 8$

25. $x^2 + 20x + 100$
26. $x^2 + 18x + 81$
27. $b^2 + 9b + 20$
28. $b^2 + 13b + 40$

29. $x^2 - 11x - 42$
30. $x^2 + 9x - 70$
31. $b^2 - b - 20$
32. $b^2 + 3b - 40$

33. $y^2 - 14y - 51$
34. $y^2 - y - 72$
35. $p^2 - 4p - 21$
36. $p^2 + 16p + 39$

37. $y^2 - 8y + 32$
38. $y^2 - 9y + 81$
39. $x^2 - 20x + 75$
40. $p^2 + 24p + 63$

41. $x^2 - 15x + 56$
42. $x^2 + 21x + 38$
43. $x^2 + x - 56$
44. $x^2 + 5x - 36$

45. $a^2 - 21a - 72$
46. $a^2 - 7a - 44$
47. $a^2 - 15a + 36$
48. $a^2 - 21a + 54$

49. $z^2 - 9z - 136$
50. $z^2 + 14z - 147$
51. $c^2 - c - 90$
52. $c^2 - 3c - 180$

2   Factor.

53. $2x^2 + 6x + 4$
54. $3x^2 + 15x + 18$
55. $3a^2 + 3a - 18$

56. $4x^2 - 4x - 8$
57. $ab^2 + 2ab - 15a$
58. $ab^2 + 7ab - 8a$

59. $xy^2 - 5xy + 6x$

60. $xy^2 + 8xy + 15x$

61. $z^3 - 7z^2 + 12z$

62. $2a^3 + 6a^2 + 4a$

63. $3y^3 - 15y^2 + 18y$

64. $4y^3 + 12y^2 - 72y$

65. $3x^2 + 3x - 36$

66. $2x^3 - 2x^2 - 4x$

67. $5z^2 - 15z - 140$

68. $6z^2 + 12z - 90$

69. $2a^3 + 8a^2 - 64a$

70. $3a^3 - 9a^2 - 54a$

71. $x^2 - 5xy + 6y^2$

72. $x^2 + 4xy - 21y^2$

73. $a^2 - 9ab + 20b^2$

74. $a^2 - 15ab + 50b^2$

75. $x^2 - 3xy - 28y^2$

76. $s^2 + 2st - 48t^2$

77. $y^2 - 15yz - 41z^2$

78. $y^2 + 85yz + 36z^2$

79. $z^4 - 12z^3 + 35z^2$

80. $z^4 + 2z^3 - 80z^2$

81. $b^4 - 22b^3 + 120b^2$

82. $b^4 - 3b^3 - 10b^2$

83. $2y^4 - 26y^3 - 96y^2$

84. $3y^4 + 54y^3 + 135y^2$

85. $x^4 + 7x^3 - 8x^2$

86. $x^4 - 11x^3 - 12x^2$

87. $4x^2y + 20xy - 56y$

88. $3x^2y - 6xy - 45y$

89. $8y^2 - 32y + 24$

90. $10y^2 - 100y + 90$

91. $c^3 + 13c^2 + 30c$

92. $c^3 + 18c^2 - 40c$

93. $3x^3 - 36x^2 + 81x$

94. $4x^3 + 4x^2 - 24x$

95. $x^2 - 8xy + 15y^2$

96. $y^2 - 7xy - 8x^2$

97. $a^2 - 13ab + 42b^2$

98. $y^2 + 4yz - 21z^2$

99. $y^2 + 8yz + 7z^2$

100. $y^2 - 16yz + 15z^2$

101. $3x^2y + 60xy - 63y$

102. $4x^2y - 68xy - 72y$

103. $3x^3 + 3x^2 - 36x$

104. $4x^3 + 12x^2 - 160x$

105. $4z^3 + 32z^2 - 132z$

106. $5z^3 - 50z^2 - 120z$

SUPPLEMENTAL EXERCISES 8.2

Factor.

107. $20 + c^2 + 9c$

108. $x^2y - 54y - 3xy$

109. $45a^2 + a^2b^2 - 14a^2b$

110. $12p^2 - 96p + 3p^3$

Find all integers k such that the trinomial can be factored over the integers.

111. $x^2 + kx + 35$

112. $x^2 + kx + 18$

113. $x^2 - kx + 21$

114. $x^2 - kx + 14$

Determine the positive integer values of k for which the following polynomials are factorable over the integers.

115. $y^2 + 4y + k$ **116.** $z^2 + 7z + k$ **117.** $a^2 - 6a + k$

118. $c^2 - 7c + k$ **119.** $x^2 - 3x + k$ **120.** $y^2 + 5y + k$

121. Exercises 115–120 included the requirement that $k > 0$. If k is allowed to be any integer, how many different values of k are possible for each polynomial? Explain your answer.

S E C T I O N   **8.3**

Factoring Polynomials of the Form $ax^2 + bx + c$

1   Factor trinomials of the form $ax^2 + bx + c$ using trial factors

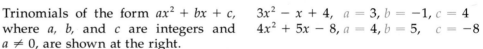

Trinomials of the form $ax^2 + bx + c$, where a, b, and c are integers and $a \neq 0$, are shown at the right.

$3x^2 - x + 4$, $a = 3, b = -1, c = 4$
$4x^2 + 5x - 8$, $a = 4, b = 5,$ $c = -8$

These trinomials differ from those in the previous section in that the coefficient of x^2 is not 1. There are various methods of factoring these trinomials. The method described in this objective is factoring trinomials using trial factors.

To **factor a trinomial of the form $ax^2 + bx + c$** means to express the polynomial as the product of two binomials. Factoring such polynomials by trial and error may require testing many trial factors. To reduce the number of trial factors, remember the following points.

Points to Remember in Factoring $ax^2 + bx + c$

1. If the terms of the trinomial do not have a common factor, then the terms of a binomial factor cannot have a common factor.
2. When the constant term of the trinomial is positive, the constant terms of the binomials have the same sign as the coefficient of x in the trinomial.
3. When the constant term of the trinomial is negative, the constant terms of the binomials have opposite signs.

Factor: $10x^2 - x - 3$

The terms of the trinomial do not have a common factor; therefore, a binomial factor will not have a common factor.

Because the constant term, c, of the trinomial is negative (-3), the constant terms of the binomial factors will have opposite signs [Point 3].

Find the factors of a (10) and the factors of c (−3).

Factors of 10	Factors of −3
1, 10	1, −3
2, 5	−1, 3

Using these factors, write trial factors, and use the Outer and Inner products of FOIL to check the middle term.

Trial Factors	Middle Term
$(x + 1)(10x - 3)$	$-3x + 10x = 7x$
$(x - 1)(10x + 3)$	$3x - 10x = -7x$
$(2x + 1)(5x - 3)$	$-6x + 5x = -x$
$(2x - 1)(5x + 3)$	$6x - 5x = x$
$(10x + 1)(x - 3)$	$-30x + x = -29x$
$(10x - 1)(x + 3)$	$30x - x = 29x$
$(5x + 1)(2x - 3)$	$-15x + 2x = -13x$
$(5x - 1)(2x + 3)$	$15x - 2x = 13x$

From the list of trial factors, $10x^2 - x - 3 = (2x + 1)(5x - 3)$.

Check: $(2x + 1)(5x - 3) = 10x^2 - 6x + 5x - 3$
$$= 10x^2 - x - 3$$

All the trial factors for this trinomial were listed in this example. However, once the correct binomial factors are found, it is not necessary to continue checking the remaining trial factors.

Factor: $4x^2 - 27x + 18$

The terms of the trinomial do not have a common factor; therefore, a binomial factor will not have a common factor.

Because the constant term, c, of the trinomial is positive (18), the constant terms of the binomial factors will have the same sign as the coefficient of x. Because the coefficient of x is −27, both signs will be negative [Point 2].

Find the factors of a (4) and the negative factors of c (18).

Factors of 4	Factors of 18
1, 4	−1, −18
2, 2	−2, −9
	−3, −6

Using these factors, write trial factors, and use the Outer and Inner products of FOIL to check the middle term.

Trial Factors	Middle Term
$(x - 1)(4x - 18)$	Common factor
$(x - 2)(4x - 9)$	$-9x - 8x = -17x$
$(x - 3)(4x - 6)$	Common factor
$(2x - 1)(2x - 18)$	Common factor
$(2x - 2)(2x - 9)$	Common factor
$(2x - 3)(2x - 6)$	Common factor
$(4x - 1)(x - 18)$	$-72x - x = -73x$
$(4x - 2)(x - 9)$	Common factor
$(4x - 3)(x - 6)$	$-24x - 3x = -27x$

The correct factors have been found. The remaining trial factors need not be checked.

$$4x^2 - 27x + 18 = (4x - 3)(x - 6)$$

The second example on page 355 illustrates that many of the trial factors may have common factors and thus need not be tried. For the remainder of this chapter, the trial factors with a common factor will not be listed.

Example 1 Factor: $3x^2 + 20x + 12$

Solution

Factors of 3	Factors of 12
1, 3	1, 12
	2, 6
	3, 4

▶ Because 20 is positive, only the positive factors of 12 need be tried.

Trial Factors	Middle Term
$(x + 3)(3x + 4)$	$4x + 9x = 13x$
$(3x + 1)(x + 12)$	$36x + x = 37x$
$(3x + 2)(x + 6)$	$18x + 2x = 20x$

▶ Write the trial factors. Use FOIL to check the middle term.

$3x^2 + 20x + 12 = (3x + 2)(x + 6)$

Check $(3x + 2)(x + 6) = 3x^2 + 18x + 2x + 12$
$= 3x^2 + 20x + 12$

Problem 1 Factor: $6x^2 - 11x + 5$

Solution See page A32.

Example 2 Factor: $6x^2 - 5x - 6$

Solution

Factors of 6	Factors of −6
1, 6	1, −6
2, 3	−1, 6
	2, −3
	−2, 3

▶ Find the factors of a (6) and the factors of c (−6).

Trial Factors	Middle Term
$(x - 6)(6x + 1)$	$x - 36x = -35x$
$(x + 6)(6x - 1)$	$-x + 36x = 35x$
$(2x - 3)(3x + 2)$	$4x - 9x = -5x$

▶ Write the trial factors. Use FOIL to check the middle term.

$6x^2 - 5x - 6 = (2x - 3)(3x + 2)$

Check $(2x - 3)(3x + 2) = 6x^2 + 4x - 9x - 6$
$= 6x^2 - 5x - 6$

Problem 2 Factor: $8x^2 + 14x - 15$

Solution See page A32.

Example 3 Factor: $15 - 2x - x^2$

Solution

Factors of 15	Factors of -1
1, 15	1, -1
3, 5	

▶ The terms have no common factor. The coefficient of x^2 is -1.

Trial Factors	Middle Term
$(1 + x)(15 - x)$	$-x + 15x = 14x$
$(1 - x)(15 + x)$	$x - 15x = -14x$
$(3 + x)(5 - x)$	$-3x + 5x = 2x$
$(3 - x)(5 + x)$	$3x - 5x = -2x$

▶ Write the trial factors. Use FOIL to check the middle term.

$$15 - 2x - x^2 = (3 - x)(5 + x)$$

Check $(3 - x)(5 + x) = 15 + 3x - 5x - x^2$
$$= 15 - 2x - x^2$$

Problem 3 Factor: $24 - 2y - y^2$

Solution See page A33.

The first step in factoring a trinomial is to determine whether there is a common factor. If there is a common factor, factor out the GCF of the terms.

Example 4 Factor: $3x^3 - 23x^2 + 14x$

Solution The GCF of $3x^3$, $23x^2$, and $14x$ is x.

▶ Find the GCF of the terms of the polynomial.

$$3x^3 - 23x^2 + 14x = x(3x^2 - 23x + 14)$$

▶ Factor out the GCF.

Factors of 3	Factors of 14
1, 3	-1, -14
	-2, -7

▶ Factor the trinomial $3x^2 - 23x + 14$.

Trial Factors	Middle Term
$(x - 1)(3x - 14)$	$-14x - 3x = -17x$
$(x - 14)(3x - 1)$	$-x - 42x = -43x$
$(x - 2)(3x - 7)$	$-7x - 6x = -13x$
$(x - 7)(3x - 2)$	$-2x - 21x = -23x$

$$3x^3 - 23x^2 + 14x = x(x - 7)(3x - 2)$$

Check $x(x - 7)(3x - 2) = x(3x^2 - 2x - 21x + 14)$
$$= x(3x^2 - 23x + 14)$$
$$= 3x^3 - 23x^2 + 14x$$

Problem 4 Factor: $4a^2b^2 - 30a^2b + 14a^2$

Solution See page A33.

2   Factor trinomials of the form $ax^2 + bx + c$
by grouping

In the previous objective, trinomials of the form $ax^2 + bx + c$ were factored using trial factors. In this objective, factoring by grouping is used.

To factor $ax^2 + bx + c$, first find two factors of $a \cdot c$ whose sum is b. Use the two factors to rewrite the middle term of the trinomial as the sum of two terms. Then use factoring by grouping to write the factorization of the trinomial.

Factor: $2x^2 + 13x + 15$

$a = 2, c = 15, a \cdot c = 2 \cdot 15 = 30$

Find two positive factors of 30 whose sum is 13.

Positive Factors of 30	Sum
1, 30	31
2, 15	17
3, 10	13
5, 6	11

The factors are 3 and 10.

Use the factors 3 and 10 to rewrite $13x$ as $3x + 10x$.

$$2x^2 + 13x + 15$$
$$= 2x^2 + 3x + 10x + 15$$

Factor by grouping.

$$= (2x^2 + 3x) + (10x + 15)$$
$$= x(2x + 3) + 5(2x + 3)$$
$$= (2x + 3)(x + 5)$$

Check: $(2x + 3)(x + 5) = 2x^2 + 10x + 3x + 15$
$$= 2x^2 + 13x + 15$$

Factor: $6x^2 - 11x - 10$

$a = 6, c = -10, a \cdot c = 6(-10) = -60$

Find two factors of -60 whose sum is -11.

Factors of −60	Sum
1, −60	−59
−1, 60	59
2, −30	−28
−2, 30	28
3, −20	−17
−3, 20	17
4, −15	−11

The required sum has been found. The remaining factors need not be checked. The factors are 4 and -15.

Use the factors 4 and -15 to rewrite $-11x$ as $4x - 15x$.

$$6x^2 - 11x - 10$$
$$= 6x^2 + 4x - 15x - 10$$

Factor by grouping.
Note: $-15x - 10 = -(15x + 10)$

$$= (6x^2 + 4x) - (15x + 10)$$
$$= 2x(3x + 2) - 5(3x + 2)$$
$$= (3x + 2)(2x - 5)$$

Check: $(3x + 2)(2x - 5) = 6x^2 - 15x + 4x - 10$
$$= 6x^2 - 11x - 10$$

Factor: $3x^2 - 2x - 4$

$a = 3, c = -4, a \cdot c = 3(-4) = -12$

Find two factors of -12 whose sum is -2.

Factors of -12	Sum
1, -12	-11
-1, 12	11
2, -6	-4
-2, 6	4
3, -4	-1
-3, 4	1

No integer factors of -12 have a sum of -2. Therefore, $3x^2 - 2x - 4$ is nonfactorable over the integers.

Example 5 Factor: $2x^2 + 19x - 10$

Solution $a \cdot c = 2(-10) = -20$ ▶ Find $a \cdot c$.

$-1(20) = -20$ ▶ Find two numbers whose product is -20 and whose sum is 19.
$-1 + 20 = 19$

$2x^2 + 19x - 10$
$= 2x^2 - x + 20x - 10$ ▶ Rewrite $19x$ as $-x + 20x$.
$= (2x^2 - x) + (20x - 10)$ ▶ Factor by grouping.
$= x(2x - 1) + 10(2x - 1)$
$= (2x - 1)(x + 10)$

Problem 5 Factor: $2a^2 + 13a - 7$

Solution See page A33.

Example 6 Factor: $8y^2 - 10y - 3$

Solution $a \cdot c = 8(-3) = -24$ ▶ Find $a \cdot c$.

$2(-12) = -24$ ▶ Find two numbers whose product is -24 and whose sum is -10.
$2 + (-12) = -10$

$8y^2 - 10y - 3$
$= 8y^2 + 2y - 12y - 3$ ▶ Rewrite $-10y$ as $2y - 12y$.
$= (8y^2 + 2y) - (12y + 3)$ ▶ Factor by grouping.
$= 2y(4y + 1) - 3(4y + 1)$
$= (4y + 1)(2y - 3)$

Problem 6 Factor: $4a^2 - 11a - 3$

Solution See page A33.

Remember that the first step in factoring a trinomial is to determine whether there is a common factor. If there is a common factor, factor out the GCF of the terms.

Example 7   Factor: $24x^2y - 76xy + 40y$

Solution $24x^2y - 76xy + 40y$ ▶ The terms of the polynomial
    $= 4y(6x^2 - 19x + 10)$ have a common factor, $4y$. Fac-
        tor out the GCF.

$a \cdot c = 6(10) = 60$ ▶ To factor $6x^2 - 19x + 10$, first
        find $a \cdot c$.

$-4(-15) = 60$ ▶ Find two numbers whose prod-
$-4 + (-15) = -19$ uct is 60 and whose sum is -19.

$6x^2 - 19x + 10$
    $= 6x^2 - 4x - 15x + 10$ ▶ Rewrite $-19x$ as $-4x - 15x$.
    $= (6x^2 - 4x) - (15x - 10)$ ▶ Factor by grouping.
    $= 2x(3x - 2) - 5(3x - 2)$
    $= (3x - 2)(2x - 5)$

$24x^2y - 76xy + 40y$ ▶ Write the complete factorization
    $= 4y(6x^2 - 19x + 10)$ of the given polynomial.
    $= 4y(3x - 2)(2x - 5)$

Problem 7   Factor: $15x^3 + 40x^2 - 80x$

Solution   See page A33.

CONCEPT REVIEW 8.3

Determine whether the statement is always true, sometimes true, or never true.

1. The value of a in the trinomial $3x^2 + 8x - 7$ is -7.

2. The factored form of $2x^2 - 7x - 15$ is $(2x + 3)(x - 5)$.

3. To check the factorization of a trinomial of the form $ax^2 + bx + c$, multiply the two binomial factors.

4. The terms of the binomial $3x - 9$ have a common factor.

5. If the terms of a trinomial of the form $ax^2 + bx + c$ do not have a common factor, then one of its binomial factors cannot have a common factor.

6. To factor a trinomial of the form $ax^2 + bx + c$ by grouping, first find two numbers whose product is ac and whose sum is b.

EXERCISES 8.3

1   Factor by the method of using trial factors.

1. $2x^2 + 3x + 1$ 2. $5x^2 + 6x + 1$ 3. $2y^2 + 7y + 3$ 4. $3y^2 + 7y + 2$

5. $2a^2 - 3a + 1$ 6. $3a^2 - 4a + 1$ 7. $2b^2 - 11b + 5$ 8. $3b^2 - 13b + 4$

9. $2x^2 + x - 1$ **10.** $4x^2 - 3x - 1$ **11.** $2x^2 - 5x - 3$ **12.** $3x^2 + 5x - 2$

13. $6z^2 - 7z + 3$ **14.** $9z^2 + 3z + 2$ **15.** $6t^2 - 11t + 4$ **16.** $10t^2 + 11t + 3$

17. $8x^2 + 33x + 4$ **18.** $7x^2 + 50x + 7$ **19.** $3b^2 - 16b + 16$ **20.** $6b^2 - 19b + 15$

21. $2z^2 - 27z - 14$ **22.** $4z^2 + 5z - 6$ **23.** $3p^2 + 22p - 16$ **24.** $7p^2 + 19p + 10$

25. $6x^2 - 17x + 12$ **26.** $15x^2 - 19x + 6$ **27.** $5b^2 + 33b - 14$ **28.** $8x^2 - 30x + 25$

29. $6a^2 + 7a - 24$ **30.** $14a^2 + 15a - 9$ **31.** $18t^2 - 9t - 5$ **32.** $12t^2 + 28t - 5$

33. $15a^2 + 26a - 21$ **34.** $6a^2 + 23a + 21$ **35.** $8y^2 - 26y + 15$ **36.** $18y^2 - 27y + 4$

37. $8z^2 + 2z - 15$ **38.** $10z^2 + 3z - 4$ **39.** $3x^2 + 14x - 5$ **40.** $15a^2 - 22a + 8$

41. $12x^2 + 25x + 12$ **42.** $10b^2 + 43b - 9$ **43.** $3z^2 + 95z + 10$ **44.** $8z^2 - 36z + 1$

45. $3x^2 + xy - 2y^2$ **46.** $6x^2 + 10xy + 4y^2$ **47.** $28 + 3z - z^2$

48. $15 - 2z - z^2$ **49.** $8 - 7x - x^2$ **50.** $12 + 11x - x^2$

51. $9x^2 + 33x - 60$ **52.** $16x^2 - 16x - 12$ **53.** $24x^2 - 52x + 24$

54. $60x^2 + 95x + 20$ **55.** $35a^4 + 9a^3 - 2a^2$ **56.** $15a^4 + 26a^3 + 7a^2$

57. $15b^2 - 115b + 70$ **58.** $25b^2 + 35b - 30$ **59.** $10x^3 + 12x^2 + 2x$

60. $9x^3 - 39x^2 + 12x$ **61.** $10y^3 - 44y^2 + 16y$ **62.** $14y^3 + 94y^2 - 28y$

63. $4yz^3 + 5yz^2 - 6yz$ **64.** $2yz^3 - 17yz^2 + 8yz$ **65.** $20b^4 + 41b^3 + 20b^2$

66. $6b^4 - 13b^3 + 6b^2$ **67.** $9x^3y + 12x^2y + 4xy$ **68.** $9a^3b - 9a^2b^2 - 10ab^3$

2   Factor by grouping.

69. $2t^2 - t - 10$ **70.** $2t^2 + 5t - 12$ **71.** $3p^2 - 16p + 5$ **72.** $6p^2 + 5p + 1$

73. $12y^2 - 7y + 1$ **74.** $6y^2 - 5y + 1$ **75.** $5x^2 - 62x - 7$ **76.** $9x^2 - 13x - 4$

77. $12y^2 + 19y + 5$ **78.** $5y^2 - 22y + 8$ **79.** $7a^2 + 47a - 14$ **80.** $11a^2 - 54a - 5$

81. $4z^2 + 11z + 6$ **82.** $6z^2 - 25z + 14$ **83.** $22p^2 + 51p - 10$ **84.** $14p^2 - 41p + 15$

85. $8y^2 + 17y + 9$ **86.** $12y^2 - 145y + 12$ **87.** $6b^2 - 13b + 6$ **88.** $20b^2 + 37b + 15$

89. $33b^2 + 34b - 35$ **90.** $15b^2 - 43b + 22$ **91.** $18y^2 - 39y + 20$ **92.** $24y^2 + 41y + 12$

93. $15x^2 - 82x + 24$ **94.** $13z^2 + 49z - 8$ **95.** $10z^2 - 29z + 10$ **96.** $15z^2 - 44z + 32$

97. $36z^2 + 72z + 35$ **98.** $16z^2 + 8z - 35$ **99.** $14y^2 - 29y + 12$ **100.** $8y^2 + 30y + 25$

101. $6x^2 + 35x - 6$ **102.** $4x^2 + 6x + 2$ **103.** $12x^2 + 33x - 9$ **104.** $15y^2 - 50y + 35$

105. $30y^2 + 10y - 20$ **106.** $2x^3 - 11x^2 + 5x$ **107.** $2x^3 - 3x^2 - 5x$ **108.** $3a^2 + 5ab - 2b^2$

109. $2a^2 - 9ab + 9b^2$ **110.** $4y^2 - 11yz + 6z^2$ **111.** $2y^2 + 7yz + 5z^2$ **112.** $12 - x - x^2$

113. $18 + 17x - x^2$ **114.** $21 - 20x - x^2$ **115.** $360y^2 + 4y - 4$ **116.** $10t^2 - 5t - 50$

117. $16t^2 + 40t - 96$ **118.** $3p^3 - 16p^2 + 5p$ **119.** $6p^3 + 5p^2 + p$ **120.** $26z^2 + 98z - 24$

121. $30z^2 - 87z + 30$ **122.** $12a^3 + 14a^2 - 48a$ **123.** $42a^3 + 45a^2 - 27a$ **124.** $36p^2 - 9p^3 - p^4$

125. $9x^2y - 30xy^2 + 25y^3$ **126.** $8x^2y - 38xy^2 + 35y^3$

127. $9x^3y - 24x^2y^2 + 16xy^3$ **128.** $45a^3b - 78a^2b^2 + 24ab^3$

SUPPLEMENTAL EXERCISES 8.3

Factor.

129. $6y + 8y^3 - 26y^2$ **130.** $22p^2 - 3p^3 + 16p$ **131.** $a^3b - 24ab - 2a^2b$

132. $3xy^2 - 14xy + 2xy^3$ **133.** $25t^2 + 60t - 10t^3$ **134.** $3xy^3 + 2x^3y - 7x^2y^2$

Factor.

135. $2(y + 2)^2 - (y + 2) - 3$ **136.** $3(a + 2)^2 - (a + 2) - 4$

137. $10(x + 1)^2 - 11(x + 1) - 6$ **138.** $4(y - 1)^2 - 7(y - 1) - 2$

Find all integers k such that the trinomial can be factored over the integers.

139. $2x^2 + kx + 3$ **140.** $2x^2 + kx - 3$

141. $3x^2 + kx + 2$ **142.** $3x^2 + kx - 2$

143. $2x^2 + kx + 5$ **144.** $2x^2 + kx - 5$

145. Given that $x + 2$ is a factor of $x^3 - 2x^2 - 5x + 6$, factor $x^3 - 2x^2 - 5x + 6$ completely.

146. In your own words, explain how the signs of the last terms of the two binomial factors of a trinomial are determined.

147. The area of a rectangle is $(3x^2 + x - 2)$ ft². Find the dimensions of the rectangle in terms of the variable x. Given that $x > 0$, specify the dimension that is the length and the dimension that is the width. Can $x < 0$? Can $x = 0$? Explain your answers.

$A = 3x^2 + x - 2$

S E C T I O N **8.4**

Special Factoring

1 Factor the difference of two squares and perfect-square trinomials

POINT OF INTEREST

See page 342 for some geometric proofs for factoring the difference of two squares and perfect-square trinomials.

Recall from Objective 4 in Section 7.3 that the product of the sum and difference of the same two terms equals the square of the first term minus the square of the second term.

$$(a + b)(a - b) = a^2 - b^2$$

The expression $a^2 - b^2$ is the difference of two squares. The pattern above suggests the following rule for factoring the difference of two squares.

Rule for Factoring the Difference of Two Squares

Difference of Two Squares		Sum and Difference of Two Terms
$a^2 - b^2$	$=$	$(a + b)(a - b)$

LOOK CLOSELY

Convince yourself that the sum of two squares is nonfactorable over the integers by trying to factor $x^2 + 4$.

$a^2 + b^2$ is the *sum* of two squares. It is nonfactorable over the integers.

Example 1   Factor.   A. $x^2 - 16$ B. $x^2 - 10$ C. $z^6 - 25$

Solution   A. $x^2 - 16 = x^2 - 4^2$
                          ► Write $x^2 - 16$ as the difference of two squares.
                 $= (x + 4)(x - 4)$ ► The factors are the sum and difference of the terms x and 4.

B. $x^2 - 10$ is nonfactorable over the integers.
                          ► Because 10 is not a square, $x^2 - 10$ cannot be written as the difference of two squares.

C. $z^6 - 25 = (z^3)^2 - 5^2$
                          ► Write $z^6 - 25$ as the difference of two squares.
                 $= (z^3 + 5)(z^3 - 5)$ ► The factors are the sum and difference of the terms z^3 and 5.

Problem 1   Factor.   A. $25a^2 - b^2$ B. $6x^2 - 1$ C. $n^8 - 36$

Solution   See page A33.

Example 2   Factor: $z^4 - 16$

Solution $z^4 - 16 = (z^2)^2 - (4)^2$ ▶ This is the difference of two squares.

$= (z^2 + 4)(z^2 - 4)$ ▶ The factors are the sum and difference of the terms z^2 and 4.

$= (z^2 + 4)(z + 2)(z - 2)$ ▶ Factor $z^2 - 4$, which is the difference of two squares. $z^2 + 4$ is nonfactorable over the integers.

Problem 2   Factor: $n^4 - 81$

Solution   See page A33.

Recall from Objective 4 in Section 7.3 the pattern for finding the square of a binomial.

$$(a + b)^2 = (a + b)(a + b) = a^2 + ab + ab + b^2$$
$$= a^2 + 2ab + b^2$$

Square of first term —
Twice the product of the two terms —
Square of last term —

The square of a binomial is a perfect-square trinomial. The pattern above suggests the following rule for factoring a perfect-square trinomial.

Rule for Factoring a Perfect-Square Trinomial

Perfect-Square Trinomial			Square of a Binomial
$a^2 + 2ab + b^2$	=	$(a + b)(a + b)$	=	$(a + b)^2$
$a^2 - 2ab + b^2$	=	$(a - b)(a - b)$	=	$(a - b)^2$

Note in these patterns that the sign in the binomial is the sign of the middle term of the trinomial.

Factor: $4x^2 - 20x + 25$

Check that the first term and the last term are squares. $4x^2 = (2x)^2, 25 = 5^2$

Use the squared terms to factor the trinomial as the square of a binomial. The sign of the binomial is the sign of the middle term of the trinomial. $(2x - 5)^2$

Check the factorization. $(2x - 5)^2$
$= (2x)^2 + 2(2x)(-5) + (-5)^2$
$= 4x^2 - 20x + 25$

The factorization is correct. $4x^2 - 20x + 25 = (2x - 5)^2$

Factor: $9x^2 + 30x + 16$

Check that the first term and the last term are squares.

$9x^2 = (3x)^2,\ 16 = 4^2$

Use the squared terms to factor the trinomial as the square of a binomial. The sign of the binomial is the sign of the middle term in the trinomial.

$(3x + 4)^2$

Check the factorization.

$(3x + 4)^2$
$= (3x)^2 + 2(3x)(4) + 4^2$
$= 9x^2 + 24x + 16$

$9x^2 + 24x + 16 \neq 9x^2 + 30x + 16$
The proposed factorization is not correct.

In this case, the polynomial is not a perfect-square trinomial. It may, however, still factor. In fact, $9x^2 + 30x + 16 = (3x + 2)(3x + 8)$. If the trinomial does not check as a perfect-square trinomial, try to factor it by another method.

A perfect-square trinomial can always be factored using either of the methods presented in Section 8.3. However, noticing that a trinomial is a perfect-square trinomial can save you a considerable amount of time.

Example 3   Factor.   A. $9x^2 - 30x + 25$ B. $4x^2 + 37x + 9$

Solution   A. $9x^2 = (3x)^2,\ 25 = 5^2$

▶ Check that the first and last terms are squares.

$(3x - 5)^2$

▶ Use the squared terms to factor the trinomial as the square of a binomial.

$(3x - 5)^2$
$= (3x)^2 + 2(3x)(-5) + (-5)^2$
$= 9x^2 - 30x + 25$

▶ Check the factorization.

▶ The factorization checks.

$9x^2 - 30x + 25 = (3x - 5)^2$

B. $4x^2 = (2x)^2,\ 9 = 3^2$

▶ Check that the first and last terms are squares.

$(2x + 3)^2$

▶ Use the squared terms to factor the trinomial as the square of a binomial.

$(2x + 3)^2$
$= (2x)^2 + 2(2x)(3) + 3^2$
$= 4x^2 + 12x + 9$

▶ Check the factorization.

▶ The factorization does not check.

$4x^2 + 37x + 9$
$= (4x + 1)(x + 9)$

▶ Use another method to factor the trinomial.

Problem 3   Factor.   A. $16y^2 + 8y + 1$ B. $x^2 + 14x + 36$

Solution   See page A34.

2   Factor completely

When factoring a polynomial completely, ask yourself the following questions about the polynomial.

1. Is there a common factor? If so, factor out the common factor.
2. Is the polynomial the difference of two squares? If so, factor.
3. Is the polynomial a perfect-square trinomial? If so, factor.
4. Is the polynomial a trinomial that is the product of two binomials? If so, factor.
5. Does the polynomial contain four terms? If so, try factoring by grouping.
6. Is each binomial factor nonfactorable over the integers? If not, factor.

Example 4   Factor.   A. $3x^2 - 48$ B. $x^3 - 3x^2 - 4x + 12$
                            C. $4x^2y^2 + 12xy^2 + 9y^2$

Solution   A. $3x^2 - 48$ ▶ The GCF of the terms is 3.
                $= 3(x^2 - 16)$ Factor out the common factor.
                $= 3(x + 4)(x - 4)$ ▶ Factor the difference of two squares.

B. $x^3 - 3x^2 - 4x + 12$ ▶ The polynomial contains four
   $= (x^3 - 3x^2) - (4x - 12)$ terms. Factor by grouping.
   $= x^2(x - 3) - 4(x - 3)$
   $= (x - 3)(x^2 - 4)$
   $= (x - 3)(x + 2)(x - 2)$ ▶ Factor the difference of two squares.

C. $4x^2y^2 + 12xy^2 + 9y^2$ ▶ The GCF of the terms is y^2.
   $= y^2(4x^2 + 12x + 9)$ Factor out the common factor.
   $= y^2(2x + 3)^2$ ▶ Factor the perfect-square trinomial.

Problem 4   Factor.   A. $12x^3 - 75x$ B. $a^2b - 7a^2 - b + 7$
                            C. $4x^3 + 28x^2 - 120x$

Solution   See page A34.

CONCEPT REVIEW 8.4

Determine whether the statement is always true, sometimes true, or never true.

1. The expression $x^2 - 12$ is an example of the difference of two squares.

2. The expression $(y + 8)(y - 8)$ is the product of the sum and difference of the same two terms. The two terms are y and 8.

3. A binomial is factorable.

4. A trinomial is factorable.

5. If a binomial is multiplied times itself, the result is a perfect-square trinomial.

6. In a perfect-square trinomial, the first and last terms are perfect squares.

7. If a polynomial contains four terms, try to factor it as a perfect-square trinomial.

8. The expression $x^2 + 9$ is the sum of two squares. It factors as $(x + 3)(x + 3)$.

EXERCISES 8.4

1 Factor.

1. $x^2 - 4$
2. $x^2 - 9$
3. $a^2 - 81$
4. $a^2 - 49$

5. $4x^2 - 1$
6. $9x^2 - 16$
7. $y^2 + 2y + 1$
8. $y^2 + 14y + 49$

9. $a^2 - 2a + 1$
10. $x^2 + 8x - 16$
11. $z^2 - 18z - 81$
12. $x^2 - 12x + 36$

13. $x^6 - 9$
14. $y^{12} - 121$
15. $25x^2 - 1$
16. $9x^2 - 1$

17. $1 - 49x^2$
18. $1 - 64x^2$
19. $x^2 + 2xy + y^2$
20. $x^2 + 6xy + 9y^2$

21. $4a^2 + 4a + 1$
22. $25x^2 + 10x + 1$
23. $64a^2 - 16a + 1$
24. $9a^2 + 6a + 1$

25. $t^2 + 36$
26. $x^2 + 64$
27. $x^4 - y^2$
28. $b^4 - 16a^2$

29. $9x^2 - 16y^2$
30. $25z^2 - y^2$
31. $16b^2 + 8b + 1$
32. $4a^2 - 20a + 25$

33. $4b^2 + 28b + 49$
34. $9a^2 - 42a + 49$
35. $25a^2 + 30ab + 9b^2$

36. $4a^2 - 12ab + 9b^2$
37. $x^2y^2 - 4$
38. $a^2b^2 - 25$

39. $16 - x^2y^2$
40. $49x^2 + 28xy + 4y^2$
41. $4y^2 - 36yz + 81z^2$

42. $64y^2 - 48yz + 9z^2$
43. $9a^2b^2 - 6ab + 1$
44. $16x^2y^2 - 24xy + 9$

45. $m^4 - 256$
46. $81 - t^4$
47. $9x^2 + 13x + 4$

48. $x^2 + 10x + 16$
49. $y^8 - 81$
50. $9 + 24a + 16a^2$

2 Factor.

51. $2x^2 - 18$
52. $y^3 - 10y^2 + 25y$
53. $x^4 + 2x^3 - 35x^2$
54. $a^4 - 11a^3 + 24a^2$

55. $5b^2 + 75b + 180$
56. $6y^2 - 48y + 72$
57. $3a^2 + 36a + 10$
58. $5a^2 - 30a + 4$

59. $2x^2y + 16xy - 66y$
60. $3a^2b + 21ab - 54b$
61. $x^3 - 6x^2 - 5x$
62. $b^3 - 8b^2 - 7b$

63. $3y^2 - 36$
64. $3y^2 - 147$
65. $20a^2 + 12a + 1$
66. $12a^2 - 36a + 27$

67. $x^2y^2 - 7xy^2 - 8y^2$ **68.** $a^2b^2 + 3a^2b - 88a^2$ **69.** $10a^2 - 5ab - 15b^2$

70. $16x^2 - 32xy + 12y^2$ **71.** $50 - 2x^2$ **72.** $72 - 2x^2$

73. $12a^3b - a^2b^2 - ab^3$ **74.** $2x^3y - 7x^2y^2 + 6xy^3$ **75.** $2ax - 2a + 2bx - 2b$

76. $4ax - 12a - 2bx + 6b$ **77.** $12a^3 - 12a^2 + 3a$ **78.** $18a^3 + 24a^2 + 8a$

79. $243 + 3a^2$ **80.** $75 + 27y^2$ **81.** $12a^3 - 46a^2 + 40a$

82. $24x^3 - 66x^2 + 15x$ **83.** $x^3 - 2x^2 - x + 2$ **84.** $ay^2 - by^2 - a + b$

85. $4a^3 + 20a^2 + 25a$ **86.** $2a^3 - 8a^2b + 8ab^2$ **87.** $27a^2b - 18ab + 3b$

88. $a^2b^2 - 6ab^2 + 9b^2$ **89.** $48 - 12x - 6x^2$ **90.** $21x^2 - 11x^3 - 2x^4$

91. $ax^2 - 4a + bx^2 - 4b$ **92.** $a^2x - b^2x - a^2y + b^2y$ **93.** $x^4 - x^2y^2$

94. $b^4 - a^2b^2$ **95.** $18a^3 + 24a^2 + 8a$ **96.** $32xy^2 - 48xy + 18x$

97. $2b + ab - 6a^2b$ **98.** $20x - 11xy - 3xy^2$ **99.** $4x - 20 - x^3 + 5x^2$

100. $ay^2 - by^2 - 9a + 9b$ **101.** $72xy^2 + 48xy + 8x$ **102.** $4x^2y + 8xy + 4y$

103. $15y^2 - 2xy^2 - x^2y^2$ **104.** $4x^4 - 38x^3 + 48x^2$ **105.** $y^3 - 9y$

106. $a^4 - 16$ **107.** $2x^4y^2 - 2x^2y^2$ **108.** $6x^5y - 6xy^5$

109. $x^9 - x^5$ **110.** $8b^5 - 2b^3$ **111.** $24x^3y + 14x^2y - 20xy$

112. $12x^3y - 60x^2y + 63xy$ **113.** $4x^4y^2 - 20x^3y^2 + 25x^2y^2$ **114.** $9x^4y^2 + 24x^3y^2 + 16x^2y^2$

115. $x^3 - 2x^2 - 4x + 8$ **116.** $24x^2y + 6x^3y - 45x^4y$ **117.** $8xy^2 - 20x^2y^2 + 12x^3y^2$

118. $45y^2 - 42y^3 - 24y^4$ **119.** $36a^3b - 62a^2b^2 + 12ab^3$ **120.** $18a^3b + 57a^2b^2 + 30ab^3$

121. $5x^2y^2 - 11x^3y^2 - 12x^4y^2$ **122.** $24x^2y^2 - 32x^3y^2 + 10x^4y^2$ **123.** $(4x - 3)^2 - y^2$

124. $(2a + 3)^2 - 25b^2$ **125.** $(x^2 - 4x + 4) - y^2$ **126.** $(4x^2 + 12x + 9) - 4y^2$

SUPPLEMENTAL EXERCISES 8.4

Find all integers k such that the trinomial is a perfect-square trinomial.

127. $4x^2 - kx + 9$ **128.** $25x^2 - kx + 1$ **129.** $36x^2 + kxy + y^2$ **130.** $64x^2 + kxy + y^2$

131. $x^2 + 6x + k$ **132.** $x^2 - 4x + k$ **133.** $x^2 - 2x + k$ **134.** $x^2 + 10x + k$

135. The prime factorization of a number is $2^3 \cdot 3^2$. How many of its whole-number factors are perfect squares?

136. The product of two numbers is 48. One of the two numbers is a perfect square. The other is a prime number. Find the sum of the two numbers.

137. What is the smallest whole number by which 300 can be multiplied so that the product will be a perfect square?

138. The area of a square is $(16x^2 + 24x + 9)$ ft^2. Find the dimensions of the square in terms of the variable x. Can $x = 0$? What are the possible values of x?

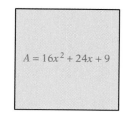

$A = 16x^2 + 24x + 9$

The cube of an integer is a **perfect cube.** Because $2^3 = 8$, 8 is a perfect cube. Because $4^3 = 64$, 64 is a perfect cube. A variable expression can be a perfect cube; the exponents on variables of perfect cubes are multiples of 3. Therefore, x^3, x^6, and x^9 are perfect cubes. The sum and the difference of two perfect cubes are factorable. They can be written as the product of a binomial and a trinomial. Their factoring patterns are shown below.

$a^3 + b^3$ is the sum of two cubes. $a^3 + b^3 = (a + b)(a^2 - ab + b^2)$
$a^3 - b^3$ is the difference of two cubes. $a^3 - b^3 = (a - b)(a^2 + ab + b^2)$

To factor $x^3 - 8$, write the binomial as the difference of two perfect cubes. Use the factoring pattern shown above. Replace a with x and b with 2.

$$x^3 - 8 = (x)^3 - (2)^3$$

$$= (x - 2)(x^2 + 2x + 4)$$

Factor.

139. $x^3 + 8$ **140.** $y^3 + 27$ **141.** $y^3 - 27$ **142.** $x^3 - 1$

143. $y^3 + 64$ **144.** $x^3 - 125$ **145.** $8x^3 - 1$ **146.** $27y^3 + 1$

147. Select any odd integer greater than 1, square it, and then subtract 1. Is the result evenly divisible by 8? Prove that this procedure always produces a number divisible by 8. (*Suggestion*: Any odd integer greater than 1 can be expressed as $2n + 1$, where n is a natural number.)

S E C T I O N **8.5**

Solving Equations

1 Solve equations by factoring

Recall that the Multiplication Property of Zero states that the product of a number and zero is zero.

If a is a real number, then $a \cdot 0 = 0$.

Consider the equation $a \cdot b = 0$. If this is a true equation, then either $a = 0$ or $b = 0$.

Principle of Zero Products

If the product of two factors is zero, then at least one of the factors must be zero.

If $a \cdot b = 0$, then $a = 0$ or $b = 0$.

The Principle of Zero Products is used in solving equations.

Solve: $(x - 2)(x - 3) = 0$

If $(x - 2)(x - 3) = 0$, then $(x - 2) = 0$ or $(x - 3) = 0$.

Solve each equation for x.

$(x - 2)(x - 3) = 0$

$$x - 2 = 0 \qquad x - 3 = 0$$
$$x = 2 \qquad x = 3$$

Check:
$$\begin{array}{c|c} (x - 2)(x - 3) = 0 & (x - 2)(x - 3) = 0 \\ \hline (2 - 2)(2 - 3) \mid 0 & (3 - 2)(3 - 3) \mid 0 \\ 0(-1) \mid 0 & 1(0) \mid 0 \\ 0 = 0 & 0 = 0 \end{array}$$

A true equation   A true equation

Write the solutions.   The solutions are 2 and 3.

An equation of the form $ax^2 + bx + c = 0$, $a \neq 0$, is a **quadratic equation**. A quadratic equation is in **standard form** when the polynomial is in descending order and equal to zero.

$3x^2 + 2x + 1 = 0$

$4x^2 - 3x + 2 = 0$

A quadratic equation can be solved by using the Principle of Zero Products when the polynomial $ax^2 + bx + c$ is factorable.

Example 1   Solve: $2x^2 + x = 6$

Solution
$$2x^2 + x = 6$$
$$2x^2 + x - 6 = 0 \qquad \blacktriangleright \text{Write the equation in standard form.}$$
$$(2x - 3)(x + 2) = 0 \qquad \blacktriangleright \text{Factor the trinomial.}$$
$$2x - 3 = 0 \qquad x + 2 = 0 \qquad \blacktriangleright \text{Set each factor equal to zero (the Principle of Zero Products).}$$
$$2x = 3 \qquad x = -2 \qquad \blacktriangleright \text{Solve each equation for } x.$$
$$x = \frac{3}{2}$$

Check

$2x^2 + x = 6$		$2x^2 + x = 6$	
$2\left(\dfrac{3}{2}\right)^2 + \dfrac{3}{2}$	6	$2(-2)^2 + (-2)$	6
$2\left(\dfrac{9}{4}\right) + \dfrac{3}{2}$	6	$2 \cdot 4 - 2$	6
$\dfrac{9}{2} + \dfrac{3}{2}$	6	$8 - 2$	6
	$6 = 6$		$6 = 6$

The solutions are $\dfrac{3}{2}$ and -2.   ▶ Write the solutions.

Problem 1   Solve: $2x^2 - 50 = 0$

Solution   See page A34.

Example 2   Solve: $(x - 3)(x - 10) = -10$

Solution
$(x - 3)(x - 10) = -10$
$x^2 - 13x + 30 = -10$ ▶ Multiply $(x - 3)(x - 10)$.
$x^2 - 13x + 40 = 0$ ▶ Write the equation in standard form.

$(x - 8)(x - 5) = 0$ ▶ Factor.

$x - 8 = 0 \qquad x - 5 = 0$ ▶ Set each factor equal to zero.
$\qquad x = 8 \qquad\quad x = 5$ ▶ Solve each equation for x.

The solutions are 8 and 5.   ▶ Write the solutions.

Problem 2   Solve: $(x + 2)(x - 7) = 52$

Solution   See page A34.

LOOK CLOSELY

The Principle of Zero Products cannot be used unless 0 is on one side of the equation.

2   Application problems

Example 3   The sum of the squares of two consecutive positive odd integers is equal to 130. Find the two integers.

Strategy   ■ First positive odd integer: n
Second positive odd integer: $n + 2$
Square of the first positive odd integer: n^2
Square of the second positive odd integer: $(n + 2)^2$
■ The sum of the square of the first positive odd integer and the square of the second positive odd integer is 130.

Solution
$\qquad n^2 + (n + 2)^2 = 130$
$\quad n^2 + n^2 + 4n + 4 = 130$
$\qquad\quad 2n^2 + 4n - 126 = 0$
$\qquad\quad 2(n^2 + 2n - 63) = 0$
$\qquad\qquad n^2 + 2n - 63 = 0$ ▶ Divide each side of the equation
$\qquad\qquad (n - 7)(n + 9) = 0$ by 2.

$$n - 7 = 0 \qquad n + 9 = 0$$
$$n = 7 \qquad\qquad n = -9$$

▶ Because -9 is not a positive odd integer, it is not a solution.

$$n = 7$$
$$n + 2 = 7 + 2 = 9$$

▶ The first positive odd integer is 7.
▶ Substitute the value of n into the variable expression for the second positive odd integer and evaluate.

The two integers are 7 and 9.

Problem 3   The sum of the squares of two consecutive positive integers is 85. Find the two integers.

Solution   See page A34.

Example 4   A stone is thrown into a well with an initial velocity of 8 ft/s. The well is 440 ft deep. How many seconds later will the stone hit the bottom of the well? Use the equation $d = vt + 16t^2$, where d is the distance in feet, v is the initial velocity in feet per second, and t is the time in seconds.

Strategy   To find the time for the stone to drop to the bottom of the well, replace the variables d and v by their given values and solve for t.

Solution

$$d = vt + 16t^2$$
$$440 = 8t + 16t^2$$
$$-16t^2 - 8t + 440 = 0$$
$$16t^2 + 8t - 440 = 0$$
$$8(2t^2 + t - 55) = 0$$
$$2t^2 + t - 55 = 0$$
$$(2t + 11)(t - 5) = 0$$

▶ Multiply each side of the equation by -1.

▶ Divide each side of the equation by 8.

$$2t + 11 = 0 \qquad\qquad t - 5 = 0$$
$$2t = -11 \qquad\qquad\quad t = 5$$

$$t = -\frac{11}{2}$$

▶ Because the time cannot be a negative number, $-\dfrac{11}{2}$ is not a solution.

The time is 5 s.

Problem 4   The length of a rectangle is 3 m more than twice the width. The area of the rectangle is 90 m². Find the length and width of the rectangle.

Solution   See page A34.

CONCEPT REVIEW 8.5

Determine whether the statement is always true, sometimes true, or never true.

1. If you multiply two numbers and the product is zero, then either one or both of the numbers must be zero.

2. The equation $2x^2 + 5x - 7$ is a quadratic equation.

3. The equation $4x^2 - 9 = 0$ is a quadratic equation.

4. The equation $3x + 1 = 0$ is a quadratic equation in standard form.

5. If $(x - 8)(x + 6) = 0$, then $x = -8$ or $x = 6$.

6. If a quadratic equation is not in standard form, the first step in solving the equation by factoring is to write it in standard form.

EXERCISES 8.5

1   Solve.

1. $(y + 3)(y + 2) = 0$
2. $(y - 3)(y - 5) = 0$
3. $(z - 7)(z - 3) = 0$

4. $(z + 8)(z - 9) = 0$
5. $x(x - 5) = 0$
6. $x(x + 2) = 0$

7. $a(a - 9) = 0$
8. $a(a + 12) = 0$
9. $y(2y + 3) = 0$

10. $t(4t - 7) = 0$
11. $2a(3a - 2) = 0$
12. $4b(2b + 5) = 0$

13. $(b + 2)(b - 5) = 0$
14. $(b - 8)(b + 3) = 0$
15. $x^2 - 81 = 0$

16. $9x^2 - 1 = 0$
17. $16x^2 - 49 = 0$
18. $x^2 + 6x + 8 = 0$

19. $x^2 - 8x + 15 = 0$
20. $z^2 + 5z - 14 = 0$
21. $z^2 + z - 72 = 0$

22. $x^2 - 5x + 6 = 0$
23. $x^2 - 3x - 10 = 0$
24. $y^2 + 4y - 21 = 0$

25. $2y^2 - y - 1 = 0$
26. $2a^2 - 9a - 5 = 0$
27. $3a^2 + 14a + 8 = 0$

28. $a^2 - 5a = 0$
29. $x^2 - 7x = 0$
30. $2a^2 - 8a = 0$

31. $a^2 + 5a = -4$
32. $a^2 - 5a = 24$
33. $y^2 - 5y = -6$

34. $y^2 - 7y = 8$
35. $2t^2 + 7t = 4$
36. $3t^2 + t = 10$

37. $3t^2 - 13t = -4$
38. $5t^2 - 16t = -12$
39. $x(x - 12) = -27$

40. $x(x - 11) = 12$
41. $y(y - 7) = 18$
42. $y(y + 8) = -15$

43. $p(p + 3) = -2$
44. $p(p - 1) = 20$
45. $y(y + 4) = 45$

46. $y(y - 8) = -15$
47. $x(x + 3) = 28$
48. $p(p - 14) = 15$

49. $(x + 8)(x - 3) = -30$
50. $(x + 4)(x - 1) = 14$
51. $(y + 3)(y + 10) = -10$

52. $(z - 5)(z + 4) = 52$

53. $(z - 8)(z + 4) = -35$

54. $(z - 6)(z + 1) = -10$

55. $(a + 3)(a + 4) = 72$

56. $(a - 4)(a + 7) = -18$

57. $(z + 3)(z - 10) = -42$

58. $(2x + 5)(x + 1) = -1$

59. $(y + 3)(2y + 3) = 5$

60. $(y + 5)(3y - 2) = -14$

2   Solve.

61. The square of a positive number is seven more than six times the positive number. Find the number.

62. The square of a negative number is fifteen more than twice the negative number. Find the number.

63. The sum of the squares of two consecutive positive integers is sixty-one. Find the two integers.

64. The sum of the squares of two consecutive positive even integers is fifty-two. Find the two integers.

65. The product of two consecutive positive integers is two hundred forty. Find the integers.

66. The product of two consecutive positive even integers is one hundred sixty-eight. Find the integers.

67. The length of the base of a triangle is three times the height. The area of the triangle is 54 ft². Find the base and height of the triangle.

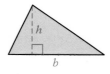

68. The height of a triangle is 4 m more than twice the length of the base. The area of the triangle is 35 m². Find the height of the triangle.

69. The length of a rectangle is four times the width. The area is 400 m². Find the length and width of the rectangle.

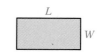

70. The length of a rectangle is two more than twice the width. The area is 144 ft². Find the length and width of the rectangle.

71. The length of each side of a square is extended 2 cm. The area of the resulting square is 64 cm². Find the length of a side of the original square.

72. The length of each side of a square is extended 4 m. The area of the resulting square is 64 m². Find the length of a side of the original square.

73. A circle has a radius of 10 in. Find the increase in area when the radius is increased by 2 in. Round to the nearest tenth.

74. The radius of a circle is increased by 3 ft, increasing the area by 100 ft^2. Find the radius of the original circle. Round to the nearest tenth.

75. The page of a book measures 6 in. by 9 in. A uniform border around the page leaves 28 in^2 for type. Find the dimensions of the type area.

76. A small garden measures 8 ft by 10 ft. A uniform border around the garden increases the total area to 168 ft^2. Find the width of the border.

Use the formula $d = vt + 16t^2$, where d is the distance in feet, v is the initial velocity in feet per second, and t is the time in seconds.

77. An object is released from a plane at an altitude of 1600 ft. The initial velocity is 0 ft/s. How many seconds later will the object hit the ground?

78. An object is released from the top of a building 320 ft high. The initial velocity is 16 ft/s. How many seconds later will the object hit the ground?

Use the formula $S = \dfrac{n^2 + n}{2}$, where S is the sum of the first n natural numbers.

79. How many consecutive natural numbers beginning with 1 will give a sum of 78?

80. How many consecutive natural numbers beginning with 1 will give a sum of 120?

Use the formula $N = \dfrac{t^2 - t}{2}$, where N is the number of football games that must be scheduled in a league with t teams if each team is to play every other team once.

81. A league has 28 games scheduled. How many teams are in the league if each team plays every other team once?

82. A league has 45 games scheduled. How many teams are in the league if each team plays every other team once?

Use the formula $h = vt - 16t^2$, where h is the height an object will attain (neglecting air resistance) in t seconds and v is the initial velocity.

83. A baseball player hits a "Baltimore chop," meaning the ball bounces off home plate after the batter hits it. The ball leaves home plate with an initial upward velocity of 64 ft/s. How many seconds after the ball hits home plate will the ball be 64 ft above the ground?

84. A golf ball is thrown onto a cement surface and rebounds straight up. The initial velocity of the rebound is 96 ft/s. How many seconds later will the golf ball return to the ground?

SUPPLEMENTAL EXERCISES 8.5

Solve.

85. $2y(y + 4) = -5(y + 3)$

86. $2y(y + 4) = 3(y + 4)$

87. $(a - 3)^2 = 36$

88. $(b + 5)^2 = 16$

89. $p^3 = 9p^2$

90. $p^3 = 7p^2$

91. $(2z - 3)(z + 5) = (z + 1)(z + 3)$

92. $(x + 3)(2x - 1) = (3 - x)(5 - 3x)$

93. Find $3n^2$ if $n(n + 5) = -4$.

94. Find $2n^3$ if $n(n + 3) = 4$.

95. The length of a rectangle is 7 cm, and the width is 4 cm. If both the length and the width are increased by equal amounts, the area of the rectangle is increased by 42 cm³. Find the length and width of the larger rectangle.

96. A rectangular piece of cardboard is 10 in. longer than it is wide. Squares 2 in. on a side are to be cut from each corner, and then the sides will be folded up to make an open box with a volume of 192 in³. Find the length and width of the piece of cardboard.

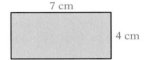

97. Explain the error made in solving the equation at the right. Solve the equation correctly.

$(x + 2)(x - 3) = 6$
$x + 2 = 6 \qquad x - 3 = 6$
$ x = 4 \qquad x = 9$

98. Explain the error made in solving the equation at the right. Solve the equation correctly.

$x^2 = x$
$\dfrac{x^2}{x} = \dfrac{x}{x}$
$x = 1$

99. In your own words, explain why it is possible to solve a quadratic equation using the Principle of Zero Products.

Project in Mathematics

Prime and Composite Numbers

A **prime number** is a natural number greater than 1 whose only natural number factors are itself and 1. The number 11 is a prime number because the only natural number factors of 11 are 11 and 1.

Eratosthenes, a Greek philosopher and astronomer who lived from 270 to 190 B.C., devised a method of identifying prime numbers. It is called the **Sieve of Eratosthenes.** The procedure is illustrated on the next page.

1	(2)	(3)	4	(5)	6	(7)	8	9	10
(11)	12	(13)	14	15	16	(17)	18	(19)	20
21	22	(23)	24	25	26	27	28	(29)	30
(31)	32	33	34	35	36	(37)	38	39	40
(41)	42	(43)	44	45	46	(47)	48	49	50
51	52	(53)	54	55	56	57	58	(59)	60
(61)	62	63	64	65	66	(67)	68	69	70
(71)	72	(73)	74	75	76	77	78	(79)	80
81	82	(83)	84	85	86	87	88	(89)	90
91	92	93	94	95	96	(97)	98	99	100

List all the natural numbers from 1 to 100. Cross out the number 1, because it is not a prime number. The number 2 is prime; circle it. Cross out all the other multiples of 2 (4, 6, 8, ...), because they are not prime. The number 3 is prime; circle it. Cross out all the other multiples of 3 (6, 9, 12, ...) that are not already crossed out. The number 4, the next consecutive number in the list, has already been crossed out. The number 5 is prime; circle it. Cross out all the other multiples of 5 that are not already crossed out. Continue in this manner until all the prime numbers less than 100 are circled.

A **composite number** is a natural number greater than 1 that has a natural-number factor other than itself and 1. The number 21 is a composite number because it has factors of 3 and 7. All the numbers crossed out in the table above, except the number 1, are composite numbers.

Solve.

1. Use the Sieve of Eratosthenes to find the prime numbers between 100 and 200.

2. How many prime numbers are even numbers?

3. Find the "twin primes" between 1 and 200. Twin primes are two prime numbers whose difference is 2. For instance, 3 and 5 are twin primes; 5 and 7 are also twin primes.

4. **a.** List two prime numbers that are consecutive natural numbers.

   b. Can there be any other pairs of prime numbers that are consecutive natural numbers?

5. Some primes are the sum of a square and 1. For example, $5 = 2^2 + 1$. Find another prime p such that $p = n^2 + 1$, where n is a natural number.

6. Find a prime p such that $p = n^2 - 1$, where n is a natural number.

7. **a.** 4! (which is read "4 factorial") is equal to $4 \cdot 3 \cdot 2 \cdot 1$. Show that $4! + 2$, $4! + 3$, and $4! + 4$ are all composite numbers.

   b. 5! (which is read "5 factorial") is equal to $5 \cdot 4 \cdot 3 \cdot 2 \cdot 1$. Will $5! + 2$, $5! + 3$, $5! + 4$, and $5! + 5$ generate four consecutive composite numbers?

   c. Use the notation 6! to represent a list of five consecutive composite numbers.

Focus on Problem Solving

The Trial-and-Error Method

A topic in Section 8.3 is factoring trinomials using trial factors. This method involves writing trial factors and then using the FOIL method to determine which pair of factors is the correct one. The **trial-and-error method** of arriving at a solution to a problem involves performing repeated tests or experiments until a satisfactory conclusion is reached. However, not all problems solved using the trial-and-error method have a strategy by which to determine the answer. Here is an example:

Explain how you could cut through a cube so that the face of the resulting solid is **a.** a square, **b.** an equilateral triangle, **c.** a trapezoid, and **d.** a hexagon.

There is no formula to apply to this problem; there is no computation to perform. This problem requires picturing a cube and the results after cutting through it at different places on its surface and at different angles. For part (a), cutting perpendicular to the top and bottom of the cube and parallel to two of its sides will result in a square. The other shapes may prove more difficult.

When solving problems of this type, keep an open mind. Sometimes when using the trial-and-error method, we are hampered by narrowness of vision; we cannot expand our thinking to include other possibilities. Then when we see someone else's solution, it appears so obvious to us! For example, for the question above, it is necessary to conceive of cutting through the cube at places other than the top surface; we need to be open to the idea of beginning the cut at one of the corner points of the cube.

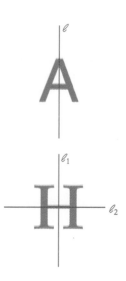

Look at the letter A printed at the left. If the letter were folded along line ℓ, the two sides of the letter would match exactly. This letter has **symmetry** with respect to line ℓ. Line ℓ is called the **axis of symmetry.** Now consider the letter H printed below at the left. Both line ℓ_1 and ℓ_2 are axes of symmetry for this letter; the letter could be folded along either line and the two sides would match exactly. Does the letter A have more than one axis of symmetry? Find axes of symmetry for other capital letters of the alphabet. Which lower-case letters have one axis of symmetry? Do any of the lower-case letters have more than one axis of symmetry?

How many axes of symmetry does a square have? In determining lines of symmetry for a square, begin by drawing a square as shown at the right. The horizontal line of symmetry and the vertical line of symmetry may be immediately obvious to you. But there are two others. Do you see that a line drawn through opposite corners of the square is also a line of symmetry?

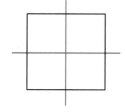

Find the number of axes of symmetry for each of the following figures: equilateral triangle, rectangle, regular pentagon, regular hexagon, and circle.

Many of the questions in this text that require an answer of "always true, sometimes true, or never true" are best solved by the trial-and-error method. For example, consider the following statement:

If two rectangles have the same area, then they have the same perimeter.

Try some numbers. Each of two rectangles, one measuring 6 units by 2 units and another measuring 4 units by 3 units, has an area of 12 square units, but the perimeter of the first is 16 units and the perimeter of the second is 14 units. So the answer "always true" has been eliminated. We still need to determine whether or not there is a case when it is true. After experimenting with a lot of numbers, you may come to realize that we are trying to determine whether it is possible for two different pairs of factors of a number to have the same sum. Is it?

Don't be afraid to make many experiments, and remember that *errors,* or tests that "don't work," are a part of the trial-and-*error* process.

Chapter Summary

Key Words

The *greatest common factor* (GCF) of two or more integers is the greatest integer that is a factor of all the integers. (Objective 8.1.1)

The greatest common factor of 12 and 18 is 6 because 6 is the greatest integer that divides evenly into both 12 and 18.

To *factor a polynomial* means to write the polynomial as a product of other polynomials.   (Objective 8.1.1)

To factor $8x + 12$ means to write it as the product $4(2x + 3)$. The expression $8x + 12$ is a sum. The expression $4(2x + 3)$ is a product (the polynomial $2x + 3$ is multiplied by 4).

To *factor a trinomial of the form* $ax^2 + bx + c$ means to express the trinomial as the product of two binomials. (Objective 8.3.1)

To factor $3x^2 + 7x + 2$ means to write it as the product $(3x + 1)(x + 2)$.

A polynomial that does not factor using only integers is *nonfactorable over the integers.*   (Objective 8.2.1)

The trinomial $x^2 + x + 2$ is nonfactorable over the integers. There are no two integers whose product is 2 and whose sum is 1.

An equation of the form $ax^2 + bx + c = 0$, $a \neq 0$, is a *quadratic equation.* A quadratic equation is in *standard form* when the polynomial is in descending order and equal to zero. (Objective 8.5.1)

The equation $2x^2 + 7x + 3 = 0$ is a quadratic equation in standard form.

Essential Rules and Procedures

Factoring the Difference of Two Squares
The difference of two squares is the product of the sum and difference of two terms.
$a^2 - b^2 = (a + b)(a - b)$ (Objective 8.4.1)

$y^2 - 81 = (y + 9)(y - 9)$

Factoring a Perfect-Square Trinomial

A perfect-square trinomial is the square of a binomial.

$a^2 + 2ab + b^2 = (a + b)^2$

$a^2 - 2ab + b^2 = (a - b)^2$ (Objective 8.4.1)

$x^2 + 10x + 25 = (x + 5)^2$
$x^2 - 10x + 25 = (x - 5)^2$

General Factoring Strategy   (Objective 8.4.2)

1. Is there a common factor? If so, factor out the common factor.

$24x + 6 = 6(4x + 1)$
$2x^2y + 6xy + 8y = 2y(x^2 + 3x + 4)$

2. Is the polynomial the difference of two squares? If so, factor.

$4x^2 - 49 = (2x + 7)(2x - 7)$

3. Is the polynomial a perfect-square trinomial? If so, factor.

$9x^2 + 6x + 1 = (3x + 1)^2$

4. Is the polynomial a trinomial that is the product of two binomials? If so, factor.

$2x^2 + 7x + 5 = (2x + 5)(x + 1)$

5. Does the polynomial contain four terms? If so, try factoring by grouping.

$x^3 - 3x^2 + 2x - 6 = (x^3 - 3x^2) + (2x - 6)$
$= x^2(x - 3) + 2(x - 3)$
$= (x - 3)(x^2 + 2)$

6. Is each binomial factor nonfactorable over the integers? If not, factor.

$x^4 - 16 = (x^2 + 4)(x^2 - 4)$
$= (x^2 + 4)(x + 2)(x - 2)$

The Principle of Zero Products

If the product of two factors is zero, then at least one of the factors must be zero.

If $a \cdot b = 0$, then $a = 0$ or $b = 0$.

The Principle of Zero Products is used to solve a quadratic equation by factoring.   (Objective 8.5.1)

$x^2 + x = 12$
$x^2 + x - 12 = 0$
$(x + 4)(x - 3) = 0$
$x + 4 = 0 \qquad x - 3 = 0$
$x = -4 \qquad\quad x = 3$

Chapter Review Exercises

1. Factor: $14y^9 - 49y^6 + 7y^3$

2. Factor: $3a^2 - 12a + ab - 4b$

3. Factor: $c^2 + 8c + 12$

4. Factor: $a^3 - 5a^2 + 6a$

5. Factor: $6x^2 - 29x + 28$

6. Factor: $3y^2 + 16y - 12$

7. Factor: $18a^2 - 3a - 10$

8. Factor: $a^2b^2 - 1$

9. Factor: $4y^2 - 16y + 16$

10. Solve: $a(5a + 1) = 0$

11. Factor: $12a^2b + 3ab^2$

12. Factor: $b^2 - 13b + 30$

13. Factor: $10x^2 + 25x + 4xy + 10y$

14. Factor: $3a^2 - 15a - 42$

15. Factor: $n^4 - 2n^3 - 3n^2$

16. Factor: $2x^2 - 5x + 6$

17. Factor: $6x^2 - 7x + 2$

18. Factor: $16x^2 + 49$

19. Solve: $(x - 2)(2x - 3) = 0$

20. Factor: $7x^2 - 7$

21. Factor: $3x^5 - 9x^4 - 4x^3$

22. Factor: $4x(x - 3) - 5(3 - x)$

23. Factor: $a^2 + 5a - 14$

24. Factor: $y^2 + 5y - 36$

25. Factor: $5x^2 - 50x - 120$

26. Solve: $(x + 1)(x - 5) = 16$

27. Factor: $7a^2 + 17a + 6$

28. Factor: $4x^2 + 83x + 60$

29. Factor: $9y^4 - 25z^2$

30. Factor: $5x^2 - 5x - 30$

31. Solve: $6 - 6y^2 = 5y$

32. Factor: $12b^3 - 58b^2 + 56b$

33. Factor: $5x^3 + 10x^2 + 35x$

34. Factor: $x^2 - 23x + 42$

35. Factor: $a(3a + 2) - 7(3a + 2)$

36. Factor: $8x^2 - 38x + 45$

37. Factor: $10a^2x - 130ax + 360x$

38. Factor: $2a^2 - 19a - 60$

39. Factor: $21ax - 35bx - 10by + 6ay$

40. Factor: $a^6 - 100$

41. Factor: $16a^2 + 8a + 1$

42. Solve: $4x^2 + 27x = 7$

43. Factor: $20a^2 + 10a - 280$

44. Factor: $6x - 18$

45. Factor: $3x^4y + 2x^3y + 6x^2y$

46. Factor: $d^2 + 3d - 40$

47. Factor: $24x^2 - 12xy + 10y - 20x$

48. Factor: $4x^3 - 20x^2 - 24x$

49. Solve: $x^2 - 8x - 20 = 0$

50. Factor: $3x^2 - 17x + 10$

51. Factor: $16x^2 - 94x + 33$

52. Factor: $9x^2 - 30x + 25$

53. Factor: $12y^2 + 16y - 3$

54. Factor: $3x^2 + 36x + 108$

55. The length of a playing field is twice the width. The area is 5000 yd². Find the length and width of the playing field.

56. The length of a hockey field is 20 yd less than twice the width. The area of the field is 6000 yd². Find the length and width of the hockey field.

57. The sum of the squares of two consecutive positive integers is forty-one. Find the two integers.

58. The size of a motion picture on the screen is given by the equation $S = d^2$, where d is the distance between the projector and the screen. Find the distance between the projector and the screen when the size of the picture is 400 ft².

59. A rectangular garden plot has dimensions 15 ft by 12 ft. A uniform path around the garden increases the total area to 270 ft^2. What is the width of the resulting area?

60. The length of each side of a square is extended 4 ft. The area of the resulting square is 576 ft^2. Find the length of a side of the original square.

Chapter Test

1. Factor: $6x^2y^2 + 9xy^2 + 12y^2$

2. Factor: $6x^3 - 8x^2 + 10x$

3. Factor: $p^2 + 5p + 6$

4. Factor: $a(x - 2) + b(2 - x)$

5. Solve: $(2a - 3)(a + 7) = 0$

6. Factor: $a^2 - 19a + 48$

7. Factor: $x^3 + 2x^2 - 15x$

8. Factor: $8x^2 + 20x - 48$

9. Factor: $ab + 6a - 3b - 18$

10. Solve: $4x^2 - 1 = 0$

11. Factor: $6x^2 + 19x + 8$

12. Factor: $x^2 - 9x - 36$

13. Factor: $2b^2 - 32$

14. Factor: $4a^2 - 12ab + 9b^2$

15. Factor: $px + x - p - 1$

16. Factor: $5x^2 - 45x - 15$

17. Factor: $2x^2 + 4x - 5$

18. Factor: $4x^2 - 49y^2$

19. Solve: $x(x - 8) = -15$

20. Factor: $p^2 + 12p + 36$

21. Factor: $18x^2 - 48xy + 32y^2$

22. Factor: $2y^4 - 14y^3 - 16y^2$

23. The length of a rectangle is 3 cm more than twice the width. The area of the rectangle is 90 cm^2. Find the length and width of the rectangle.

24. The length of the base of a triangle is three times the height. The area of the triangle is 24 in^2. Find the length of the base of the triangle.

25. The product of two consecutive negative integers is one hundred fifty-six. Find the two integers.

Cumulative Review Exercises

1. Subtract: $4 - (-5) - 6 - 11$

2. Divide: $0.372 \div (-0.046)$
Round to the nearest tenth.

3. Simplify: $(3 - 7)^2 \div (-2) - 3 \cdot (-4)$

4. Evaluate $-2a^2 \div 2b - c$ when $a = -4$, $b = 2$, and $c = -1$.

5. Identify the property that justifies the statement.
$(3 + 8) + 7 = 3 + (8 + 7)$

6. Multiply: $-\frac{3}{4}(-24x^2)$

7. Simplify: $-2[3x - 4(3 - 2x) - 8x]$

8. Solve: $-\frac{5}{7}x = -\frac{10}{21}$

9. Solve: $4 + 3(x - 2) = 13$

10. Solve: $3x - 2 = 12 - 5x$

11. Solve: $-2 + 4[3x - 2(4 - x) - 3] = 4x + 2$

12. 120% of what number is 42?

13. Solve: $-4x - 2 \geq 10$

14. Solve: $9 - 2(4x - 5) < 3(7 - 6x)$

15. Graph: $y = \frac{3}{4}x - 2$

16. Graph: $f(x) = -3x - 3$

17. Find the domain and range of the relation $\{(-5, -4), (-3, -2), (-1, 0), (1, 2), (3, 4)\}$. Is the relation a function?

18. Evaluate $f(x) = 6x - 5$ at $x = 11$.

19. Graph the solution set of $x + 3y > 2$.

20. Solve by substitution: $6x + y = 7$
$x - 3y = 17$

21. Solve by the addition method: $2x - 3y = -4$
$5x + y = 7$

22. Add: $(3y^3 - 5y^2 - 6) + (2y^2 - 8y + 1)$

23. Simplify: $(-3a^4b^2)^3$

24. Multiply: $(x + 2)(x^2 - 5x + 4)$

25. Divide: $(8x^2 + 4x - 3) \div (2x - 3)$

26. Simplify: $(x^{-4}y^2)^3$

27. Factor: $3a - 3b - ax + bx$

28. Factor: $x^2 + 3xy - 10y^2$

29. Factor: $6a^4 + 22a^3 + 12a^2$

30. Factor: $25a^2 - 36b^2$

31. Factor: $12x^2 - 36xy + 27y^2$

32. Solve: $3x^2 + 11x - 20 = 0$

33. Find the range of the function given by the equation $f(x) = \frac{4}{5}x - 3$ if the domain is $\{-10, -5, 0, 5, 10\}$.

34. The daily high temperatures, in degrees Celsius, during one week were recorded as follows: $-4°, -7°, 2°, 0°, -1°, -6°, -5°$. Find the average daily high temperature for the week.

35. The width of a rectangle is 40% of the length. The perimeter of the rectangle is 42 cm. Find the length and width of the rectangle.

36. A board 10 ft long is cut into two pieces. Four times the length of the shorter piece is 2 ft less than three times the length of the longer piece. Find the length of each piece.

37. Company A rents cars for $6 a day and 25¢ for every mile driven. Company B rents cars for $15 a day and 10¢ per mile. You want to rent a car for 6 days. What is the maximum number of miles you can drive a Company A car if it is to cost you less than a Company B car?

38. An investment of $4000 is made at an annual simple interest rate of 8%. How much additional money must be invested at an annual simple interest rate of 11% so that the total interest earned is $1035?

39. A stereo that regularly sells for $165 is on sale for $99. Find the discount rate.

40. Find three consecutive even integers such that five times the middle integer is twelve more than twice the sum of the first and third.

9

Algebraic Fractions

$$\frac{7\frac{1}{2}^{\circ}}{360^{\circ}} = \frac{520 \text{ miles}}{C}$$

Objectives

Measurement of the Circumference of the Earth

Distances on the earth, the circumference of the earth, and the distance to the moon and stars are known to great precision. Eratosthenes, the fifth librarian of Alexandria (230 B.C.), laid the foundation of scientific geography with his determination of the circumference of the earth.

Eratosthenes was familiar with certain astronomical data that enabled him to calculate the circumference of the earth by using a proportion statement.

Eratosthenes knew that on a mid-summer day, the sun was directly overhead at Syrene, as shown in the diagram.

At the same time, at Alexandria the sun was at a $7\frac{1}{2}^{\circ}$ angle from the zenith. The distance from Syrene to Alexandria was 5000 stadia (about 520 mi).

Knowing that the ratio of the $7\frac{1}{2}^{\circ}$ angle to one revolution (360°) is equal to the ratio of the arc length (520 mi) to the circumference, Eratosthenes was able to write and solve a proportion.

This result, calculated over 2000 years ago, is very close to the accepted value of 24,800 miles.

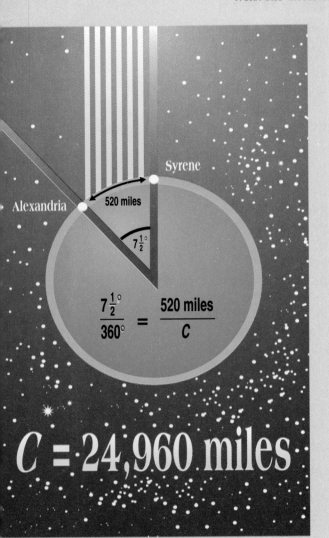

$$\frac{7\frac{1}{2}^{\circ}}{360^{\circ}} = \frac{520 \text{ miles}}{C}$$

$$C = 24,960 \text{ miles}$$

SECTION 9.1

Multiplication and Division of Algebraic Fractions

1 Simplify algebraic fractions

A fraction in which the numerator and denominator are polynomials is called an **algebraic fraction.** Examples of algebraic fractions are shown at the right.

$$\frac{5}{z} \qquad \frac{x^2 + 1}{2x - 1} \qquad \frac{y^2 - 3}{3xy + 1}$$

Care must be exercised with algebraic fractions to ensure that when the variables are replaced with numbers, the resulting denominator is not zero.

Consider the algebraic fraction at the right. The value of x cannot be 2 because the denominator would then be zero.

$$\frac{3x + 1}{2x - 4}$$

$$\frac{3 \cdot 2 + 1}{2 \cdot 2 - 4} = \frac{7}{0} \leftarrow \text{Not a real number}$$

A fraction is in simplest form when the numerator and denominator have no common factors other than 1. The Multiplication Property of One is used to write an algebraic fraction in simplest form.

Simplify: $\dfrac{x^2 - 4}{x^2 - 2x - 8}$

Factor the numerator and denominator.

$$\frac{x^2 - 4}{x^2 - 2x - 8} = \frac{(x - 2)(x + 2)}{(x - 4)(x + 2)}$$

$$= \frac{x - 2}{x - 4} \cdot \boxed{\frac{x + 2}{x + 2}}$$

$$= \frac{x - 2}{x - 4} \cdot 1$$

The restrictions $x \neq -2$ and $x \neq 4$ are necessary to prevent division by zero.

$$= \frac{x - 2}{x - 4}, \; x \neq -2, 4$$

This simplification is usually shown with slashes through the common factors. The last simplification would be shown as follows.

$$\frac{x^2 - 4}{x^2 - 2x - 8} = \frac{(x - 2)\overset{1}{\cancel{(x + 2)}}}{(x - 4)\underset{1}{\cancel{(x + 2)}}} = \frac{x - 2}{x - 4}, \; x \neq -2, 4$$

In this problem, it is stated that $x \neq -2, 4$. Look at the factored form of the original algebraic fraction.

$$\frac{(x - 2)(x + 2)}{(x - 4)(x + 2)}$$

If either of the factors in the denominator is zero, then the denominator is zero.

$$x - 4 = 0 \qquad\qquad x + 2 = 0$$
$$x = 4 \qquad\qquad x = -2$$

When x is 4 or -2, the denominator is 0. Therefore, for this fraction, $x \neq -2, 4$.

For the remaining examples, we will omit the restrictions on the variables that prevent division by zero and assume the values of the variables are such that division by zero is not possible.

To simplify $\dfrac{10 + 3x - x^2}{x^2 - 4x - 5}$, factor the numerator and denominator.

$$\frac{10 + 3x - x^2}{x^2 - 4x - 5} = \frac{(5 - x)(2 + x)}{(x - 5)(x + 1)}$$

Divide by the common factors. Remember that $5 - x = -(x - 5)$. Therefore,

$$\frac{5 - x}{x - 5} = \frac{-(x - 5)}{x - 5} = \frac{-1}{1} = -1$$

$$= \frac{\overset{-1}{\cancel{(5 - x)}}(2 + x)}{\underset{1}{\cancel{(x - 5)}}(x + 1)}$$

Write the answer in simplest form.

$$= -\frac{x + 2}{x + 1}$$

Example 1   Simplify.   A. $\dfrac{4x^3 y^4}{6x^4 y}$ B. $\dfrac{9 - x^2}{x^2 + x - 12}$

Solution   A. $\dfrac{4x^3 y^4}{6x^4 y} = \dfrac{\overset{1}{\cancel{2}} \cdot 2x^3 y^4}{\underset{1}{\cancel{2}} \cdot 3x^4 y} = \dfrac{2y^3}{3x}$ ▶ Simplify using the rules of exponents.

   B. $\dfrac{9 - x^2}{x^2 + x - 12} = \dfrac{\overset{-1}{\cancel{(3 - x)}}(3 + x)}{\underset{1}{\cancel{(x - 3)}}(x + 4)} = -\dfrac{x + 3}{x + 4}$

Problem 1   Simplify.   A. $\dfrac{6x^5 y}{12x^2 y^3}$ B. $\dfrac{x^2 + 2x - 24}{16 - x^2}$

Solution   See page A35.

---

2   Multiply algebraic fractions

The product of two fractions is a fraction whose numerator is the product of the numerators of the two fractions and whose denominator is the product of the denominators of the two fractions.

$$\frac{a}{b} \cdot \frac{c}{d} = \frac{ac}{bd}$$

$$\frac{2}{3} \cdot \frac{4}{5} = \frac{8}{15}$$

$$\frac{3x}{y} \cdot \frac{2}{z} = \frac{6x}{yz}$$

$$\frac{x + 2}{x} \cdot \frac{3}{x - 2} = \frac{3(x + 2)}{x(x - 2)}$$

Multiply: $\dfrac{x^2 + 3x}{x^2 - 3x - 4} \cdot \dfrac{x^2 - 5x + 4}{x^2 + 2x - 3}$

$$\dfrac{x^2 + 3x}{x^2 - 3x - 4} \cdot \dfrac{x^2 - 5x + 4}{x^2 + 2x - 3}$$

Factor the numerator and denominator of each fraction.

$$= \dfrac{x(x + 3)}{(x - 4)(x + 1)} \cdot \dfrac{(x - 4)(x - 1)}{(x + 3)(x - 1)}$$

Multiply. Divide by the common factors.

$$= \dfrac{x\cancel{(x+3)}\cancel{(x-4)}\cancel{(x-1)}}{\cancel{(x-4)}(x+1)\cancel{(x+3)}\cancel{(x-1)}}$$

Write the answer in simplest form.

$$= \dfrac{x}{x + 1}$$

Example 2 Multiply.

A. $\dfrac{10x^2 - 15x}{12x - 8} \cdot \dfrac{3x - 2}{20x - 25}$ B. $\dfrac{x^2 + x - 6}{x^2 + 7x + 12} \cdot \dfrac{x^2 + 3x - 4}{4 - x^2}$

Solution A. $\dfrac{10x^2 - 15x}{12x - 8} \cdot \dfrac{3x - 2}{20x - 25}$

$$= \dfrac{5x(2x - 3)}{4(3x - 2)} \cdot \dfrac{(3x - 2)}{5(4x - 5)}$$ ▶ Factor the numerator and denominator of each fraction.

$$= \dfrac{\cancel{5}x(2x - 3)\cancel{(3x-2)}}{2 \cdot 2\cancel{(3x-2)}\cancel{5}(4x - 5)}$$ ▶ Multiply. Divide by the common factors.

$$= \dfrac{x(2x - 3)}{4(4x - 5)}$$ ▶ Write the answer in simplest form.

B. $\dfrac{x^2 + x - 6}{x^2 + 7x + 12} \cdot \dfrac{x^2 + 3x - 4}{4 - x^2}$

$$= \dfrac{(x + 3)(x - 2)}{(x + 3)(x + 4)} \cdot \dfrac{(x + 4)(x - 1)}{(2 - x)(2 + x)}$$

$$= \dfrac{\cancel{(x+3)}\overset{-1}{\cancel{(x-2)}}\cancel{(x+4)}(x - 1)}{\cancel{(x+3)}\cancel{(x+4)}\cancel{(2-x)}(2 + x)}$$

$$= -\dfrac{x - 1}{x + 2}$$

Problem 2 Multiply.

A. $\dfrac{12x^2 + 3x}{10x - 15} \cdot \dfrac{8x - 12}{9x + 18}$ B. $\dfrac{x^2 + 2x - 15}{9 - x^2} \cdot \dfrac{x^2 - 3x - 18}{x^2 - 7x + 6}$

Solution See page A35.

3   Divide algebraic fractions

The **reciprocal** of a fraction is a fraction with the numerator and denominator interchanged.

$$\text{Fraction}\begin{cases}\dfrac{a}{b} & \dfrac{b}{a} \\[2mm] x^2 = \dfrac{x^2}{1} & \dfrac{1}{x^2} \\[2mm] \dfrac{x+2}{x} & \dfrac{x}{x+2}\end{cases}\text{Reciprocal}$$

To divide two fractions, multiply by the reciprocal of the divisor.

$$\frac{a}{b} \div \frac{c}{d} = \frac{a}{b} \cdot \frac{d}{c} = \frac{ad}{bc}$$

$$\frac{4}{x} \div \frac{y}{5} = \frac{4}{x} \cdot \frac{5}{y} = \frac{20}{xy}$$

$$\frac{x+4}{x} \div \frac{x-2}{4} = \frac{x+4}{x} \cdot \frac{4}{x-2} = \frac{4(x+4)}{x(x-2)}$$

The basis for the division rule is shown below.

$$\underbrace{\frac{a}{b} \div \frac{c}{d}}_{} = \frac{\dfrac{a}{b}}{\dfrac{c}{d}} = \frac{\dfrac{a}{b} \cdot \dfrac{d}{c}}{\dfrac{c}{d} \cdot \dfrac{d}{c}} = \frac{\dfrac{a}{b} \cdot \dfrac{d}{c}}{1} = \underbrace{\frac{a}{b} \cdot \frac{d}{c}}_{}$$

Example 3   Divide.

A. $\dfrac{xy^2 - 3x^2y}{z^2} \div \dfrac{6x^2 - 2xy}{z^3}$ B. $\dfrac{2x^2 + 5x + 2}{2x^2 + 3x - 2} \div \dfrac{3x^2 + 13x + 4}{2x^2 + 7x - 4}$

Solution   A. $\dfrac{xy^2 - 3x^2y}{z^2} \div \dfrac{6x^2 - 2xy}{z^3} = \dfrac{xy^2 - 3x^2y}{z^2} \cdot \dfrac{z^3}{6x^2 - 2xy}$

$$= \frac{xy(y - 3x) \cdot \overset{-1}{z^3}}{z^2 \cdot 2x(3x - y)}$$

$$= -\frac{yz}{2}$$

B. $\dfrac{2x^2 + 5x + 2}{2x^2 + 3x - 2} \div \dfrac{3x^2 + 13x + 4}{2x^2 + 7x - 4} = \dfrac{2x^2 + 5x + 2}{2x^2 + 3x - 2} \cdot \dfrac{2x^2 + 7x - 4}{3x^2 + 13x + 4}$

$$= \frac{(2x+1)(x+2)}{(2x-1)(x+2)} \cdot \frac{(2x-1)(x+4)}{(3x+1)(x+4)}$$

$$= \frac{2x+1}{3x+1}$$

Problem 3   Divide.

A. $\dfrac{a^2}{4bc^2 - 2b^2c} \div \dfrac{a}{6bc - 3b^2}$ B. $\dfrac{3x^2 + 26x + 16}{3x^2 - 7x - 6} \div \dfrac{2x^2 + 9x - 5}{x^2 + 2x - 15}$

Solution   See page A35.

CONCEPT REVIEW 9.1

Determine whether the statement is always true, sometimes true, or never true.

1. Algebraic fractions are rational expressions.

2. A fraction is in simplest form when the only factor common to both the numerator and the denominator is 1.

3. Before multiplying two algebraic fractions, we must write the fractions in terms of a common denominator.

4. The expression $\dfrac{x}{x + 2}$ is not a real number if $x = 0$.

5. The procedure for multiplying algebraic fractions is the same as that for multiplying arithmetic fractions.

6. When an algebraic fraction is rewritten in simplest form, its value is less than it was before it was rewritten in simplest form.

7. To divide two algebraic fractions, multiply the reciprocal of the first fraction by the second fraction.

EXERCISES 9.1

1 Simplify.

1. $\dfrac{9x^3}{12x^4}$

2. $\dfrac{16x^2y}{24xy^3}$

3. $\dfrac{(x + 3)^2}{(x + 3)^3}$

4. $\dfrac{(2x - 1)^5}{(2x - 1)^4}$

5. $\dfrac{3n - 4}{4 - 3n}$

6. $\dfrac{5 - 2x}{2x - 5}$

7. $\dfrac{6y(y + 2)}{9y^2(y + 2)}$

8. $\dfrac{12x^2(3 - x)}{18x(3 - x)}$

9. $\dfrac{6x(x - 5)}{8x^2(5 - x)}$

10. $\dfrac{14x^3(7 - 3x)}{21x(3x - 7)}$

11. $\dfrac{a^2 + 4a}{ab + 4b}$

12. $\dfrac{x^2 - 3x}{2x - 6}$

13. $\dfrac{4 - 6x}{3x^2 - 2x}$

14. $\dfrac{5xy - 3y}{9 - 15x}$

15. $\dfrac{y^2 - 3y + 2}{y^2 - 4y + 3}$

16. $\dfrac{x^2 + 5x + 6}{x^2 + 8x + 15}$

17. $\dfrac{x^2 + 3x - 10}{x^2 + 2x - 8}$

18. $\dfrac{a^2 + 7a - 8}{a^2 + 6a - 7}$

19. $\dfrac{x^2 + x - 12}{x^2 - 6x + 9}$

20. $\dfrac{x^2 + 8x + 16}{x^2 - 2x - 24}$

21. $\dfrac{x^2 - 3x - 10}{25 - x^2}$

22. $\dfrac{4 - y^2}{y^2 - 3y - 10}$

23. $\dfrac{2x^3 + 2x^2 - 4x}{x^3 + 2x^2 - 3x}$

24. $\dfrac{3x^3 - 12x}{6x^3 - 24x^2 + 24x}$

25. $\dfrac{6x^2 - 7x + 2}{6x^2 + 5x - 6}$

26. $\dfrac{2n^2 - 9n + 4}{2n^2 - 5n - 12}$

2 Multiply.

27. $\dfrac{8x^2}{9y^3} \cdot \dfrac{3y^2}{4x^3}$

28. $\dfrac{4a^2b^3}{15x^5y^2} \cdot \dfrac{25x^3y}{16ab}$

29. $\dfrac{12x^3y^4}{7a^2b^3} \cdot \dfrac{14a^3b^4}{9x^2y^2}$

30. $\dfrac{18a^4b^2}{25x^2y^3} \cdot \dfrac{50x^5y^6}{27a^6b^2}$

31. $\dfrac{3x - 6}{5x - 20} \cdot \dfrac{10x - 40}{27x - 54}$

32. $\dfrac{8x - 12}{14x + 7} \cdot \dfrac{42x + 21}{32x - 48}$

33. $\dfrac{3x^2 + 2x}{2xy - 3y} \cdot \dfrac{2xy^3 - 3y^3}{3x^3 + 2x^2}$

34. $\dfrac{4a^2x - 3a^2}{2by + 5b} \cdot \dfrac{2b^3y + 5b^3}{4ax - 3a}$

35. $\dfrac{x^2 + 5x + 4}{x^3y^2} \cdot \dfrac{x^2y^3}{x^2 + 2x + 1}$

36. $\dfrac{x^2 + x - 2}{xy^2} \cdot \dfrac{x^3y}{x^2 + 5x + 6}$

37. $\dfrac{x^4y^2}{x^2 + 3x - 28} \cdot \dfrac{x^2 - 49}{xy^4}$

38. $\dfrac{x^5y^3}{x^2 + 13x + 30} \cdot \dfrac{x^2 + 2x - 3}{x^7y^2}$

39. $\dfrac{2x^2 - 5x}{2xy + y} \cdot \dfrac{2xy^2 + y^2}{5x^2 - 2x^3}$

40. $\dfrac{3a^3 + 4a^2}{5ab - 3b} \cdot \dfrac{3b^3 - 5ab^3}{3a^2 + 4a}$

41. $\dfrac{x^2 - 2x - 24}{x^2 - 5x - 6} \cdot \dfrac{x^2 + 5x + 6}{x^2 + 6x + 8}$

42. $\dfrac{x^2 - 8x + 7}{x^2 + 3x - 4} \cdot \dfrac{x^2 + 3x - 10}{x^2 - 9x + 14}$

43. $\dfrac{x^2 + 2x - 35}{x^2 + 4x - 21} \cdot \dfrac{x^2 + 3x - 18}{x^2 + 9x + 18}$

44. $\dfrac{y^2 + y - 20}{y^2 + 2y - 15} \cdot \dfrac{y^2 + 4y - 21}{y^2 + 3y - 28}$

45. $\dfrac{x^2 - 3x - 4}{x^2 + 6x + 5} \cdot \dfrac{x^2 + 5x + 6}{8 + 2x - x^2}$

46. $\dfrac{25 - n^2}{n^2 - 2n - 35} \cdot \dfrac{n^2 - 8n - 20}{n^2 - 3n - 10}$

47. $\dfrac{12x^2 - 6x}{x^2 + 6x + 5} \cdot \dfrac{2x^4 + 10x^3}{4x^2 - 1}$

48. $\dfrac{8x^3 + 4x^2}{x^2 - 3x + 2} \cdot \dfrac{x^2 - 4}{16x^2 + 8x}$

49. $\dfrac{16 + 6x - x^2}{x^2 - 10x - 24} \cdot \dfrac{x^2 - 6x - 27}{x^2 - 17x + 72}$

50. $\dfrac{x^2 - 11x + 28}{x^2 - 13x + 42} \cdot \dfrac{x^2 + 7x + 10}{20 - x - x^2}$

51. $\dfrac{2x^2 + 5x + 2}{2x^2 + 7x + 3} \cdot \dfrac{x^2 - 7x - 30}{x^2 - 6x - 40}$

52. $\dfrac{x^2 - 4x - 32}{x^2 - 8x - 48} \cdot \dfrac{3x^2 + 17x + 10}{3x^2 - 22x - 16}$

53. $\dfrac{2x^2 + x - 3}{2x^2 - x - 6} \cdot \dfrac{2x^2 - 9x + 10}{2x^2 - 3x + 1}$

54. $\dfrac{3y^2 + 14y + 8}{2y^2 + 7y - 4} \cdot \dfrac{2y^2 + 9y - 5}{3y^2 + 16y + 5}$

55. $\dfrac{6x^2 - 11x + 4}{6x^2 + x - 2} \cdot \dfrac{12x^2 + 11x + 2}{8x^2 + 14x + 3}$

56. $\dfrac{6 - x - 2x^2}{4x^2 + 3x - 10} \cdot \dfrac{3x^2 + 7x - 20}{2x^2 + 5x - 12}$

3   Divide.

57. $\dfrac{4x^2y^3}{15a^2b^3} \div \dfrac{6xy}{5a^3b^5}$

58. $\dfrac{9x^3y^4}{16a^4b^2} \div \dfrac{45x^4y^2}{14a^7b}$

59. $\dfrac{6x - 12}{8x + 32} \div \dfrac{18x - 36}{10x + 40}$

60. $\dfrac{28x + 14}{45x - 30} \div \dfrac{14x + 7}{30x - 20}$

61. $\dfrac{6x^3 + 7x^2}{12x - 3} \div \dfrac{6x^2 + 7x}{36x - 9}$

62. $\dfrac{5a^2y + 3a^2}{2x^3 + 5x^2} \div \dfrac{10ay + 6a}{6x^3 + 15x^2}$

63. $\dfrac{x^2 + 4x + 3}{x^2y} \div \dfrac{x^2 + 2x + 1}{xy^2}$

64. $\dfrac{x^3y^2}{x^2 - 3x - 10} \div \dfrac{xy^4}{x^2 - x - 20}$

65. $\dfrac{x^2 - 49}{x^4y^3} \div \dfrac{x^2 - 14x + 49}{x^4y^3}$

66. $\dfrac{x^2y^5}{x^2 - 11x + 30} \div \dfrac{xy^6}{x^2 - 7x + 10}$

67. $\dfrac{4ax - 8a}{c^2} \div \dfrac{2y - xy}{c^3}$

68. $\dfrac{3x^2y - 9xy}{a^2b} \div \dfrac{3x^2 - x^3}{ab^2}$

69. $\dfrac{x^2 - 5x + 6}{x^2 - 9x + 18} \div \dfrac{x^2 - 6x + 8}{x^2 - 9x + 20}$

70. $\dfrac{x^2 + 3x - 40}{x^2 + 2x - 35} \div \dfrac{x^2 + 2x - 48}{x^2 + 3x - 18}$

71. $\dfrac{x^2 + 2x - 15}{x^2 - 4x - 45} \div \dfrac{x^2 + x - 12}{x^2 - 5x - 36}$

72. $\dfrac{y^2 - y - 56}{y^2 + 8y + 7} \div \dfrac{y^2 - 13y + 40}{y^2 - 4y - 5}$

73. $\dfrac{8 + 2x - x^2}{x^2 + 7x + 10} \div \dfrac{x^2 - 11x + 28}{x^2 - x - 42}$

74. $\dfrac{x^2 - x - 2}{x^2 - 7x + 10} \div \dfrac{x^2 - 3x - 4}{40 - 3x - x^2}$

75. $\dfrac{2x^2 - 3x - 20}{2x^2 - 7x - 30} \div \dfrac{2x^2 - 5x - 12}{4x^2 + 12x + 9}$

76. $\dfrac{6n^2 + 13n + 6}{4n^2 - 9} \div \dfrac{6n^2 + n - 2}{4n^2 - 1}$

77. $\dfrac{9x^2 - 16}{6x^2 - 11x + 4} \div \dfrac{6x^2 + 11x + 4}{8x^2 + 10x + 3}$

78. $\dfrac{15 - 14x - 8x^2}{4x^2 + 4x - 15} \div \dfrac{4x^2 + 13x - 12}{3x^2 + 13x + 4}$

79. $\dfrac{8x^2 + 18x - 5}{10x^2 - 9x + 2} \div \dfrac{8x^2 + 22x + 15}{10x^2 + 11x - 6}$

80. $\dfrac{10 + 7x - 12x^2}{8x^2 - 2x - 15} \div \dfrac{6x^2 - 13x + 5}{10x^2 - 13x + 4}$

SUPPLEMENTAL EXERCISES 9.1

For what values of x is the algebraic fraction undefined? (*Hint:* Set the denominator equal to zero and solve for x.)

81. $\dfrac{x}{(x + 6)(x - 1)}$

82. $\dfrac{x}{(x - 2)(x + 5)}$

83. $\dfrac{8}{x^2 - 1}$

84. $\dfrac{7}{x^2 - 16}$

85. $\dfrac{x - 4}{x^2 - x - 6}$

86. $\dfrac{x + 5}{x^2 - 4x - 5}$

87. $\dfrac{3x}{x^2 + 6x + 9}$

88. $\dfrac{3x - 8}{3x^2 - 10x - 8}$

89. $\dfrac{4x + 7}{6x^2 - 5x - 4}$

Simplify.

90. $\dfrac{y^2}{x} \cdot \dfrac{x}{2} \div \dfrac{y}{x}$

91. $\dfrac{ab}{3} \cdot \dfrac{a}{b^2} \div \dfrac{a}{4}$

92. $\left(\dfrac{2x}{y}\right)^3 \div \left(\dfrac{x}{3y}\right)^2$

93. $\left(\dfrac{c}{3}\right)^2 \div \left(\dfrac{c}{2} \cdot \dfrac{c}{4}\right)$

94. $\left(\dfrac{a - 3}{b}\right)^2 \left(\dfrac{b}{3 - a}\right)^3$

95. $\left(\dfrac{x - 4}{y^2}\right)^3 \cdot \left(\dfrac{y}{4 - x}\right)^2$

96. $\dfrac{x^2 + 3x - 40}{x^2 + 2x - 35} \div \dfrac{x^2 + 2x - 48}{x^2 + 3x - 18} \cdot \dfrac{x^2 - 36}{x^2 - 9}$

97. $\dfrac{x^2 + x - 6}{x^2 + 7x + 12} \cdot \dfrac{x^2 + 3x - 4}{x^2 + x - 2} \div \dfrac{x^2 - 16}{x^2 - 4}$

98. Given the expression $\dfrac{9}{x^2 + 1}$, choose some values of x and evalute the expression for those values. Is it possible to choose a value of x for which the value of the expression is greater than 10? If so, what is that value of x? If not, explain why it is not possible.

99. Given the expression $\dfrac{1}{y - 3}$, choose some values of y and evaluate the expression for those values. Is it possible to choose a value of y for which the value of the expression is greater than 10,000,000? If so, what is that value of y? Explain your answer.

SECTION **9.2**

Expressing Fractions in Terms of the Least Common Multiple (LCM)

1 ▶ **Find the least common multiple (LCM) of two or more polynomials**

The **least common multiple (LCM)** of two or more numbers is the smallest number that contains the prime factorization of each number.

The LCM of 12 and 18 is 36. 36 contains the prime factors of 12 and the prime factors of 18.

$$12 = 2 \cdot 2 \cdot 3$$
$$18 = 2 \cdot 3 \cdot 3$$

$$\text{LCM} = 36 = \overbrace{2 \cdot \underbrace{2 \cdot 3 \cdot 3}}^{\text{Factors of 12}}$$
$$\underbrace{}_{\text{Factors of 18}}$$

The least common multiple of two or more polynomials is the simplest polynomial that contains the factors of each polynomial.

To find the LCM of two or more polynomials, first factor each polynomial completely. The LCM is the product of each factor the greatest number of times it occurs in any one factorization.

LOOK CLOSELY

The LCM must contain the factors of each polynomial. As shown with braces at the right, the LCM contains the factors of $4x^2 + 4x$ and the factors of $x^2 + 2x + 1$.

The LCM of $4x^2 + 4x$ and $x^2 + 2x + 1$ is the product of the LCM of the numerical coefficients and each variable factor the greatest number of times it occurs in any one factorization.

$$4x^2 + 4x = 4x(x + 1) = 2 \cdot 2 \cdot x(x + 1)$$
$$x^2 + 2x + 1 = (x + 1)(x + 1)$$

$$\text{LCM} = \overbrace{2 \cdot 2 \cdot x(x + 1)}^{\text{Factors of } 4x^2 + 4x}(x + 1) = 4x(x + 1)(x + 1)$$
$$\underbrace{}_{\text{Factors of } x^2 + 2x + 1}$$

Example 1   Find the LCM of $4x^2y$ and $6xy^2$.

Solution $4x^2y = 2 \cdot 2 \cdot x \cdot x \cdot y$
$6xy^2 = 2 \cdot 3 \cdot x \cdot y \cdot y$

▶ Factor each polynomial completely.

$\text{LCM} = 2 \cdot 2 \cdot 3 \cdot x \cdot x \cdot y \cdot y$
$\quad\quad = 12x^2y^2$

▶ Write the product of the LCM of the numerical coefficients and each variable factor the greatest number of times it occurs in any one factorization.

Problem 1   Find the LCM of $8uv^2$ and $12uw$.

Solution   See page A35.

> **Example 2** Find the LCM of $x^2 - x - 6$ and $9 - x^2$.
>
> **Solution** $x^2 - x - 6 = (x - 3)(x + 2)$
> $9 - x^2 = -(x^2 - 9) = -(x + 3)(x - 3)$
>
> LCM $= (x - 3)(x + 2)(x + 3)$
>
> **Problem 2** Find the LCM of $m^2 - 6m + 9$ and $m^2 - 2m - 3$.
>
> **Solution** See page A35.

2 ## Express two fractions in terms of the LCM of their denominators

When adding and subtracting fractions, it is often necessary to express two or more fractions in terms of a common denominator. This common denominator is the LCM of the denominators of the fractions.

Write the fractions $\dfrac{x + 1}{4x^2}$ and $\dfrac{x - 3}{6x^2 - 12x}$ in terms of the LCM of the denominators.

Find the LCM of the denominators.

The LCM is $12x^2(x - 2)$.

LOOK CLOSELY
$\dfrac{3(x - 2)}{3(x - 2)} = 1$ and $\dfrac{2x}{2x} = 1$. We are multiplying each fraction by 1, so we are not changing the value of either fraction.

For each fraction, multiply the numerator and denominator by the factors whose product with the denominator is the LCM.

$\dfrac{x + 1}{4x^2} = \dfrac{x + 1}{4x^2} \cdot \dfrac{3(x - 2)}{3(x - 2)} = \dfrac{3x^2 - 3x - 6}{12x^2(x - 2)}$ ⟵
$\dfrac{x - 3}{6x^2 - 12x} = \dfrac{x - 3}{6x(x - 2)} \cdot \dfrac{2x}{2x} = \dfrac{2x^2 - 6x}{12x^2(x - 2)}$ ⟵ } LCM

> **Example 3** Write the fractions $\dfrac{x + 2}{3x^2}$ and $\dfrac{x - 1}{8xy}$ in terms of the LCM of the denominators.
>
> **Solution** The LCM is $24x^2y$.
>
> $\dfrac{x + 2}{3x^2} = \dfrac{x + 2}{3x^2} \cdot \dfrac{8y}{8y} = \dfrac{8xy + 16y}{24x^2y}$ ▶ The product of $3x^2$ and $8y$ is the LCM.
>
> $\dfrac{x - 1}{8xy} = \dfrac{x - 1}{8xy} \cdot \dfrac{3x}{3x} = \dfrac{3x^2 - 3x}{24x^2y}$ ▶ The product of $8xy$ and $3x$ is the LCM.
>
> **Problem 3** Write the fractions $\dfrac{x - 3}{4xy^2}$ and $\dfrac{2x + 1}{9y^2z}$ in terms of the LCM of the denominators.
>
> **Solution** See page A35.

Example 4   Write the fractions $\dfrac{2x-1}{2x-x^2}$ and $\dfrac{x}{x^2+x-6}$ in terms of the LCM of the denominators.

Solution $\dfrac{2x-1}{2x-x^2} = \dfrac{2x-1}{-(x^2-2x)} = -\dfrac{2x-1}{x^2-2x}$ ▶ Rewrite $\dfrac{2x-1}{2x-x^2}$ with a denominator of x^2-2x.

The LCM is $x(x-2)(x+3)$.

$\dfrac{2x-1}{2x-x^2} = -\dfrac{2x-1}{x(x-2)} \cdot \dfrac{x+3}{x+3} = -\dfrac{2x^2+5x-3}{x(x-2)(x+3)}$

$\dfrac{x}{x^2+x-6} = \dfrac{x}{(x-2)(x+3)} \cdot \dfrac{x}{x} = \dfrac{x^2}{x(x-2)(x+3)}$

Problem 4   Write the fractions $\dfrac{x+4}{x^2-3x-10}$ and $\dfrac{2x}{25-x^2}$ in terms of the LCM of the denominators.

Solution   See page A35.

CONCEPT REVIEW 9.2

Determine whether the statement is always true, sometimes true, or never true.

1. The least common multiple of two numbers is the smallest number that contains all the prime factors of both numbers.

2. The LCD is the least common multiple of the denominators of two or more fractions.

3. The LCM of x^2, x^5, and x^8 is x^2.

4. We can rewrite $\dfrac{x}{y}$ as $\dfrac{4x}{4y}$ by using the Multiplication Property of One.

5. Rewriting two fractions in terms of the LCM of their denominators is the reverse process of simplifying the fractions.

6. To rewrite an algebraic fraction in terms of a common denominator, determine what factor you must multiply the denominator by so that the denominator will be the common denominator. Then multiply the numerator and denominator of the fraction by that factor.

EXERCISES 9.2

1   Find the LCM of the expressions.

1. $8x^3y$
    $12xy^2$

2. $6ab^2$
    $18ab^3$

3. $10x^4y^2$
    $15x^3y$

4. $12a^2b$
    $18ab^3$

5. $8x^2$
    $4x^2+8x$

6. $6y^2$
    $4y+12$

7. $2x^2y$
    $3x^2+12x$

8. $4xy^2$
    $6xy^2+12y^2$

9. $8x^2(x-1)^2$
       $10x^3(x-1)$

10. $3x + 3$
       $2x^2 + 4x + 2$

11. $4x - 12$
       $2x^2 - 12x + 18$

12. $(x-1)(x+2)$
       $(x-1)(x+3)$

13. $(2x-1)(x+4)$
       $(2x+1)(x+4)$

14. $(2x+3)^2$
       $(2x+3)(x-5)$

15. $(x-7)(x+2)$
       $(x-7)^2$

16. $(x-1)$
       $(x-2)$
       $(x-1)(x-2)$

17. $(x+4)(x-3)$
       $x + 4$
       $x - 3$

18. $x^2 - x - 6$
       $x^2 + x - 12$

19. $x^2 + 3x - 10$
       $x^2 + 5x - 14$

20. $x^2 + 5x + 4$
       $x^2 - 3x - 28$

21. $x^2 - 10x + 21$
       $x^2 - 8x + 15$

22. $x^2 - 2x - 24$
       $x^2 - 36$

23. $x^2 + 7x + 10$
       $x^2 - 25$

24. $x^2 - 7x - 30$
       $x^2 - 5x - 24$

25. $2x^2 - 7x + 3$
       $2x^2 + x - 1$

26. $3x^2 - 11x + 6$
       $3x^2 + 4x - 4$

27. $2x^2 - 9x + 10$
       $2x^2 + x - 15$

28. $6 + x - x^2$
       $x + 2$
       $x - 3$

29. $15 + 2x - x^2$
       $x - 5$
       $x + 3$

30. $5 + 4x - x^2$
       $x - 5$
       $x + 1$

31. $x^2 + 3x - 18$
       $3 - x$
       $x + 6$

32. $x^2 - 5x + 6$
       $1 - x$
       $x - 6$

2   Write each fraction in terms of the LCM of the denominators.

33. $\dfrac{4}{x} ; \dfrac{3}{x^2}$

34. $\dfrac{5}{ab^2} ; \dfrac{6}{ab}$

35. $\dfrac{x}{3y^2} ; \dfrac{z}{4y}$

36. $\dfrac{5y}{6x^2} ; \dfrac{7}{9xy}$

37. $\dfrac{y}{x(x-3)} ; \dfrac{6}{x^2}$

38. $\dfrac{a}{y^2} ; \dfrac{6}{y(y+5)}$

39. $\dfrac{9}{(x-1)^2} ; \dfrac{6}{x(x-1)}$

40. $\dfrac{a^2}{y(y+7)} ; \dfrac{a}{(y+7)^2}$

41. $\dfrac{3}{x-3} ; -\dfrac{5}{x(3-x)}$

42. $\dfrac{b}{y(y-4)} ; \dfrac{b^2}{4-y}$

43. $\dfrac{3}{(x-5)^2} ; \dfrac{2}{5-x}$

44. $\dfrac{3}{7-y} ; \dfrac{2}{(y-7)^2}$

45. $\dfrac{3}{x^2+2x} ; \dfrac{4}{x^2}$

46. $\dfrac{2}{y-3} ; \dfrac{3}{y^3-3y^2}$

47. $\dfrac{x-2}{x+3} ; \dfrac{x}{x-4}$

48. $\dfrac{x^2}{2x-1} ; \dfrac{x+1}{x+4}$

49. $\dfrac{3}{x^2+x-2} ; \dfrac{x}{x+2}$

50. $\dfrac{3x}{x-5} ; \dfrac{4}{x^2-25}$

51. $\dfrac{5}{2x^2-9x+10} ; \dfrac{x-1}{2x-5}$

52. $\dfrac{x-3}{3x^2+4x-4}; \dfrac{2}{x+2}$

53. $\dfrac{x}{x^2+x-6}; \dfrac{2x}{x^2-9}$

54. $\dfrac{x-1}{x^2+2x-15}; \dfrac{x}{x^2+6x+5}$

55. $\dfrac{x}{9-x^2}; \dfrac{x-1}{x^2-6x+9}$

56. $\dfrac{2x}{10+3x-x^2}; \dfrac{x+2}{x^2-8x+15}$

57. $\dfrac{3x}{x-5}; \dfrac{x}{x+4}; \dfrac{3}{20+x-x^2}$

58. $\dfrac{x+1}{x+5}; \dfrac{x+2}{x-7}; \dfrac{3}{35+2x-x^2}$

SUPPLEMENTAL EXERCISES 9.2

Write each expression in terms of the LCM of the denominators.

59. $\dfrac{3}{10^2}; \dfrac{5}{10^4}$

60. $\dfrac{8}{10^3}; \dfrac{9}{10^5}$

61. $b; \dfrac{5}{b}$

62. $3; \dfrac{2}{n}$

63. $1; \dfrac{y}{y-1}$

64. $x; \dfrac{x}{x^2-1}$

65. $\dfrac{x^2+1}{(x-1)^3}; \dfrac{x+1}{(x-1)^2}; \dfrac{1}{x-1}$

66. $\dfrac{a^2+a}{(a+1)^3}; \dfrac{a+1}{(a+1)^2}; \dfrac{1}{a+1}$

67. $\dfrac{b}{4a^2-4b^2}; \dfrac{a}{8a-8b}$

68. $\dfrac{c}{6c^2+7cd+d^2}; \dfrac{d}{3c^2-3d^2}$

69. $\dfrac{1}{x^2+2x+xy+2y}; \dfrac{1}{x^2+xy-2x-2y}$

70. $\dfrac{1}{ab+3a-3b-b^2}; \dfrac{1}{ab+3a+3b+b^2}$

71. When is the LCM of two expressions equal to their product?

SECTION 9.3

Addition and Subtraction of Algebraic Fractions

1 Add and subtract algebraic fractions with the same denominator

When adding algebraic fractions in which the denominators are the same, add the numerators. The denominator of the sum is the common denominator.

$$\frac{a}{b} + \frac{c}{b} = \frac{a+c}{b}$$

$$\frac{5x}{18} + \frac{7x}{18} = \frac{12x}{18} = \frac{2x}{3}$$

$$\frac{x}{x^2-1} + \frac{1}{x^2-1} = \frac{x+1}{x^2-1} = \frac{\overset{1}{\cancel{(x+1)}}}{(x-1)\underset{1}{\cancel{(x+1)}}} = \frac{1}{x-1}$$

Note that the sum is written in simplest form.

When subtracting algebraic fractions in which the denominators are the same, subtract the numerators. The denominator of the difference is the common denominator. Write the answer in simplest form.

$$\frac{2x}{x-2} - \frac{4}{x-2} = \frac{2x-4}{x-2} = \frac{2\overset{1}{\cancel{(x-2)}}}{\underset{1}{\cancel{x-2}}} = 2$$

$$\frac{3x-1}{x^2-5x+4} - \frac{2x+3}{x^2-5x+4} = \frac{(3x-1)-(2x+3)}{x^2-5x+4}$$

$$= \frac{x-4}{x^2-5x+4}$$

$$= \frac{\overset{1}{\cancel{(x-4)}}}{\underset{1}{\cancel{(x-4)}}(x-1)}$$

$$= \frac{1}{x-1}$$

LOOK CLOSELY

Be careful with signs when subtracting algebraic fractions. Note that we must subtract the *entire* numerator $2x+3$.
$(3x-1)-(2x+3)$
$\quad = 3x-1-2x-3$

Example 1 Add or subtract.   A. $\frac{7}{x^2} + \frac{9}{x^2}$ B. $\frac{3x^2}{x^2-1} - \frac{x+4}{x^2-1}$

Solution A. $\frac{7}{x^2} + \frac{9}{x^2} = \frac{7+9}{x^2}$

$$= \frac{16}{x^2}$$

▶ The denominators are the same. Add the numerators.

B. $\frac{3x^2}{x^2-1} - \frac{x+4}{x^2-1} = \frac{3x^2-(x+4)}{x^2-1}$

$$= \frac{3x^2-x-4}{x^2-1}$$

▶ The denominators are the same. Subtract the numerators.

$$= \frac{(3x-4)\overset{1}{\cancel{(x+1)}}}{(x-1)\underset{1}{\cancel{(x+1)}}}$$

▶ Write the answer in simplest form.

$$= \frac{3x-4}{x-1}$$

Problem 1 Add or subtract.   A. $\frac{3}{xy} + \frac{12}{xy}$ B. $\frac{2x^2}{x^2-x-12} - \frac{7x+4}{x^2-x-12}$

Solution See page A36.

2   Add and subtract algebraic fractions with different denominators

Before two fractions with different denominators can be added or subtracted, each fraction must be expressed in terms of a common denominator. This common denominator is the LCM of the denominators of the fractions.

Add: $\dfrac{x-3}{x^2-2x} + \dfrac{6}{x^2-4}$

Find the LCM of the denominators.

$$x^2 - 2x = x(x-2)$$
$$x^2 - 4 = (x-2)(x+2)$$

The LCM is $x(x-2)(x+2)$.

$$\frac{x-3}{x^2-2x} + \frac{6}{x^2-4}$$

Write each fraction in terms of the LCM of the denominators.

$$= \frac{x-3}{x(x-2)} \cdot \frac{x+2}{x+2} + \frac{6}{(x-2)(x+2)} \cdot \frac{x}{x}$$

Multiply the factors in the numerator.

$$= \frac{x^2-x-6}{x(x-2)(x+2)} + \frac{6x}{x(x-2)(x+2)}$$

Add the fractions.

$$= \frac{x^2-x-6+6x}{x(x-2)(x+2)}$$

$$= \frac{x^2+5x-6}{x(x-2)(x+2)}$$

Factor the numerator to determine whether there are common factors in the numerator and denominator.

$$= \frac{(x+6)(x-1)}{x(x-2)(x+2)}$$

Example 2   Add or subtract.

A. $\dfrac{y}{x} - \dfrac{4y}{3x} + \dfrac{3y}{4x}$ B. $\dfrac{2x}{x-3} - \dfrac{5}{3-x}$ C. $x - \dfrac{3}{5x}$

Solution   A. The LCM of the denominators is $12x$.   ▶ Find the LCM of the denominators.

$$\frac{y}{x} - \frac{4y}{3x} + \frac{3y}{4x}$$

$$= \frac{y}{x} \cdot \frac{12}{12} - \frac{4y}{3x} \cdot \frac{4}{4} + \frac{3y}{4x} \cdot \frac{3}{3}$$ ▶ Write each fraction in terms of the LCM.

$$= \frac{12y}{12x} - \frac{16y}{12x} + \frac{9y}{12x}$$

$$= \frac{12y - 16y + 9y}{12x}$$

$$= \frac{5y}{12x}$$

B. The LCM of $x - 3$ and   ► $3 - x = -(x - 3)$
$3 - x$ is $x - 3$.

$$\frac{2x}{x - 3} - \frac{5}{3 - x}$$

$$= \frac{2x}{x - 3} - \frac{5}{-(x - 3)} \cdot \frac{-1}{-1}$$ ► Multiply $\dfrac{5}{-(x - 3)}$ by $\dfrac{-1}{-1}$ so that the denominator will be $x - 3$.

$$= \frac{2x}{x - 3} - \frac{-5}{x - 3}$$

$$= \frac{2x - (-5)}{x - 3}$$

$$= \frac{2x + 5}{x - 3}$$

C. The LCM of the denominators is $5x$.

$$x - \frac{3}{5x} = \frac{x}{1} - \frac{3}{5x}$$

$$= \frac{x}{1} \cdot \frac{5x}{5x} - \frac{3}{5x}$$

$$= \frac{5x^2}{5x} - \frac{3}{5x}$$

$$= \frac{5x^2 - 3}{5x}$$

Problem 2   Add or subtract.

A. $\dfrac{z}{8y} - \dfrac{4z}{3y} + \dfrac{5z}{4y}$ B. $\dfrac{5x}{x - 2} - \dfrac{3}{2 - x}$ C. $y + \dfrac{5}{y - 7}$

Solution   See page A36.

Example 3   Add or subtract.   A. $\dfrac{2x}{2x - 3} - \dfrac{1}{x + 1}$ B. $\dfrac{x + 3}{x^2 - 2x - 8} + \dfrac{3}{4 - x}$

Solution   A. The LCM is $(2x - 3)(x + 1)$.

$$\frac{2x}{2x - 3} - \frac{1}{x + 1} = \frac{2x}{2x - 3} \cdot \frac{x + 1}{x + 1} - \frac{1}{x + 1} \cdot \frac{2x - 3}{2x - 3}$$

$$= \frac{2x^2 + 2x}{(2x - 3)(x + 1)} - \frac{2x - 3}{(2x - 3)(x + 1)}$$

$$= \frac{(2x^2 + 2x) - (2x - 3)}{(2x - 3)(x + 1)}$$

$$= \frac{2x^2 + 3}{(2x - 3)(x + 1)}$$

B. The LCM is $(x - 4)(x + 2)$.

$$\frac{x + 3}{x^2 - 2x - 8} + \frac{3}{4 - x} = \frac{x + 3}{(x - 4)(x + 2)} + \frac{3}{-(x - 4)} \cdot \frac{-1 \cdot (x + 2)}{-1 \cdot (x + 2)}$$

$$= \frac{x + 3}{(x - 4)(x + 2)} + \frac{-3(x + 2)}{(x - 4)(x + 2)}$$

$$= \frac{(x + 3) + (-3)(x + 2)}{(x - 4)(x + 2)}$$

$$= \frac{x + 3 - 3x - 6}{(x - 4)(x + 2)}$$

$$= \frac{-2x - 3}{(x - 4)(x + 2)}$$

Problem 3   Add or subtract.   A. $\dfrac{4x}{3x - 1} - \dfrac{9}{x + 4}$ B. $\dfrac{2x - 1}{x^2 - 25} + \dfrac{2}{5 - x}$

Solution   See page A36.

CONCEPT REVIEW 9.3

Determine whether the statement is always true, sometimes true, or never true.

1. To add two fractions, add the numerators and the denominators.

2. The procedure for subtracting two algebraic fractions is the same as that for subtracting two arithmetic fractions.

3. To add two algebraic fractions, first multiply both fractions by the LCM of their denominators.

4. If $x \neq -2$ and $x \neq 0$, then $\dfrac{x}{x + 2} + \dfrac{3}{x + 2} = \dfrac{x + 3}{x + 2} = \dfrac{3}{2}$.

5. If $y \neq 8$, then $\dfrac{y}{y - 8} - \dfrac{8}{y - 8} = \dfrac{y - 8}{y - 8} = 1$.

6. If $x \neq 0$, then $\dfrac{4x - 3}{x} - \dfrac{3x + 1}{x} = \dfrac{4x - 3 - 3x + 1}{x} = \dfrac{x - 2}{x}$.

EXERCISES 9.3

1   Add or subtract.

1. $\dfrac{3}{y^2} + \dfrac{8}{y^2}$

2. $\dfrac{6}{ab} - \dfrac{2}{ab}$

3. $\dfrac{3}{x + 4} - \dfrac{10}{x + 4}$

4. $\dfrac{x}{x + 6} - \dfrac{2}{x + 6}$

5. $\dfrac{3x}{2x + 3} + \dfrac{5x}{2x + 3}$

6. $\dfrac{6y}{4y + 1} - \dfrac{11y}{4y + 1}$

7. $\dfrac{2x + 1}{x - 3} + \dfrac{3x + 6}{x - 3}$

8. $\dfrac{4x + 3}{2x - 7} + \dfrac{3x - 8}{2x - 7}$

9. $\dfrac{5x - 1}{x + 9} - \dfrac{3x + 4}{x + 9}$

10. $\dfrac{6x - 5}{x - 10} - \dfrac{3x - 4}{x - 10}$

11. $\dfrac{x - 7}{2x + 7} - \dfrac{4x - 3}{2x + 7}$

12. $\dfrac{2n}{3n + 4} - \dfrac{5n - 3}{3n + 4}$

13. $\dfrac{x}{x^2 + 2x - 15} - \dfrac{3}{x^2 + 2x - 15}$

14. $\dfrac{3x}{x^2 + 3x - 10} - \dfrac{6}{x^2 + 3x - 10}$

15. $\dfrac{2x + 3}{x^2 - x - 30} - \dfrac{x - 2}{x^2 - x - 30}$

16. $\dfrac{3x - 1}{x^2 + 5x - 6} - \dfrac{2x - 7}{x^2 + 5x - 6}$

17. $\dfrac{4y + 7}{2y^2 + 7y - 4} - \dfrac{y - 5}{2y^2 + 7y - 4}$

18. $\dfrac{x + 1}{2x^2 - 5x - 12} + \dfrac{x + 2}{2x^2 - 5x - 12}$

19. $\dfrac{2x^2 + 3x}{x^2 - 9x + 20} + \dfrac{2x^2 - 3}{x^2 - 9x + 20} - \dfrac{4x^2 + 2x + 1}{x^2 - 9x + 20}$

20. $\dfrac{2x^2 + 3x}{x^2 - 2x - 63} - \dfrac{x^2 - 3x + 21}{x^2 - 2x - 63} - \dfrac{x - 7}{x^2 - 2x - 63}$

2   Add or subtract.

21. $\dfrac{4}{x} + \dfrac{5}{y}$

22. $\dfrac{7}{a} + \dfrac{5}{b}$

23. $\dfrac{12}{x} - \dfrac{5}{2x}$

24. $\dfrac{5}{3a} - \dfrac{3}{4a}$

25. $\dfrac{1}{2x} - \dfrac{5}{4x} + \dfrac{7}{6x}$

26. $\dfrac{7}{4y} + \dfrac{11}{6y} - \dfrac{8}{3y}$

27. $\dfrac{5}{3x} - \dfrac{2}{x^2} + \dfrac{3}{2x}$

28. $\dfrac{6}{y^2} + \dfrac{3}{4y} - \dfrac{2}{5y}$

29. $\dfrac{2}{x} - \dfrac{3}{2y} + \dfrac{3}{5x} - \dfrac{1}{4y}$

30. $\dfrac{5}{2a} + \dfrac{7}{3b} - \dfrac{2}{b} - \dfrac{3}{4a}$

31. $\dfrac{2x + 1}{3x} + \dfrac{x - 1}{5x}$

32. $\dfrac{4x - 3}{6x} + \dfrac{2x + 3}{4x}$

33. $\dfrac{x - 3}{6x} + \dfrac{x + 4}{8x}$

34. $\dfrac{2x - 3}{2x} + \dfrac{x + 3}{3x}$

35. $\dfrac{2x + 9}{9x} - \dfrac{x - 5}{5x}$

36. $\dfrac{3y - 2}{12y} - \dfrac{y - 3}{18y}$

37. $\dfrac{x + 4}{2x} - \dfrac{x - 1}{x^2}$

38. $\dfrac{x - 2}{3x^2} - \dfrac{x + 4}{x}$

39. $\dfrac{x - 10}{4x^2} + \dfrac{x + 1}{2x}$

40. $\dfrac{x + 5}{3x^2} + \dfrac{2x + 1}{2x}$

41. $y + \dfrac{8}{3y}$

42. $\dfrac{7}{2n} - n$

43. $\dfrac{4}{x + 4} + x$

44. $x + \dfrac{3}{x + 2}$

45. $5 - \dfrac{x - 2}{x + 1}$

46. $3 + \dfrac{x - 1}{x + 1}$

47. $\dfrac{2x + 1}{6x^2} - \dfrac{x - 4}{4x}$

48. $\dfrac{x + 3}{6x} - \dfrac{x - 3}{8x^2}$

49. $\dfrac{x + 2}{xy} - \dfrac{3x - 2}{x^2 y}$

50. $\dfrac{3x - 1}{xy^2} - \dfrac{2x + 3}{xy}$

51. $\dfrac{4x - 3}{3x^2 y} + \dfrac{2x + 1}{4xy^2}$

52. $\dfrac{5x + 7}{6xy^2} - \dfrac{4x - 3}{8x^2 y}$

53. $\dfrac{x - 2}{8x^2} - \dfrac{x + 7}{12xy}$

54. $\dfrac{3x - 1}{6y^2} - \dfrac{x + 5}{9xy}$

55. $\dfrac{4}{x - 2} + \dfrac{5}{x + 3}$

56. $\dfrac{2}{x - 3} + \dfrac{5}{x - 4}$

57. $\dfrac{6}{x - 7} - \dfrac{4}{x + 3}$

58. $\dfrac{3}{y + 6} - \dfrac{4}{y - 3}$

59. $\dfrac{2x}{x + 1} + \dfrac{1}{x - 3}$

60. $\dfrac{3x}{x - 4} + \dfrac{2}{x + 6}$

61. $\dfrac{4x}{2x - 1} - \dfrac{5}{x - 6}$

62. $\dfrac{6x}{x + 5} - \dfrac{3}{2x + 3}$

63. $\dfrac{2a}{a - 7} + \dfrac{5}{7 - a}$

64. $\dfrac{4x}{6 - x} + \dfrac{5}{x - 6}$

65. $\dfrac{x}{x^2 - 9} + \dfrac{3}{x - 3}$

66. $\dfrac{y}{y^2 - 16} + \dfrac{1}{y - 4}$

67. $\dfrac{2x}{x^2 - x - 6} - \dfrac{3}{x + 2}$

68. $\dfrac{5x}{x^2 + 2x - 8} - \dfrac{2}{x + 4}$

69. $\dfrac{3x - 1}{x^2 - 10x + 25} - \dfrac{3}{x - 5}$

70. $\dfrac{2a + 3}{a^2 - 7a + 12} - \dfrac{2}{a - 3}$

71. $\dfrac{x + 4}{x^2 - x - 42} + \dfrac{3}{7 - x}$

72. $\dfrac{x + 3}{x^2 - 3x - 10} + \dfrac{2}{5 - x}$

73. $\dfrac{x}{2x + 4} - \dfrac{2}{x^2 + 2x}$

74. $\dfrac{x + 2}{4x + 16} - \dfrac{2}{x^2 + 4x}$

75. $\dfrac{x - 1}{x^2 - x - 2} + \dfrac{3}{x^2 - 3x + 2}$

76. $\dfrac{a + 2}{a^2 + a - 2} + \dfrac{3}{a^2 + 2a - 3}$

77. $\dfrac{1}{x + 1} + \dfrac{x}{x - 6} - \dfrac{5x - 2}{x^2 - 5x - 6}$

78. $\dfrac{x}{x - 4} + \dfrac{5}{x + 5} - \dfrac{11x - 8}{x^2 + x - 20}$

79. $\dfrac{3x + 1}{x - 1} - \dfrac{x - 1}{x - 3} + \dfrac{x + 1}{x^2 - 4x + 3}$

80. $\dfrac{4x + 1}{x - 8} - \dfrac{3x + 2}{x + 4} - \dfrac{49x + 4}{x^2 - 4x - 32}$

SUPPLEMENTAL EXERCISES 9.3

Simplify.

81. $\dfrac{a}{a - b} + \dfrac{b}{b - a} + 1$

82. $\dfrac{y}{x - y} + 2 - \dfrac{x}{y - x}$

83. $b - 3 + \dfrac{5}{b + 4}$

84. $2y - 1 + \dfrac{6}{y + 5}$

85. $\dfrac{(n + 1)^2}{(n - 1)^2} - 1$

86. $1 - \dfrac{(y - 2)^2}{(y + 2)^2}$

87. $\dfrac{2x + 9}{3 - x} + \dfrac{x + 5}{x + 7} - \dfrac{2x^2 + 3x - 3}{x^2 + 4x - 21}$

88. $\dfrac{3x + 5}{x + 5} - \dfrac{x + 1}{2 - x} - \dfrac{4x^2 - 3x - 1}{x^2 + 3x - 10}$

89. $\dfrac{x^2 + x - 6}{x^2 + 2x - 8} \cdot \dfrac{x^2 + 5x + 4}{x^2 + 2x - 3} - \dfrac{2}{x - 1}$

90. $\dfrac{x^2 + 9x + 20}{x^2 + 4x - 5} \div \dfrac{x^2 - 49}{x^2 + 6x - 7} - \dfrac{x}{x - 7}$

91. $\dfrac{x^2 - 9}{x^2 + 6x + 9} \div \dfrac{x^2 + x - 20}{x^2 - x - 12} + \dfrac{1}{x + 1}$

92. $\dfrac{x^2 - 25}{x^2 + 10x + 25} \cdot \dfrac{x^2 - 7x + 10}{x^2 - x - 2} + \dfrac{1}{x + 1}$

93. Find the sum of the following: $\dfrac{1}{1 \cdot 2} + \dfrac{1}{2 \cdot 3}$

$$\dfrac{1}{1 \cdot 2} + \dfrac{1}{2 \cdot 3} + \dfrac{1}{3 \cdot 4}$$

$$\dfrac{1}{1 \cdot 2} + \dfrac{1}{2 \cdot 3} + \dfrac{1}{3 \cdot 4} + \dfrac{1}{4 \cdot 5}$$

Note the pattern in these sums, and then find the sum of 50 terms, of 100 terms, and of 1000 terms.

Rewrite the expression as the sum of two fractions in simplest form.

94. $\dfrac{5b + 4a}{ab}$

95. $\dfrac{6x + 7y}{xy}$

96. $\dfrac{3x^2 + 4xy}{x^2 y^2}$

97. $\dfrac{2mn^2 + 8m^2 n}{m^3 n^3}$

98. In your own words, explain the procedure for adding algebraic fractions with different denominators.

S E C T I O N 9.4

Complex Fractions

1 Simplify complex fractions

POINT OF INTEREST

There are many instances of complex fractions in application problems. For example, the fraction $\dfrac{1}{\dfrac{1}{r_1} + \dfrac{1}{r_2}}$ is used to determine the total resistance in certain electric circuits.

A **complex fraction** is a fraction whose numerator or denominator contains one or more fractions. Examples of complex fractions are shown at the right.

$$\dfrac{3}{2 - \dfrac{1}{2}} \qquad \dfrac{4 + \dfrac{1}{x}}{3 + \dfrac{2}{x}} \qquad \dfrac{\dfrac{1}{x - 1} + x + 3}{x - 3 + \dfrac{1}{x + 4}}$$

To simplify $\dfrac{1 - \dfrac{4}{x^2}}{1 + \dfrac{2}{x}}$, find the LCM of the denominators of the fractions in the numerator and denominator. (The LCM of x^2 and x is x^2.) Multiply the numerator and denominator of the complex fraction by the LCM. Then simplify.

LOOK CLOSELY

First, we are multiplying the complex fraction by $\dfrac{x^2}{x^2}$, which equals 1, so we are not changing the value of the fraction. Second, we are using the Distributive Property to multiply $\left(1 - \dfrac{4}{x^2}\right)x^2$ and $\left(1 + \dfrac{2}{x}\right)x^2$.

$$\dfrac{1 - \dfrac{4}{x^2}}{1 + \dfrac{2}{x}} = \dfrac{1 - \dfrac{4}{x^2}}{1 + \dfrac{2}{x}} \cdot \dfrac{x^2}{x^2}$$

$$= \dfrac{1 \cdot x^2 - \dfrac{4}{x^2} \cdot x^2}{1 \cdot x^2 + \dfrac{2}{x} \cdot x^2}$$

$$= \dfrac{x^2 - 4}{x^2 + 2x}$$

$$= \dfrac{(x - 2)\overset{1}{\cancel{(x + 2)}}}{x\underset{1}{\cancel{(x + 2)}}}$$

$$= \dfrac{x - 2}{x}$$

Example 1   Simplify.   A. $\dfrac{\dfrac{1}{x} + \dfrac{1}{2}}{\dfrac{1}{x^2} - \dfrac{1}{4}}$ B. $\dfrac{1 - \dfrac{2}{x} - \dfrac{15}{x^2}}{1 - \dfrac{11}{x} + \dfrac{30}{x^2}}$

Solution   A. The LCM of x, 2, x^2, and 4 is $4x^2$.

$$\dfrac{\dfrac{1}{x} + \dfrac{1}{2}}{\dfrac{1}{x^2} - \dfrac{1}{4}} = \dfrac{\dfrac{1}{x} + \dfrac{1}{2}}{\dfrac{1}{x^2} - \dfrac{1}{4}} \cdot \dfrac{4x^2}{4x^2}$$

$$= \dfrac{\dfrac{1}{x} \cdot 4x^2 + \dfrac{1}{2} \cdot 4x^2}{\dfrac{1}{x^2} \cdot 4x^2 - \dfrac{1}{4} \cdot 4x^2}$$

$$= \dfrac{4x + 2x^2}{4 - x^2}$$

$$= \dfrac{2x(2 \overset{1}{\cancel{+ x}})}{(2 - x)(2 \underset{1}{\cancel{+ x}})}$$

$$= \dfrac{2x}{2 - x}$$

B. The LCM of x and x^2 is x^2.

$$\dfrac{1 - \dfrac{2}{x} - \dfrac{15}{x^2}}{1 - \dfrac{11}{x} + \dfrac{30}{x^2}} = \dfrac{1 - \dfrac{2}{x} - \dfrac{15}{x^2}}{1 - \dfrac{11}{x} + \dfrac{30}{x^2}} \cdot \dfrac{x^2}{x^2}$$

$$= \dfrac{1 \cdot x^2 - \dfrac{2}{x} \cdot x^2 - \dfrac{15}{x^2} \cdot x^2}{1 \cdot x^2 - \dfrac{11}{x} \cdot x^2 + \dfrac{30}{x^2} \cdot x^2}$$

$$= \dfrac{x^2 - 2x - 15}{x^2 - 11x + 30}$$

$$= \dfrac{(\overset{1}{\cancel{x - 5}})(x + 3)}{(\underset{1}{\cancel{x - 5}})(x - 6)}$$

$$= \dfrac{x + 3}{x - 6}$$

Problem 1   Simplify.   A. $\dfrac{\dfrac{1}{3} - \dfrac{1}{x}}{\dfrac{1}{9} - \dfrac{1}{x^2}}$ B. $\dfrac{1 + \dfrac{4}{x} + \dfrac{3}{x^2}}{1 + \dfrac{10}{x} + \dfrac{21}{x^2}}$

Solution   See page A36.

CONCEPT REVIEW 9.4

Determine whether the statement is always true, sometimes true, or never true.

1. A complex fraction is a fraction that has fractions in the numerator and denominator.

2. The first step in simplifying a complex fraction is to find the least common multiple of the denominators of the fractions in the numerator and denominator of the complex fraction.

3. To simplify a complex fraction, multiply the complex fraction by the LCM of the denominators of the fractions in the numerator and denominator of the complex fraction.

4. When we multiply the numerator and denominator of a complex fraction by the same expression, we are using the Multiplication Property of One.

5. Our goal in simplifying a complex fraction is to rewrite it so that there are no fractions in the numerator or in the denominator. We then express the fraction in simplest form.

6. For the complex fraction $\dfrac{1 + \dfrac{4}{x} + \dfrac{3}{x^2}}{1 + \dfrac{10}{x} + \dfrac{21}{x^2}}$, x can be any real number except 0.

EXERCISES 9.4

1 Simplify.

1. $\dfrac{1 + \dfrac{3}{x}}{1 - \dfrac{9}{x^2}}$

2. $\dfrac{1 + \dfrac{4}{x}}{1 - \dfrac{16}{x^2}}$

3. $\dfrac{2 - \dfrac{8}{x + 4}}{3 - \dfrac{12}{x + 4}}$

4. $\dfrac{5 - \dfrac{25}{x + 5}}{1 - \dfrac{3}{x + 5}}$

5. $\dfrac{1 + \dfrac{5}{y - 2}}{1 - \dfrac{2}{y - 2}}$

6. $\dfrac{2 - \dfrac{11}{2x - 1}}{3 - \dfrac{17}{2x - 1}}$

7. $\dfrac{4 - \dfrac{2}{x + 7}}{5 + \dfrac{1}{x + 7}}$

8. $\dfrac{5 + \dfrac{3}{x - 8}}{2 - \dfrac{1}{x - 8}}$

9. $\dfrac{\dfrac{3}{x - 2} + 3}{\dfrac{4}{x - 2} + 4}$

10. $\dfrac{\dfrac{3}{2x + 1} - 3}{2 - \dfrac{4x}{2x + 1}}$

11. $\dfrac{2 - \dfrac{3}{x} - \dfrac{2}{x^2}}{2 + \dfrac{5}{x} + \dfrac{2}{x^2}}$

12. $\dfrac{2 + \dfrac{5}{x} - \dfrac{12}{x^2}}{4 - \dfrac{4}{x} - \dfrac{3}{x^2}}$

13. $\dfrac{1 - \dfrac{1}{x} - \dfrac{6}{x^2}}{1 - \dfrac{9}{x^2}}$

14. $\dfrac{1 + \dfrac{4}{x} + \dfrac{4}{x^2}}{1 - \dfrac{2}{x} - \dfrac{8}{x^2}}$

15. $\dfrac{1 - \dfrac{5}{x} - \dfrac{6}{x^2}}{1 + \dfrac{6}{x} + \dfrac{5}{x^2}}$

16. $\dfrac{1 - \dfrac{7}{a} + \dfrac{12}{a^2}}{1 + \dfrac{1}{a} - \dfrac{20}{a^2}}$

17. $\dfrac{1 - \dfrac{6}{x} + \dfrac{8}{x^2}}{\dfrac{4}{x^2} + \dfrac{3}{x} - 1}$

18. $\dfrac{1 + \dfrac{3}{x} - \dfrac{18}{x^2}}{\dfrac{21}{x^2} - \dfrac{4}{x} - 1}$

19. $\dfrac{x - \dfrac{4}{x+3}}{1 + \dfrac{1}{x+3}}$

20. $\dfrac{y + \dfrac{1}{y-2}}{1 + \dfrac{1}{y-2}}$

21. $\dfrac{1 - \dfrac{x}{2x+1}}{x - \dfrac{1}{2x+1}}$

22. $\dfrac{1 - \dfrac{2x-2}{3x-1}}{x - \dfrac{4}{3x-1}}$

23. $\dfrac{x - 5 + \dfrac{14}{x+4}}{x + 3 - \dfrac{2}{x+4}}$

24. $\dfrac{a + 4 + \dfrac{5}{a-2}}{a + 6 + \dfrac{15}{a-2}}$

25. $\dfrac{x + 3 - \dfrac{10}{x-6}}{x + 2 - \dfrac{20}{x-6}}$

26. $\dfrac{x - 7 + \dfrac{5}{x-1}}{x - 3 + \dfrac{1}{x-1}}$

27. $\dfrac{y - 6 + \dfrac{22}{2y+3}}{y - 5 + \dfrac{11}{2y+3}}$

28. $\dfrac{x + 2 - \dfrac{12}{2x-1}}{x + 1 - \dfrac{9}{2x-1}}$

29. $\dfrac{x - \dfrac{2}{2x-3}}{2x - 1 - \dfrac{8}{2x-3}}$

30. $\dfrac{x + 3 - \dfrac{18}{2x+1}}{x - \dfrac{6}{2x+1}}$

31. $\dfrac{1 - \dfrac{2}{x+1}}{1 + \dfrac{1}{x-2}}$

32. $\dfrac{1 - \dfrac{1}{x+2}}{1 + \dfrac{2}{x-1}}$

33. $\dfrac{1 - \dfrac{2}{x+4}}{1 + \dfrac{3}{x-1}}$

34. $\dfrac{1 + \dfrac{1}{x-2}}{1 - \dfrac{3}{x+2}}$

35. $\dfrac{\dfrac{1}{x} - \dfrac{2}{x-1}}{\dfrac{3}{x} + \dfrac{1}{x-1}}$

36. $\dfrac{\dfrac{3}{n+1} + \dfrac{1}{n}}{\dfrac{2}{n+1} + \dfrac{3}{n}}$

37. $\dfrac{\dfrac{3}{2x-1} - \dfrac{1}{x}}{\dfrac{4}{x} + \dfrac{2}{2x-1}}$

38. $\dfrac{\dfrac{4}{3x+1} + \dfrac{3}{x}}{\dfrac{6}{x} - \dfrac{2}{3x+1}}$

39. $\dfrac{\dfrac{3}{b-4} - \dfrac{2}{b+1}}{\dfrac{5}{b+1} - \dfrac{1}{b-4}}$

40. $\dfrac{\dfrac{5}{x-5} - \dfrac{3}{x-1}}{\dfrac{6}{x-1} + \dfrac{2}{x-5}}$

SUPPLEMENTAL EXERCISES 9.4

Simplify.

41. $1 + \dfrac{1}{1 + \dfrac{1}{2}}$

42. $1 + \dfrac{1}{1 + \dfrac{1}{1 + \dfrac{1}{2}}}$

43. $1 - \dfrac{1}{1 - \dfrac{1}{x}}$

44. $1 - \dfrac{1}{1 - \dfrac{1}{y+1}}$

45. $\dfrac{a^{-1} - b^{-1}}{a^{-2} - b^{-2}}$

46. $\dfrac{x^{-2} - y^{-2}}{x^{-2}y^{-2}}$

47. $\left(\dfrac{y}{4} - \dfrac{4}{y}\right) \div \left(\dfrac{4}{y} - 3 + \dfrac{y}{2}\right)$

48. $\left(\dfrac{b}{8} - \dfrac{8}{b}\right) \div \left(\dfrac{8}{b} - 5 + \dfrac{b}{2}\right)$

49. How would you explain to a classmate why we multiply the numerator and denominator of a complex fraction by the LCM of the denominators of the fractions in the numerator and denominator?

SECTION 9.5

Equations Containing Fractions

1   Solve equations containing fractions

In Chapter 3, equations containing fractions were solved by the method of clearing denominators. Recall that to **clear denominators,** we multiply each side of an equation by the LCM of the denominators. The result is an equation that contains no fractions. In this section, we will again solve equations containing fractions by multiplying each side of the equation by the LCM of the denominators. The difference between this section and Chapter 3 is that the fractions in these equations contain variables in the denominators.

Solve: $\dfrac{3x-1}{4x} + \dfrac{2}{3x} = \dfrac{7}{6x}$

LOOK CLOSELY

Note that we are now solving *equations*, not operating on *expressions*. We are not writing each fraction in terms of the LCM of the denominators; we are multiplying both sides of the equation by the LCM of the denominators.

Find the LCM of the denominators.

The LCM of $4x$, $3x$, and $6x$ is $12x$.

Multiply each side of the equation by the LCM of the denominators.

Simplify using the Distributive Property.

Solve for x.

$$\frac{3x-1}{4x} + \frac{2}{3x} = \frac{7}{6x}$$

$$12x\left(\frac{3x-1}{4x} + \frac{2}{3x}\right) = 12x\left(\frac{7}{6x}\right)$$

$$12x\left(\frac{3x-1}{4x}\right) + 12x\left(\frac{2}{3x}\right) = 12x\left(\frac{7}{6x}\right)$$

$$\frac{12x}{1}\left(\frac{3x-1}{4x}\right) + \frac{12x}{1}\left(\frac{2}{3x}\right) = \frac{12x}{1}\left(\frac{7}{6x}\right)$$

$$3(3x-1) + 4(2) = 2(7)$$
$$9x - 3 + 8 = 14$$
$$9x + 5 = 14$$
$$9x = 9$$
$$x = 1$$

1 checks as a solution.
The solution is 1.

▎**Example 1**   Solve: $\dfrac{4}{x} - \dfrac{x}{2} = \dfrac{7}{2}$

Solution

$$\frac{4}{x} - \frac{x}{2} = \frac{7}{2}$$ ▶ The LCM of x and 2 is $2x$.

$$2x\left(\frac{4}{x} - \frac{x}{2}\right) = 2x\left(\frac{7}{2}\right)$$

$$\frac{2x}{1}\cdot\frac{4}{x} - \frac{2x}{1}\cdot\frac{x}{2} = \frac{2x}{1}\cdot\frac{7}{2}$$

$$8 - x^2 = 7x$$ ▶ This is a quadratic equation.
$$0 = x^2 + 7x - 8$$
$$0 = (x+8)(x-1)$$

$$x + 8 = 0 \qquad x - 1 = 0$$
$$x = -8 \qquad\quad x = 1$$

Both -8 and 1 check as solutions.
The solutions are -8 and 1.

Problem 1   Solve: $x + \dfrac{1}{3} = \dfrac{4}{3x}$

Solution   See page A37.

Occasionally, a value of a variable in a fractional equation makes one of the denominators zero. In this case, that value of the variable is not a solution of the equation.

Solve: $\dfrac{2x}{x-2} = 1 + \dfrac{4}{x-2}$

Find the LCM of the denominators.

The LCM is $x - 2$.

$$\frac{2x}{x-2} = 1 + \frac{4}{x-2}$$

Multiply each side of the equation by the LCM of the denominators.

$$(x-2)\frac{2x}{x-2} = (x-2)\left(1 + \frac{4}{x-2}\right)$$

Simplify using the Distributive Property.

$$(x-2)\left(\frac{2x}{x-2}\right) = (x-2)\cdot 1 + (x-2)\cdot \frac{4}{x-2}$$

Solve for x.

$$2x = x - 2 + 4$$
$$2x = x + 2$$
$$x = 2$$

When x is replaced by 2, the denominators of $\dfrac{2x}{x-2}$ and $\dfrac{4}{x-2}$ are zero.

Therefore, 2 is not a solution of the equation. The equation has no solution.

Example 2   Solve: $\dfrac{3x}{x-4} = 5 + \dfrac{12}{x-4}$

Solution

$$\frac{3x}{x-4} = 5 + \frac{12}{x-4}$$ ▶ The LCM is $x - 4$.

$$\frac{(x-4)}{1} \cdot \frac{3x}{x-4} = \frac{(x-4)}{1}\left(5 + \frac{12}{x-4}\right)$$

$$3x = (x-4)5 + 12$$
$$3x = 5x - 20 + 12$$
$$3x = 5x - 8$$
$$-2x = -8$$
$$x = 4$$

4 does not check as a solution.
The equation has no solution.

Problem 2   Solve: $\dfrac{5x}{x + 2} = 3 - \dfrac{10}{x + 2}$

Solution   See page A37.

Quantities such as 4 meters, 15 seconds, and 8 gallons are number quantities written with units. In these examples, the units are meters, seconds, and gallons.

A **ratio** is the quotient of two quantities that have the same unit.

The length of a living room is 16 ft, and the width is 12 ft. The ratio of the length to the width is written

$\dfrac{16 \text{ ft}}{12 \text{ ft}} = \dfrac{16}{12} = \dfrac{4}{3}$ A ratio is in simplest form when the two numbers do not have a common factor. Note that the units are not written.

A **rate** is the quotient of two quantities that have different units.

There are 2 lb of salt in 8 gal of water. The salt-to-water rate is

$\dfrac{2 \text{ lb}}{8 \text{ gal}} = \dfrac{1 \text{ lb}}{4 \text{ gal}}$ A rate is in simplest form when the two numbers do not have a common factor. The units are written as part of the rate.

A **proportion** is an equation that states the equality of two ratios or rates.

Examples of proportions are shown below.

$$\frac{30 \text{ mi}}{4 \text{ h}} = \frac{15 \text{ mi}}{2 \text{ h}} \qquad \frac{4}{6} = \frac{8}{12} \qquad \frac{3}{4} = \frac{x}{8}$$

Because a proportion is an equation containing fractions, the same method used to solve an equation containing fractions is used to solve a proportion. Multiply each side of the equation by the LCM of the denominators. Then solve for the variable.

To solve the proportion $\dfrac{4}{x} = \dfrac{2}{3}$, multiply each side of the proportion by the LCM of the denominators.

$$\frac{4}{x} = \frac{2}{3}$$

Solve the equation.

$$3x\left(\frac{4}{x}\right) = 3x\left(\frac{2}{3}\right)$$

$$12 = 2x$$

$$6 = x$$

The solution is 6.

Example 3   Solve.   A. $\dfrac{8}{x+3} = \dfrac{4}{x}$ B. $\dfrac{6}{x+4} = \dfrac{12}{5x-13}$

Solution   A. $\dfrac{8}{x+3} = \dfrac{4}{x}$

$$x(x+3)\,\dfrac{8}{x+3} = x(x+3)\,\dfrac{4}{x}$$ ▶ Multiply each side of the pro-
$$8x = (x+3)4$$ portion by the LCM of the de-
$$8x = 4x + 12$$ nominators.
$$4x = 12$$
$$x = 3$$ ▶ Remember to check the solution because it is a fractional
equation.

The solution is 3.

B. $\dfrac{6}{x+4} = \dfrac{12}{5x-13}$

$$(5x-13)(x+4)\,\dfrac{6}{x+4} = (5x-13)(x+4)\,\dfrac{12}{5x-13}$$
$$(5x-13)6 = (x+4)12$$
$$30x - 78 = 12x + 48$$
$$18x - 78 = 48$$
$$18x = 126$$
$$x = 7$$

The solution is 7.

Problem 3   Solve.   A. $\dfrac{2}{x+3} = \dfrac{6}{5x+5}$ B. $\dfrac{5}{2x-3} = \dfrac{10}{x+3}$

Solution   See page A37.

2   Application problems: proportions

Example 4   The monthly loan payment for a car is $29.50 for each $1000 borrowed. At this rate, find the monthly payment for a $9000 car loan.

Strategy   To find the monthly payment, write and solve a proportion using P to represent the monthly car payment.

LOOK CLOSELY

It is also correct to write the proportion with the loan amounts in the numerators and the monthly payments in the denominators. The solution will be the same.

Solution $\dfrac{29.50}{1000} = \dfrac{P}{9000}$ ▶ The monthly payments are in the numerators. The loan amounts are in the denominators.

$$9000\left(\dfrac{29.50}{1000}\right) = 9000\left(\dfrac{P}{9000}\right)$$
$$265.50 = P$$

The monthly payment is $265.50.

Problem 4 Nine ceramic tiles are required to tile a 4 ft² area. At this rate, how many square feet can be tiled with 270 ceramic tiles?

4 square feet

Solution See page A38.

Example 5 An investment of $1200 earns $96 each year. At the same rate, how much additional money must be invested to earn $128 each year?

Strategy To find the additional amount of money that must be invested, write and solve a proportion using x to represent the additional money. Then $1200 + x$ is the total amount invested.

LOOK CLOSELY

It is also correct to write the proportion with the amounts earned in the numerators and the amounts invested in the denominators. The solution will be the same.

Solution

$$\frac{1200}{96} = \frac{1200 + x}{128}$$

▶ The amounts invested are in the numerators. The amounts earned are in the denominators.

$$\frac{25}{2} = \frac{1200 + x}{128}$$

▶ Simplify $\frac{1200}{96}$.

$$128\left(\frac{25}{2}\right) = 128\left(\frac{1200 + x}{128}\right)$$

$$1600 = 1200 + x$$

$$400 = x$$

An additional $400 must be invested.

Problem 5 Three ounces of a medication are required for a 120 lb adult. At the same rate, how many additional ounces of the medication are required for a 180 lb adult?

Solution See page A38.

CONCEPT REVIEW 9.5

Determine whether the statement is always true, sometimes true, or never true.

1. The process of clearing denominators in an equation containing fractions is an application of the Multiplication Property of Equations.

2. If the denominator of a fraction is $x + 3$, then $x \neq 3$ for that fraction.

3. The first step in solving an equation containing fractions is to find the LCM of the denominators.

4. The solutions to an equation containing algebraic fractions must be checked because a solution might make one of the denominators equal to zero.

5. If, after the clearing of denominators, a fractional equation results in a quadratic equation, the equation has two solutions.

6. Both sides of an equation containing fractions can be multiplied by the same number without changing the solution of the equation.

7. To solve a proportion, multiply each side of the proportion by the LCM of the denominators.

EXERCISES 9.5

1   Solve.

1. $\dfrac{6}{2a + 1} = 2$

2. $\dfrac{12}{3x - 2} = 3$

3. $\dfrac{9}{2x - 5} = -2$

4. $\dfrac{6}{4 - 3x} = 3$

5. $2 + \dfrac{5}{x} = 7$

6. $3 + \dfrac{8}{n} = 5$

7. $1 - \dfrac{9}{x} = 4$

8. $3 - \dfrac{12}{x} = 7$

9. $\dfrac{2}{y} + 5 = 9$

10. $\dfrac{6}{x} + 3 = 11$

11. $\dfrac{3}{x - 2} = \dfrac{4}{x}$

12. $\dfrac{5}{x + 3} = \dfrac{3}{x - 1}$

13. $\dfrac{2}{3x - 1} = \dfrac{3}{4x + 1}$

14. $\dfrac{5}{3x - 4} = \dfrac{-3}{1 - 2x}$

15. $\dfrac{-3}{2x + 5} = \dfrac{2}{x - 1}$

16. $\dfrac{4}{5y - 1} = \dfrac{2}{2y - 1}$

17. $\dfrac{4x}{x - 4} + 5 = \dfrac{5x}{x - 4}$

18. $\dfrac{2x}{x + 2} - 5 = \dfrac{7x}{x + 2}$

19. $2 + \dfrac{3}{a - 3} = \dfrac{a}{a - 3}$

20. $\dfrac{x}{x + 4} = 3 - \dfrac{4}{x + 4}$

21. $\dfrac{x}{x - 1} = \dfrac{8}{x + 2}$

22. $\dfrac{x}{x + 12} = \dfrac{1}{x + 5}$

23. $\dfrac{2x}{x + 4} = \dfrac{3}{x - 1}$

24. $\dfrac{5}{3n - 8} = \dfrac{n}{n + 2}$

25. $x + \dfrac{6}{x - 2} = \dfrac{3x}{x - 2}$

26. $x - \dfrac{6}{x - 3} = \dfrac{2x}{x - 3}$

27. $\dfrac{x}{x + 2} + \dfrac{2}{x - 2} = \dfrac{x + 6}{x^2 - 4}$

28. $\dfrac{x}{x + 4} = \dfrac{11}{x^2 - 16} + 2$

29. $\dfrac{8}{y} = \dfrac{2}{y - 2} + 1$

30. $\dfrac{8}{r} + \dfrac{3}{r - 1} = 3$

31. $\dfrac{4}{5} = \dfrac{12}{x}$

32. $\dfrac{6}{x} = \dfrac{2}{3}$

33. $\dfrac{16}{9} = \dfrac{64}{x}$

34. $\dfrac{18}{x + 4} = \dfrac{9}{5}$

35. $\dfrac{2}{11} = \dfrac{20}{x - 3}$

36. $\dfrac{2}{x} = \dfrac{4}{x + 1}$

37. $\dfrac{16}{x - 2} = \dfrac{8}{x}$

38. $\dfrac{2}{x - 1} = \dfrac{6}{2x + 1}$

39. $\dfrac{9}{x + 2} = \dfrac{3}{x - 2}$

40. $\dfrac{2x}{7} = \dfrac{x - 2}{14}$

2   Solve.

41. An exit poll showed that 4 out of every 7 voters cast a ballot in favor of an amendment to a city charter. At this rate, how many people voted in favor of the amendment if 35,000 people voted?

42. A quality control inspector found 3 defective transistors in a shipment of 500 transistors. At this rate, how many transistors would be defective in a shipment of 2000 transistors?

43. An air conditioning specialist recommends 2 air vents for every 300 ft² of floor space. At this rate, how many air vents are required for an office building of 21,000 ft²?

44. In a city of 25,000 homes, a survey was taken to determine the number with cable television. Of the 300 homes surveyed, 210 had cable television. Estimate the number of homes in the city that have cable television.

45. A simple syrup is made by dissolving 2 c of sugar in $\frac{2}{3}$ c of boiling water. At this rate, how many cups of sugar are required for 2 c of boiling water?

46. The lighting for a billboard is provided by solar energy. If three energy panels generate 10 watts of power, how many panels are needed to provide 600 watts of power?

47. As part of a conservation effort for a lake, 40 fish were caught, tagged, and then released. Later, 80 fish were caught from the lake. Four of these 80 fish were found to have tags. Estimate the number of fish in the lake.

48. In a wildlife preserve, 10 elk were captured, tagged, and then released. Later, 15 elk were captured and 2 were found to have tags. Estimate the number of elk in the preserve.

49. A health department estimates that 8 vials of a malaria serum will treat 100 people. At this rate, how many vials are required to treat 175 people?

50. A company will accept a shipment of 10,000 computer disks if there are 2 or fewer defects in a sample of 100 randomly chosen disks. Assume that there are 300 defective disks in the shipment and that the rate of defective disks in the sample is the same as the rate in the shipment. Will the shipment be accepted?

51. A company will accept a shipment of 20,000 precision bearings if there are 3 or fewer defects in a sample of 100 randomly chosen bearings. Assume that there are 400 defective bearings in the shipment and that the rate of defective bearings in the sample is the same as the rate in the shipment. Will the shipment be accepted?

52. On a map, two cities are $2\frac{5}{8}$ in. apart. If $\frac{3}{8}$ in. on the map represents 25 mi, find the number of miles in the distance between the two cities.

53. On a map, two cities are $5\frac{5}{8}$ in. apart. If $\frac{3}{4}$ in. on the map represents 100 mi, find the number of miles in the distance between the two cities.

54. A painter estimates that 5 gal of paint will cover 1200 ft² of wall surface. How many additional gallons are required to cover 1680 ft²?

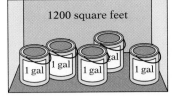

1200 square feet

55. A 50-acre field yields 1100 bushels of wheat annually. How many additional acres must be planted so that the annual yield will be 1320 bushels?

56. The sales tax on a car that sold for $12,000 is $780. At the same rate, how much higher is the sales tax on a car that sells for $13,500?

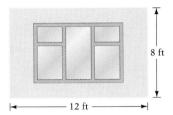

57. To conserve energy and still allow for as much natural lighting as possible, an architect suggests that the ratio of the area of a window to the area of the total wall surface be 5 to 12. Using this ratio, determine the recommended area of a window to be installed in a wall that measures 8 ft by 12 ft.

8 ft

12 ft

58. The engine of a small rocket burns 170,000 lb of fuel in 1 min. At this rate, how many pounds of fuel does the rocket burn in 45 s?

59. A caterer estimates that 5 gal of coffee will serve 50 people. How much additional coffee is necessary to serve 70 people?

60. A green paint is created by mixing 3 parts of yellow with every 5 parts of blue. How many gallons of yellow paint are needed to make 60 gal of this green paint?

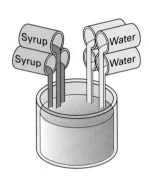

61. A soft drink is made by mixing 4 parts of carbonated water with every 3 parts of syrup. How many milliliters of water are in 280 ml of soft drink?

SUPPLEMENTAL EXERCISES 9.5

Solve.

62. $\frac{2}{3}(x + 2) - \frac{x + 1}{6} = \frac{1}{2}$

63. $\frac{3}{4}a = \frac{1}{2}(3 - a) + \frac{a - 2}{4}$

64. $\frac{x}{2x^2 - x - 1} = \frac{3}{x^2 - 1} + \frac{3}{2x + 1}$

65. $\frac{x + 1}{x^2 + x - 2} = \frac{x + 2}{x^2 - 1} + \frac{3}{x + 2}$

66. The sum of a number and its reciprocal is $\frac{26}{5}$. Find the number.

67. The sum of the multiplicative inverses of two consecutive integers is $-\frac{7}{12}$. Find the integers.

68. The denominator of a fraction is 2 more than the numerator. If both the numerator and denominator of the fraction are increased by 3, the new fraction is $\frac{4}{5}$. Find the original fraction.

69. Three people put their money together to buy lottery tickets. The first person put in $25, the second person put in $30, and the third person put in $35. One of their tickets was a winning ticket. If they won $4.5 million, what was the first person's share of the winnings?

70. A basketball player has made 5 out of every 6 foul shots attempted. If 42 foul shots were missed in the player's career, how many foul shots were made in the player's career?

71. The "sitting fee" for school pictures is $4. If 10 photos cost $10, including the sitting fee, what would 24 photos cost, including the sitting fee?

72. No one belongs to both the Math Club and the Photography Club, but the two clubs join to hold a car wash. Ten members of the Math Club and 6 members of the Photography Club participate. The profits from the car wash are $120. If each club's profits are proportional to the number of members participating, what share of the profits does the Math Club receive?

73. Explain the procedure for solving an equation containing fractions. Include in your discussion an explanation of how the LCM of the denominators is used to eliminate fractions in the equation.

74. According to the graph shown below, is the ratio of PCs to Macintosh computers greater in universities than in 2-year colleges? Is the ratio of PCs to Macintosh computers greater in 4-year colleges or in 2-year colleges? Write a short paragraph about the relative use of PCs and Macintosh computers for all institutions.

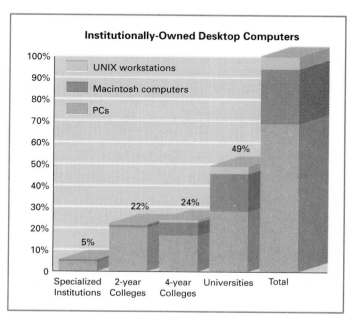

75. The graph and table shown below are based on ones that were printed in the *Boston Globe*. According to the data presented, is the proportion

$$\frac{\text{Undeclared corporate income in 1981}}{\text{Undeclared individual income in 1981}} = \frac{\text{Undeclared corporate income in 1992}}{\text{Undeclared individual income in 1992}}$$

a true proportion? If not, which segment is growing at a faster rate than the other? To the nearest $10 billion, how much individual income was undeclared in 1992? What was the percent increase in math errors in individual returns from 1981 to 1992? Write a few sentences about the change in tax forms filed without full payment from 1981 to 1992.

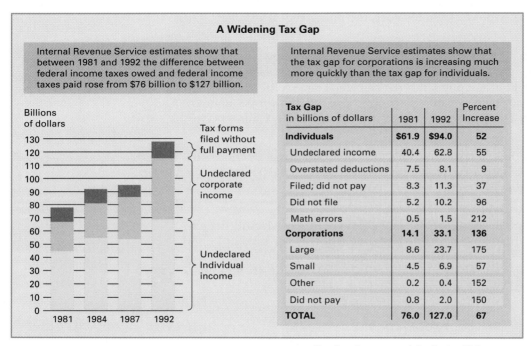

A Widening Tax Gap

Internal Revenue Service estimates show that between 1981 and 1992 the difference between federal income taxes owed and federal income taxes paid rose from $76 billion to $127 billion.

Internal Revenue Service estimates show that the tax gap for corporations is increasing much more quickly than the tax gap for individuals.

Tax Gap in billions of dollars	1981	1992	Percent Increase
Individuals	**$61.9**	**$94.0**	**52**
Undeclared income	40.4	62.8	55
Overstated deductions	7.5	8.1	9
Filed; did not pay	8.3	11.3	37
Did not file	5.2	10.2	96
Math errors	0.5	1.5	212
Corporations	**14.1**	**33.1**	**136**
Large	8.6	23.7	175
Small	4.5	6.9	57
Other	0.2	0.4	152
Did not pay	0.8	2.0	150
TOTAL	**76.0**	**127.0**	**67**

Reprinted courtesy of the *Boston Globe*.

S E C T I O N **9.6**

Literal Equations

1 Solve a literal equation for one of the variables

A **literal equation** is an equation that contains more than one variable. Examples of literal equations are shown at the right.

$$2x + 3y = 6$$
$$4w - 2x + z = 0$$

Formulas are used to express a relationship among physical quantities. A **formula** is a literal equation that states rules about measurements. Examples of formulas are shown below.

$$\frac{1}{R_1} + \frac{1}{R_2} = \frac{1}{R} \qquad \text{(Physics)}$$

$$s = a + (n - 1)d \qquad \text{(Mathematics)}$$

$$A = P + Prt \qquad \text{(Business)}$$

The Addition and Multiplication Properties can be used to solve a literal equation for one of the variables. The goal is to rewrite the equation so that the letter being solved for is alone on one side of the equation and all numbers and other variables are on the other side.

In solving $A = P(1 + i)$ for i, the goal is to rewrite the equation so that i is on one side of the equation and all other variables are on the other side.

Use the Distributive Property to remove parentheses.

$$A = P(1 + i)$$
$$A = P + Pi$$

Subtract P from each side of the equation.

$$A - P = P - P + Pi$$
$$A - P = Pi$$

Divide each side of the equation by P.

$$\frac{A - P}{P} = \frac{Pi}{P}$$

$$\frac{A - P}{P} = i$$

Example 1   A. Solve $I = \dfrac{E}{R + r}$ for R.   B. Solve $L = a(1 + ct)$ for c.

Solution   A.

$$I = \frac{E}{R + r}$$
$$(R + r)I = (R + r)\frac{E}{R + r}$$
$$RI + rI = E$$
$$RI + rI - rI = E - rI$$
$$RI = E - rI$$
$$\frac{RI}{I} = \frac{E - rI}{I}$$
$$R = \frac{E - rI}{I}$$

B.

$$L = a(1 + ct)$$
$$L = a + act$$
$$L - a = a - a + act$$
$$L - a = act$$
$$\frac{L - a}{at} = \frac{act}{at}$$
$$\frac{L - a}{at} = c$$

Problem 1   A. Solve $s = \dfrac{A + L}{2}$ for L.   B. Solve $S = a + (n - 1)d$ for n.

Solution   See page A38.

Example 2   Solve $S = C - rC$ for C.

Solution
$$S = C - rC$$
$$S = C(1 - r) \quad \blacktriangleright \text{ Factor } C \text{ from } C - rC.$$
$$\frac{S}{1 - r} = \frac{C(1 - r)}{1 - r} \quad \blacktriangleright \text{ Divide each side of the equation by } 1 - r.$$
$$\frac{S}{1 - r} = C$$

Problem 2   Solve $S = C + rC$ for C.

Solution   See page A39.

CONCEPT REVIEW 9.6

Determine whether the statement is always true, sometimes true, or never true.

1. An equation that contains more than one variable is a literal equation.

2. The linear equation $y = 4x - 5$ is not a literal equation.

3. Literal equations are solved using the same properties of equations that are used to solve equations in one variable.

4. In solving a literal equation, the goal is to get the variable being solved for alone on one side of the equation and all numbers and other variables on the other side of the equation.

5. In solving $L = a(1 + ct)$ for c, you need to use the Distributive Property. In solving $S = C - rC$ for C, you need to factor.

6. We can divide both sides of a literal equation by the same nonzero expression.

EXERCISES 9.6

1   Solve the formula for the variable given.

1. $A = \frac{1}{2}bh; h$ (Geometry)

2. $P = a + b + c; b$ (Geometry)

3. $d = rt; t$ (Physics)

4. $E = IR; R$ (Physics)

5. $PV = nRT; T$ (Chemistry)

6. $A = bh; h$ (Geometry)

7. $P = 2L + 2W; L$ (Geometry)

8. $F = \dfrac{9}{5}C + 32; C$ (Temperature conversion)

9. $A = \dfrac{1}{2}h(b_1 + b_2); b_1$ (Geometry)

10. $C = \dfrac{5}{9}(F - 32); F$ (Temperature conversion)

11. $V = \dfrac{1}{3}Ah; h$ (Geometry)

12. $P = R - C; C$ (Business)

13. $R = \dfrac{C - S}{t}; S$ (Business)

14. $P = \dfrac{R - C}{n}; R$ (Business)

15. $A = P + Prt; P$ (Business)

16. $T = fm - gm; m$ (Engineering)

17. $A = Sw + w; w$ (Physics)

18. $a = S - Sr; S$ (Mathematics)

SUPPLEMENTAL EXERCISES 9.6

The surface area of a right circular cylinder is given by the formula $S = 2\pi rh + 2\pi r^2$, where r is the radius of the base and h is the height of the cylinder.

19. **a.** Solve the formula $S = 2\pi rh + 2\pi r^2$ for h.

    b. Use your answer to part (a) to find the height of a right circular cylinder when the surface area is 12π in^2 and the radius is 1 in.

    c. Use your answer to part (a) to find the height of a right circular cylinder when the surface area is 24π in^2 and the radius is 2 in.

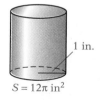

1 in.

$S = 12\pi$ in^2

Break-even analysis is a method used to determine the sales volume required for a company to break even, or experience neither a profit nor a loss, on the sale of its product. The break-even point represents the number of units that must be made and sold for income from sales to equal the cost of producing the product. The break-even point can be calculated using the formula $B = \dfrac{F}{S - V}$, where F is the fixed costs, S is the selling price per unit, and V is the variable costs per unit.

20. **a.** Solve the formula $B = \dfrac{F}{S - V}$ for S.

    b. Use your answer to part (a) to find the selling price per desk required for a company to break even. The fixed costs are $20,000, the variable costs per desk are $80, and the company plans to make and sell 200 desks.

    c. Use your answer to part (a) to find the selling price per camera required for a company to break even. The fixed costs are $15,000, the variable costs per camera are $50, and the company plans to make and sell 600 cameras.

When markup is based on selling price, the selling price of a product is given by the formula $S = \dfrac{C}{1 - r}$, where C is the cost of the product and r is the markup rate.

21.   **a.** Solve the formula $S = \dfrac{C}{1 - r}$ for r.

    b. Use your answer to part (a) to find the markup rate on a tennis racket when the cost is \$112 and the selling price is \$140.

    c. Use your answer to part (a) to find the markup rate on a radio when the cost is \$50.40 and the selling price is \$72.

Resistors are used to control the flow of current. The total resistance of two resistors in a circuit can be given by the formula $R = \dfrac{1}{\dfrac{1}{R_1} + \dfrac{1}{R_2}}$, where R_1 and R_2 are the resistances of the two resistors in the circuit. Resistance is measured in ohms.

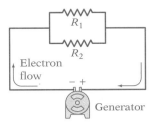

22.   **a.** Solve the formula $R = \dfrac{1}{\dfrac{1}{R_1} + \dfrac{1}{R_2}}$ for R_1.

    b. Use your answer to part (a) to find the resistance in R_1 if the resistance in R_2 is 30 ohms and the total resistance is 12 ohms.

    c. Use your answer to part (a) to find the resistance in R_1 if the resistance in R_2 is 15 ohms and the total resistance is 6 ohms.

SECTION 9.7

Application Problems

1   **Work problems**

If a painter can paint a room in 4 h, then in 1 h the painter can paint $\dfrac{1}{4}$ of the room. The painter's rate of work is $\dfrac{1}{4}$ of the room each hour. The **rate of work** is that part of a task that is completed in 1 unit of time.

A pipe can fill a tank in 30 min. This pipe can fill $\dfrac{1}{30}$ of the tank in 1 min. The rate of work is $\dfrac{1}{30}$ of the tank each minute. If a second pipe can fill the tank in x min, the rate of work for the second pipe is $\dfrac{1}{x}$ of the tank each minute.

In solving a work problem, the goal is to determine the time it takes to complete a task. The basic equation that is used to solve work problems is

Rate of work · Time worked = Part of task completed

For example, if a faucet can fill a sink in 6 min, then in 5 min the faucet will fill $\frac{1}{6} \cdot 5 = \frac{5}{6}$ of the sink. In 5 min, the faucet completes $\frac{5}{6}$ of the task.

Solve: A painter can paint a ceiling in 60 min. The painter's apprentice can paint the same ceiling in 90 min. How long will it take to paint the ceiling if they work together.

STRATEGY for solving a work problem

◼ For each person or machine, write a numerical or variable expression for the rate of work, the time worked, and the part of the task completed. The results can be recorded in a table.

Unknown time to paint the ceiling working together: t

	Rate of work	·	Time worked	=	Part of task completed
Painter	$\frac{1}{60}$	·	t	=	$\frac{t}{60}$
Apprentice	$\frac{1}{90}$	·	t	=	$\frac{t}{90}$

◼ Determine how the parts of the task completed are related. Use the fact that the sum of the parts of the task completed must equal 1, the complete task.

$$\frac{t}{60} + \frac{t}{90} = 1$$

▶ The part of the task completed by the painter plus the part of the task completed by the apprentice must equal 1.

$$180\left(\frac{t}{60} + \frac{t}{90}\right) = 180 \cdot 1$$
$$3t + 2t = 180$$
$$5t = 180$$
$$t = 36$$

Working together, they will paint the ceiling in 36 min.

Example 1 A small water pipe takes four times longer to fill a tank than does a large water pipe. With both pipes open, it takes 3 h to fill the tank. Find the time it would take the small pipe, working alone, to fill the tank.

Strategy
- Time for large pipe to fill the tank: t
- Time for small pipe to fill the tank: $4t$

	Rate	Time	Part
Small pipe	$\frac{1}{4t}$	3	$\frac{3}{4t}$
Large pipe	$\frac{1}{t}$	3	$\frac{3}{t}$

- The sum of the parts of the task completed must equal 1.

Solution

$$\frac{3}{4t} + \frac{3}{t} = 1$$

▶ The part of the task completed by the small pipe plus the part of the task completed by the large pipe must equal 1.

$$4t\left(\frac{3}{4t} + \frac{3}{t}\right) = 4t \cdot 1$$
$$3 + 12 = 4t$$
$$15 = 4t$$
$$\frac{15}{4} = t$$

▶ t is the time for the large pipe to fill the tank.

$$4t = 4\left(\frac{15}{4}\right) = 15$$

▶ Substitute the value of t into the variable expression for the time for the small pipe to fill the tank.

The small pipe, working alone, takes 15 h to fill the tank.

Problem 1 Two computer printers that work at the same rate are working together to print the payroll checks for a corporation. After they work together for 3 h, one of the printers quits. The second requires 2 more hours to complete the checks. Find the time it would take one printer, working alone, to print the checks.

Solution See page A39.

Fills tank in $4t$ hours

Fills tank in t hours

Fills $\frac{3}{4t}$ of the tank in 3 hours

Fills $\frac{3}{t}$ of the tank in 3 hours

2 Uniform motion problems

A car that travels constantly in a straight line at 30 mph is in uniform motion. When an object is in **uniform motion,** its speed does not change.

The basic equation used to solve uniform motion problems is

Distance = Rate · Time

An alternative form of this equation is written by solving the equation for time:

$$\frac{\textbf{Distance}}{\textbf{Rate}} = \textbf{Time}$$

This form of the equation is useful when the total time of travel for two objects is known or the times of travel for two objects are equal.

Solve: The speed of a boat in still water is 20 mph. The boat traveled 120 mi down a river in the same amount of time it took the boat to travel 80 mi up the river. Find the rate of the river's current.

STRATEGY *for solving a uniform motion problem*

■ For each object, write a numerical or variable expression for the distance, rate, and time. The results can be recorded in a table.

The unknown rate of the river's current: r

	Distance	÷	Rate	=	Time
Down river	120	÷	$20 + r$	=	$\frac{120}{20 + r}$
Up river	80	÷	$20 - r$	=	$\frac{80}{20 - r}$

■ Determine how the times traveled by the two objects are related. For example, it may be known that the times are equal, or the total time may be known.

$$\frac{120}{20 + r} = \frac{80}{20 - r}$$

$$(20 + r)(20 - r)\frac{120}{20 + r} = (20 + r)(20 - r)\frac{80}{20 - r}$$

$$(20 - r)120 = (20 + r)80$$

$$2400 - 120r = 1600 + 80r$$

$$-200r = -800$$

$$r = 4$$

▶ The time spent traveling down the river is equal to the time spent traveling up the river.

The rate of the river's current is 4 mph.

Example 2   A cyclist rode the first 20 mi of a trip at a constant rate. For the next 16 mi, the cyclist reduced the speed by 2 mph. The total time for the 36 mi was 4 h. Find the rate of the cyclist for each leg of the trip.

Strategy   ■ Rate for the first 20 mi: r

20 mi   16 mi

36 mi in 4 h

	Distance	Rate	Time
First 20 mi	20	r	$\frac{20}{r}$
Next 16 mi	16	$r - 2$	$\frac{16}{r - 2}$

■ The total time for the trip was 4 h.

Solution

$$\frac{20}{r} + \frac{16}{r-2} = 4$$

$$r(r-2)\left[\frac{20}{r} + \frac{16}{r-2}\right] = r(r-2) \cdot 4$$

$$(r-2)20 + 16r = 4r^2 - 8r$$

$$20r - 40 + 16r = 4r^2 - 8r$$

$$36r - 40 = 4r^2 - 8r$$

$$0 = 4r^2 - 44r + 40$$

$$0 = 4(r^2 - 11r + 10)$$

$$0 = r^2 - 11r + 10$$

$$0 = (r-10)(r-1)$$

▶ The time spent riding the first 20 mi plus the time spent riding the next 16 mi is equal to 4 h.

▶ This is a quadratic equation.

▶ Divide each side by 4.

▶ Solve by factoring.

$$r - 10 = 0 \qquad\qquad r - 1 = 0$$
$$r = 10 \qquad\qquad r = 1$$

$$r - 2 = 10 - 2 \qquad r - 2 = 1 - 2$$
$$= 8 \qquad\qquad = -1$$

▶ Find the rate for the last 16 mi. The solution $r = 1$ mph is not possible, because the rate on the last 16 mi would be -1 mph.

10 mph was the rate for the first 20 mi.
8 mph was the rate for the next 16 mi.

Problem 2   The total time for a sailboat to sail back and forth across a lake 6 km wide was 3 h. The rate sailing back was twice the rate sailing across the lake. Find the rate of the sailboat going across the lake.

Solution   See page A39.

CONCEPT REVIEW 9.7

Determine whether the statement is always true, sometimes true, or never true.

1. The work problems in this section require use of the equation
    Rate of work · Time worked = 1.

2. If it takes a janitorial crew 5 hours to clean a company's offices, then in x hours the crew has completed $\frac{x}{5}$ of the job.

3. If it takes an automotive crew x min to service a car, then the rate of work is $\frac{1}{x}$ of the job each minute.

4. If only two people worked on a job and together they completed it, and one person completed $\frac{t}{30}$ of the job and the other person completed $\frac{t}{20}$ of the job, then $\frac{t}{30} + \frac{t}{20} = 1$.

5. Uniform motion problems are based on the equation
   Distance · Rate = Time.

6. If a plane flies 300 mph in calm air and the rate of the wind is r mph, then the rate of the plane flying with the wind can be represented as $300 - r$ and the rate of the plane flying against the wind can be represented as $300 + r$.

EXERCISES 9.7

1   Solve.

1. An experienced painter can paint a garage twice as fast as an apprentice. Working together, the painters require 4 h to paint the garage. How long would it take the experienced painter, working alone, to paint the garage?

2. One grocery clerk can stock a shelf in 20 min. A second clerk requires 30 min to stock the same shelf. How long would it take to stock the shelf if the two clerks worked together?

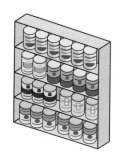

3. One person with a skiploader requires 12 h to remove a large quantity of earth. With a larger skiploader, the same amount of earth can be removed in 4 h. How long would it take to remove the earth if both skiploaders were operated together?

4. One worker can dig the trenches for a sprinkler system in 3 h. A second worker requires 6 h to do the same task. How long would it take to dig the trenches with both people working together?

5. One computer can solve a complex prime factorization problem in 75 h. A second computer can solve the same problem in 50 h. How long would it take both computers, working together, to solve the problem?

6. A new machine makes 10,000 aluminum cans three times faster than an older machine. With both machines operating, it takes 9 h to make 10,000 cans. How long would it take the new machine, working alone, to make 10,000 cans?

7. A small air conditioner will cool a room 2° in 15 min. A larger air conditioner will cool the room 2° in 10 min. How long would it take to cool the room 2° with both air conditioners operating?

8. One printing press can print the first edition of a book in 55 min. A second printing press requires 66 min to print the same number of copies. How long would it take to print the first edition of the book with both presses operating?

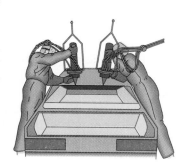

9. Two welders working together can complete a job in 6 h. One of the welders, working alone, can complete the task in 10 h. How long would it take the second welder, working alone, to complete the task?

10. Two pipelines can fill a small tank in 30 min. Working alone, the larger pipeline can fill the tank in 45 min. How long would it take the smaller pipeline, working alone, to fill the tank?

11. Working together, two dock workers can load a crate in 6 min. One of the dock workers, working alone, can load the crate in 15 min. How long would it take the other dock worker, working alone, to load the crate?

12. With two harvesters operating, a plot of land can be harvested in 1 h. If only the newer harvester is used, the land can be harvested in 1.5 h. How long would it take to harvest the field using only the older harvester?

13. A cement mason can build a barbecue in 8 h. A second mason requires 12 h to do the same task. After working alone for 4 h, the first mason quits. How long will it take the second mason to complete the job?

14. A mechanic requires 2 h to repair a transmission. An apprentice requires 6 h to make the same repairs. If the mechanic works alone on a transmission for 1 h and then stops, how long will it take the apprentice to complete the repairs?

15. One computer technician can wire a modem in 4 h. A second technician requires 6 h to do the same job. After working alone for 2 h, the first technician quits. How long will it take the second technician to complete the wiring?

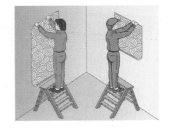

16. A wallpaper hanger requires 2 h to hang the wallpaper on one wall of a room. A second wallpaper hanger requires 4 h to hang the same amount of wallpaper. The first wallpaper hanger works alone for 1 h and then quits. How long will it take the second hanger, working alone, to finish papering the wall?

17. A large heating unit and a small heating unit are being used to heat the water in a pool. The large unit, working alone, requires 8 h to heat the pool. After both units have been operating for 2 h, the large unit is turned off. The small unit requires 9 more hours to heat the pool. How long would it take the small unit, working alone, to heat the pool?

18. Two machines fill cereal boxes at the same rate. After the two machines work together for 7 h, one machine breaks down. The second machine requires 14 more hours to finish filling the boxes. How long would it have taken one of the machines, working alone, to fill the boxes?

19. Two welders who work at the same rate are riveting the girders of a building. After they work together for 10 h, one of the welders quits. The second welder requires 20 more hours to complete the welds. Find the time it would have taken one of the welders, working alone, to complete the welds.

20. A large drain and a small drain are opened to drain a pool. The large drain can empty the pool in 6 h. After both drains have been open for 1 h, the large drain becomes clogged and is closed. The smaller drain remains open and requires 9 more hours to empty the pool. How long would it have taken the small drain, working alone, to empty the pool?

2 Solve.

21. A camper drove 90 mi to a recreational area and then hiked 5 mi into the woods. The rate of the camper while driving was nine times the rate while hiking. The time spent hiking and driving was 3 h. Find the rate at which the camper hiked.

22. The president of a company traveled 1800 mi by jet and 300 mi on a prop plane. The rate of the jet was four times the rate of the prop plane. The entire trip took 5 h. Find the rate of the jet plane.

23. A jogger ran 8 mi in the same amount of time as it took a cyclist to ride 20 mi. The rate of the cyclist was 12 mph greater than the rate of the jogger. Find the rate of the jogger and the rate of the cyclist.

24. An express train traveled 600 mi in the same amount of time as it took a freight train to travel 360 mi. The rate of the express train was 20 mph greater than the rate of the freight train. Find the rate of each train.

25. To assess the damage done by a fire, a forest ranger traveled 1080 mi by jet and then an additional 180 mi by helicopter. The rate of the jet was four times the rate of the helicopter. The entire trip took 5 h. Find the rate of the jet.

26. A twin-engine plane flies 800 mi in the same amount of time as it takes a single-engine plane to fly 600 mi. The rate of the twin-engine plane is 50 mph greater than the rate of the single-engine plane. Find the rate of the twin-engine plane.

27. Two planes leave an airport and head for another airport 900 mi away. The rate of the first plane is twice the rate of the second plane. The second plane arrives at the airport 3 h after the first plane. Find the rate of the second plane.

28. A car and a bus leave a town at 1 P.M. and head for a town 300 mi away. The rate of the car is twice the rate of the bus. The car arrives 5 h ahead of the bus. Find the rate of the car.

29. A car is traveling at a rate that is 36 mph greater than the rate of a cyclist. The car travels 384 mi in the same amount of time it takes the cyclist to travel 96 mi. Find the rate of the car.

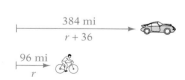

30. An engineer traveled 165 mi by car and then an additional 660 mi by plane. The rate of the plane was four times the rate of the car. The total trip took 6 h. Find the rate of the car.

31. A backpacker hiking into a wilderness area walked 9 mi at a constant rate and then reduced this rate by 1 mph. Another 4 mi was hiked at the reduced rate. The time required to hike the last 4 mi was 1 h less than the time required to walk the first 9 mi. Find the rate at which the hiker walked the first 9 mi.

32. After sailing 15 mi, a sailor changed direction and increased the boat's speed by 2 mph. An additional 19 mi was sailed at the increased speed. The total sailing time was 4 h. Find the rate of the boat for the first 15 mi.

33. A small motor on a fishing boat can move the boat at a rate of 6 mph in calm water. Traveling with the current, the boat can travel 24 mi in the same amount of time it takes to travel 12 mi against the current. Find the rate of the current.

34. A commercial jet can fly 550 mph in calm air. Traveling with the jet stream, the plane can fly 2400 mi in the same amount of time it takes to fly 2000 mi against the jet stream. Find the rate of the jet stream.

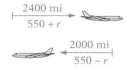

35. A cruise ship can sail 28 mph in calm water. Sailing with the gulf stream, the ship can sail 170 mi in the same amount of time it takes to sail 110 mi against the gulf stream. Find the rate of the gulf stream.

36. Paddling in calm water, a canoeist can paddle at a rate of 8 mph. Traveling with the current, the canoeist went 30 mi in the same amount of time it took to travel 18 mi against the current. Find the rate of the current.

37. On a recent trip, a trucker traveled 330 mi at a constant rate. Because of road conditions, the trucker then reduced the speed by 25 mph. An additional 30 mi was traveled at the reduced rate. The entire trip took 7 h. Find the rate of the trucker for the first 330 mi.

38. Commuting from work to home, a lab technician traveled 10 mi at a constant rate through congested traffic. Upon reaching the expressway, the technician increased the speed by 20 mph. An additional 20 mi was traveled at the increased speed. The total time for the trip was 1 h. Find the rate of the technician through the congested traffic.

39. Rowing with the current of a river, a rowing team can row 25 mi in the same amount of time it takes to row 15 mi against the current. The rate of the rowing team in calm water is 20 mph. Find the rate of the current.

40. A plane can fly 180 mph in calm air. Flying with the wind, the plane can fly 600 mi in the same amount of time it takes to fly 480 mi against the wind. Find the rate of the wind.

SUPPLEMENTAL EXERCISES 9.7

Solve.

41. One pipe can fill a tank in 2 h, a second pipe can fill the tank in 4 h, and a third pipe can fill the tank in 5 h. How long would it take to fill the tank with all three pipes operating?

42. A mason can construct a retaining wall in 10 h. The mason's more experienced apprentice can do the same job in 15 h. How long would it take the mason's less experienced apprentice to do the job if, working together, all three can complete the wall in 5 h?

43. An Outing Club traveled 32 mi by canoe and then hiked 4 mi. The rate by boat was four times the rate on foot. The time spent walking was 1 h less than the time spent canoeing. Find the amount of time spent traveling by canoe.

44. A motorist drove 120 mi before running out of gas and walking 4 mi to a gas station. The rate of the motorist in the car was ten times the rate walking. The time spent walking was 2 h less than the time spent driving. How long did it take for the motorist to drive the 120 mi?

45. Because of bad weather, a bus driver reduced the usual speed along a 150 mi bus route by 10 mph. The bus arrived only 30 min later than its usual arrival time. How fast does the bus usually travel?

Project in Mathematics

Intensity of Illumination

You are already aware that the standard unit of length in the metric system is the meter (m) and that the standard unit of mass in the metric system is the kilogram (kg). You may not know that the standard unit of light intensity is the candela.

The rate at which light falls upon a one-square-unit area of surface is called the **intensity of illumination**. Intensity of illumination is measured in **lumens** (lm). A lumen is defined in the following illustration.

Picture a source of light equal to 1 candela positioned at the center of a hollow sphere that has a radius of 1 m. The rate at which light falls upon 1 m^2 of the inner surface of the sphere is equal to one lumen (1 lm). If a light source equal to 4 candelas is positioned at the center of the sphere, each square meter of the inner surface receives four times as much illumination, or 4 lm.

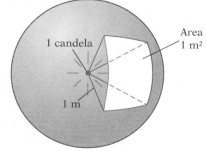

Light rays diverge as they leave a light source. The light that falls upon an area of 1 m² at a distance of 1 m from the source of light spreads out over an area of 4 m² when it is 2 m from the source. The same light spreads out over an area of 9 m² when it is 3 m from the light source and over an area of 16 m² when it is 4 m from the light source. Therefore, as a surface moves farther away from the source of light, the intensity of illumination on the surface decreases from its value at 1 m to $\left(\frac{1}{2}\right)^2$, or $\frac{1}{4}$, that value at 2 m; to $\left(\frac{1}{3}\right)^2$, or $\frac{1}{9}$, that value at 3 m; and to $\left(\frac{1}{4}\right)^2$, or $\frac{1}{16}$, that value at 4 m.

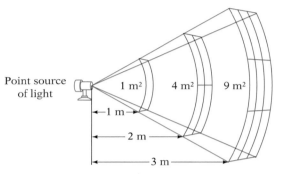

The formula for the intensity of illumination is

$$I = \frac{s}{r^2}$$

where I is the intensity of illumination in lumens, s is the strength of the light source in candelas, and r is the distance in meters between the light source and the illuminated surface.

A 30-candela lamp is 0.5 m above a desk. Find the illumination on the desk.

$$I = \frac{s}{r^2}$$

$$I = \frac{30}{(0.5)^2} = 120$$

The illumination on the desk is 120 lm.

Solve.

1. A 100-candela light is hanging 5 m above a floor. What is the intensity of illumination on the floor beneath it?

2. A 25-candela source of light is 2 m above a desk. Find the intensity of illumination on the desk.

3. How strong a light source is needed to cast 20 lm of light on a surface 4 m from the source?

4. How strong a light source is needed to cast 80 lm of light on a surface 5 m from the source?

5. How far from the desk surface must a 40-candela light source be positioned if the desired intensity of illumination is 10 lm?

6. Find the distance between a 36-candela light source and a surface if the intensity of illumination on the surface is 0.01 lm.

7. Two lights cast the same intensity of illumination on a wall. One light is 6 m from the wall and has a rating of 36 candelas. The second light is 8 m from the wall. Find the candela rating of the second light.

8. A 40-candela light source and a 10-candela light source both throw the same intensity of illumination on a wall. The 10-candela light is 6 m from the wall. Find the distance from the 40-candela light to the wall.

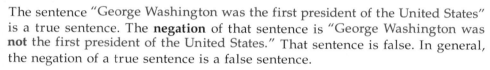

Focus on Problem Solving

Negations and *If... then...* Sentences

The sentence "George Washington was the first president of the United States" is a true sentence. The **negation** of that sentence is "George Washington was **not** the first president of the United States." That sentence is false. In general, the negation of a true sentence is a false sentence.

The negation of a false sentence is a true sentence. For instance, the sentence "The moon is made of green cheese" is a false sentence. The negation of that sentence, "The moon is **not** made of green cheese," is true.

The words *all, no* (or *none*) and *some* are called **quantifiers**. Writing the negation of a sentence that contains these words requires special attention. Consider the sentence "All pets are dogs." This sentence is not true because there are pets that are not dogs—cats, for example, are pets. Because the sentence is false, its negation must be true. You might be tempted to write "All pets are not dogs," but that sentence is not true because some pets are dogs. The correct negation of "All pets are dogs," is "Some pets are not dogs." Note the use of the word *some* in the negation.

Now consider the sentence "Some computers are portable." Because that sentence is true, its negation must be false. Writing "Some computers are not portable" as the negation is not correct, because that sentence is true. The negation of "Some computers are portable" is "No computers are portable."

The sentence "No flowers have red blooms" is false, because there is at least one flower (roses, for example) that has red blooms. Because the sentence is false, its negation must be true. The negation is "Some flowers have red blooms."

Statement	Negation
All *A* are *B*.	Some *A* are not *B*.
No *A* are *B*.	Some *A* are *B*.
Some *A* are *B*.	No *A* are *B*.
Some *A* are not *B*.	All *A* are *B*.

A **premise** is a known or assumed fact. A premise can be stated using one of the quantifiers (*all, no, none,* or *some*) or can be stated using an *If... then...* sentence. For instance, the sentence "All triangles have three sides" can be written "*If* the figure is a triangle, *then* it has three sides."

We can write the sentence "No whole numbers are negative numbers" as an *If...then...* sentence: If a number is a whole number, then it is not a negative number.

Write the negation of the sentence.

1. All cats like milk.

2. All computers need people.

3. Some trees are tall.

4. Some vegetables are good for you to eat.

5. No politicians are honest.

6. No houses have kitchens.

7. All police officers are tall.

8. All lakes are not polluted.

9. Some banks are not open on Sunday.

10. Some colleges do not offer night classes.

11. Some drivers are unsafe.

12. Some speeches are interesting.

13. All laws are good.

14. All Mark Twain books are funny.

15. All businesses are not profitable.

16. All motorcycles are not large.

Write the sentence as an *If... then...* sentence.

17. All students at Barlock College must take a life science course.

18. All baseballs are round.

19. All computers need people.

20. All cats like milk.

21. No odd number is evenly divisible by 2.

22. No rectangles have five sides.

23. All roads lead to Rome.

24. All dogs have fleas.

25. No triangle has four angles.

26. No prime number greater than 2 is an even number.

Chapter Summary

Key Words

An *algebraic fraction* is a fraction in which the numerator or denominator is a variable expression. An algebraic fraction is in simplest form when the numerator and denominator have no common factors other than 1.   (Objective 9.1.1)

$\dfrac{x+4}{x-3}$ is an algebraic fraction in simplest form.

The *reciprocal* of a fraction is that fraction with the numerator and denominator interchanged.   (Objective 9.1.3)

The reciprocal of $\dfrac{x-7}{y}$ is $\dfrac{y}{x-7}$.

The *least common multiple* (LCM) of two or more numbers is the smallest number that contains the prime factorization of each number.   (Objective 9.2.1)

The LCM of 8 and 12 is 24 because 24 is the smallest number that is a multiple of both 8 and 12.

A *complex fraction* is a fraction whose numerator or denominator contains one or more fractions.   (Objective 9.4.1)

$\dfrac{x+\dfrac{1}{x}}{2-\dfrac{3}{x}}$ is a complex fraction.

A *ratio* is the quotient of two quantities that have the same unit. A *rate* is the quotient of two quantities that have different units. A *proportion* is an equation that states the equality of two ratios or rates.   (Objective 9.5.1)

$\dfrac{3}{8}=\dfrac{15}{40}$ and $\dfrac{20\text{ m}}{5\text{ s}}=\dfrac{80\text{ m}}{20\text{ s}}$ are examples of proportions.

A *literal equation* is an equation that contains more than one variable. A *formula* is a literal equation that states rules about measurements.   (Objective 9.6.1)

$2x+3y=6$ is an example of a literal equation. $A=\pi r^2$ is a literal equation that is also a formula.

Essential Rules and Procedures

Simplifying Algebraic Fractions
Factor the numerator and denominator. Divide by the common factors.   (Objective 9.1.1)

$\dfrac{x^2-3x-10}{x^2-25}=\dfrac{(x+2)(x-5)}{(x+5)(x-5)}=\dfrac{x+2}{x+5}$

Multiplying Algebraic Fractions
Multiply the numerators. Multiply the denominators. Write the answer in simplest form.

$\dfrac{a}{b}\cdot\dfrac{c}{d}=\dfrac{ac}{bd}$ (Objective 9.1.2)

$\dfrac{x^2-3x}{x^2+x}\cdot\dfrac{x^2+5x+4}{x^2-4x+3}$
$=\dfrac{x(x-3)}{x(x+1)}\cdot\dfrac{(x+4)(x+1)}{(x-3)(x-1)}$
$=\dfrac{x(x-3)(x+4)(x+1)}{x(x+1)(x-3)(x-1)}=\dfrac{x+4}{x-1}$

Dividing Algebraic Fractions
To divide two fractions, multiply by the reciprocal of the divisor.

$\dfrac{a}{b}\div\dfrac{c}{d}=\dfrac{a}{b}\cdot\dfrac{d}{c}=\dfrac{ad}{bc}$ (Objective 9.1.3)

$\dfrac{3x^3y^2}{8a^4b^5}\div\dfrac{6x^4y}{4a^3b^2}=\dfrac{3x^3y^2}{8a^4b^5}\cdot\dfrac{4a^3b^2}{6x^4y}$
$=\dfrac{3x^3y^2\cdot 4a^3b^2}{8a^4b^5\cdot 6x^4y}=\dfrac{y}{4ab^3x}$

Essential Rules and Procedures

Adding and Subtracting Algebraic Fractions
1. Find the LCM of the denominators.
2. Write each fraction with the LCM as the denominator.
3. Add or subtract the numerators. The denominator of the sum or difference is the common denominator.
4. Write the answer in simplest form.
(Objective 9.3.2)

$$\frac{4x-2}{3x+12} - \frac{x-2}{x+4} = \frac{4x-2}{3(x+4)} - \frac{x-2}{x+4}$$

$$= \frac{4x-2}{3(x+4)} - \frac{x-2}{x+4} \cdot \frac{3}{3}$$

$$= \frac{4x-2}{3(x+4)} - \frac{3x-6}{3(x+4)}$$

$$= \frac{4x-2-(3x-6)}{3(x+4)}$$

$$= \frac{x+4}{3(x+4)} = \frac{1}{3}$$

Simplifying Complex Fractions
Multiply the numerator and denominator of the complex fraction by the LCM of the denominators of the fractions in the numerator and denominator. (Objective 9.4.1)

$$\frac{\frac{1}{x}+\frac{1}{y}}{\frac{1}{x}-\frac{1}{y}} = \frac{\frac{1}{x}+\frac{1}{y}}{\frac{1}{x}-\frac{1}{y}} \cdot \frac{xy}{xy} = \frac{\frac{1}{x}\cdot xy + \frac{1}{y}\cdot xy}{\frac{1}{x}\cdot xy - \frac{1}{y}\cdot xy}$$

$$= \frac{y+x}{y-x}$$

Solving Equations Containing Fractions
Clear denominators by multiplying each side of the equation by the LCM of the denominators. Then solve for the variable. (Objective 9.5.1)

$$\frac{1}{2a} = \frac{2}{a} - \frac{3}{8}$$

$$8a\left(\frac{1}{2a}\right) = 8a\left(\frac{2}{a} - \frac{3}{8}\right)$$

$$4 = 16 - 3a$$

$$-12 = -3a$$

$$4 = a$$

Solving Literal Equations
Rewrite the equation so that the letter being solved for is alone on one side of the equation and all numbers and other variables are on the other side. (Objective 9.6.1)

Solve $A = \frac{1}{2}bh$ for b.

$$2(A) = 2\left(\frac{1}{2}bh\right)$$

$$2A = bh$$

$$\frac{2A}{h} = \frac{bh}{h}$$

$$\frac{2A}{h} = b$$

Work Problems
Rate of work · Time worked = Part of task completed
(Objective 9.7.1)

Pat can do a certain job in 3 h. Chris can do the same job in 5 h. How long would it take them, working together, to get the job done?

$$\frac{t}{3} + \frac{t}{5} = 1$$

Uniform Motion Problems
Distance ÷ Rate = Time   (Objective 9.7.2)

Train A's speed is 15 mph faster than Train B's speed. Train A travels 150 mi in the same amount of time it takes Train B to travel 120 mi. Find the rate of Train B.

$$\frac{120}{r} = \frac{150}{r + 15}$$

Chapter Review Exercises

1. Multiply: $\dfrac{8ab^2}{15x^3y} \cdot \dfrac{5xy^4}{16a^2b}$

2. Add: $\dfrac{5}{3x - 4} + \dfrac{4}{2x + 3}$

3. Solve $4x + 3y = 12$ for x.

4. Simplify: $\dfrac{16x^5y^3}{24xy^{10}}$

5. Divide: $\dfrac{20x^2 - 45x}{6x^3 + 4x^2} \div \dfrac{40x^3 - 90x^2}{12x^2 + 8x}$

6. Simplify: $\dfrac{x - \dfrac{16}{5x - 2}}{3x - 4 - \dfrac{88}{5x - 2}}$

7. Find the LCM of $24a^2b^5$ and $36a^3b$.

8. Subtract: $\dfrac{5x}{3x + 7} - \dfrac{x}{3x + 7}$

9. Write each fraction in terms of the LCM of the denominators.

$\dfrac{3}{16x}; \dfrac{5}{8x^2}$

10. Simplify: $\dfrac{2x^2 - 13x - 45}{2x^2 - x - 15}$

11. Divide: $\dfrac{x^2 - 5x - 14}{x^2 - 3x - 10} \div \dfrac{x^2 - 4x - 21}{x^2 - 9x + 20}$

12. Add: $\dfrac{2y}{5y - 7} + \dfrac{3}{7 - 5y}$

13. Multiply: $\dfrac{3x^3 + 10x^2}{10x - 2} \cdot \dfrac{20x - 4}{6x^4 + 20x^3}$

14. Subtract: $\dfrac{5x + 3}{2x^2 + 5x - 3} - \dfrac{3x + 4}{2x^2 + 5x - 3}$

15. Find the LCM of $5x^4(x - 7)^2$ and $15x(x - 7)$.

16. Solve: $\dfrac{6}{x - 7} = \dfrac{8}{x - 6}$

17. Solve: $\dfrac{x + 8}{x + 4} = 1 + \dfrac{5}{x + 4}$

18. Simplify: $\dfrac{12a^2b(4x - 7)}{15ab^2(7 - 4x)}$

19. Simplify: $\dfrac{5x - 1}{x^2 - 9} + \dfrac{4x - 3}{x^2 - 9} - \dfrac{8x - 1}{x^2 - 9}$

20. Write each fraction in terms of the LCM of the denominators.

$\dfrac{2}{x + 3}; \dfrac{7}{x - 3}$

21. Solve: $\dfrac{20}{2x+3} = \dfrac{17x}{2x+3} - 5$

22. Simplify: $\dfrac{\dfrac{5}{x-1} - \dfrac{3}{x+3}}{\dfrac{6}{x+3} + \dfrac{2}{x-1}}$

23. Add: $\dfrac{x-1}{x+2} + \dfrac{3x-2}{5-x} + \dfrac{5x^2+15x-11}{x^2-3x-10}$

24. Divide: $\dfrac{18x^2+25x-3}{9x^2-28x+3} \div \dfrac{2x^2+9x+9}{x^2-6x+9}$

25. Simplify: $\dfrac{x^2+x-30}{15+2x-x^2}$

26. Solve: $\dfrac{5}{7} + \dfrac{x}{2} = 2 - \dfrac{x}{7}$

27. Simplify: $\dfrac{x + \dfrac{6}{x-5}}{1 + \dfrac{2}{x-5}}$

28. Multiply: $\dfrac{3x^2+4x-15}{x^2-11x+28} \cdot \dfrac{x^2-5x-14}{3x^2+x-10}$

29. Solve: $\dfrac{x}{5} = \dfrac{x+12}{9}$

30. Solve: $\dfrac{3}{20} = \dfrac{x}{80}$

31. Simplify: $\dfrac{1 - \dfrac{1}{x}}{1 - \dfrac{8x-7}{x^2}}$

32. Solve $x - 2y = 15$ for x.

33. Add: $\dfrac{6}{a} + \dfrac{9}{b}$

34. Find the LCM of $10x^2 - 11x + 3$ and $20x^2 - 17x + 3$.

35. Solve $i = \dfrac{100m}{c}$ for c.

36. Solve: $\dfrac{15}{x} = \dfrac{3}{8}$

37. Solve: $\dfrac{22}{2x+5} = 2$

38. Add: $\dfrac{x+7}{15x} + \dfrac{x-2}{20x}$

39. Multiply: $\dfrac{16a^2-9}{16a^2-24a+9} \cdot \dfrac{8a^2-13a-6}{4a^2-5a-6}$

40. Write each fraction in terms of the LCM of the denominators.

$$\dfrac{x}{12x^2+16x-3}, \dfrac{4x^2}{6x^2+7x-3}$$

41. Add: $\dfrac{3}{4ab} + \dfrac{5}{4ab}$

42. Solve: $\dfrac{20}{x+2} = \dfrac{5}{16}$

43. Solve: $\dfrac{5x}{3} - \dfrac{2}{5} = \dfrac{8x}{5}$

44. Divide: $\dfrac{6a^2b^7}{25x^3y} \div \dfrac{12a^3b^4}{5x^2y^2}$

45. A brick mason can construct a patio in 3 h. If the mason works with an apprentice, they can construct the patio in 2 h. How long would it take the apprentice, working alone, to construct the patio?

46. A weight of 21 lb stretches a spring 14 in. At the same rate, how far would a weight of 12 lb stretch the spring?

47. The rate of a jet is 400 mph in calm air. Traveling with the wind, the jet can fly 2100 mi in the same amount of time it takes to fly 1900 mi against the wind. Find the rate of the wind.

48. A gardener uses 4 oz of insecticide to make 2 gal of garden spray. At this rate, how much additional insecticide is necessary to make 10 gal of the garden spray?

49. One hose can fill a pool in 15 h. A second hose can fill the pool in 10 h. How long would it take to fill the pool using both hoses?

50. A car travels 315 mi in the same amount of time as it takes a bus to travel 245 mi. The rate of the car is 10 mph greater than that of the bus. Find the rate of the car.

C hapter Test

1. Divide: $\dfrac{x^2 + 3x + 2}{x^2 + 5x + 4} \div \dfrac{x^2 - x - 6}{x^2 + 2x - 15}$

2. Subtract: $\dfrac{2x}{x^2 + 3x - 10} - \dfrac{4}{x^2 + 3x - 10}$

3. Find the LCM of $6x - 3$ and $2x^2 + x - 1$.

4. Solve: $\dfrac{3}{x + 4} = \dfrac{5}{x + 6}$

5. Multiply: $\dfrac{x^3 y^4}{x^2 - 4x + 4} \cdot \dfrac{x^2 - x - 2}{x^6 y^4}$

6. Simplify: $\dfrac{1 + \dfrac{1}{x} - \dfrac{12}{x^2}}{1 + \dfrac{2}{x} - \dfrac{8}{x^2}}$

7. Write each fraction in terms of the LCM of the denominators.

$\dfrac{3}{x^2 - 2x}, \dfrac{x}{x^2 - 4}$

8. Solve $3x + 5y + 15 = 0$ for x.

9. Solve: $\dfrac{6}{x} - 2 = 1$

10. Subtract: $\dfrac{2}{2x - 1} - \dfrac{3}{3x + 1}$

11. Divide: $\dfrac{x^2 - x - 56}{x^2 + 8x + 7} \div \dfrac{x^2 - 13x + 40}{x^2 - 4x - 5}$

12. Subtract: $\dfrac{3x}{x^2 + 5x - 24} - \dfrac{9}{x^2 + 5x - 24}$

13. Find the LCM of $3x^2 + 6x$ and $2x^2 + 8x + 8$.

14. Simplify: $\dfrac{x^2 - 7x + 10}{25 - x^2}$

15. Solve: $\dfrac{3x}{x - 3} - 2 = \dfrac{10}{x - 3}$

16. Solve $f = v + at$ for t.

17. Simplify: $\dfrac{12x^4 y^2}{18xy^7}$

18. Subtract: $\dfrac{2}{2x - 1} - \dfrac{1}{x + 1}$

19. Solve: $\dfrac{2}{x - 2} = \dfrac{12}{x + 3}$

20. Multiply: $\dfrac{x^5 y^3}{x^2 - x - 6} \cdot \dfrac{x^2 - 9}{x^2 y^4}$

21. Write each fraction in terms of the LCM of the denominators.

$$\dfrac{3y}{x(1 - x)}, \dfrac{x}{(x + 1)(x - 1)}$$

22. Simplify: $\dfrac{1 - \dfrac{2}{x} - \dfrac{15}{x^2}}{1 - \dfrac{25}{x^2}}$

23. A salt solution is formed by mixing 4 lb of salt with 10 gal of water. At this rate, how many additional pounds of salt are required for 15 gal of water?

24. A small plane can fly at 110 mph in calm air. Flying with the wind, the plane can fly 260 mi in the same amount of time it takes to fly 180 mi against the wind. Find the rate of the wind.

25. One pipe can fill a tank in 9 min. A second pipe requires 18 min to fill the tank. How long would it take both pipes, working together, to fill the tank?

Cumulative Review Exercises

1. Evaluate: $-|-17|$

2. Evaluate: $-\dfrac{3}{4} \cdot (2)^3$

3. Simplify: $\left(\dfrac{2}{3}\right)^2 \div \left(\dfrac{3}{2} - \dfrac{2}{3}\right) + \dfrac{1}{2}$

4. Evaluate $-a^2 + (a - b)^2$ when $a = -2$ and $b = 3$.

5. Simplify: $-2x - (-3y) + 7x - 5y$

6. Simplify: $2[3x - 7(x - 3) - 8]$

7. Solve: $3 - \dfrac{1}{4}x = 8$

8. Solve: $3[x - 2(x - 3)] = 2(3 - 2x)$

9. Find $16\dfrac{2}{3}\%$ of 60.

10. Solve: $\dfrac{5}{9}x < 1$

11. Solve: $x - 2 \geq 4x - 47$

12. Graph: $y = 2x - 1$

13. Graph: $f(x) = 3x + 2$

14. Graph: $x = 3$

15. Graph the solution set of $5x + 2y < 6$.

16. Find the range of the function given by the equation $f(x) = -2x + 11$ if the domain is $\{-8, -4, 0, 3, 7\}$.

17. Evaluate $f(x) = 4x - 3$ at $x = 10$.

18. Solve by substitution: $6x - y = 1$
$\qquad\qquad\qquad\qquad\qquad\quad y = 3x + 1$

19. Solve by the addition method: $2x - 3y = 4$
$\qquad\qquad\qquad\qquad\qquad\qquad\quad 4x + y = 1$

20. Multiply: $(3xy^4)(-2x^3y)$

21. Simplify: $(a^4b^3)^5$

22. Simplify: $\dfrac{a^2b^{-5}}{a^{-1}b^{-3}}$

23. Multiply: $(a - 3b)(a + 4b)$

24. Divide: $\dfrac{15b^4 - 5b^2 + 10b}{5b}$

25. Divide: $(x^3 - 8) \div (x - 2)$

26. Factor: $12x^2 - x - 1$

27. Factor: $y^2 - 7y + 6$

28. Factor: $2a^3 + 7a^2 - 15a$

29. Factor: $4b^2 - 100$

30. Solve: $(x + 3)(2x - 5) = 0$

31. Simplify: $\dfrac{x^2 + 3x - 28}{16 - x^2}$

32. Divide: $\dfrac{x^2 - 3x - 10}{x^2 - 4x - 12} \div \dfrac{x^2 - x - 20}{x^2 - 2x - 24}$

33. Subtract: $\dfrac{6}{3x - 1} - \dfrac{2}{x + 1}$

34. Solve: $\dfrac{4x}{x - 3} - 2 = \dfrac{8}{x - 3}$

35. Solve $f = v + at$ for a.

36. Translate "the difference between five times a number and eighteen is the opposite of three" into an equation and solve.

37. An investment of $5000 is made at an annual simple interest rate of 7%. How much additional money must be invested at an annual simple interest rate of 11% so that the total interest earned is 9% of the total investment?

38. A silversmith mixes 60 g of an alloy that is 40% silver with 120 g of another silver alloy. The resulting alloy is 60% silver. Find the percent of silver in the 120 g alloy.

39. The length of the base of a triangle is 2 in. less than twice the height. The area of the triangle is 30 in². Find the base and height of the triangle.

40. One water pipe can fill a tank in 12 min. A second pipe requires 24 min to fill the tank. How long would it take both pipes, working together, to fill the tank?

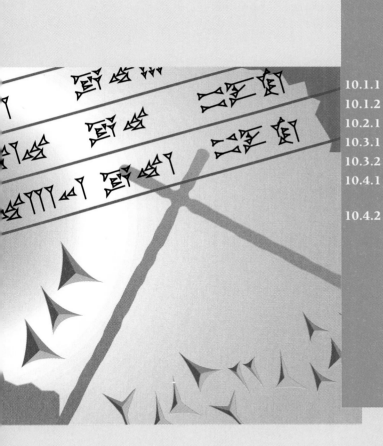

10

Radical Expressions

Objectives

A Table of Square Roots

The practice of finding the square root of a number has existed for at least two thousand years. The process of finding a square root is tedious and time-consuming. Therefore, before the advent of the hand-held calculator, it was convenient to have tables of square roots.

The table shown at the left is part of an old Babylonian clay table that was written around 350 B.C. It is an incomplete table of square roots written in a style called *cuneiform*.

The number base of the Babylonians was 60 instead of the 10 we use today. The symbol ∇ was used for 1, and 10 was written as ◁. Examples of numbers using this system are given below.

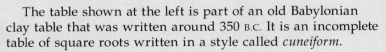

A translation of the first two lines of the table is given below. The number in parentheses is the equivalent base-10 number that would be used today. You might try to translate the third line. The answer appears at the bottom of this page.

40 × 60 + 1 (= 2401), which is the square of 49

41 × 60 + 40 (= 2500), which is the square of 50

Answer: 43 × 60 + 21 (= 2601), which is the square of 51.

S E C T I O N **10.1**

Introduction to Radical Expressions

1 Simplify numerical radical expressions

A **square root** of a positive number x is a number whose square is x.

A square root of 16 is 4 because $4^2 = 16$.

A square root of 16 is -4 because $(-4)^2 = 16$.

Every positive number has two square roots, one a positive number and one a negative number. The symbol $\sqrt{}$, called a **radical sign**, is used to indicate the positive or **principal square root** of a number. For example, $\sqrt{16} = 4$ and $\sqrt{25} = 5$. The number or variable expression under the radical sign is called the **radicand**.

When the negative square root of a number is to be found, a negative sign is placed in front of the radical. For example, $-\sqrt{16} = -4$ and $-\sqrt{25} = -5$.

The square of an integer is a **perfect square.** 49, 81, and 144 are examples of perfect squares.

$$7^2 = 49$$
$$9^2 = 81$$
$$12^2 = 144$$

An integer that is a perfect square can be written as the product of prime factors, each of which has an even exponent when expressed in exponential form.

$$49 = 7 \cdot 7 = 7^2$$
$$81 = 3 \cdot 3 \cdot 3 \cdot 3 = 3^4$$
$$144 = 2 \cdot 2 \cdot 2 \cdot 2 \cdot 3 \cdot 3 = 2^4 3^2$$

To find the square root of a perfect square written in exponential form, remove the radical sign, and divide each exponent by 2.

To simplify $\sqrt{625}$, write the prime factorization of the radicand in exponential form.

$$\sqrt{625} = \sqrt{5^4}$$

Remove the radical sign, and divide the exponent by 2.

$$= 5^2$$

Simplify.

$$= 25$$

The square root of a negative number is not a real number because the square of a real number is always positive or zero.

$\sqrt{-4}$ is not a real number.

$\sqrt{-25}$ is not a real number.

If a number is not a perfect square, its square root can only be approximated. For example, 2 and 7 are not perfect squares. Thus their square roots can only be approximated. These numbers are **irrational numbers**. Their decimal representations never terminate or repeat.

$$\sqrt{2} \approx 1.4142135\ldots$$

$$\sqrt{7} \approx 2.6457513\ldots$$

The approximate square roots of numbers that are not perfect squares can be found using a calculator. The square roots can be rounded to any given place value.

A radical expression is in simplest form when the radicand contains no factor greater than 1 that is a perfect square. The Product Property of Square Roots is used to simplify radical expressions.

The Product Property of Square Roots

If a and b are positive real numbers, then $\sqrt{ab} = \sqrt{a} \cdot \sqrt{b}$.

To simplify $\sqrt{96}$, write the prime factorization of the radicand in exponential form.

$$\sqrt{96} = \sqrt{2^5 \cdot 3}$$

Write the radicand as a product of a perfect square and factors that do not contain a perfect square. Remember that a perfect square has an even exponent when expressed in exponential form.

$$= \sqrt{2^4(2 \cdot 3)}$$

Use the Product Property of Square Roots. Write perfect squares under the first radical sign and all remaining factors under the second radical sign.

$$= \sqrt{2^4}\sqrt{2 \cdot 3}$$

Simplify.

$$= 2^2\sqrt{2 \cdot 3}$$
$$= 4\sqrt{6}$$

This last example shows that $\sqrt{96} = 4\sqrt{6}$. The two expressions are different representations of the same number. Using a calculator to evaluate each expression, we find that $\sqrt{96} \approx 9.79796$ and $4\sqrt{6} \approx 4(2.44949) = 9.79796$.

Example 1   Simplify: $3\sqrt{90}$

Solution $3\sqrt{90} = 3\sqrt{2 \cdot 3^2 \cdot 5}$ ▶ Write the prime factorization of the radicand in exponential form.

$\qquad = 3\sqrt{3^2(2 \cdot 5)}$ ▶ Write the radicand as a product of a perfect square and factors that do not contain a perfect square.

$\qquad = 3\sqrt{3^2}\sqrt{2 \cdot 5}$ ▶ Use the Product Property of Square Roots.

$\qquad = 3 \cdot 3\sqrt{10}$ ▶ Simplify.

$\qquad = 9\sqrt{10}$

Problem 1   Simplify: $-5\sqrt{32}$

Solution   See page A39.

Example 2   Simplify: $\sqrt{252}$

Solution $\sqrt{252} = \sqrt{2^2 \cdot 3^2 \cdot 7}$

$\qquad = \sqrt{2^2 \cdot 3^2}\sqrt{7}$

$\qquad = 2 \cdot 3\sqrt{7}$

$\qquad = 6\sqrt{7}$

Problem 2   Simplify: $\sqrt{216}$

   Solution   See page A39.

2   Simplify variable radical expressions

Variable expressions that contain radicals do not always represent real numbers.

The variable expression at the right does not represent a real number when x is a negative number, such as -4.

$\sqrt{x^3}$

$\sqrt{(-4)^3} = \sqrt{-64}$ ← Not a real number

Now consider the expression $\sqrt{x^2}$ and evaluate this expression for $x = -2$ and $x = 2$.

$\sqrt{x^2}$

$\sqrt{(-2)^2} = \sqrt{4} = 2 = |-2|$

$\sqrt{2^2} \quad = \sqrt{4} = 2 = |2|$

This suggests the following.

The Square Root of a^2

For any real number a, $\sqrt{a^2} = |a|$.

If $a \geq 0$, then $\sqrt{a^2} = a$.

In order to avoid variable expressions that do not represent real numbers, and so that absolute value signs are not needed for certain expressions, the variables in this chapter will represent *positive* numbers unless otherwise stated.

A variable or a product of variables written in exponential form is a **perfect square** when each exponent is an even number.

To find the square root of a perfect square, remove the radical sign, and divide each exponent by 2.

For example, to simplify $\sqrt{a^6}$, remove the radical sign, and divide the exponent by 2.

$\sqrt{a^6} = a^3$

A variable radical expression is in simplest form when the radicand contains no factor greater than 1 that is a perfect square.

To simplify $\sqrt{x^7}$, write x^7 as the product of x and a perfect square.

$\sqrt{x^7} = \sqrt{x^6 \cdot x}$

Use the Product Property of Square Roots.

$= \sqrt{x^6}\sqrt{x}$

Simplify the square root of the perfect square.

$= x^3\sqrt{x}$

Example 3   Simplify: $\sqrt{b^{15}}$

   Solution $\sqrt{b^{15}} = \sqrt{b^{14} \cdot b} = \sqrt{b^{14}} \cdot \sqrt{b} = b^7\sqrt{b}$

Problem 3   Simplify: $\sqrt{y^{19}}$

   Solution   See page A39.

Simplify: $3x\sqrt{8x^3y^{13}}$

Write the prime factorization of the radicand in exponential form.

$$3x\sqrt{8x^3y^{13}} = 3x\sqrt{2^3x^3y^{13}}$$

Write the radicand as a product of a perfect square and factors that do not contain a perfect square.

$$= 3x\sqrt{2^2x^2y^{12}(2xy)}$$

Use the Product Property of Square Roots.

$$= 3x\sqrt{2^2x^2y^{12}}\,\sqrt{2xy}$$

Simplify.

$$= 3x \cdot 2xy^6\sqrt{2xy}$$
$$= 6x^2y^6\sqrt{2xy}$$

Example 4   Simplify.   A. $\sqrt{24x^5}$ B. $2a\sqrt{18a^3b^{10}}$

Solution   A. $\sqrt{24x^5} = \sqrt{2^3 \cdot 3 \cdot x^5}$ B. $2a\sqrt{18a^3b^{10}} = 2a\sqrt{2 \cdot 3^2 \cdot a^3b^{10}}$
$\qquad\qquad\quad = \sqrt{2^2x^4(2 \cdot 3x)}$ $\qquad\qquad\qquad\qquad\quad = 2a\sqrt{3^2a^2b^{10}(2a)}$
$\qquad\qquad\quad = \sqrt{2^2x^4}\,\sqrt{2 \cdot 3x}$ $\qquad\qquad\qquad\qquad\quad = 2a\sqrt{3^2a^2b^{10}}\,\sqrt{2a}$
$\qquad\qquad\quad = 2x^2\sqrt{6x}$ $\qquad\qquad\qquad\qquad\qquad\quad = 2a \cdot 3ab^5\sqrt{2a}$
$\qquad\qquad\qquad\qquad\qquad\qquad\qquad\qquad\qquad\quad = 6a^2b^5\sqrt{2a}$

Problem 4   Simplify.   A. $\sqrt{45b^7}$ B. $3a\sqrt{28a^9b^{18}}$

Solution   See page A39.

Simplify: $\sqrt{25(x + 2)^2}$

Write the prime factorization of the radicand in exponential form.

$$\sqrt{25(x + 2)^2} = \sqrt{5^2(x + 2)^2}$$

Simplify.

$$= 5(x + 2)$$
$$= 5x + 10$$

Example 5   Simplify: $\sqrt{16(x + 5)^2}$

Solution $\sqrt{16(x + 5)^2} = \sqrt{2^4(x + 5)^2} = 2^2(x + 5) = 4(x + 5) = 4x + 20$

Problem 5   Simplify: $\sqrt{25(a + 3)^2}$

Solution   See page A39.

CONCEPT REVIEW 10.1

Determine whether the statement is always true, sometimes true, or never true.

1. The square root of a positive number is a positive number, and the square root of a negative number is a negative number.

2. Every positive number has two square roots, one of which is the additive inverse of the other.

3. The square root of a number that is not a perfect square is an irrational number.

4. If the radicand of a radical expression is evenly divisible by a perfect square greater than 1, then the radical expression is not in simplest form.

5. If a and b are real numbers, then $\sqrt{ab} = \sqrt{a} \cdot \sqrt{b}$.

6. When a perfect square is written in exponential form, the exponents are multiples of 2.

EXERCISES 10.1

1 Simplify.

1. $\sqrt{16}$	**2.** $\sqrt{64}$	**3.** $\sqrt{49}$	**4.** $\sqrt{144}$	**5.** $\sqrt{32}$	**6.** $\sqrt{50}$
7. $\sqrt{8}$	**8.** $\sqrt{12}$	**9.** $6\sqrt{18}$	**10.** $-3\sqrt{48}$	**11.** $5\sqrt{40}$	**12.** $2\sqrt{28}$
13. $\sqrt{15}$	**14.** $\sqrt{21}$	**15.** $\sqrt{29}$	**16.** $\sqrt{13}$	**17.** $-9\sqrt{72}$	**18.** $11\sqrt{80}$
19. $\sqrt{45}$	**20.** $\sqrt{225}$	**21.** $\sqrt{0}$	**22.** $\sqrt{210}$	**23.** $6\sqrt{128}$	**24.** $9\sqrt{288}$
25. $\sqrt{105}$	**26.** $\sqrt{55}$	**27.** $\sqrt{900}$	**28.** $\sqrt{300}$	**29.** $5\sqrt{180}$	**30.** $7\sqrt{98}$
31. $\sqrt{250}$	**32.** $\sqrt{120}$	**33.** $\sqrt{96}$	**34.** $\sqrt{160}$	**35.** $\sqrt{324}$	**36.** $\sqrt{444}$

Find the decimal approximation to the nearest thousandth.

37. $\sqrt{240}$	**38.** $\sqrt{300}$	**39.** $\sqrt{288}$	**40.** $\sqrt{600}$
41. $\sqrt{245}$	**42.** $\sqrt{525}$	**43.** $\sqrt{352}$	**44.** $\sqrt{363}$

2 Simplify.

45. $\sqrt{x^6}$	**46.** $\sqrt{x^{12}}$	**47.** $\sqrt{y^{15}}$	**48.** $\sqrt{y^{11}}$
49. $\sqrt{a^{20}}$	**50.** $\sqrt{a^{16}}$	**51.** $\sqrt{x^4 y^4}$	**52.** $\sqrt{x^{12} y^8}$
53. $\sqrt{4x^4}$	**54.** $\sqrt{25y^8}$	**55.** $\sqrt{24x^2}$	**56.** $\sqrt{x^3 y^{15}}$
57. $\sqrt{x^3 y^7}$	**58.** $\sqrt{a^{15} b^5}$	**59.** $\sqrt{a^3 b^{11}}$	**60.** $\sqrt{24y^7}$
61. $\sqrt{60x^5}$	**62.** $\sqrt{72y^7}$	**63.** $\sqrt{49a^4 b^8}$	**64.** $\sqrt{144x^2 y^8}$

65. $\sqrt{18x^5y^7}$ **66.** $\sqrt{32a^5b^{15}}$ **67.** $\sqrt{40x^{11}y^7}$

68. $\sqrt{72x^9y^3}$ **69.** $\sqrt{80a^9b^{10}}$ **70.** $\sqrt{96a^5b^7}$

71. $2\sqrt{16a^2b^3}$ **72.** $5\sqrt{25a^4b^7}$ **73.** $x\sqrt{x^4y^2}$

74. $y\sqrt{x^3y^6}$ **75.** $4\sqrt{20a^4b^7}$ **76.** $5\sqrt{12a^3b^4}$

77. $3x\sqrt{12x^2y^7}$ **78.** $4y\sqrt{18x^5y^4}$ **79.** $2x^2\sqrt{8x^2y^3}$

80. $3y^2\sqrt{27x^4y^3}$ **81.** $\sqrt{25(a+4)^2}$ **82.** $\sqrt{81(x+y)^4}$

83. $\sqrt{4(x+2)^4}$ **84.** $\sqrt{9(x+2)^2}$ **85.** $\sqrt{x^2+4x+4}$

86. $\sqrt{b^2+8b+16}$ **87.** $\sqrt{y^2+2y+1}$ **88.** $\sqrt{a^2+6a+9}$

SUPPLEMENTAL EXERCISES 10.1

Simplify.

89. $\sqrt{0.0025a^3b^5}$ **90.** $-\frac{3y}{4}\sqrt{64x^4y^2}$ **91.** $\sqrt{x^2y^3+x^3y^2}$ **92.** $\sqrt{4a^5b^4-4a^4b^5}$

93. Given $f(x) = \sqrt{2x-1}$, find each of the following. Write your answer in simplest form.
   a. $f(1)$ **b.** $f(5)$ **c.** $f(14)$

94. Use the roster method to list the whole numbers between $\sqrt{8}$ and $\sqrt{90}$.

95. The area of a square is 76 cm². Find the length of a side of the square. Round to the nearest tenth.

Assuming x can be any real number, for what values of x is the radical expression a real number? Write the answer as an inequality or write "all real numbers."

96. $\sqrt{x}$ **97.** $\sqrt{4x}$ **98.** $\sqrt{x-2}$ **99.** $\sqrt{x+5}$

100. $\sqrt{6-4x}$ **101.** $\sqrt{5-2x}$ **102.** $\sqrt{x^2+7}$ **103.** $\sqrt{x^2+1}$

104. Describe in your own words how to simplify a radical expression.

105. Explain why $2\sqrt{2}$ is in simplest form and $\sqrt{8}$ is not in simplest form.

SECTION **10.2**

Addition and Subtraction of Radical Expressions

1 Add and subtract radical expressions

The Distributive Property is used to simplify the sum or difference of radical expressions with the same radicand.

$$5\sqrt{2} + 3\sqrt{2} = (5 + 3)\sqrt{2} = 8\sqrt{2}$$
$$6\sqrt{2x} - 4\sqrt{2x} = (6 - 4)\sqrt{2x} = 2\sqrt{2x}$$

Radical expressions that are in simplest form and have different radicands cannot be simplified by the Distributive Property.

$2\sqrt{3} + 4\sqrt{2}$ cannot be simplified by the Distributive Property.

To simplify the sum or difference of radical expressions, simplify each term. Then use the Distributive Property.

Subtract: $4\sqrt{8} - 10\sqrt{2}$

Simplify each term.

$$\begin{aligned}
4\sqrt{8} - 10\sqrt{2} &= 4\sqrt{2^3} - 10\sqrt{2} \\
&= 4\sqrt{2^2 \cdot 2} - 10\sqrt{2} \\
&= 4\sqrt{2^2}\sqrt{2} - 10\sqrt{2} \\
&= 4 \cdot 2\sqrt{2} - 10\sqrt{2} \\
&= 8\sqrt{2} - 10\sqrt{2}
\end{aligned}$$

Subtract by using the Distributive Property.
$$\begin{aligned}
&= (8 - 10)\sqrt{2} \\
&= -2\sqrt{2}
\end{aligned}$$

Example 1   Simplify.   A. $5\sqrt{2} - 3\sqrt{2} + 12\sqrt{2}$ B. $3\sqrt{12} - 5\sqrt{27}$

Solution   A. $5\sqrt{2} - 3\sqrt{2} + 12\sqrt{2}$
$$\begin{aligned}
&= (5 - 3 + 12)\sqrt{2} \quad \blacktriangleright \text{Use the Distributive Property.} \\
&= 14\sqrt{2}
\end{aligned}$$

B. $\begin{aligned}[t] 3\sqrt{12} - 5\sqrt{27} &= 3\sqrt{2^2 \cdot 3} - 5\sqrt{3^3} \\
&= 3\sqrt{2^2 \cdot 3} - 5\sqrt{3^2 \cdot 3} \\
&= 3\sqrt{2^2}\sqrt{3} - 5\sqrt{3^2}\sqrt{3} \\
&= 3 \cdot 2\sqrt{3} - 5 \cdot 3\sqrt{3} \\
&= 6\sqrt{3} - 15\sqrt{3} \\
&= -9\sqrt{3}
\end{aligned}$

Problem 1   Simplify.   A. $9\sqrt{3} + 3\sqrt{3} - 18\sqrt{3}$ B. $2\sqrt{50} - 5\sqrt{32}$

Solution   See page A40.

To subtract $8\sqrt{18x} - 2\sqrt{32x}$, simplify each term. Then subtract the radical expressions.

$$
\begin{aligned}
8\sqrt{18x} - 2\sqrt{32x} &= 8\sqrt{2 \cdot 3^2 x} - 2\sqrt{2^5 x} \\
&= 8\sqrt{3^2}\sqrt{2x} - 2\sqrt{2^4}\sqrt{2x} \\
&= 8 \cdot 3\sqrt{2x} - 2 \cdot 2^2\sqrt{2x} \\
&= 24\sqrt{2x} - 8\sqrt{2x} \\
&= 16\sqrt{2x}
\end{aligned}
$$

Example 2   Simplify.
A. $3\sqrt{12x^3} - 2x\sqrt{3x}$ B. $2x\sqrt{8y} - 3\sqrt{2x^2 y} + 2\sqrt{32x^2 y}$

Solution   A. $3\sqrt{12x^3} - 2x\sqrt{3x} = 3\sqrt{2^2 \cdot 3 \cdot x^3} - 2x\sqrt{3x}$
$$
\begin{aligned}
&= 3\sqrt{2^2 \cdot x^2}\sqrt{3x} - 2x\sqrt{3x} \\
&= 3 \cdot 2x\sqrt{3x} - 2x\sqrt{3x} \\
&= 6x\sqrt{3x} - 2x\sqrt{3x} \\
&= 4x\sqrt{3x}
\end{aligned}
$$

B. $2x\sqrt{8y} - 3\sqrt{2x^2 y} + 2\sqrt{32x^2 y}$
$$
\begin{aligned}
&= 2x\sqrt{2^3 y} - 3\sqrt{2x^2 y} + 2\sqrt{2^5 x^2 y} \\
&= 2x\sqrt{2^2}\sqrt{2y} - 3\sqrt{x^2}\sqrt{2y} + 2\sqrt{2^4 x^2}\sqrt{2y} \\
&= 2x \cdot 2\sqrt{2y} - 3 \cdot x\sqrt{2y} + 2 \cdot 2^2 x\sqrt{2y} \\
&= 4x\sqrt{2y} - 3x\sqrt{2y} + 8x\sqrt{2y} \\
&= 9x\sqrt{2y}
\end{aligned}
$$

Problem 2   Simplify.
A. $y\sqrt{28y} + 7\sqrt{63y^3}$ B. $2\sqrt{27a^5} - 4a\sqrt{12a^3} + a^2\sqrt{75a}$

Solution   See page A40.

CONCEPT REVIEW 10.2

Determine whether the statement is always true, sometimes true, or never true.

1. $5\sqrt{2} + 6\sqrt{3} = 11\sqrt{5}$

2. The Distributive Property is used to add or subtract radical expressions with the same radicand.

3. $4\sqrt{7}$ and $\sqrt{7}$ are examples of like radical expressions.

4. The expressions $3\sqrt{12}$ and $5\sqrt{3}$ cannot be subtracted until $3\sqrt{12}$ is rewritten in simplest form.

5. $5\sqrt{32x} + 4\sqrt{18x} = 5(4\sqrt{2x}) + 4(3\sqrt{2x}) = 20\sqrt{2x} + 12\sqrt{2x} = 32\sqrt{2x}$

6. The expressions $9\sqrt{11}$ and $10\sqrt{13}$ can be added after each radical is written in simplest form.

EXERCISES 10.2

1   Simplify.

1. $2\sqrt{2} + \sqrt{2}$

2. $3\sqrt{5} + 8\sqrt{5}$

3. $-3\sqrt{7} + 2\sqrt{7}$

4. $4\sqrt{5} - 10\sqrt{5}$

5. $-3\sqrt{11} - 8\sqrt{11}$

6. $-3\sqrt{3} - 5\sqrt{3}$

7. $2\sqrt{x} + 8\sqrt{x}$

8. $3\sqrt{y} + 2\sqrt{y}$

9. $8\sqrt{y} - 10\sqrt{y}$

10. $-5\sqrt{2a} + 2\sqrt{2a}$

11. $-2\sqrt{3b} - 9\sqrt{3b}$

12. $-7\sqrt{5a} - 5\sqrt{5a}$

13. $3x\sqrt{2} - x\sqrt{2}$

14. $2y\sqrt{3} - 9y\sqrt{3}$

15. $2a\sqrt{3a} - 5a\sqrt{3a}$

16. $-5b\sqrt{3x} - 2b\sqrt{3x}$

17. $3\sqrt{xy} - 8\sqrt{xy}$

18. $-4\sqrt{xy} + 6\sqrt{xy}$

19. $\sqrt{45} + \sqrt{125}$

20. $\sqrt{32} - \sqrt{98}$

21. $2\sqrt{2} + 3\sqrt{8}$

22. $4\sqrt{128} - 3\sqrt{32}$

23. $5\sqrt{18} - 2\sqrt{75}$

24. $5\sqrt{75} - 2\sqrt{18}$

25. $5\sqrt{4x} - 3\sqrt{9x}$

26. $-3\sqrt{25y} + 8\sqrt{49y}$

27. $3\sqrt{3x^2} - 5\sqrt{27x^2}$

28. $-2\sqrt{8y^2} + 5\sqrt{32y^2}$

29. $2x\sqrt{xy^2} - 3y\sqrt{x^2y}$

30. $4a\sqrt{b^2a} - 3b\sqrt{a^2b}$

31. $3x\sqrt{12x} - 5\sqrt{27x^3}$

32. $2a\sqrt{50a} + 7\sqrt{32a^3}$

33. $4y\sqrt{8y^3} - 7\sqrt{18y^5}$

34. $2a\sqrt{8ab^2} - 2b\sqrt{2a^3}$

35. $b^2\sqrt{a^5b} + 3a^2\sqrt{ab^5}$

36. $y^2\sqrt{x^5y} + x\sqrt{x^3y^5}$

37. $4\sqrt{2} - 5\sqrt{2} + 8\sqrt{2}$

38. $3\sqrt{3} + 8\sqrt{3} - 16\sqrt{3}$

39. $5\sqrt{x} - 8\sqrt{x} + 9\sqrt{x}$

40. $\sqrt{x} - 7\sqrt{x} + 6\sqrt{x}$

41. $8\sqrt{2} - 3\sqrt{y} - 8\sqrt{2}$

42. $8\sqrt{3} - 5\sqrt{2} - 5\sqrt{3}$

43. $8\sqrt{8} - 4\sqrt{32} - 9\sqrt{50}$

44. $2\sqrt{12} - 4\sqrt{27} + \sqrt{75}$

45. $-2\sqrt{3} + 5\sqrt{27} - 4\sqrt{45}$

46. $-2\sqrt{8} - 3\sqrt{27} + 3\sqrt{50}$

47. $4\sqrt{75} + 3\sqrt{48} - \sqrt{99}$

48. $2\sqrt{75} - 5\sqrt{20} + 2\sqrt{45}$

49. $\sqrt{25x} - \sqrt{9x} + \sqrt{16x}$

50. $\sqrt{4x} - \sqrt{100x} - \sqrt{49x}$

51. $3\sqrt{3x} + \sqrt{27x} - 8\sqrt{75x}$

52. $5\sqrt{5x} + 2\sqrt{45x} - 3\sqrt{80x}$

53. $2a\sqrt{75b} - a\sqrt{20b} + 4a\sqrt{45b}$

54. $2b\sqrt{75a} - 5b\sqrt{27a} + 2b\sqrt{20a}$

55. $x\sqrt{3y^2} - 2y\sqrt{12x^2} + xy\sqrt{3}$

56. $a\sqrt{27b^2} + 3b\sqrt{147a^2} - ab\sqrt{3}$

57. $3\sqrt{ab^3} + 4a\sqrt{a^2b} - 5b\sqrt{4ab}$

58. $5\sqrt{a^3b} + a\sqrt{4ab} - 3\sqrt{49a^3b}$

59. $3a\sqrt{2ab^2} - \sqrt{a^2b^2} + 4b\sqrt{3a^2b}$

60. $2\sqrt{4a^2b^2} - 3a\sqrt{9ab^2} + 4b\sqrt{a^2b}$

SUPPLEMENTAL EXERCISES 10.2

Add or subtract.

61. $5\sqrt{x + 2} + 3\sqrt{x + 2}$

62. $8\sqrt{a + 5} - 4\sqrt{a + 5}$

63. $\dfrac{1}{2}\sqrt{8x^2y} + \dfrac{1}{3}\sqrt{18x^2y}$

64. $\dfrac{1}{4}\sqrt{48ab^2} + \dfrac{1}{5}\sqrt{75ab^2}$

65. $\dfrac{a}{3}\sqrt{54ab^3} + \dfrac{b}{4}\sqrt{96a^3b}$

66. $\dfrac{x}{6}\sqrt{72xy^5} + \dfrac{y}{7}\sqrt{98x^3y^3}$

67. $2\sqrt{8x + 4y} - 5\sqrt{18x + 9y}$

68. $3\sqrt{a^3 + a^2} + 5\sqrt{4a^3 + 4a^2}$

Solve.

69. The lengths of the sides of a triangle are $4\sqrt{3}$ cm, $2\sqrt{3}$ cm, and $2\sqrt{15}$ cm. Find the perimeter of the triangle.

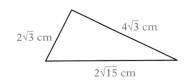

70. The length of a rectangle is $3\sqrt{2}$ cm. The width is $\sqrt{2}$ cm. Find the perimeter of the rectangle.

71. The length of a rectangle is $4\sqrt{5}$ cm. The width is $\sqrt{5}$ cm. Find the decimal approximation of the perimeter. Round to the nearest tenth.

72. Given $G(x) = \sqrt{x + 5} + \sqrt{5x + 3}$, write $G(3)$ in simplest form.

73. Is the equation $\sqrt{a^2 + b^2} = \sqrt{a} + \sqrt{b}$ true for all real numbers a and b?

74. Use complete sentences to explain the steps in simplifying the radical expression $4\sqrt{2a^3b} + 5\sqrt{5a^3b}$.

S E C T I O N **10.3**

Multiplication and Division of Radical Expressions

1 ## Multiply radical expressions

The Product Property of Square Roots is used to multiply variable radical expressions.

$$\sqrt{2x}\,\sqrt{3y} = \sqrt{2x \cdot 3y}$$
$$= \sqrt{6xy}$$

Multiply: $\sqrt{2x^2}\,\sqrt{32x^5}$

Use the Product Property of Square Roots.

$$\sqrt{2x^2}\,\sqrt{32x^5} = \sqrt{2x^2 \cdot 32x^5}$$

Multiply the radicands.

$$= \sqrt{64x^7}$$

Simplify.

$$= \sqrt{2^6 x^7}$$
$$= \sqrt{2^6 x^6}\,\sqrt{x}$$
$$= 2^3 x^3 \sqrt{x}$$
$$= 8x^3 \sqrt{x}$$

Example 1   Multiply: $\sqrt{3x^4}\,\sqrt{2x^2 y}\,\sqrt{6xy^2}$

Solution $\sqrt{3x^4}\,\sqrt{2x^2 y}\,\sqrt{6xy^2} = \sqrt{3x^4 \cdot 2x^2 y \cdot 6xy^2}$
$$= \sqrt{36x^7 y^3}$$
$$= \sqrt{2^2 3^2 x^7 y^3}$$
$$= \sqrt{2^2 3^2 x^6 y^2}\,\sqrt{xy}$$
$$= 2 \cdot 3x^3 y \sqrt{xy}$$
$$= 6x^3 y \sqrt{xy}$$

Problem 1   Multiply: $\sqrt{5a}\,\sqrt{15a^3 b^4}\,\sqrt{3b^5}$

Solution   See page A40.

When the expression $(\sqrt{x})^2$ is simplified by using the Product Property of Square Roots, the result is x.

$$(\sqrt{x})^2 = \sqrt{x}\,\sqrt{x}$$
$$= \sqrt{x \cdot x}$$
$$= \sqrt{x^2}$$
$$= x$$

For $a > 0$, $(\sqrt{a})^2 = \sqrt{a^2} = a$.

Multiply: $\sqrt{2x}(x + \sqrt{2x})$

Use the Distributive Property to remove parentheses.

$$\sqrt{2x}(x + \sqrt{2x}) = \sqrt{2x}(x) + \sqrt{2x}\,\sqrt{2x}$$
$$= x\sqrt{2x} + (\sqrt{2x})^2$$
$$= x\sqrt{2x} + 2x$$

Example 2   Multiply: $\sqrt{3ab}(\sqrt{3a} + \sqrt{9b})$

Solution $\sqrt{3ab}(\sqrt{3a} + \sqrt{9b}) = \sqrt{3ab}(\sqrt{3a}) + \sqrt{3ab}(\sqrt{9b})$
$$= \sqrt{3^2a^2b} + \sqrt{3^3ab^2}$$
$$= \sqrt{3^2a^2}\sqrt{b} + \sqrt{3^2b^2}\sqrt{3a}$$
$$= 3a\sqrt{b} + 3b\sqrt{3a}$$

Problem 2   Multiply: $\sqrt{5x}(\sqrt{5x} - \sqrt{25y})$

Solution   See page A40.

To multiply $(\sqrt{2} - 3x)(\sqrt{2} + x)$, use the FOIL method to multiply the two binomials.

$$(\sqrt{2} - 3x)(\sqrt{2} + x) = (\sqrt{2})^2 + x\sqrt{2} - 3x\sqrt{2} - 3x^2$$
$$= 2 + (x - 3x)\sqrt{2} - 3x^2$$
$$= 2 - 2x\sqrt{2} - 3x^2$$

Example 3   Multiply: $(2\sqrt{x} - \sqrt{y})(5\sqrt{x} - 2\sqrt{y})$

Solution $(2\sqrt{x} - \sqrt{y})(5\sqrt{x} - 2\sqrt{y})$
$$= 10(\sqrt{x})^2 - 4\sqrt{xy} - 5\sqrt{xy} + 2(\sqrt{y})^2$$
$$= 10x - 9\sqrt{xy} + 2y$$

Problem 3   Multiply: $(3\sqrt{x} - \sqrt{y})(5\sqrt{x} - 2\sqrt{y})$

Solution   See page A40.

The expressions $a + b$ and $a - b$, which are the sum and difference of two terms, are called **conjugates** of each other.

The product of conjugates is the difference of two squares.

$$(a + b)(a - b) = a^2 - b^2$$
$$(2 + \sqrt{7})(2 - \sqrt{7}) = 2^2 - (\sqrt{7})^2 = 4 - 7 = -3$$
$$(3 + \sqrt{y})(3 - \sqrt{y}) = 3^2 - (\sqrt{y})^2 = 9 - y$$

Example 4   Multiply: $(\sqrt{a} - \sqrt{b})(\sqrt{a} + \sqrt{b})$

Solution $(\sqrt{a} - \sqrt{b})(\sqrt{a} + \sqrt{b}) = (\sqrt{a})^2 - (\sqrt{b})^2 = a - b$

Problem 4   Multiply: $(2\sqrt{x} + 7)(2\sqrt{x} - 7)$

Solution   See page A40.

2   Divide radical expressions

The square root of a quotient is equal to the quotient of the square roots.

The Quotient Property of Square Roots

If a and b are positive real numbers, then $\sqrt{\dfrac{a}{b}} = \dfrac{\sqrt{a}}{\sqrt{b}}$.

POINT OF INTEREST

A radical expression that occurs in Einstein's Theory of Relativity is

$$\frac{1}{\sqrt{1 - \dfrac{v^2}{c^2}}}$$

where v is the velocity of an object and c is the speed of light.

To simplify $\sqrt{\dfrac{4x^2}{z^6}}$, rewrite the radical expression as the quotient of the square roots.

Simplify.

$$\sqrt{\frac{4x^2}{z^6}} = \frac{\sqrt{4x^2}}{\sqrt{z^6}} = \frac{\sqrt{2^2x^2}}{\sqrt{z^6}}$$

$$= \frac{2x}{z^3}$$

To simplify $\sqrt{\dfrac{24x^3y^7}{3x^7y^2}}$, simplify the radicand.

Rewrite the radical expression as the quotient of the square roots.

Simplify.

$$\sqrt{\frac{24x^3y^7}{3x^7y^2}} = \sqrt{\frac{8y^5}{x^4}}$$

$$= \frac{\sqrt{8y^5}}{\sqrt{x^4}}$$

$$= \frac{\sqrt{2^3y^5}}{\sqrt{x^4}}$$

$$= \frac{\sqrt{2^2y^4}\sqrt{2y}}{\sqrt{x^4}}$$

$$= \frac{2y^2\sqrt{2y}}{x^2}$$

To simplify $\dfrac{\sqrt{4x^2y}}{\sqrt{xy}}$, use the Quotient Property of Square Roots.

Simplify the radicand.

Simplify the radical expression.

$$\frac{\sqrt{4x^2y}}{\sqrt{xy}} = \sqrt{\frac{4x^2y}{xy}}$$

$$= \sqrt{4x}$$

$$= \sqrt{2^2}\sqrt{x}$$

$$= 2\sqrt{x}$$

A radical expression is not in simplest form if a radical remains in the denominator. The procedure used to remove a radical from the denominator is called **rationalizing the denominator.**

The expression $\dfrac{2}{\sqrt{3}}$ has a radical expression in the denominator. Multiply the expression by $\dfrac{\sqrt{3}}{\sqrt{3}}$, which equals 1.

Simplify.

$$\frac{2}{\sqrt{3}} = \frac{2}{\sqrt{3}} \cdot \frac{\sqrt{3}}{\sqrt{3}}$$

$$= \frac{2\sqrt{3}}{(\sqrt{3})^2}$$

$$= \frac{2\sqrt{3}}{3}$$

Thus $\dfrac{2}{\sqrt{3}} = \dfrac{2\sqrt{3}}{3}$, but $\dfrac{2}{\sqrt{3}}$ is not in simplest form. $\dfrac{2\sqrt{3}}{3}$ is in simplest form, because no radical remains in the denominator and the radical in the numerator contains no perfect-square factors other than 1.

When the denominator contains a radical expression with two terms, simplify the radical expression by multiplying the numerator and denominator by the conjugate of the denominator.

Simplify: $\dfrac{\sqrt{2y}}{\sqrt{y} + 3}$

Multiply the numerator and denominator by $\sqrt{y} - 3$, the conjugate of $\sqrt{y} + 3$.

$$\dfrac{\sqrt{2y}}{\sqrt{y} + 3} = \dfrac{\sqrt{2y}}{\sqrt{y} + 3} \cdot \dfrac{\sqrt{y} - 3}{\sqrt{y} - 3}$$

Simplify.

$$= \dfrac{\sqrt{2y^2} - 3\sqrt{2y}}{(\sqrt{y})^2 - 3^2}$$

$$= \dfrac{y\sqrt{2} - 3\sqrt{2y}}{y - 9}$$

The following list summarizes the discussions about radical expressions in simplest form.

Radical Expressions in Simplest Form

A radical expression is in simplest form if:

1. The radicand contains no factor greater than 1 that is a perfect square.
2. There is no fraction under the radical sign.
3. There is no radical in the denominator of a fraction.

Example 5   Simplify.   A. $\dfrac{\sqrt{4x^2y^5}}{\sqrt{3x^4y}}$ B. $\dfrac{\sqrt{2}}{\sqrt{2} - \sqrt{x}}$ C. $\dfrac{3 - \sqrt{5}}{2 + 3\sqrt{5}}$

Solution   A. $\dfrac{\sqrt{4x^2y^5}}{\sqrt{3x^4y}} = \sqrt{\dfrac{2^2x^2y^5}{3x^4y}} = \sqrt{\dfrac{2^2y^4}{3x^2}} = \dfrac{2y^2}{x\sqrt{3}} = \dfrac{2y^2}{x\sqrt{3}} \cdot \dfrac{\sqrt{3}}{\sqrt{3}} = \dfrac{2y^2\sqrt{3}}{3x}$

B. $\dfrac{\sqrt{2}}{\sqrt{2} - \sqrt{x}} = \dfrac{\sqrt{2}}{\sqrt{2} - \sqrt{x}} \cdot \dfrac{\sqrt{2} + \sqrt{x}}{\sqrt{2} + \sqrt{x}} = \dfrac{2 + \sqrt{2x}}{2 - x}$

C. $\dfrac{3 - \sqrt{5}}{2 + 3\sqrt{5}} = \dfrac{3 - \sqrt{5}}{2 + 3\sqrt{5}} \cdot \dfrac{2 - 3\sqrt{5}}{2 - 3\sqrt{5}}$

$\qquad = \dfrac{6 - 9\sqrt{5} - 2\sqrt{5} + 3 \cdot 5}{4 - 9 \cdot 5}$

$\qquad = \dfrac{21 - 11\sqrt{5}}{-41}$

$\qquad = -\dfrac{21 - 11\sqrt{5}}{41}$

Problem 5   Simplify.   A. $\sqrt{\dfrac{15x^6y^7}{3x^7y^9}}$ B. $\dfrac{\sqrt{y}}{\sqrt{y} + 3}$ C. $\dfrac{5 + \sqrt{y}}{1 - 2\sqrt{y}}$

Solution   See page A40.

CONCEPT REVIEW 10.3

Determine whether the statement is always true, sometimes true, or never true.

1. By the Product Property of Square Roots, if $a > 0$ and $b > 0$, then $\sqrt{a} \cdot \sqrt{b} = \sqrt{ab}$.

2. When we square a square root, the result is the radicand.

3. The procedure for rationalizing the denominator is used when a fraction has a radical expression in the denominator.

4. The square root of a fraction is equal to the square root of the numerator over the square root of the denominator.

5. A radical expression is in simplest form if the radicand contains no factor other than 1 that is a perfect square.

6. The conjugate of $5 - \sqrt{3}$ is $\sqrt{3} - 5$.

EXERCISES 10.3

1   Multiply.

1. $\sqrt{5}\sqrt{5}$

2. $\sqrt{11}\sqrt{11}$

3. $\sqrt{3}\sqrt{12}$

4. $\sqrt{2}\sqrt{8}$

5. $\sqrt{x}\sqrt{x}$

6. $\sqrt{y}\sqrt{y}$

7. $\sqrt{xy^3}\sqrt{x^5y}$

8. $\sqrt{a^3b^5}\sqrt{ab^5}$

9. $\sqrt{3a^2b^5}\sqrt{6ab^7}$

10. $\sqrt{5x^3y}\sqrt{10x^2y}$

11. $\sqrt{6a^3b^2}\sqrt{24a^5b}$

12. $\sqrt{8ab^5}\sqrt{12a^7b}$

13. $\sqrt{2}(\sqrt{2} - \sqrt{3})$

14. $3(\sqrt{12} - \sqrt{3})$

15. $\sqrt{x}(\sqrt{x} - \sqrt{y})$

16. $\sqrt{b}(\sqrt{a} - \sqrt{b})$

17. $\sqrt{5}(\sqrt{10} - \sqrt{x})$

18. $\sqrt{6}(\sqrt{y} - \sqrt{18})$

19. $\sqrt{8}(\sqrt{2} - \sqrt{5})$

20. $\sqrt{10}(\sqrt{20} - \sqrt{a})$

21. $(\sqrt{x} - 3)^2$

22. $(2\sqrt{a} - y)^2$

23. $\sqrt{3a}(\sqrt{3a} - \sqrt{3b})$

24. $\sqrt{5x}(\sqrt{10x} - \sqrt{x})$

25. $\sqrt{2ac}\sqrt{5ab}\sqrt{10cb}$

26. $\sqrt{3xy}\sqrt{6x^3y}\sqrt{2y^2}$

27. $(3\sqrt{x} - 2y)(5\sqrt{x} - 4y)$

28. $(5\sqrt{x} + 2\sqrt{y})(3\sqrt{x} - \sqrt{y})$

29. $(\sqrt{x} - \sqrt{y})(\sqrt{x} + \sqrt{y})$

30. $(\sqrt{3x} + y)(\sqrt{3x} - y)$

31. $(2\sqrt{x} + \sqrt{y})(5\sqrt{x} + 4\sqrt{y})$

32. $(5\sqrt{x} - 2\sqrt{y})(3\sqrt{x} - 4\sqrt{y})$

2   Simplify.

33. $\dfrac{\sqrt{32}}{\sqrt{2}}$

34. $\dfrac{\sqrt{45}}{\sqrt{5}}$

35. $\dfrac{\sqrt{98}}{\sqrt{2}}$

36. $\dfrac{\sqrt{48}}{\sqrt{3}}$

37. $\dfrac{\sqrt{27a}}{\sqrt{3a}}$

38. $\dfrac{\sqrt{72x^5}}{\sqrt{2x}}$

39. $\dfrac{\sqrt{15x^3y}}{\sqrt{3xy}}$

40. $\dfrac{\sqrt{40x^5y^2}}{\sqrt{5xy}}$

41. $\dfrac{\sqrt{2a^5b^4}}{\sqrt{98ab^4}}$

42. $\dfrac{\sqrt{48x^5y^2}}{\sqrt{3x^3y}}$

43. $\dfrac{1}{\sqrt{3}}$

44. $\dfrac{1}{\sqrt{8}}$

45. $\dfrac{15}{\sqrt{75}}$

46. $\dfrac{6}{\sqrt{72}}$

47. $\dfrac{6}{\sqrt{12x}}$

48. $\dfrac{14}{\sqrt{7y}}$

49. $\dfrac{8}{\sqrt{32x}}$

50. $\dfrac{15}{\sqrt{50x}}$

51. $\dfrac{3}{\sqrt{x}}$

52. $\dfrac{4}{\sqrt{2x}}$

53. $\dfrac{\sqrt{8x^2y}}{\sqrt{2x^4y^2}}$

54. $\dfrac{\sqrt{4x^2}}{\sqrt{9y}}$

55. $\dfrac{\sqrt{16a}}{\sqrt{49ab}}$

56. $\dfrac{5\sqrt{8}}{4\sqrt{50}}$

57. $\dfrac{5\sqrt{18}}{9\sqrt{27}}$

58. $\dfrac{\sqrt{12a^3b}}{\sqrt{24a^2b^2}}$

59. $\dfrac{\sqrt{3xy}}{\sqrt{27x^3y^2}}$

60. $\dfrac{\sqrt{9xy^2}}{\sqrt{27x}}$

61. $\dfrac{\sqrt{4x^2y}}{\sqrt{3xy^3}}$

62. $\dfrac{\sqrt{16x^3y^2}}{\sqrt{8x^3y}}$

63. $\dfrac{1}{\sqrt{2}-3}$

64. $\dfrac{5}{\sqrt{7}-3}$

65. $\dfrac{3}{5+\sqrt{5}}$

66. $\dfrac{7}{\sqrt{2}-7}$

67. $\dfrac{\sqrt{xy}}{\sqrt{x}-\sqrt{y}}$

68. $\dfrac{\sqrt{x}}{\sqrt{x}-\sqrt{y}}$

69. $\dfrac{5\sqrt{x^2y}}{\sqrt{75xy^2}}$

70. $\dfrac{3\sqrt{ab}}{a\sqrt{6b}}$

71. $\dfrac{\sqrt{2}}{\sqrt{2}-\sqrt{3}}$

72. $\dfrac{1+\sqrt{2}}{1-\sqrt{2}}$

73. $\dfrac{\sqrt{5}}{\sqrt{2}-\sqrt{5}}$

74. $\dfrac{\sqrt{6}}{\sqrt{3}-\sqrt{2}}$

75. $\dfrac{\sqrt{x}}{\sqrt{x}+3}$

76. $\dfrac{\sqrt{y}}{2-\sqrt{y}}$

77. $\dfrac{5\sqrt{3}-7\sqrt{3}}{4\sqrt{3}}$

78. $\dfrac{10\sqrt{7}-2\sqrt{7}}{2\sqrt{7}}$

79. $\dfrac{5\sqrt{8}-3\sqrt{2}}{\sqrt{2}}$

80. $\dfrac{5\sqrt{12}-\sqrt{3}}{\sqrt{27}}$

81. $\dfrac{3\sqrt{2}-8\sqrt{2}}{\sqrt{2}}$

82. $\dfrac{5\sqrt{3}-2\sqrt{3}}{2\sqrt{3}}$

83. $\dfrac{2\sqrt{8}+3\sqrt{2}}{\sqrt{32}}$

84. $\dfrac{3-\sqrt{6}}{5-2\sqrt{6}}$

85. $\dfrac{6-2\sqrt{3}}{4+3\sqrt{3}}$

86. $\dfrac{\sqrt{2}+2\sqrt{6}}{2\sqrt{2}-3\sqrt{6}}$

87. $\dfrac{2\sqrt{3}-\sqrt{6}}{5\sqrt{3}+2\sqrt{6}}$

88. $\dfrac{3+\sqrt{x}}{2-\sqrt{x}}$

89. $\dfrac{\sqrt{a}-4}{2\sqrt{a}+2}$

90. $\dfrac{3+2\sqrt{y}}{2-\sqrt{y}}$

91. $\dfrac{2+\sqrt{y}}{\sqrt{y}-3}$

92. $\dfrac{\sqrt{x}+\sqrt{y}}{\sqrt{x}-\sqrt{y}}$

SUPPLEMENTAL EXERCISES 10.3

Simplify.

93. $-\sqrt{1.3}\,\sqrt{1.3}$

94. $\sqrt{\dfrac{5}{8}}\,\sqrt{\dfrac{5}{8}}$

95. $-\sqrt{\dfrac{16}{81}}$

96. $\sqrt{1\dfrac{9}{16}}$

97. $\sqrt{2\dfrac{1}{4}}$

98. $-\sqrt{6\dfrac{1}{4}}$

99. Answer true or false. If the equation is false, correct the right side.
   a. $(\sqrt{y})^4 = y^2$ **b.** $(2\sqrt{x})^3 = 8x\sqrt{x}$
   c. $(\sqrt{x} + 1)^2 = x + 1$ **d.** $\dfrac{1}{2 - \sqrt{3}} = 2 + \sqrt{3}$

Solve.

100. Show that 2 is a solution of the equation $\sqrt{x + 2} + \sqrt{x - 1} = 3$.

101. Is 16 a solution of the equation $\sqrt{x} - \sqrt{x + 9} = 1$?

102. Show that $(1 + \sqrt{6})$ and $(1 - \sqrt{6})$ are solutions of the equation $x^2 - 2x - 5 = 0$.

103. In your own words, describe the process of rationalizing the denominator.

104. The number $\dfrac{\sqrt{5} + 1}{2}$ is called the golden ratio. Research the golden ratio and write a few paragraphs about this number and its applications.

SECTION **10.4**

Solving Equations Containing Radical Expressions

1 Solve equations containing one or more radical expressions

An equation that contains a variable expression in a radicand is a **radical equation.**

$$\left.\begin{array}{l}\sqrt{x} = 4 \\[4pt] \sqrt{x + 2} = \sqrt{x - 7}\end{array}\right\} \begin{array}{l}\text{Radical} \\ \text{equations}\end{array}$$

The following property of equality states that if two numbers are equal, then the squares of the numbers are equal. This property is used to solve radical equations.

Property of Squaring Both Sides of an Equation

If a and b are real numbers and $a = b$, then $a^2 = b^2$.

Solve: $\sqrt{x - 2} - 7 = 0$

Rewrite the equation with the radical on one side of the equation and the constant on the other side.

$$\sqrt{x - 2} - 7 = 0$$
$$\sqrt{x - 2} = 7$$

Square both sides of the equation.

$$(\sqrt{x - 2})^2 = 7^2$$

Solve the resulting equation.

$$x - 2 = 49$$
$$x = 51$$

LOOK CLOSELY

Any time each side of an equation is squared, you must check the proposed solution of the equation.

Check the solution. When both sides of an equation are squared, the resulting equation may have a solution that is not a solution of the original equation.

Check:
$$\begin{array}{c|c} \sqrt{x - 2} - 7 = 0 & \\ \hline \sqrt{51 - 2} - 7 & 0 \\ \sqrt{49} - 7 & 0 \\ \sqrt{7^2} - 7 & 0 \\ 7 - 7 & 0 \\ 0 = 0 & \end{array}$$

The solution is 51.

Example 1   Solve: $\sqrt{3x} + 2 = 5$

Solution $\sqrt{3x} + 2 = 5$
$\sqrt{3x} = 3$ ▸ Rewrite the equation so that the radical is alone on one side of the equation.
$(\sqrt{3x})^2 = 3^2$ ▸ Square both sides of the equation.
$3x = 9$ ▸ Solve for x.
$x = 3$

Check
$$\begin{array}{c|c} \sqrt{3x} + 2 = 5 & \\ \hline \sqrt{3 \cdot 3} + 2 & 5 \\ \sqrt{3^2} + 2 & 5 \\ 3 + 2 & 5 \\ 5 = 5 & \end{array}$$
▸ Both sides of the equation were squared. The solution must be checked.

▸ This is a true equation. The solution checks.

The solution is 3.

Problem 1   Solve: $\sqrt{4x} + 3 = 7$

Solution   See page A40.

Example 2   Solve.   A. $0 = 3 - \sqrt{2x - 3}$ B. $\sqrt{2x - 5} + 3 = 0$

Solution   A. $0 = 3 - \sqrt{2x - 3}$
$\sqrt{2x - 3} = 3$ ▸ Rewrite the equation so that the radical is alone on one side of the equation.
$(\sqrt{2x - 3})^2 = 3^2$ ▸ Square both sides of the equation.
$2x - 3 = 9$
$2x = 12$
$x = 6$

Check $0 = 3 - \sqrt{2x - 3}$

$$\begin{array}{c|l}
0 & 3 - \sqrt{2 \cdot 6 - 3} \\
0 & 3 - \sqrt{12 - 3} \\
0 & 3 - \sqrt{9} \\
0 & 3 - \sqrt{3^2} \\
0 & 3 - 3 \\
\end{array}$$

$0 = 0$ ▶ This is a true equation. The solution checks.

The solution is 6.

B. $\sqrt{2x - 5} + 3 = 0$

$\sqrt{2x - 5} = -3$ ▶ Rewrite the equation so that the radical is alone on one side of the equation.

$(\sqrt{2x - 5})^2 = (-3)^2$ ▶ Square each side of the equation.

$2x - 5 = 9$ ▶ Solve for x.

$2x = 14$

$x = 7$

Check $\sqrt{2x - 5} + 3 = 0$

$$\begin{array}{l|c}
\sqrt{2 \cdot 7 - 5} + 3 & 0 \\
\sqrt{14 - 5} + 3 & 0 \\
\sqrt{9} + 3 & 0 \\
3 + 3 & 0 \\
\end{array}$$

$6 \neq 0$ ▶ This is not a true equation. The solution does not check.

There is no solution.

Problem 2   Solve.   A. $\sqrt{3x - 2} - 5 = 0$ B. $\sqrt{4x - 7} + 5 = 0$

Solution   See page A40.

The following example illustrates the procedure for solving a radical equation containing two radical expressions. Note that the process of squaring both sides of the equation is performed twice.

Solve: $\sqrt{5 + x} + \sqrt{x} = 5$

Solve for one of the radical expressions. $\sqrt{5 + x} + \sqrt{x} = 5$

$\sqrt{5 + x} = 5 - \sqrt{x}$

Square each side. $(\sqrt{5 + x})^2 = (5 - \sqrt{x})^2$

Recall that $(a - b)^2 = a^2 - 2ab + b^2$. $5 + x = 25 - 10\sqrt{x} + x$

Simplify. $-20 = -10\sqrt{x}$

This is still a radical equation. $2 = \sqrt{x}$

Square each side. $2^2 = (\sqrt{x})^2$

$4 = x$

4 checks as the solution.   The solution is 4.

Example 3   Solve: $\sqrt{x} - \sqrt{x-5} = 1$

Solution
$$\sqrt{x} - \sqrt{x-5} = 1$$
$$\sqrt{x} = 1 + \sqrt{x-5}$$ ▶ Solve for one of the radical expressions.
$$(\sqrt{x})^2 = (1 + \sqrt{x-5})^2$$ ▶ Square each side.
$$x = 1 + 2\sqrt{x-5} + (x-5)$$ ▶ Simplify.
$$4 = 2\sqrt{x-5}$$
$$2 = \sqrt{x-5}$$ ▶ This is a radical equation.
$$2^2 = (\sqrt{x-5})^2$$ ▶ Square each side.
$$4 = x - 5$$ ▶ Simplify.
$$9 = x$$

Check
$$\sqrt{x} - \sqrt{x-5} = 1$$

$\sqrt{9} - \sqrt{9-5}$	1
$3 - \sqrt{4}$	1
$3 - 2$	1
$1 = 1$	

The solution is 9.

Problem 3   Solve: $\sqrt{x} + \sqrt{x+9} = 9$

Solution   See page A41.

---

2   ## Application problems

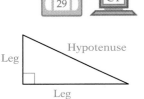

A right triangle contains one 90° angle. The side opposite the 90° angle is called the **hypotenuse**. The other two sides are called **legs**.

The angles in a right triangle are usually labeled with the capital letters A, B, and C, with C reserved for the right angle. The side opposite angle A is side a, the side opposite angle B is side b, and c is the hypotenuse.

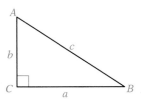

The Greek mathematician Pythagoras is generally credited with the discovery that the square of the hypotenuse of a right triangle is equal to the sum of the squares of the two legs. This is called the **Pythagorean Theorem**.

The figure at the right is a right triangle with legs measuring 3 units and 4 units and a hypotenuse measuring 5 units. Each side of the triangle is also the side of a square. The number of square units in the area of the largest square is equal to the sum of the numbers of square units in the areas of the smaller squares.

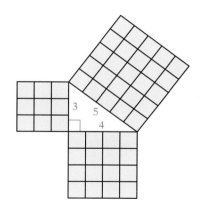

| square of the hypotenuse | = | sum of the squares of the two legs |

$$5^2 = 3^2 + 4^2$$
$$25 = 9 + 16$$
$$25 = 25$$

Pythagorean Theorem

> If a and b are the lengths of the legs of a right triangle and c is the length of the hypotenuse, then $c^2 = a^2 + b^2$.

If the lengths of two sides of a right triangle are known, the Pythagorean Theorem can be used to find the length of the third side.

The Pythagorean Theorem is used to find the hypotenuse when the two legs are known.

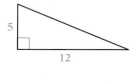

$$\text{Hypotenuse} = \sqrt{(\text{leg})^2 + (\text{leg})^2}$$
$$c = \sqrt{a^2 + b^2}$$
$$c = \sqrt{(5)^2 + (12)^2}$$
$$c = \sqrt{25 + 144}$$
$$c = \sqrt{169}$$
$$c = 13$$

LOOK CLOSELY

If we let $a = 12$ and $b = 5$, the result is the same.

The Pythagorean Theorem is used to find the length of a leg when one leg and the hypotenuse are known.

$$\text{Leg} = \sqrt{(\text{hypotenuse})^2 - (\text{leg})^2}$$
$$a = \sqrt{c^2 - b^2}$$
$$a = \sqrt{(25)^2 - (20)^2}$$
$$a = \sqrt{625 - 400}$$
$$a = \sqrt{225}$$
$$a = 15$$

Example 4 and Problem 4 illustrate the use of the Pythagorean Theorem. Example 5 and Problem 5 illustrate other applications of radical equations.

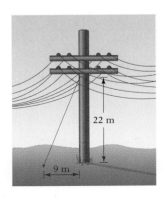

Example 4   A guy wire is attached to a point 22 m above the ground on a telephone pole that is perpendicular to the ground. The wire is anchored to the ground at a point 9 m from the base of the pole. Find the length of the guy wire. Round to the nearest hundredth.

Strategy   To find the length of the guy wire, use the Pythagorean Theorem. One leg is the distance from the bottom of the wire to the base of the telephone pole. The other leg is the distance from the top of the wire to the base of the telephone pole. The guy wire is the hypotenuse. Solve the Pythagorean Theorem for the hypotenuse.

Solution $c = \sqrt{a^2 + b^2}$
$c = \sqrt{(22)^2 + (9)^2}$
$c = \sqrt{484 + 81}$
$c = \sqrt{565}$
$c \approx 23.77$

The guy wire has a length of 23.77 m.

Problem 4   A ladder 12 ft long is resting against a building. How high on the building will the ladder reach when the bottom of the ladder is 5 ft from the building? Round to the nearest hundredth.

Solution   See page A41.

Example 5   How far would a submarine periscope have to be above the water for the lookout to locate a ship 5 mi away? The equation for the distance in miles that the lookout can see is $d = \sqrt{1.5h}$, where h is the height in feet above the surface of the water. Round to the nearest hundredth.

Strategy   To find the height above water, replace d in the equation with the given value. Then solve for h.

Solution $d = \sqrt{1.5h}$
$5 = \sqrt{1.5h}$
$5^2 = (\sqrt{1.5h})^2$
$25 = 1.5h$
$\dfrac{25}{1.5} = h$
$16.67 \approx h$

The periscope must be 16.67 ft above the water.

Problem 5   Find the length of a pendulum that makes one swing in 1.5 s. The equation for the time for one swing is $T = 2\pi \sqrt{\dfrac{L}{32}}$, where T is the time in seconds and L is the length in feet. Round to the nearest hundredth.

Solution   See page A41.

CONCEPT REVIEW 10.4

Determine whether the statement is always true, sometimes true, or never true.

1. A radical equation is an equation that contains a radical.

2. We can square both sides of an equation without changing the solutions of the equation.

3. We use the Property of Squaring Both Sides of an Equation in order to eliminate a radical expression from an equation.

4. The first step in solving a radical equation is to square both sides of the equation.

5. If $a^2 = b^2$, then $a = b$.

6. The Pythagorean Theorem is used to find the length of a side of a triangle.

EXERCISES 10.4

1   Solve and check.

1. $\sqrt{x} = 5$
2. $\sqrt{y} = 7$
3. $\sqrt{a} = 12$
4. $\sqrt{a} = 9$

5. $\sqrt{5x} = 5$
6. $\sqrt{3x} = 4$
7. $\sqrt{4x} = 8$
8. $\sqrt{6x} = 3$

9. $\sqrt{2x} - 4 = 0$
10. $3 - \sqrt{5x} = 0$
11. $\sqrt{4x} + 5 = 2$
12. $\sqrt{3x} + 9 = 4$

13. $\sqrt{3x - 2} = 4$
14. $\sqrt{5x + 6} = 1$
15. $\sqrt{2x + 1} = 7$
16. $\sqrt{5x + 4} = 3$

17. $0 = 2 - \sqrt{3 - x}$
18. $0 = 5 - \sqrt{10 + x}$
19. $\sqrt{5x + 2} = 0$
20. $\sqrt{3x - 7} = 0$

21. $\sqrt{3x} - 6 = -4$
22. $\sqrt{5x} + 8 = 23$
23. $0 = \sqrt{3x - 9} - 6$
24. $0 = \sqrt{2x + 7} - 3$

25. $\sqrt{5x - 1} = \sqrt{3x + 9}$
26. $\sqrt{3x + 4} = \sqrt{12x - 14}$

27. $\sqrt{5x - 3} = \sqrt{4x - 2}$
28. $\sqrt{5x - 9} = \sqrt{2x - 3}$

29. $\sqrt{x^2 - 5x + 6} = \sqrt{x^2 - 8x + 9}$

30. $\sqrt{x^2 - 2x + 4} = \sqrt{x^2 + 5x - 12}$

31. $\sqrt{x} = \sqrt{x + 3} - 1$

32. $\sqrt{x + 5} = \sqrt{x} + 1$

33. $\sqrt{2x + 5} = 5 - \sqrt{2x}$

34. $\sqrt{2x} + \sqrt{2x + 9} = 9$

35. $\sqrt{3x} - \sqrt{3x + 7} = 1$

36. $\sqrt{x} - \sqrt{x + 9} = 1$

2   Solve. Round to the nearest hundredth.

37. Five added to the square root of the product of four and a number is equal to seven. Find the number.

38. The product of a number and the square root of three is equal to the square root of twenty-seven. Find the number.

39. Two added to the square root of the sum of a number and five is equal to six. Find the number.

40. The product of a number and the square root of seven is equal to the square root of twenty-eight. Find the number.

41. The two legs of a right triangle measure 5 cm and 9 cm. Find the length of the hypotenuse.

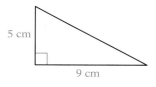

42. The two legs of a right triangle measure 8 in. and 4 in. Find the length of the hypotenuse.

43. The hypotenuse of a right triangle measures 12 ft. One leg of the triangle measures 7 ft. Find the length of the other leg of the triangle.

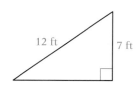

44. The hypotenuse of a right triangle measures 20 cm. One leg of the triangle measures 16 cm. Find the length of the other leg of the triangle.

45. How far would a submarine periscope have to be above the water for the lookout to locate a ship 4 mi away? The equation for the distance in miles that the lookout can see is $d = \sqrt{1.5h}$, where h is the height in feet above the surface of the water.

46. How far would a submarine periscope have to be above the water for the lookout to locate a ship 7 mi away? The equation for the distance in miles that the lookout can see is $d = \sqrt{1.5h}$, where h is the height in feet above the surface of the water.

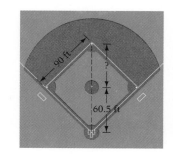

47. The infield of a baseball diamond is a square. The distance between successive bases is 90 ft. The pitcher's mound is on the diagonal between home plate and second base at a distance of 60.5 ft from home plate. Is the pitcher's mound more or less than halfway between home plate and second base?

48. The infield of a softball diamond is a square. The distance between successive bases is 60 ft. The pitcher's mound is on the diagonal between home plate and second base at a distance of 46 ft from home plate. Is the pitcher's mound more or less than halfway between home plate and second base?

49. Find the length of a pendulum that makes one swing in 3 s. The equation for the time of one swing is $T = 2\pi\sqrt{\dfrac{L}{32}}$, where T is the time in seconds and L is the length in feet.

50. Find the length of a pendulum that makes one swing in 2 s. The equation for the time of one swing is $T = 2\pi\sqrt{\dfrac{L}{32}}$, where T is the time in seconds and L is the length in feet.

51. An L-shaped sidewalk from the parking lot to a memorial is shown in the figure at the right. The distance directly across the grass to the memorial is 650 ft. The distance to the corner is 600 ft. Find the distance from the corner to the memorial.

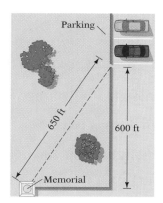

52. A commuter plane leaves an airport traveling due south at 400 mph. Another plane leaving at the same time travels due east at 300 mph. Find the distance between the two planes after 2 h.

53. A stone is dropped from a bridge and hits the water 2 s later. How high is the bridge? The equation for the distance an object falls in T seconds is $T = \sqrt{\dfrac{d}{16}}$, where d is the distance in feet.

54. A stone is dropped into a mine shaft and hits the bottom 3.5 s later. How deep is the mine shaft? The equation for the distance an object falls in T seconds is $T = \sqrt{\dfrac{d}{16}}$, where d is the distance in feet.

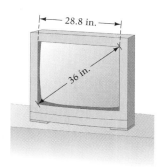

55. The measure of a television screen is given by the length of a diagonal across the screen. A 36 in. television has a width of 28.8 in. Find the height of the screen.

56. The measure of a television screen is given by the length of a diagonal across the screen. A 33 in. big-screen television has a width of 26.4 in. Find the height of the screen.

SUPPLEMENTAL EXERCISES 10.4

Solve.

57. $\sqrt{\dfrac{5y + 2}{3}} = 3$

58. $\sqrt{\dfrac{3y}{5}} - 1 = 2$

59. $\sqrt{9x^2 + 49} + 1 = 3x + 2$

60. In the coordinate plane, a triangle is formed by drawing lines between the points (0, 0) and (5, 0), (5, 0) and (5, 12), and (5, 12) and (0, 0). Find the number of units in the perimeter of the triangle.

61. The hypotenuse of a right triangle is $5\sqrt{2}$ cm, and one leg is $4\sqrt{2}$ cm.
   a. Find the perimeter of the triangle.   **b.** Find the area of the triangle.

62. Write an expression in factored form for the shaded region in the diagram at the right.

63. A circular fountain is being designed for a triangular plaza in a cultural center. The fountain is placed so that each side of the triangle touches the fountain as shown in the diagram at the right. Find the area of the fountain. The formula for the radius of the circle is given by

$$r = \sqrt{\frac{(s - a)(s - b)(s - c)}{s}}$$

where $s = \frac{1}{2}(a + b + c)$ and a, b, and c are the lengths of the sides of the triangle. Round to the nearest hundredth.

64. Can the Pythagorean Theorem be used to find the length of side c of the triangle at the right? If so, determine c. If not, explain why the theorem cannot be used.

65. What is a Pythagorean triple? Provide at least three examples of Pythagorean triples.

Project in Mathematics

Mean and Standard Deviation

An automotive engineer tests the miles-per-gallon ratings of 15 cars and records the results as follows:

25   22   21   27   25   35   29   31   25   26   21   39   34   32   28

The **mean** of the data is the sum of the measurements divided by the number of measurements. The symbol for the mean is $\bar{x}$.

$$\text{Mean} = \bar{x} = \frac{\text{Sum of all data values}}{\text{Number of data values}}$$

To find the mean for the data above, add the numbers and then divide by 15.

$$\bar{x} = \frac{25 + 22 + 21 + 27 + 25 + 35 + 29 + 31 + 25 + 26 + 21 + 39 + 34 + 32 + 28}{15}$$

$$= \frac{420}{15} = 28$$

The mean number of miles per gallon for the 15 cars tested was 28 mi/gal.

The mean is one of the most frequently computed averages. It is the one that is commonly used to calculate a student's performance in a class.

The scores for a history student on 5 tests were 78, 82, 91, 87, and 93. What was the mean score for this student?

To find the mean, add the numbers. Then divide by 5.

$$\bar{x} = \frac{78 + 82 + 91 + 87 + 93}{5}$$

$$= \frac{431}{5} = 86.2$$

The mean score for the history student was 86.2.

Consider two students, each of whom has taken 5 exams.

Scores for Student A

84	86	83	85	87

Scores for Student B

90	75	94	68	98

$$\bar{x} = \frac{84 + 86 + 83 + 85 + 87}{5} = \frac{425}{5} = 85$$

$$\bar{x} = \frac{90 + 75 + 94 + 68 + 98}{5} = \frac{425}{5} = 85$$

The mean for Student A is 85.   The mean for Student B is 85.

For each of these students, the mean (average) for the 5 exams is 85. However, Student A has a more consistent record of scores than Student B. One way to measure the consistency or "clustering" of data near the mean is to use the **standard deviation.**

To calculate the standard deviation:

1. Sum the squares of the differences between each value of data and the mean.
2. Divide the result in Step 1 by the number of items in the set of data.
3. Take the square root of the result in Step 2.

The calculation for Student A is shown at the right.

Step 1:

x	$(x - \bar{x})$	$(x - \bar{x})^2$
84	$(84 - 85)$	$(-1)^2 = 1$
86	$(86 - 85)$	$1^2 = 1$
83	$(83 - 85)$	$(-2)^2 = 4$
85	$(85 - 85)$	$0^2 = 0$
87	$(87 - 85)$	$2^2 = 4$
		Total $= 10$

Step 2: $\frac{10}{5} = 2$

The symbol for standard deviation is the lower-case Greek letter *sigma, σ.*   Step 3: $\sigma = \sqrt{2} \approx 1.414$

The standard deviation for Student A's scores is approximately 1.414.

Following a similar procedure for Student B shows that the standard deviation for Student B's scores is approximately 11.524. Because the standard deviation of Student B's scores is greater than that of Student A's (11.524 > 1.414), Student B's scores are not as consistent as those of Student A.

Solve.

1. An airline recorded the times for a ground crew to unload the baggage from 5 airplanes. The recorded times (in minutes) were 12, 18, 20, 14, and 16. Find the standard deviation of these times.

2. The weights in ounces of 6 newborn infants were recorded by a hospital. The weights were 96, 105, 84, 90, 102, and 99. Find the standard deviation of the weights.

3. The numbers of rooms occupied in a hotel on 6 consecutive days were 234, 321, 222, 246, 312, and 396. Find the standard deviation for the number of rooms.

4. Seven coins were tossed 100 times. The numbers of heads recorded for each coin were 56, 63, 49, 50, 48, 53, and 52. Find the standard deviation of the number of heads.

5. The temperatures for 11 consecutive days at a desert resort were 95°, 98°, 98°, 104°, 97°, 100°, 96°, 97°, 108°, 93°, and 104°. For the same days, the temperatures in Antarctica were 27°, 28°, 28°, 30°, 28°, 27°, 30°, 25°, 24°, 26°, and 21°. Which location has the greater standard deviation of temperatures?

6. The scores for 5 college basketball games were 56, 68, 60, 72, and 64. The scores for 5 professional basketball games were 106, 118, 110, 122, and 114. Which scores have the greater standard deviation?

7. The weights in pounds of the 5-man front line of a college football team are 210, 245, 220, 230, and 225. Find the standard deviation of the weights.

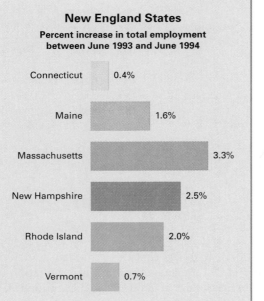

New England States

Percent increase in total employment between June 1993 and June 1994

Connecticut	0.4%
Maine	1.6%
Massachusetts	3.3%
New Hampshire	2.5%
Rhode Island	2.0%
Vermont	0.7%

8. Use the information given in the graph at the left to determine the mean percent increase in total employment in the New England states from June 1993 to June 1994. Find the standard deviation of the percent increases.

9. One student received test scores of 85, 92, 86, and 89. A second student received scores of 90, 97, 91, and 94 (exactly 5 points more on each test). Are the means of the two students the same? If not, what is the relationship between the means of the two students? Are the standard deviations of the scores of the two students the same? If not, what is the relationship between the standard deviations of the scores of the two students?

10. Grade-point average (GPA) is a *weighted* mean. It is called a weighted mean because a grade in a 5-unit course has more influence on your GPA than a grade in a 2-unit course. GPA is calculated by multiplying the numerical equivalent of each grade by the number of units, adding those products, and then dividing by the total number of units. Calculate your GPA for the last quarter or semester.

11. If you average 40 mph for 1 h and then 50 mph for 1 h, is your average speed $\frac{40 + 50}{2} = 45$ mph? Why or why not?

12. A company is negotiating with its employees the terms of a raise in salary. One proposal would add $500 a year to each employee's salary. The second proposal would give each employee a 4% raise. Explain how each of these proposals would affect the current mean and standard deviation of salaries for the company.

Focus on Problem Solving

Working Backward

Sometimes the solution to a problem can be found by *working backward*. This problem-solving technique can be used to find a winning strategy for a game called Nim.

There are many variations of this game. For our game, there are two players, Player A and Player B, who alternately place 1, 2, or 3 matchsticks in a pile. The object of the game is to place the 32nd matchstick in the pile. Is there a strategy that Player A can use to guarantee winning the game?

Working backward, if there are 29, 30, or 31 matchsticks in the pile when it is A's turn to play, A can win by placing 3 (29 + 3 = 32), 2 (30 + 2 = 32), or 1 (31 + 1 = 32) matchsticks on the pile. If there are to be 29, 30, or 31 matchsticks in the pile when it is A's turn, there must be 28 matchsticks in the pile when it is B's turn.

Working backward from 28, if there are to be 28 matches in the pile at B's turn, there must be 25, 26, or 27 at A's turn. Player A can then add 3, 2, or 1 matchsticks to the pile to bring the number to 28. For there to be 25, 26, or 27 matchsticks in the pile at A's turn, there must be 24 matchsticks at B's turn.

Now working backward from 24, if there are to be 24 matches in the pile at B's turn, there must be 21, 22, or 23 at A's turn. Player A can then add 3, 2, or 1 matchsticks to the pile to bring the number to 24. For there to be 21, 22, or 23 matchsticks in the pile at A's turn, there must be 20 matchsticks at B's turn.

So far, we have found that for Player A to win, there must be 28, 24, or 20 matchsticks in the pile when it is B's turn to play. Note that each time, the number is decreasing by 4. Continuing this pattern, Player A will win if there are 16, 12, 8, or 4 matchsticks in the pile when it is B's turn.

Player A can guarantee winning by making sure the number of matchsticks in the pile is a multiple of 4. To ensure this, Player A allows Player B to go first and then adds exactly enough matchsticks to the pile to bring the total to a multiple of 4.

For example, suppose B places 3 matchsticks in the pile; then A places 1 matchstick (3 + 1 = 4). Now B places 2 matchsticks in the pile. The total is now 6 matchsticks. Player A then places 2 matchsticks in the pile to bring the total to 8, a multiple of 4. If play continues in this way, Player A will win.

Here are some variations of Nim. See if you can develop a winning strategy for Player A. (*Hint:* It may not be possible.)

1. Suppose the goal is to place the last matchstick in a pile of 30 matches.

2. Suppose the players make two piles of matchsticks, with the maximum number of matchsticks in each pile to be 20.

3. In this variation of Nim, there are 40 matchsticks in a pile. Each player alternately selects 1, 2, or 3 matches from the pile. The player who selects the last match wins.

Chapter Summary

Key Words

A *square root* of a positive number x is a number whose square is x. The square root of a negative number is not a real number.   (Objective 10.1.1)

A square root of 25 is 5 becuase $5^2 = 25$.
A square root of 25 is -5 because $(-5)^2 = 25$.
$\sqrt{-9}$ is not a real number.

The *principal square root* of a number is the positive square root. The symbol $\sqrt{}$ is called a *radical sign* and is used to indicate the principal square root of a number. The negative square root of a number is indicated by placing a negative sign in front of the radical. The *radicand* is the expression under the radical symbol.   (Objective 10.1.1)

$\sqrt{25} = 5$

$-\sqrt{25} = -5$

In the expression $\sqrt{25}$, 25 is the radicand.

The square of an integer is a *perfect square*. If a number is not a perfect square, its square root can only be approximated. Such numbers are *irrational numbers*. Their decimal representations never terminate or repeat.   (Objective 10.1.1)

$2^2 = 4, 3^2 = 9, 4^2 = 16, 5^2 = 25, 6^2 = 36, \ldots$
4, 9, 16, 25, 36, ... are perfect squares.
5 is not a perfect square. $\sqrt{5}$ is an irrational number.

Conjugates are binomials that differ only in the sign of the second term. The expressions $a + b$ and $a - b$ are conjugates. (Objective 10.3.1)

$3 + \sqrt{7}$ and $3 - \sqrt{7}$ are conjugates.
$\sqrt{x} + 2$ and $\sqrt{x} - 2$ are conjugates.

A *radical equation* is an equation that contains a variable expression in a radicand. (Objective 10.4.1)

$\sqrt{2x} + 8 = 12$ is a radical equation.
$2x + \sqrt{8} = 12$ is not a radical equation.

Essential Rules and Procedures

The Product Property of Square Roots
If a and b are positive real numbers, then $\sqrt{ab} = \sqrt{a} \cdot \sqrt{b}$.
(Objective 10.1.1)

$\sqrt{12} = \sqrt{4 \cdot 3} = \sqrt{4}\sqrt{3} = 2\sqrt{3}$

Simplifying Radical Expressions

A radical is in simplest form when the radicand contains no factor greater than 1 that is a perfect square.
(Objective 10.1.1)

$\sqrt{12}$ is not in simplest form.

$2\sqrt{3}$ is in simplest form.

Adding and Subtracting Radical Expressions

The Distributive Property is used to simplify the sum or difference of like radical expressions.   (Objective 10.2.1)

$16\sqrt{3x} - 4\sqrt{3x} = (16 - 4)\sqrt{3x} = 12\sqrt{3x}$

Multiplying Radical Expressions

The Product Property of Square Roots is used to multiply radical expressions.

$\sqrt{5x}\,\sqrt{7y} = \sqrt{5x \cdot 7y} = \sqrt{35xy}$

Use the Distributive Property to remove parentheses.

$\sqrt{y}(2 + \sqrt{3x}) = 2\sqrt{y} + \sqrt{3xy}$

Use the FOIL method to multiply two binomial radical expressions.
(Objective 10.3.1)

$(5 - \sqrt{x})(11 + \sqrt{x})$
$= 55 + 5\sqrt{x} - 11\sqrt{x} - (\sqrt{x})^2$
$= 55 - 6\sqrt{x} - x$

The Quotient Property of Square Roots

If a and b are positive real numbers, then $\sqrt{\dfrac{a}{b}} = \dfrac{\sqrt{a}}{\sqrt{b}}$.
(Objective 10.3.2)

$\sqrt{\dfrac{9x^2}{y^8}} = \dfrac{\sqrt{9x^2}}{\sqrt{y^8}} = \dfrac{3x}{y^4}$

Dividing Radical Expressions

The Quotient Property of Square Roots is used to divide radical expressions.   (Objective 10.3.2)

$\dfrac{\sqrt{3x^5y}}{\sqrt{75xy^3}} = \sqrt{\dfrac{3x^5y}{75xy^3}} = \sqrt{\dfrac{x^4}{25y^2}} = \dfrac{\sqrt{x^4}}{\sqrt{25y^2}} = \dfrac{x^2}{5y}$

Rationalizing the Denominator

A radical is not in simplest form if there is a radical in the denominator. The procedure used to remove a radical from the denominator is called rationalizing the denominator.
(Objective 10.3.2)

$\dfrac{5}{\sqrt{7}} = \dfrac{5}{\sqrt{7}} \cdot \dfrac{\sqrt{7}}{\sqrt{7}} = \dfrac{5\sqrt{7}}{7}$

Property of Squaring Both Sides of an Equation

If a and b are real numbers and $a = b$, then $a^2 = b^2$.
(Objective 10.4.1)

$\sqrt{x + 3} = 4$
$(\sqrt{x + 3})^2 = 4^2$
$x + 3 = 16$
$x = 13$

Solving Radical Equations

When an equation contains one radical expression, write the equation with the radical alone on one side of the equation. Square both sides of the equation. Solve for the variable. Whenever both sides of an equation are squared, the solutions must be checked.   (Objective 10.4.1)

$\sqrt{2x} - 1 = 5$
$\sqrt{2x} = 6$
$(\sqrt{2x})^2 = 6^2$
$2x = 36$
$x = 18$ The solution checks.

Pythagorean Theorem

If a and b are the lengths of the legs of a right triangle and c is the hypotenuse, then $c^2 = a^2 + b^2$.   (Objective 10.4.2)

Two legs of a right triangle measure 4 cm and 7 cm. To find the length of the hypotenuse, use the equation

$c = \sqrt{a^2 + b^2}$
$c = \sqrt{4^2 + 7^2} = \sqrt{16 + 49} = \sqrt{65}$

The hypotenuse measures $\sqrt{65}$ cm.

Chapter Review Exercises

1. Subtract: $5\sqrt{3} - 16\sqrt{3}$

2. Simplify: $\dfrac{2x}{\sqrt{3} - \sqrt{5}}$

3. Simplify: $\sqrt{x^2 + 16x + 64}$

4. Solve: $3 - \sqrt{7x} = 5$

5. Simplify: $\dfrac{5\sqrt{y} - 2\sqrt{y}}{3\sqrt{y}}$

6. Add: $6\sqrt{7} + \sqrt{7}$

7. Multiply: $(6\sqrt{a} + 5\sqrt{b})(2\sqrt{a} + 3\sqrt{b})$

8. Simplify: $\sqrt{49(x + 3)^4}$

9. Simplify: $2\sqrt{36}$

10. Solve: $\sqrt{b} = 4$

11. Subtract: $9x\sqrt{5} - 5x\sqrt{5}$

12. Multiply: $(\sqrt{5ab} - \sqrt{7})(\sqrt{5ab} + \sqrt{7})$

13. Solve: $\sqrt{2x + 9} = \sqrt{8x - 9}$

14. Simplify: $\sqrt{35}$

15. Add: $2x\sqrt{60x^3y^3} + 3x^2y\sqrt{15xy}$

16. Simplify: $\dfrac{\sqrt{3x^3y}}{\sqrt{27xy^5}}$

17. Simplify: $(3\sqrt{x} - \sqrt{y})^2$

18. Simplify: $\sqrt{(a + 4)^2}$

19. Simplify: $5\sqrt{48}$

20. Add: $3\sqrt{12x} + 5\sqrt{48x}$

21. Simplify: $\dfrac{8}{\sqrt{x} - 3}$

22. Multiply: $\sqrt{6a}(\sqrt{3a} + \sqrt{2a})$

23. Simplify: $3\sqrt{18a^5b}$

24. Simplify: $-3\sqrt{120}$

25. Subtract: $\sqrt{20a^5b^9} - 2ab^2\sqrt{45a^3b^5}$

26. Simplify: $\dfrac{\sqrt{98x^7y^9}}{\sqrt{2x^3y}}$

27. Solve: $\sqrt{5x + 1} = \sqrt{20x - 8}$

28. Simplify: $\sqrt{c^{18}}$

29. Simplify: $\sqrt{450}$

30. Simplify: $6a\sqrt{80b} - \sqrt{180a^2b} + 5a\sqrt{b}$

31. Simplify: $\dfrac{16}{\sqrt{a}}$

32. Solve: $6 - \sqrt{2y} = 2$

33. Multiply: $\sqrt{a^3b^4c}\,\sqrt{a^7b^2c^3}$

34. Simplify: $7\sqrt{630}$

35. Simplify: $y\sqrt{24y^6}$

36. Simplify: $\dfrac{\sqrt{250}}{\sqrt{10}}$

37. Solve: $\sqrt{x^2 + 5x + 4} = \sqrt{x^2 + 7x - 6}$

38. Multiply: $(4\sqrt{y} - \sqrt{5})(2\sqrt{y} + 3\sqrt{5})$

39. Find the decimal approximation of $\sqrt{9900}$ to the nearest thousandth.

40. Multiply: $\sqrt{7}\sqrt{7}$

41. Simplify: $5x\sqrt{150x^7}$

42. Simplify:
$2x^2\sqrt{18x^2y^5} + 6y\sqrt{2x^6y^3} - 9xy^2\sqrt{8x^4y}$

43. Simplify: $\dfrac{\sqrt{54a^3}}{\sqrt{6a}}$

44. Simplify: $4\sqrt{250}$

45. Solve: $\sqrt{5x} = 10$

46. Simplify: $\dfrac{3a\sqrt{3} + 2\sqrt{12a^2}}{\sqrt{27}}$

47. Multiply: $\sqrt{2}\sqrt{50}$

48. Simplify: $\sqrt{36x^4y^5}$

49. Simplify: $4y\sqrt{243x^{17}y^9}$

50. Solve: $\sqrt{x+1} - \sqrt{x-2} = 1$

51. Simplify: $\sqrt{400}$

52. Simplify: $-4\sqrt{8x} + 7\sqrt{18x} - 3\sqrt{50x}$

53. Solve: $0 = \sqrt{10x+4} - 8$

54. Multiply: $\sqrt{3}(\sqrt{12} - \sqrt{3})$

55. The square root of the sum of two consecutive odd integers is equal to 10. Find the larger integer.

56. The weight of an object is related to the object's distance from the surface of the earth. An equation for this relationship is $d = 4000\sqrt{\dfrac{W_0}{W_a}}$, where W_0 is the object's weight on the surface of the earth and W_a is the object's weight at a distance of d miles above the earth's surface. A space explorer weighs 36 lb when 8000 mi above the earth's surface. Find the explorer's weight on the surface of the earth.

57. The hypotenuse of a right triangle measures 18 cm. One leg of the triangle measures 11 cm. Find the length of the other leg of the triangle. Round to the nearest hundredth.

58. A bicycle will overturn if it rounds a corner too sharply or too quickly. The equation for the maximum velocity at which a cyclist can turn a corner without tipping over is given by the equation $v = 4\sqrt{r}$, where v is the velocity of the bicycle in miles per hour and r is the radius of the corner in feet. Find the radius of the sharpest corner that a cyclist can safely turn when riding at a speed of 20 mph.

59. A tsunami is a great sea wave produced by underwater earthquakes or volcanic eruption. The velocity of a tsunami as it approaches land is approximated by the equation $v = 3\sqrt{d}$, where v is the velocity in feet per second and d is the depth of the water in feet. Find the depth of the water when the velocity of a tsunami is 30 ft/s.

60. A guy wire is attached to a point 25 ft above the ground on a telephone pole that is perpendicular to the ground. The wire is anchored to the ground at a point 8 ft from the base of the pole. Find the length of the guy wire. Round to the nearest hundredth.

Chapter Test

1. Simplify: $\sqrt{121x^8y^2}$

2. Subtract: $5\sqrt{8} - 3\sqrt{50}$

3. Multiply: $\sqrt{3x^2y}\,\sqrt{6x^2}\,\sqrt{2x}$

4. Simplify: $\sqrt{45}$

5. Simplify: $\sqrt{72x^7y^2}$

6. Simplify: $3\sqrt{8y} - 2\sqrt{72x} + 5\sqrt{18x}$

7. Multiply: $(\sqrt{y} + 3)(\sqrt{y} + 5)$

8. Simplify: $\dfrac{4}{\sqrt{8}}$

9. Simplify: $\dfrac{\sqrt{162}}{\sqrt{2}}$

10. Solve: $\sqrt{5x - 6} = 7$

11. Find the decimal approximation of $\sqrt{500}$ to the nearest thousandth.

12. Simplify: $\sqrt{32a^5b^{11}}$

13. Multiply: $\sqrt{a}(\sqrt{a} - \sqrt{b})$

14. Multiply: $\sqrt{8x^3y}\,\sqrt{10xy^4}$

15. Simplify: $\dfrac{\sqrt{98a^6b^4}}{\sqrt{2a^3b^2}}$

16. Solve: $\sqrt{9x} + 3 = 18$

17. Simplify: $\sqrt{192x^{13}y^5}$

18. Simplify: $2a\sqrt{2ab^3} + b\sqrt{8a^3b} - 5ab\sqrt{ab}$

19. Multiply: $(\sqrt{a} - 2)(\sqrt{a} + 2)$

20. Simplify: $\dfrac{3}{2 - \sqrt{5}}$

21. Solve: $3 = 8 - \sqrt{5x}$

22. Multiply: $\sqrt{3}(\sqrt{6} - \sqrt{x^2})$

23. Subtract: $3\sqrt{a} - 9\sqrt{a}$

24. Simplify: $\sqrt{108}$

25. Find the decimal approximation of $\sqrt{63}$ to the nearest thousandth.

26. Simplify: $\dfrac{\sqrt{108a^7b^3}}{\sqrt{3a^4b}}$

27. Solve: $\sqrt{x} - \sqrt{x + 3} = 1$

28. The square root of the sum of two consecutive integers is equal to 9. Find the smaller integer.

29. Find the length of a pendulum that makes one swing in 2.5 s. The equation for the time of one swing of a pendulum is $T = 2\pi\sqrt{\dfrac{L}{32}}$, where T is the time in seconds and L is the length in feet. Round to the nearest hundredth.

30. A ladder 10 ft long is resting against a building. How high on the building will the ladder reach when the bottom of the ladder is 2 ft from the building? Round to the nearest hundredth.

Cumulative Review Exercises

1. Simplify: $\left(\frac{2}{3}\right)^2\left(\frac{3}{4} - \frac{3}{2}\right) + \left(\frac{1}{2}\right)^2$

2. Simplify: $-3[x - 2(3 - 2x) - 5x] + 2x$

3. Solve: $2x - 4[3x - 2(1 - 3x)] = 2(3 - 4x)$

4. Solve: $3(x - 7) \geq 5x - 12$

5. Find the slope of the line that contains the points whose coordinates are $(2, -5)$ and $(-4, 3)$.

6. Find the equation of the line that contains the point whose coordinates are $(-2, -3)$ and has slope $\frac{1}{2}$.

7. Evaluate $f(x) = \frac{5}{2}x - 8$ at $x = -4$.

8. Graph: $f(x) = -4x + 2$

9. Graph the solution set of $2x + y < -2$.

10. Solve by graphing: $3x - 2y = 8$
$4x + 5y = 3$

11. Solve by substitution: $4x - 3y = 1$
$2x + y = 3$

12. Solve by the addition method: $5x + 4y = 7$
$3x - 2y = 13$

13. Simplify: $(-3x^2y)(-2x^3y^{-4})$

14. Simplify: $\dfrac{12b^4 - 6b^2 + 2}{-6b^2}$

15. Factor: $12x^3y^2 - 9x^2y^3$

16. Factor: $9b^2 + 3b - 20$

17. Factor: $2a^3 - 16a^2 + 30a$

18. Multiply: $\dfrac{3x^3 - 6x^2}{4x^2 + 4x} \cdot \dfrac{3x - 9}{9x^3 - 45x^2 + 54x}$

19. Simplify: $\dfrac{1 - \dfrac{2}{x} - \dfrac{15}{x^2}}{1 - \dfrac{9}{x^2}}$

20. Subtract: $\dfrac{x + 2}{x - 4} - \dfrac{6}{(x - 4)(x - 3)}$

21. Solve: $\dfrac{x}{2x - 5} - 2 = \dfrac{3x}{2x - 5}$

22. Simplify: $2\sqrt{27a} - 5\sqrt{49a} + 8\sqrt{48a}$

23. Simplify: $\dfrac{\sqrt{320}}{\sqrt{5}}$

24. Solve: $\sqrt{2x - 3} - 5 = 0$

25. Three-eighths of a number is less than negative twelve. Find the largest integer that satisfies the inequality.

26. The selling price of a book is $29.40. The markup rate used by the bookstore is 20% of the cost. Find the cost of the book.

27. How many ounces of pure water must be added to 40 oz of a 12% salt solution to make a salt solution that is 5% salt?

28. The sum of two numbers is twenty-one. The product of the two numbers is one hundred four. Find the two numbers.

29. A small water pipe takes twice as long to fill a tank as does a larger water pipe. With both pipes open, it takes 16 h to fill the tank. Find the time it would take the small pipe, working alone, to fill the tank.

30. The square root of the sum of two consecutive integers is equal to 7. Find the smaller integer.

11

Quadratic Equations

Objectives

11.1.1	Solve quadratic equations by factoring
11.1.2	Solve quadratic equations by taking square roots
11.2.1	Solve quadratic equations by completing the square
11.3.1	Solve quadratic equations by using the quadratic formula
11.4.1	Graph a quadratic equation of the form $y = ax^2 + bx + c$
11.5.1	Application problems

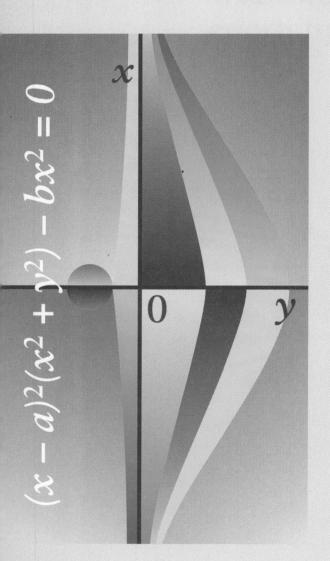

$(x - a)^2(x^2 + y^2) - bx^2 = 0$

Algebraic Symbolism

The way in which an algebraic expression or equation is written has gone through several stages of development. First there was the *rhetoric*, which was in vogue until the late 13th century. In this method, an expression would be written out in sentences. The word *res* was used to represent an unknown.

Rhetoric: From the additive *res* in the additive *res* results in a square *res*. From the three in an additive *res* comes three additive *res*, and from the subtractive four in the additive *res* comes subtractive four *res*. From three in subtractive four comes subtractive twelve.

Modern: $(x + 3)(x - 4) = x^2 - x - 12$

The second stage was the *syncoptic*, which was a shorthand in which abbreviations were used for words.

Syncoptic: *a* 6 in *b* quad $-$ *c* plano 4 in *b* + *b* cub

Modern: $6ab^2 - 4cb + b^3$

The current modern stage, called the *symbolic* stage, began with the use of exponents rather than words to symbolize exponential expressions. This occurred near the beginning of the 17th century with the publication of the book *La Geometrie* by René Descartes. Modern notation is still evolving as mathematicians continue to search for convenient methods to symbolize concepts.

S E C T I O N **11.1**

Solving Quadratic Equations by Factoring or by Taking Square Roots

1 Solve quadratic equations by factoring

In Section 5 of Chapter 8, we solved quadratic equations by factoring. In this section, we will review that material and then solve quadratic equations by taking square roots.

An equation of the form $ax^2 + bx + c = 0$, $a \neq 0$, is a **quadratic equation.**

$$4x^2 - 3x + 1 = 0, \quad a = 4, b = -3, c = 1$$

$$3x^2 - 4 = 0, \quad\quad a = 3, b = 0, \quad c = -4$$

A quadratic equation is also called a **second-degree equation.**

A quadratic equation is in **standard form** when the polynomial is in descending order and equal to zero.

Recall that the Principle of Zero Products states that if the product of two factors is zero, then at least one of the factors must be zero.

If $a \cdot b = 0$, then $a = 0$ or $b = 0$.

The Principle of Zero Products can be used in solving quadratic equations.

Solve by factoring: $2x^2 - x = 1$

Write the equation in standard form.

$$\begin{aligned} 2x^2 - x &= 1 \\ 2x^2 - x - 1 &= 0 \end{aligned}$$

Factor the left side of the equation.

$$(2x + 1)(x - 1) = 0$$

If $(2x + 1)(x - 1) = 0$, then either $2x + 1 = 0$ or $x - 1 = 0$.

$$2x + 1 = 0 \qquad\qquad x - 1 = 0$$

Solve each equation for x.

$$2x = -1 \qquad\qquad x = 1$$
$$x = -\frac{1}{2}$$

The solutions are $-\frac{1}{2}$ and 1.

A graphing calculator can be used to check the solutions of a quadratic equation. Consider the example above. The solutions appear to be $-\frac{1}{2}$ and 1. To check these solutions, store one value of x, $-\frac{1}{2}$, in the calculator. Note that after you store $-\frac{1}{2}$ in the calculator's memory, your calculator rewrites the fraction as the decimal -0.5. Evaluate the expression on the left side of the original equation: $2(-0.5)^2 - (-0.5)$. The result is 1, which is the number on the right side of the original equation. The solution -0.5 checks. Now store the other value of x, 1, in the calculator. Evaluate the expression on the left side of the original equation: $2(1)^2 - 1$. The result is 1, which is the number on the right side of the equation. The solution 1 checks.

Solve by factoring: $x^2 = 10x - 25$

	$x^2 = 10x - 25$
Write the equation in standard form.	$x^2 - 10x + 25 = 0$
Factor the left side of the equation.	$(x - 5)(x - 5) = 0$
Let each factor equal 0.	$x - 5 = 0 \qquad x - 5 = 0$
Solve each equation for x.	$x = 5 \qquad\qquad x = 5$

The factorization in this example produced two identical factors. Because $x - 5$ occurs twice in the factored form of the equation, 5 is a **double root** of the equation.

Example 1   Solve by factoring: $\dfrac{z^2}{2} - \dfrac{z}{4} - \dfrac{1}{4} = 0$

Solution

$$\frac{z^2}{2} - \frac{z}{4} - \frac{1}{4} = 0$$

$$4\left(\frac{z^2}{2} - \frac{z}{4} - \frac{1}{4}\right) = 4(0) \qquad \blacktriangleright \text{To eliminate the fractions, multi-}$$
ply each side of the equation by 4.

$$2z^2 - z - 1 = 0 \qquad \blacktriangleright \text{The quadratic equation is in stan-}$$
dard form.

$$(2z + 1)(z - 1) = 0 \qquad \blacktriangleright \text{Factor the left side of the equation.}$$

$$2z + 1 = 0 \qquad z - 1 = 0 \quad \blacktriangleright \text{Let each factor equal 0.}$$
$$2z = -1 \qquad\quad z = 1 \quad \blacktriangleright \text{Solve each equation for } z.$$
$$z = -\frac{1}{2}$$

The solutions are $-\dfrac{1}{2}$ and 1. $\blacktriangleright$ $-\dfrac{1}{2}$ and 1 check as solutions.

Problem 1   Solve by factoring: $\dfrac{3y^2}{2} + y - \dfrac{1}{2} = 0$

Solution   See page A42.

2 Solve quadratic equations by taking square roots

Consider a quadratic equation of the form $x^2 = a$. This equation can be solved by factoring.	$x^2 = 25$ $x^2 - 25 = 0$ $(x + 5)(x - 5) = 0$

$$x + 5 = 0 \qquad x - 5 = 0$$
$$x = -5 \qquad\qquad x = 5$$

The solutions are -5 and 5.

Solutions that are plus or minus the same number are frequently written using $\pm$. For the example at the bottom of page 484, this would be written: "The solutions are ± 5." Because the solutions ± 5 can be written as $\pm\sqrt{25}$, an alternative method of solving this equation is suggested.

Principle of Taking the Square Root of Each Side of an Equation

If $x^2 = a$, then $x = \pm\sqrt{a}$.

Solve by taking square roots: $x^2 = 25$

Take the square root of each side of the equation.	$x^2 = 25$ $\sqrt{x^2} = \pm\sqrt{25}$
Simplify.	$x = \pm 5$
Write the solutions.	The solutions are 5 and -5.

Solve by taking square roots: $3x^2 = 36$

Solve for x^2.	$3x^2 = 36$ $x^2 = 12$
Take the square root of each side of the equation.	$\sqrt{x^2} = \pm\sqrt{12}$
Simplify.	$x = \pm 2\sqrt{3}$

LOOK CLOSELY

You should always check your solutions by substituting the proposed solutions back into the *original* equation.

Check:

$3x^2 = 36$		$3x^2 = 36$	
$3(2\sqrt{3})^2$	36	$3(-2\sqrt{3})^2$	36
$3(12)$	36	$3(12)$	36
$36 = 36$		$36 = 36$	

These are true equations. The solutions check.

Write the solutions.

The solutions are $2\sqrt{3}$ and $-2\sqrt{3}$.

A graphing calculator can be used to check irrational solutions of a quadratic equation. Consider the example above; the solutions $2\sqrt{3}$ and $-2\sqrt{3}$ checked. To check these solutions on a calculator, store one value of x, $2\sqrt{3}$, in the calculator. Note that after you store this number in the calculator's memory, your calculator rewrites the fraction as the decimal 3.464101615, which is an approximation of $2\sqrt{3}$. Evaluate the expression on the left side of the original equation. The result is 36, which is the number on the right side of the original equation. The solution $2\sqrt{3}$ checks. Now store the other value of x, $-2\sqrt{3}$, in the calculator. Note that after you store this number in the calculator's memory, your calculator rewrites the fraction as the decimal -3.464101615, which is an approximation of $-2\sqrt{3}$. Evaluate the expression on the left side of the original equation. The result is 36, which is the number on the right side of the equation. The solution $-2\sqrt{3}$ checks.

Example 2   Solve by taking square roots.

A. $2x^2 - 72 = 0$ B. $x^2 + 16 = 0$

Solution   A. $2x^2 - 72 = 0$
$$2x^2 = 72 \qquad \blacktriangleright \text{Solve for } x^2.$$
$$x^2 = 36$$
$$\sqrt{x^2} = \pm\sqrt{36} \qquad \blacktriangleright \text{Take the square root of each side of the}$$
$$x = \pm 6 \qquad\quad \text{equation.}$$

The solutions are 6 and -6.

B. $x^2 + 16 = 0$
$$x^2 = -16$$
$$\sqrt{x^2} = \pm\sqrt{-16} \qquad \blacktriangleright \sqrt{-16} \text{ is not a real number.}$$

The equation has no real number solution.

Problem 2   Solve by taking square roots.

A. $4x^2 - 96 = 0$ B. $x^2 + 81 = 0$

Solution   See page A42.

An equation containing the square of a binomial can be solved by taking square roots.

Solve by taking square roots: $2(x - 1)^2 - 36 = 0$

Solve for $(x - 1)^2$.	$2(x - 1)^2 - 36 = 0$
	$2(x - 1)^2 = 36$
	$(x - 1)^2 = 18$
Take the square root of each side of the equation.	$\sqrt{(x - 1)^2} = \pm\sqrt{18}$
Simplify.	$x - 1 = \pm 3\sqrt{2}$
Solve for x.	$x - 1 = 3\sqrt{2} \qquad x - 1 = -3\sqrt{2}$
	$x = 1 + 3\sqrt{2} \qquad x = 1 - 3\sqrt{2}$

Check:

$$\begin{array}{r|l} 2(x - 1)^2 - 36 = 0 & \\ \hline 2(1 + 3\sqrt{2} - 1)^2 - 36 & 0 \\ 2(3\sqrt{2})^2 - 36 & 0 \\ 2(18) - 36 & 0 \\ 36 - 36 & 0 \\ 0 = 0 & \end{array}$$

$$\begin{array}{r|l} 2(x - 1)^2 - 36 = 0 & \\ \hline 2(1 - 3\sqrt{2} - 1)^2 - 36 & 0 \\ 2(-3\sqrt{2})^2 - 36 & 0 \\ 2(18) - 36 & 0 \\ 36 - 36 & 0 \\ 0 = 0 & \end{array}$$

Write the solutions.   The solutions are $1 + 3\sqrt{2}$ and $1 - 3\sqrt{2}$.

Example 3   Solve by taking square roots: $(x - 6)^2 = 12$

Solution $(x - 6)^2 = 12$
$\sqrt{(x - 6)^2} = \pm\sqrt{12}$ ▶ Take the square root of each side of the equation.
$x - 6 = \pm 2\sqrt{3}$

$x - 6 = 2\sqrt{3}$ $x - 6 = -2\sqrt{3}$ ▶ Solve for x.
$x = 6 + 2\sqrt{3}$ $x = 6 - 2\sqrt{3}$

The solutions are $6 + 2\sqrt{3}$ and $6 - 2\sqrt{3}$.

Problem 3   Solve by taking square roots: $(x + 5)^2 = 20$

Solution   See page A42.

CONCEPT REVIEW 11.1

Determine whether the statement is always true, sometimes true, or never true.

1. $2x^2 - 3x + 9$ is a quadratic equation.

2. By the Principle of Zero Products, if $(3x + 4)(x - 7) = 0$, then $3x + 4 = 0$ or $x - 7 = 0$.

3. A quadratic equation has two distinct solutions.

4. To solve the equation $3x^2 - 26x = 9$ by factoring, first write the equation in standard form.

5. To say that the solutions are ± 6 means that the solutions are 6 and -6.

6. If we take the square root of the square of a binomial, the result is the binomial.

EXERCISES 11.1

1   For the given quadratic equation, find the values of a, b, and c.

1. $3x^2 - 4x + 1 = 0$
2. $x^2 + 2x - 5 = 0$
3. $2x^2 - 5 = 0$

4. $4x^2 + 1 = 0$
5. $6x^2 - 3x = 0$
6. $-x^2 + 7x = 0$

Write the quadratic equation in standard form.

7. $x^2 - 8 = 3x$
8. $2x^2 = 4x - 1$
9. $x^2 = 16$

10. $x + 5 = x(x - 3)$
11. $2(x + 3)^2 = 5$
12. $4(x - 1)^2 = 3$

Solve by factoring.

13. $x^2 + 2x - 15 = 0$

14. $t^2 + 3t - 10 = 0$

15. $z^2 - 4z + 3 = 0$

16. $s^2 - 5s + 4 = 0$

17. $p^2 + 3p + 2 = 0$

18. $v^2 + 6v + 5 = 0$

19. $x^2 - 6x + 9 = 0$

20. $y^2 - 8y + 16 = 0$

21. $6x^2 - 9x = 0$

22. $12y^2 + 8y = 0$

23. $r^2 - 10 = 3r$

24. $t^2 - 12 = 4t$

25. $3v^2 - 5v + 2 = 0$

26. $2p^2 - 3p - 2 = 0$

27. $3s^2 + 8s = 3$

28. $3x^2 + 5x = 12$

29. $6r^2 = 12 - r$

30. $4t^2 = 4t + 3$

31. $5y^2 + 11y = 12$

32. $4v^2 - 4v + 1 = 0$

33. $9s^2 - 6s + 1 = 0$

34. $x^2 - 9 = 0$

35. $t^2 - 16 = 0$

36. $4y^2 - 1 = 0$

37. $9z^2 - 4 = 0$

38. $x + 15 = x(x - 1)$

39. $\dfrac{3x^2}{2} = 4x - 2$

40. $\dfrac{2x^2}{5} = 3x - 5$

41. $\dfrac{2x^2}{9} + x = 2$

42. $\dfrac{3x^2}{8} - x = 2$

43. $\dfrac{3}{4}z^2 - z = -\dfrac{1}{3}$

44. $\dfrac{r^2}{2} = 1 - \dfrac{r}{12}$

45. $p + 18 = p(p - 2)$

46. $r^2 - r - 2 = (2r - 1)(r - 3)$

47. $s^2 + 5s - 4 = (2s + 1)(s - 4)$

48. $x^2 + x + 5 = (3x + 2)(x - 4)$

2 Solve by taking square roots.

49. $x^2 = 36$

50. $y^2 = 49$

51. $v^2 - 1 = 0$

52. $z^2 - 64 = 0$

53. $4x^2 - 49 = 0$

54. $9w^2 - 64 = 0$

55. $9y^2 = 4$

56. $4z^2 = 25$

57. $16v^2 - 9 = 0$

58. $25x^2 - 64 = 0$

59. $y^2 + 81 = 0$

60. $z^2 + 49 = 0$

61. $w^2 - 24 = 0$

62. $v^2 - 48 = 0$

63. $(x - 1)^2 = 36$

64. $(y + 2)^2 = 49$

65. $2(x + 5)^2 = 8$

66. $4(z - 3)^2 = 100$

67. $2(x + 1)^2 = 50$

68. $3(x - 4)^2 = 27$

69. $4(x + 5)^2 = 64$

70. $9(x - 3)^2 = 81$

71. $2(x - 9)^2 = 98$

72. $5(x + 5)^2 = 125$

73. $12(x + 3)^2 = 27$

74. $8(x - 4)^2 = 50$

75. $9(x - 1)^2 - 16 = 0$

76. $4(y + 3)^2 - 81 = 0$

77. $49(v + 1)^2 - 25 = 0$ **78.** $81(y - 2)^2 - 64 = 0$ **79.** $(x - 4)^2 - 20 = 0$

80. $(y + 5)^2 - 50 = 0$ **81.** $(x + 1)^2 + 36 = 0$ **82.** $(y - 7)^2 + 49 = 0$

83. $2\left(z - \dfrac{1}{2}\right)^2 = 12$ **84.** $3\left(v + \dfrac{3}{4}\right)^2 = 36$ **85.** $4\left(x - \dfrac{2}{3}\right)^2 = 16$

SUPPLEMENTAL EXERCISES 11.1

Solve for x.

86. $(x^2 - 1)^2 = 9$ **87.** $(x^2 + 3)^2 = 25$

88. $(6x^2 - 5)^2 = 1$ **89.** $x^2 = x$

90. $ax^2 - bx = 0, a > 0$ and $b > 0$ **91.** $ax^2 - b = 0, a > 0$ and $b > 0$

92. The value P of an initial investment of A dollars after 2 years is given by $P = A(1 + r)^2$, where r is the annual percentage rate earned by the investment. If an initial investment of \$1500 grew to a value of \$1782.15 in 2 years, what was the annual percentage rate?

93. The kinetic energy of a moving body is given by $E = \dfrac{1}{2}mv^2$, where E is the kinetic energy, m is the mass, and v is the velocity. What is the velocity of a moving body whose mass is 5 kg and whose kinetic energy is 250 newton-meters?

94. On a certain type of street surface, the equation $d = 0.0074v^2$ can be used to approximate the distance d a car traveling v miles per hour will slide when its brakes are applied. After applying the brakes, the owner of a car involved in an accident skidded 40 ft. Did the traffic officer investigating the accident issue the car owner a ticket for speeding if the speed limit was 65 mph?

SECTION 11.2

Solving Quadratic Equations by Completing the Square

1 Solve quadratic equations by completing the square

Recall that a perfect-square trinomial is the square of a binomial.

Perfect-square trinomial		Square of a binomial
$x^2 + 6x + 9$	$=$	$(x + 3)^2$
$x^2 - 10x + 25$	$=$	$(x - 5)^2$
$x^2 + 8x + 16$	$=$	$(x + 4)^2$

For each perfect-square trinomial, the square of $\frac{1}{2}$ of the coefficient of x equals the constant term.

$$x^2 + 6x + 9, \qquad \left(\frac{1}{2} \cdot 6\right)^2 = 9$$

$$x^2 - 10x + 25, \qquad \left[\frac{1}{2}(-10)\right]^2 = 25 \qquad \left(\frac{1}{2} \text{ coefficient of } x\right)^2 = \text{Constant term}$$

$$x^2 + 8x + 16, \qquad \left(\frac{1}{2} \cdot 8\right)^2 = 16$$

This relationship can be used to write the constant term for a perfect-square trinomial. Adding to a binomial the constant term that makes it a perfect-square trinomial is called **completing the square.**

Complete the square of $x^2 - 8x$. Write the resulting perfect-square trinomial as the square of a binomial.

$$x^2 - 8x$$

Find the constant term.

$$\left[\frac{1}{2}(-8)\right]^2 = 16$$

Complete the square of $x^2 - 8x$ by adding the constant term.

$$x^2 - 8x + 16$$

Write the resulting perfect-square trinomial as the square of a binomial.

$$x^2 - 8x + 16 = (x - 4)^2$$

Complete the square of $y^2 + 5y$. Write the resulting perfect-square trinomial as the square of a binomial.

$$y^2 + 5y$$

Find the constant term.

$$\left(\frac{1}{2} \cdot 5\right)^2 = \left(\frac{5}{2}\right)^2 = \frac{25}{4}$$

Complete the square of $y^2 + 5y$ by adding the constant term.

$$y + 5y + \frac{25}{4}$$

Write the resulting perfect-square trinomial as the square of a binomial.

$$y^2 + 5y + \frac{25}{4} = \left(y + \frac{5}{2}\right)^2$$

A quadratic equation that cannot be solved by factoring can be solved by completing the square. The procedure is:

1. Write the equation in the form $x^2 + bx = c$.
2. Add to each side of the equation the term that completes the square on $x^2 + bx$.
3. Factor the perfect-square trinomial. Write it as the square of a binomial.
4. Take the square root of each side of the equation.
5. Solve for x.

POINT OF INTEREST

Early mathematicians solved quadratic equations by literally *completing the square*. For these mathematicians, all equations had geometric interpretations. They found that a quadratic equation could be solved by making certain figures into squares. See the Project in Mathematics at the end of this chapter for an idea of how this was done.

Solve by completing the square: $x^2 - 6x - 3 = 0$

Add the opposite of the constant term to each side of the equation.

$$x^2 - 6x - 3 = 0$$
$$x^2 - 6x = 3$$

Find the constant term that completes the square of $x^2 - 6x$.

$$\left[\frac{1}{2}(-6)\right]^2 = 9$$

Add this term to each side of the equation.

$$x^2 - 6x + 9 = 3 + 9$$

Factor the perfect-square trinomial.

$$(x - 3)^2 = 12$$

Take the square root of each side of the equation.

$$\sqrt{(x-3)^2} = \pm\sqrt{12}$$

Simplify.

$$x - 3 = \pm 2\sqrt{3}$$

Solve for x.

$$x - 3 = 2\sqrt{3} \qquad x - 3 = -2\sqrt{3}$$
$$x = 3 + 2\sqrt{3} \qquad\qquad x = 3 - 2\sqrt{3}$$

Check:

$$
\begin{array}{r|l}
x^2 - 6x - 3 = 0 & \\
\hline
(3 + 2\sqrt{3})^2 - 6(3 + 2\sqrt{3}) - 3 & 0 \\
9 + 12\sqrt{3} + 12 - 18 - 12\sqrt{3} - 3 & 0 \\
0 = 0 &
\end{array}
$$

$$
\begin{array}{r|l}
x^2 - 6x - 3 = 0 & \\
\hline
(3 - 2\sqrt{3})^2 - 6(3 - 2\sqrt{3}) - 3 & 0 \\
9 - 12\sqrt{3} + 12 - 18 + 12\sqrt{3} - 3 & 0 \\
0 = 0 &
\end{array}
$$

Write the solutions.

The solutions are $3 + 2\sqrt{3}$ and $3 - 2\sqrt{3}$.

Solve by completing the square: $2x^2 - x - 1 = 0$

Add the opposite of the constant term to each side of the equation.

$$2x^2 - x - 1 = 0$$
$$2x^2 - x = 1$$

For the method of completing the square to be used, the coefficient of the x^2 term must be 1. Multiply each term by the reciprocal of the coefficient of x^2.

$$\frac{1}{2}(2x^2 - x) = \frac{1}{2}\cdot 1$$
$$x^2 - \frac{1}{2}x = \frac{1}{2}$$

Find the constant term that completes the square of $x^2 - \frac{1}{2}x$.

$$\left[\frac{1}{2}\left(-\frac{1}{2}\right)\right]^2 = \left(-\frac{1}{4}\right)^2 = \frac{1}{16}$$

Add this term to each side of the equation.

$$x^2 - \frac{1}{2}x + \frac{1}{16} = \frac{1}{2} + \frac{1}{16}$$

Factor the perfect-square trinomial.

$$\left(x - \frac{1}{4}\right)^2 = \frac{9}{16}$$

Take the square root of each side of the equation.

$$\sqrt{\left(x - \frac{1}{4}\right)^2} = \pm\sqrt{\frac{9}{16}}$$

Simplify.

$$x - \frac{1}{4} = \pm\frac{3}{4}$$

Solve for x.

$$x - \frac{1}{4} = \frac{3}{4} \qquad x - \frac{1}{4} = -\frac{3}{4}$$

$$x = 1 \qquad\qquad x = -\frac{1}{2}$$

Check:

$$\begin{array}{c|c}
2x^2 - x - 1 = 0 & \\
\hline
2(1)^2 - 1 - 1 & 0 \\
2(1) - 1 - 1 & 0 \\
2 - 1 - 1 & 0 \\
0 = 0 &
\end{array}$$

$$\begin{array}{c|c}
2x^2 - x - 1 = 0 & \\
\hline
2\left(-\frac{1}{2}\right)^2 - \left(-\frac{1}{2}\right) - 1 & 0 \\[2mm]
2\left(\frac{1}{4}\right) - \left(-\frac{1}{2}\right) - 1 & 0 \\[2mm]
\frac{1}{2} + \frac{1}{2} - 1 & 0 \\[2mm]
0 = 0 &
\end{array}$$

Write the solutions.

The solutions are 1 and $-\frac{1}{2}$.

Example 1   Solve by completing the square.

A. $2x^2 - 4x - 1 = 0$ B. $p^2 - 4p + 6 = 0$

Solution   A. $2x^2 - 4x - 1 = 0$

$$2x^2 - 4x = 1$$ ▸ Add the opposite of -1 to each side of the equation.

$$\frac{1}{2}(2x^2 - 4x) = \frac{1}{2} \cdot 1$$ ▸ The coefficient of the x^2 term must be 1.

$$x^2 - 2x = \frac{1}{2}$$

$$x^2 - 2x + 1 = \frac{1}{2} + 1$$ ▸ Complete the square of $x^2 - 2x$. Add 1 to each side of the equation.

$$(x - 1)^2 = \frac{3}{2}$$ ▸ Factor the perfect-square trinomial.

$$\sqrt{(x - 1)^2} = \pm\sqrt{\frac{3}{2}}$$ ▸ Take the square root of each side of the equation.

$$x - 1 = \pm\frac{\sqrt{3}}{\sqrt{2}}$$

$$x - 1 = \pm\frac{\sqrt{6}}{2}$$ ▸ Rationalize the denominator. $\frac{\sqrt{3}}{\sqrt{2}} = \frac{\sqrt{3}}{\sqrt{2}} \cdot \frac{\sqrt{2}}{\sqrt{2}} = \frac{\sqrt{6}}{2}$

$$x - 1 = \frac{\sqrt{6}}{2} \qquad\qquad x - 1 = -\frac{\sqrt{6}}{2}$$

$$x = 1 + \frac{\sqrt{6}}{2} \qquad\qquad x = 1 - \frac{\sqrt{6}}{2}$$

$$= \frac{2}{2} + \frac{\sqrt{6}}{2} \qquad\qquad = \frac{2}{2} - \frac{\sqrt{6}}{2}$$

$$= \frac{2 + \sqrt{6}}{2} \qquad\qquad = \frac{2 - \sqrt{6}}{2}$$

The solutions are $\dfrac{2 + \sqrt{6}}{2}$ and $\dfrac{2 - \sqrt{6}}{2}$.

B. $p^2 - 4p + 6 = 0$

$p^2 - 4p = -6$ ▶ Add the opposite of the constant term to each side of the equation.

$p^2 - 4p + 4 = -6 + 4$ ▶ Complete the square of $p^2 - 4p$. Add 4 to each side of the equation.

$(p - 2)^2 = -2$ ▶ Factor the perfect-square trinomial.

$\sqrt{(p - 2)^2} = \pm\sqrt{-2}$ ▶ Take the square root of each side of the equation.

$\sqrt{-2}$ is not a real number.
The equation has no real number solution.

Problem 1 Solve by completing the square.
A. $3x^2 - 6x - 2 = 0$ B. $y^2 - 6y + 10 = 0$

Solution See page A42.

Example 2 Solve by completing the square: $x^2 + 6x + 4 = 0$
Approximate the solutions to the nearest thousandth.

Solution $x^2 + 6x + 4 = 0$

$x^2 + 6x = -4$

$x^2 + 6x + 9 = -4 + 9$ ▶ Complete the square of $x^2 + 6x$. Add 9 to each side of the equation.

$(x + 3)^2 = 5$

$\sqrt{(x + 3)^2} = \pm\sqrt{5}$

$x + 3 = \pm\sqrt{5}$

$$x + 3 = \sqrt{5} \qquad\qquad x + 3 = -\sqrt{5}$$
$$x = -3 + \sqrt{5} \qquad\qquad x = -3 - \sqrt{5}$$
$$\approx -3 + 2.236 \qquad\qquad \approx -3 - 2.236$$
$$\approx -0.764 \qquad\qquad \approx -5.236$$

The solutions are approximately -0.764 and -5.236.

Problem 2 Solve by completing the square: $x^2 + 8x + 8 = 0$
Approximate the solutions to the nearest thousandth.

Solution See page A43.

CONCEPT REVIEW 11.2

Determine whether the statement is always true, sometimes true, or never true.

1. When we square a binomial, the result is a perfect-square trinomial.

2. The constant term in a perfect-square trinomial is equal to the square of one-half the coefficient of x.

3. To "complete the square" means to add to $x^2 + bx$ the constant term that will make it a perfect-square trinomial.

4. To complete the square of $x^2 + 4x$, add 16 to it.

5. To complete the square of $3x^2 + 6x$, add 9 to it.

6. In solving a quadratic equation by completing the square, the next step after writing the equation in the form $(x + a)^2 = b$ is to take the square root of each side of the equation.

EXERCISES 11.2

1   Complete the square. Write the resulting perfect-square trinomial as the square of a binomial.

1. $x^2 + 12x$

2. $x^2 - 4x$

3. $x^2 + 10x$

4. $x^2 + 3x$

5. $x^2 - x$

6. $x^2 + 5x$

Solve by completing the square.

7. $x^2 + 2x - 3 = 0$

8. $y^2 + 4y - 5 = 0$

9. $v^2 + 4v + 1 = 0$

10. $y^2 - 2y - 5 = 0$

11. $v^2 - 6v + 13 = 0$

12. $x^2 + 4x + 13 = 0$

13. $x^2 + 6x = 5$

14. $w^2 - 8w = 3$

15. $x^2 = 4x - 4$

16. $z^2 = 8z - 16$

17. $z^2 = 2z + 1$

18. $y^2 = 10y - 20$

Solve. First try to solve the equation by factoring. If you are unable to solve the equation by factoring, solve the equation by completing the square.

19. $p^2 + 3p = 1$

20. $r^2 + 5r = 2$

21. $w^2 + 7w = 8$

22. $y^2 + 5y = -4$

23. $x^2 + 6x + 4 = 0$

24. $y^2 - 8y - 1 = 0$

25. $r^2 - 8r = -2$

26. $s^2 + 6s = 5$

27. $t^2 - 3t = -2$

28. $y^2 = 4y + 12$

29. $w^2 = 3w + 5$

30. $x^2 = 1 - 3x$

31. $x^2 - x - 1 = 0$

32. $x^2 - 7x = -3$

33. $y^2 - 5y + 3 = 0$

34. $z^2 - 5z = -2$

35. $v^2 + v - 3 = 0$

36. $x^2 - x = 1$

37. $y^2 = 7 - 10y$

38. $v^2 = 14 + 16v$

39. $s^2 + 3s = -1$

40. $r^2 - 3r = 5$

41. $t^2 - t = 4$

42. $y^2 + y - 4 = 0$

43. $x^2 - 3x + 5 = 0$

44. $z^2 + 5z + 7 = 0$

45. $2t^2 - 3t + 1 = 0$

46. $2x^2 - 7x + 3 = 0$

47. $2r^2 + 5r = 3$

48. $2y^2 - 3y = 9$

49. $4v^2 - 4v - 1 = 0$

50. $2s^2 - 4s - 1 = 0$

51. $4z^2 - 8z = 1$

52. $3r^2 - 2r = 2$

53. $3y - 5 = (y - 1)(y - 2)$

54. $4p + 2 = (p - 1)(p + 3)$

55. $\dfrac{x^2}{4} - \dfrac{x}{2} = 3$

56. $\dfrac{x^2}{6} - \dfrac{x}{3} = 1$

57. $\dfrac{2x^2}{3} = 2x + 3$

58. $\dfrac{3x^2}{2} = 3x + 2$

59. $\dfrac{x}{3} + \dfrac{3}{x} = \dfrac{8}{3}$

60. $\dfrac{x}{4} - \dfrac{2}{x} = \dfrac{3}{4}$

Solve by completing the square. Approximate the solutions to the nearest thousandth.

61. $y^2 + 3y = 5$

62. $w^2 + 5w = 2$

63. $2z^2 - 3z = 7$

64. $2x^2 + 3x = 11$

65. $4x^2 + 6x - 1 = 0$

66. $4x^2 + 2x - 3 = 0$

SUPPLEMENTAL EXERCISES 11.2

Solve.

67. $\sqrt{2x + 7} - 4 = x$

68. $\dfrac{x + 1}{2} + \dfrac{3}{x - 1} = 4$

69. $\dfrac{x - 2}{3} + \dfrac{2}{x + 2} = 4$

70. Explain why the equation $(x - 2)^2 = -4$ does not have a real number solution.

SECTION 11.3

Solving Quadratic Equations by Using the Quadratic Formula

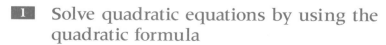

1 Solve quadratic equations by using the quadratic formula

Any quadratic equation can be solved by completing the square. Applying this method to the standard form of a quadratic equation produces a formula that can be used to solve any quadratic equation.

To solve $ax^2 + bx + c = 0$, $a \neq 0$, by completing the square, subtract the constant term from each side of the equation.

$$ax^2 + bx + c = 0$$
$$ax^2 + bx + c - c = 0 - c$$
$$ax^2 + bx = -c$$

Multiply each side of the equation by the reciprocal of a, the coefficient of x^2.

$$\frac{1}{a}(ax^2 + bx) = \frac{1}{a}(-c)$$

$$x^2 + \frac{b}{a}x = -\frac{c}{a}$$

Complete the square by adding $\left(\frac{1}{2} \cdot \frac{b}{a}\right)^2$ to each side of the equation.

$$x^2 + \frac{b}{a}x + \left(\frac{1}{2} \cdot \frac{b}{a}\right)^2 = \left(\frac{1}{2} \cdot \frac{b}{a}\right)^2 - \frac{c}{a}$$

$$x^2 + \frac{b}{a}x + \frac{b^2}{4a^2} = \frac{b^2}{4a^2} - \frac{c}{a}$$

Simplify the right side of the equation.

$$x^2 + \frac{b}{a}x + \frac{b^2}{4a^2} = \frac{b^2}{4a^2} - \left(\frac{c}{a} \cdot \frac{4a}{4a}\right)$$

$$x^2 + \frac{b}{a}x + \frac{b^2}{4a^2} = \frac{b^2}{4a^2} - \frac{4ac}{4a^2}$$

$$x^2 + \frac{b}{a}x + \frac{b^2}{4a^2} = \frac{b^2 - 4ac}{4a^2}$$

Factor the perfect-square trinomial on the left side of the equation.

$$\left(x + \frac{b}{2a}\right)^2 = \frac{b^2 - 4ac}{4a^2}$$

Take the square root of each side of the equation.

$$\sqrt{\left(x + \frac{b}{2a}\right)^2} = \pm\sqrt{\frac{b^2 - 4ac}{4a^2}}$$

$$x + \frac{b}{2a} = \pm\frac{\sqrt{b^2 - 4ac}}{2a}$$

Solve for x.

$$x + \frac{b}{2a} = \frac{\sqrt{b^2 - 4ac}}{2a} \qquad x + \frac{b}{2a} = -\frac{\sqrt{b^2 - 4ac}}{2a}$$

$$x = -\frac{b}{2a} + \frac{\sqrt{b^2 - 4ac}}{2a} \qquad x = -\frac{b}{2a} - \frac{\sqrt{b^2 - 4ac}}{2a}$$

$$x = \frac{-b + \sqrt{b^2 - 4ac}}{2a} \qquad x = \frac{-b - \sqrt{b^2 - 4ac}}{2a}$$

The Quadratic Formula

The solution of $ax^2 + bx + c = 0$, $a \neq 0$, is

$$x = \frac{-b + \sqrt{b^2 - 4ac}}{2a} \quad \text{or} \quad x = \frac{-b - \sqrt{b^2 - 4ac}}{2a}$$

The quadratic formula is frequently written in the form

$$x = \frac{-b \pm \sqrt{b^2 - 4ac}}{2a}$$

Solve by using the quadratic formula: $2x^2 = 4x - 1$

Write the equation in standard form. $a = 2$, $b = -4$, and $c = 1$.

$$2x^2 = 4x - 1$$
$$2x^2 - 4x + 1 = 0$$

Replace a, b, and c in the quadratic formula by their values.

$$x = \frac{-b \pm \sqrt{b^2 - 4ac}}{2a}$$

Simplify.

$$= \frac{-(-4) \pm \sqrt{(-4)^2 - 4 \cdot 2 \cdot 1}}{2 \cdot 2}$$

$$= \frac{4 \pm \sqrt{16 - 8}}{4}$$

$$= \frac{4 \pm \sqrt{8}}{4}$$

$$= \frac{4 \pm 2\sqrt{2}}{4} = \frac{2(2 \pm \sqrt{2})}{2 \cdot 2} = \frac{2 \pm \sqrt{2}}{2}$$

Check:

$2x^2$	$= 4x - 1$
$2\left(\dfrac{2 + \sqrt{2}}{2}\right)^2$	$4\left(\dfrac{2 + \sqrt{2}}{2}\right) - 1$
$2\left(\dfrac{4 + 4\sqrt{2} + 2}{4}\right)$	$2(2 + \sqrt{2}) - 1$
$2\left(\dfrac{6 + 4\sqrt{2}}{4}\right)$	$4 + 2\sqrt{2} - 1$
$2\left(\dfrac{3 + 2\sqrt{2}}{2}\right)$	$3 + 2\sqrt{2}$
$3 + 2\sqrt{2}$	$= 3 + 2\sqrt{2}$

$2x^2$	$= 4x - 1$
$2\left(\dfrac{2 - \sqrt{2}}{2}\right)^2$	$4\left(\dfrac{2 - \sqrt{2}}{2}\right) - 1$
$2\left(\dfrac{4 - 4\sqrt{2} + 2}{4}\right)$	$2(2 - \sqrt{2}) - 1$
$2\left(\dfrac{6 - 4\sqrt{2}}{4}\right)$	$4 - 2\sqrt{2} - 1$
$2\left(\dfrac{3 - 2\sqrt{2}}{2}\right)$	$3 - 2\sqrt{2}$
$3 - 2\sqrt{2}$	$= 3 - 2\sqrt{2}$

Write the solutions.

The solutions are $\dfrac{2 + \sqrt{2}}{2}$ and $\dfrac{2 - \sqrt{2}}{2}$.

Example 1   Solve by using the quadratic formula.

A. $2x^2 - 3x + 1 = 0$ B. $2x^2 = 8x - 5$

C. $t^2 - 3t + 7 = 0$

Solution   A. $2x^2 - 3x + 1 = 0$ ▸ $a = 2, b = -3, c = 1$

$$x = \frac{-(-3) \pm \sqrt{(-3)^2 - 4(2)(1)}}{2 \cdot 2}$$

$$= \frac{3 \pm \sqrt{9 - 8}}{4} = \frac{3 \pm \sqrt{1}}{4} = \frac{3 \pm 1}{4}$$

$$x = \frac{3 + 1}{4} \qquad x = \frac{3 - 1}{4}$$

$$= \frac{4}{4} = 1 \qquad = \frac{2}{4} = \frac{1}{2}$$

The solutions are 1 and $\frac{1}{2}$.

B. $$2x^2 = 8x - 5$$
$$2x^2 - 8x + 5 = 0$$ ▸ Write the equation in standard form. $a = 2, b = -8, c = 5$

$$x = \frac{-(-8) \pm \sqrt{(-8)^2 - 4(2)(5)}}{2 \cdot 2}$$

$$= \frac{8 \pm \sqrt{64 - 40}}{4}$$

$$= \frac{8 \pm \sqrt{24}}{4}$$

$$= \frac{8 \pm 2\sqrt{6}}{4}$$

$$= \frac{2(4 \pm \sqrt{6})}{2 \cdot 2} = \frac{4 \pm \sqrt{6}}{2}$$

The solutions are $\frac{4 + \sqrt{6}}{2}$ and $\frac{4 - \sqrt{6}}{2}$.

C. $t^2 - 3t + 7 = 0$ ▸ $a = 1, b = -3, c = 7$

$$t = \frac{-(-3) \pm \sqrt{(-3)^2 - 4(1)(7)}}{2(1)}$$

$$= \frac{3 \pm \sqrt{9 - 28}}{2} = \frac{3 \pm \sqrt{-19}}{2}$$

$\sqrt{-19}$ is not a real number.

The equation has no real number solution.

Problem 1   Solve by using the quadratic formula.
     A. $3x^2 + 4x - 4 = 0$ B. $x^2 + 2x = 1$
     C. $z^2 + 2z + 6 = 0$

Solution   See page A43.

CONCEPT REVIEW 11.3

Determine whether the statement is always true, sometimes true, or never true.

1.   Any quadratic equation can be solved by using the quadratic formula.

2.   The solutions of a quadratic equation can be written as $x = \dfrac{-b + \sqrt{b^2 - 4ac}}{2a}$

   and $x = \dfrac{-b - \sqrt{b^2 - 4ac}}{2a}$.

3.   The equation $4x^2 - 3x = 9$ is a quadratic equation in standard form.

4.   In the quadratic formula $x = \dfrac{-b \pm \sqrt{b^2 - 4ac}}{2a}$, b is the coefficient of x^2.

5. $\dfrac{6 \pm 2\sqrt{3}}{2}$ simplifies to $3 \pm 2\sqrt{3}$.

EXERCISES 11.3

1.   Solve by using the quadratic formula.

1. $z^2 + 6z - 7 = 0$ **2.** $s^2 + 3s - 10 = 0$ **3.** $w^2 = 3w + 18$ **4.** $r^2 = 5 - 4r$

5. $t^2 - 2t = 5$ **6.** $y^2 - 4y = 6$ **7.** $t^2 + 6t - 1 = 0$ **8.** $z^2 + 4z + 1 = 0$

9. $w^2 + 3w + 5 = 0$ **10.** $x^2 - 2x + 6 = 0$ **11.** $w^2 = 4w + 9$ **12.** $y^2 = 8y + 3$

Solve. First try to solve the equation by factoring. If you are unable to solve the equation by factoring, solve the equation by using the quadratic formula.

13. $p^2 - p = 0$ **14.** $2v^2 + v = 0$ **15.** $4t^2 - 4t - 1 = 0$

16. $4x^2 - 8x - 1 = 0$ **17.** $4t^2 - 9 = 0$ **18.** $4s^2 - 25 = 0$

19. $3x^2 - 6x + 2 = 0$ **20.** $5x^2 - 6x = 3$ **21.** $3t^2 = 2t + 3$

22. $4n^2 = 7n - 2$ **23.** $2x^2 + x + 1 = 0$ **24.** $3r^2 - r + 2 = 0$

25. $2y^2 + 3 = 8y$ **26.** $5x^2 - 1 = x$ **27.** $3t^2 = 7t + 6$

28. $3x^2 = 10x + 8$ **29.** $3y^2 - 4 = 5y$ **30.** $6x^2 - 5 = 3x$

31. $3x^2 = x + 3$ **32.** $2n^2 = 7 - 3n$ **33.** $5d^2 - 2d - 8 = 0$

34. $x^2 - 7x - 10 = 0$ **35.** $5z^2 + 11z = 12$ **36.** $4v^2 = v + 3$

37. $v^2 + 6v + 1 = 0$ **38.** $s^2 + 4s - 8 = 0$ **39.** $4t^2 - 12t - 15 = 0$

40. $4w^2 - 20w + 5 = 0$ **41.** $2x^2 = 4x - 5$ **42.** $3r^2 = 5r - 6$

43. $9y^2 + 6y - 1 = 0$ **44.** $9s^2 - 6s - 2 = 0$ **45.** $6s^2 - s - 2 = 0$

46. $6y^2 + 5y - 4 = 0$

47. $4p^2 + 16p = -11$

48. $4y^2 - 12y = -1$

49. $4x^2 = 4x + 11$

50. $4s^2 + 12s = 3$

51. $4p^2 = -12p - 9$

52. $3y^2 + 6y = -3$

53. $9v^2 = -30v - 23$

54. $9t^2 = 30t + 17$

55. $\dfrac{x^2}{2} - \dfrac{x}{3} = 1$

56. $\dfrac{x^2}{4} - \dfrac{x}{2} = 5$

57. $\dfrac{2x^2}{5} = x + 1$

58. $\dfrac{3x^2}{2} + 2x = 1$

59. $\dfrac{x}{5} + \dfrac{5}{x} = \dfrac{12}{5}$

60. $\dfrac{x}{4} + \dfrac{3}{x} = \dfrac{5}{2}$

Solve by using the quadratic formula. Approximate the solutions to the nearest thousandth.

61. $x^2 - 2x - 21 = 0$

62. $y^2 + 4y - 11 = 0$

63. $s^2 - 6s - 13 = 0$

64. $w^2 + 8w - 15 = 0$

65. $2p^2 - 7p - 10 = 0$

66. $3t^2 - 8t - 1 = 0$

67. $4z^2 + 8z - 1 = 0$

68. $4x^2 + 7x + 1 = 0$

69. $5v^2 - v - 5 = 0$

SUPPLEMENTAL EXERCISES 11.3

Solve.

70. $\sqrt{x^2 + 2x + 1} = x - 1$

71. $\dfrac{x + 2}{3} - \dfrac{4}{x - 2} = 2$

72. $\dfrac{x + 1}{5} - \dfrac{3}{x - 1} = 2$

73. A baseball player hits a ball. The height h of the ball above the ground at time t can be approximated by the equation $h = -16t^2 + 76t + 5$. When will the ball hit the ground? Round to the nearest hundredth. (*Hint*: The ball strikes the ground when $h = 0$ ft.)

5 ft

74. A basketball player shoots at a basket 25 ft away. The height h of the ball above the ground at time t is given by $h = -16t^2 + 32t + 6.5$. How many seconds after the ball is released does it hit the basket? Round to the nearest hundredth. (*Hint*: When it hits the basket, $h = 10$ ft.)

75. True or false?
  a. The equations $x = \sqrt{12 - x}$ and $x^2 = 12 - x$ have the same solution.
  b. If $\sqrt{a} + \sqrt{b} = c$, then $a + b = c^2$.
  c. $\sqrt{9} = \pm 3$
  d. $\sqrt{x^2} = |x|$

76. Factoring, completing the square, and using the quadratic formula are three methods of solving quadratic equations. Describe each method, and cite the advantages and disadvantages of using each.

77. Explain why the equation $0x^2 + 3x + 4 = 0$ cannot be solved by the quadratic formula.

S E C T I O N **11.4**

Graphing Quadratic Equations in Two Variables

1 Graph a quadratic equation of the form $y = ax^2 + bx + c$

An equation of the form $y = ax^2 + bx + c$, $a \neq 0$, is a **quadratic equation in two variables.** Examples of quadratic equations in two variables are shown at the right.

$$y = 3x^2 - x + 1$$
$$y = -x^2 - 3$$
$$y = 2x^2 - 5x$$

For these equations, y is a function of x, and we can write $f(x) = ax^2 + bx + c$. This represents a **quadratic function.**

The graph of a quadratic equation in two variables is a **parabola.** The graph is "cup-shaped" and opens either up or down. The graphs of two parabolas are shown below.

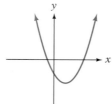

Parabola that opens up

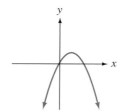

Parabola that opens down

Graph: $y = x^2 - 2x - 3$

Find several solutions of the equation. Because the graph is not a straight line, several solutions must be found in order to determine the cup shape.

Display the ordered-pair solutions in a table.

x	y
0	−3
1	−4
−1	0
2	−3
3	0

Graph the ordered-pair solutions on a rectangular coordinate system.

Draw a parabola through the points.

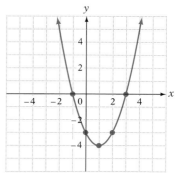

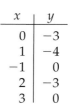 Using a graphing utility, enter the equation $y = x^2 - 2x - 3$ and verify the graph shown above. (Refer to the Appendix on page A1 for instructions on

using a graphing calculator to graph a quadratic equation.) Trace along the graph and verify that $(0, -3)$, $(1, -4)$, $(-1, 0)$, $(2, -3)$, and $(3, 0)$ are coordinates of points on the graph.

The graph of $y = -2x^2 + 1$ is shown below.

x	y
0	1
1	-1
-1	-1
2	-7
-2	-7

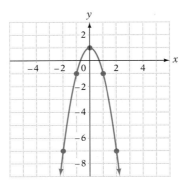

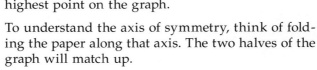 Use a graphing utility to verify the graph of $y = -2x^2 + 1$ shown above.

Note that the graph of $y = x^2 - 2x - 3$, shown on the previous page, opens up and that the coefficient of x^2 is positive. The graph of $y = -2x^2 + 1$ opens down, and the coefficient of x^2 is negative. For any quadratic equation in two variables, the coefficient of x^2 determines whether the parabola opens up or down. **When a is positive, the parabola opens up. When a is negative, the parabola opens down.**

POINT OF INTEREST

Mirrors in some telescopes are ground into the shape of a parabola. The mirror at the Palomar Mountain Observatory is 2 ft thick at the ends and weighs 14.75 tons. The mirror has been ground to a true paraboloid (the three-dimensional version of a parabola) to within 0.0000015 in. An equation of the mirror is $y = 2640x^2$.

Every parabola has an axis of symmetry and a **vertex** that is on the axis of symmetry. If the parabola opens up, the vertex is the lowest point on the graph. If the parabola opens down, the vertex is the highest point on the graph.

To understand the axis of symmetry, think of folding the paper along that axis. The two halves of the graph will match up.

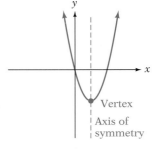

When graphing a quadratic equation in two variables, use the value of a to determine whether the parabola opens up or down. After graphing ordered-pair solutions of the equation, use symmetry to help you draw the parabola.

Example 1 Graph.   A. $y = x^2 - 2x$ B. $y = -x^2 + 4x - 4$

Solution   A. $a = 1$. a is positive.
The parabola opens up.

x	y
0	0
1	-1
-1	3
2	0
3	3

B. $a = -1$. a is negative.
The parabola opens down.

x	y
0	-4
1	-1
2	0
3	-1
4	-4

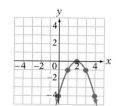

Problem 1   Graph.   A. $y = x^2 + 2$ B. $y = -x^2 - 2x + 3$

Solution   See page A44.

CONCEPT REVIEW 11.4

Determine whether the statement is always true, sometimes true, or never true.

1. The equation $y = 3x^2 - 2x + 1$ is an example of a quadratic equation in two variables.

2. The graph of an equation of the form $y = ax^2 + bx + c, a \neq 0$, is a parabola.

3. The graph of an equation of the form $y = ax^2 + c, a \neq 0$, is a straight line.

4. For the equation $y = -x^2 + 2x - 1$, when $x = 1$, $y = 2$.

5. The graph of the equation $y = -\frac{1}{2}x^2 + 5x$ is a parabola opening up.

6. If a parabola opens down, then the vertex is the highest point on the graph.

EXERCISES 11.4

1   Determine whether the graph of the equation opens up or down.

1. $y = x^2 - 3$

2. $y = -x^2 + 4$

3. $y = \frac{1}{2}x^2 - 2$

4. $y = -\frac{1}{3}x^2 + 5$

5. $y = x^2 - 2x + 3$

6. $y = -x^2 + 4x - 1$

Graph.

7. $y = x^2$

8. $y = -x^2$

9. $y = -x^2 + 1$

10. $y = x^2 - 1$

11. $y = 2x^2$

12. $y = \frac{1}{2}x^2$

13. $y = -\frac{1}{2}x^2 + 1$

14. $y = 2x^2 - 1$

15. $y = x^2 - 4x$

16. $y = x^2 + 4x$

17. $y = x^2 - 2x + 3$

18. $y = x^2 - 4x + 2$

19. $y = -x^2 + 2x + 3$

20. $y = x^2 + 2x + 1$

21. $y = -x^2 + 3x - 4$

22. $y = -x^2 + 6x - 9$

23. $y = 2x^2 + x - 3$

24. $y = -2x^2 - 3x + 3$

Graph using a graphing utility. Verify that the graph is a graph of a parabola opening up if a is positive or opening down if a is negative.

25. $y = x^2 - 2$

26. $y = -x^2 + 3$

27. $y = x^2 + 2x$

28. $y = -2x^2 + 4x$

29. $y = \frac{1}{2}x^2 - x$

30. $y = -\frac{1}{2}x^2 + 2$

31. $y = x^2 - x - 2$

32. $y = x^2 - 3x + 2$

33. $y = 2x^2 - x - 5$

34. $y = -2x^2 - 3x + 2$

35. $y = -x^2 - 2x - 1$

36. $y = \dfrac{1}{2}x^2 - 2x - 1$

SUPPLEMENTAL EXERCISES 11.4

Show that the equation is a quadratic equation in two variables by writing it in the form $y = ax^2 + bx + c$.

37. $y + 1 = (x - 4)^2$

38. $y - 2 = 3(x + 1)^2$

39. $y - 4 = 2(x - 3)^2$

The x-intercepts of a parabola are the points at which the graph crosses the x-axis. Because any point on the x-axis has y-coordinate 0, the x-intercepts of a parabola occur when $y = 0$. Therefore, the x-coordinate of an x-intercept is a solution of the equation $ax^2 + bx + c = 0$. For example, the solutions of the equation $x^2 - 1 = 0$ are 1 and -1, and the graph of the equation $y = x^2 - 1$ crosses the x-axis at $(1, 0)$ and $(-1, 0)$. Determine the x-intercepts of the graphs of the following equations.

40. $y = x^2 - 4$

41. $y = -x^2 + 4$

42. $y = x^2 - x$

43. $y = x^2 - 4x$

44. $y = x^2 - 4x + 3$

45. $y = 2x^2 - x - 1$

Evaluate the function.

46. Find $f(3)$ when $f(x) = x^2 + 3$.

47. Find $g(2)$ when $g(x) = x^2 - 2x - 3$.

48. Find $S(-2)$ when $S(t) = 2t^2 - 3t - 1$.

49. Find $P(-3)$ when $P(x) = 3x^2 - 6x - 7$.

Graph.

50. $f(x) = 2x^2 - 3$

51. $f(x) = \dfrac{1}{3}x^2$

52. $f(x) = x^2 + 2x - 4$

53. $f(x) = -x^2 + 2x - 1$

54. $f(x) = 2x^2 - 4x - 3$

55. $f(x) = \dfrac{1}{2}x^2 + x - 4$

56. 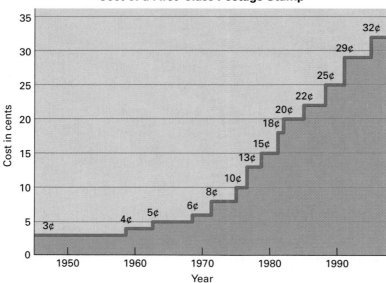 The graph below shows the cost of a first-class postage stamp from the 1950s to 1995. A quadratic function that approximately models these data is $y = 0.01572x^2 - 1.59527x + 43.07884$, where $x \geq 55$ and $x = 55$ for the year 1955, and y is the cost in cents of a first-class stamp. Using the model equation, determine what the model predicts the cost of a first-class stamp will be in the year 2000. Round to the nearest cent.

Cost of a First–Class Postage Stamp

Solve.

57. The point whose coordinates are (x_1, y_1) lies in quadrant I and is a point on the graph of the equation $y = 2x^2 - 2x + 1$. Given $y_1 = 13$, find x_1.

58. The point whose coordinates are (x_1, y_1) lies in quadrant II and is a point on the graph of the equation $y = 2x^2 - 3x - 2$. Given $y_1 = 12$, find x_1.

59. Graph $y = x^2 + 4x + 5$. By examining the graph, determine whether the solutions of the equation $x^2 + 4x + 5 = 0$ are real numbers.

60. Graph $y = -2x^2 - 4x + 5$. By examining the graph, determine whether the solutions of the equation $-2x^2 - 4x + 5 = 0$ are real numbers.

S E C T I O N **11.5**

Application Problems

1 Application problems

The application problems in this section are varieties of those problems solved earlier in the text. Each of the strategies for the problems in this section results in a quadratic equation.

Solve: It took a motorboat a total of 7 h to travel 48 mi down a river and then 48 mi back again. The rate of the current was 2 mph. Find the rate of the motorboat in calm water.

STRATEGY for solving an application problem

- Determine the type of problem. For example, is it a uniform motion problem, a geometry problem, a work problem, or an age problem?

The problem is a uniform motion problem.

- Choose a variable to represent the unknown quantity. Write numerical or variable expressions for all the remaining quantities. These results can be recorded in a table.

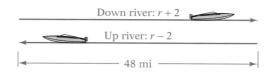

Down river: $r + 2$

Up river: $r - 2$

48 mi

The unknown rate of the motorboat: r

	Distance	÷	Rate	=	Time
Down river	48	÷	$r + 2$	=	$\dfrac{48}{r + 2}$
Up river	48	÷	$r - 2$	=	$\dfrac{48}{r - 2}$

- Determine how the quantities are related. If necessary, review the strategies presented in Chapter 4.

The total time of the trip was 7 h.

$$\frac{48}{r+2} + \frac{48}{r-2} = 7$$

$$(r+2)(r-2)\left(\frac{48}{r+2} + \frac{48}{r-2}\right) = (r+2)(r-2)7$$

$$(r-2)48 + (r+2)48 = 7r^2 - 28$$

$$48r - 96 + 48r + 96 = 7r^2 - 28$$

$$96r = 7r^2 - 28$$

$$0 = 7r^2 - 96r - 28$$

$$0 = (7r+2)(r-14)$$

$$7r + 2 = 0 \qquad\qquad r - 14 = 0$$
$$7r = -2 \qquad\qquad r = 14$$
$$r = -\frac{2}{7}$$

The solution $-\frac{2}{7}$ is not possible, because the rate cannot be negative.

The rate of the motorboat in calm water is 14 mph.

Example 1 Working together, a painter and the painter's apprentice can paint a room in 4 h. Working alone, the apprentice requires 6 more hours to paint the room than the painter requires working alone. How long does it take the painter, working alone, to paint the room?

Strategy
- This is a work problem.
- Time for the painter to paint the room: t
  Time for the apprentice to paint the room: $t + 6$

	Rate	Time	Part
Painter	$\frac{1}{t}$	4	$\frac{4}{t}$
Apprentice	$\frac{1}{t+6}$	4	$\frac{4}{t+6}$

- The sum of the parts of the task completed must equal 1.

Solution

$$\frac{4}{t} + \frac{4}{t+6} = 1$$

$$t(t+6)\left(\frac{4}{t} + \frac{4}{t+6}\right) = t(t+6) \cdot 1$$

$$(t+6)4 + t(4) = t(t+6)$$

$$4t + 24 + 4t = t^2 + 6t$$

$$0 = t^2 - 2t - 24$$

$$0 = (t-6)(t+4)$$

$$t - 6 = 0 \qquad t + 4 = 0$$
$$t = 6 \qquad\qquad t = -4 \quad \blacktriangleright \text{The solution } t = -4 \text{ is not possible.}$$

Working alone, the painter requires 6 h to paint the room.

Problem 1 The length of a rectangle is 3 m more than the width. The area is 40 m². Find the width.

Solution See page A44.

CONCEPT REVIEW 11.5

Determine whether the statement is always true, sometimes true, or never true.

1. If the length of a rectangle is three more than twice the width and the width is represented by W, then the length is represented by $3W + 2$.

2. The sum of the squares of two consecutive odd integers can be represented as $n^2 + (n + 1)^2$.

3. If it takes one pipe 15 min longer to fill a tank than it does a second pipe, then the rate of work for the second pipe can be represented by $\frac{1}{t}$ and the rate of work for the first pipe can be represented by $\frac{1}{t + 15}$.

4. If a plane's rate of speed is r and the rate of the wind is 30 mph, then the plane's rate of speed flying with the wind is $30 + r$ and the plane's rate of speed flying against the wind is $30 - r$.

5. The graph of the equation $h = 48t - 16t^2$ is the graph of a parabola opening down.

6. A number that is a solution to a quadratic equation may not be a possible answer to the application problem it models. Such a solution is disregarded.

EXERCISES 11.5

1 Solve.

1. The area of the batter's box on a major-league baseball field is 24 ft². The length of the batter's box is 2 ft more than the width. Find the length and width of the rectangular batter's box.

2. The length of the batter's box on a softball field is 1 ft more than twice the width. The area of the batter's box is 21 ft². Find the length and width of the rectangular batter's box.

3. The length of a children's playground is twice the width. The area is 5000 ft². Find the length and width of the rectangular playground.

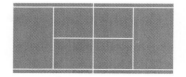

4. The length of a singles tennis court is 24 ft more than twice the width. The area is 2106 ft². Find the length and width of the singles tennis court.

5. The sum of the squares of two positive consecutive odd integers is 130. Find the two integers.

6. The sum of the squares of two consecutive positive even integers is 164. Find the two integers.

7. The sum of two integers is 12. The product of the two integers is 35. Find the two integers.

8. The difference between two integers is 4. The product of the two integers is 60. Find the integers.

9. Twice an integer equals the square of the integer. Find the integer.

10. The square of an integer equals the integer. Find the integer.

11. One computer takes 21 min longer to calculate the value of a complex equation than a second computer. Working together, these computers can complete the calculation in 10 min. How long would it take each computer, working alone, to calculate the value?

12. A tank has two drains. One drain takes 16 min longer to empty the tank than does a second drain. With both drains open, the tank is emptied in 6 min. How long would it take each drain, working alone, to empty the tank?

13. It takes 6 h longer to cross a channel in a ferryboat when one engine of the boat is used alone than when a second engine is used alone. Using both engines, the ferryboat can make the crossing in 4 h. How long would it take each engine, working alone, to power the ferryboat across the channel?

14. An apprentice mason takes 8 h longer to build a small fireplace than an experienced mason. Working together, they can build a fireplace in 3 h. How long would it take the experienced mason, working alone, to build the fireplace?

15. It took a small plane 2 h more to fly 375 mi against the wind than it took the plane to fly the same distance with the wind. The rate of the wind was 25 mph. Find the rate of the plane in calm air.

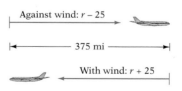

Against wind: $r - 25$

375 mi

With wind: $r + 25$

16. It took a motorboat 1 h longer to travel 36 mi against the current than it took the boat to travel 36 mi with the current. The rate of the current was 3 mph. Find the rate of the boat in calm water.

17. A motorcyclist traveled 150 mi at a constant rate before decreasing the speed by 15 mph. Another 35 mi was driven at the decreased speed. The total time for the 185 mi trip was 4 h. Find the cyclist's rate during the first 150 mi.

18. A cruise ship sailed through a 20 mi inland passageway at a constant rate before increasing its speed by 15 mph. Another 75 mi was traveled at the increased rate. The total time for the 95 mi trip was 5 h. Find the rate of the ship during the last 75 mi.

19. An arrow is projected into the air with an initial velocity of 48 ft/s. At what times will the arrow be 32 ft above the ground? Use the equation $h = 48t - 16t^2$, where h is the height, in feet, above the ground after t seconds.

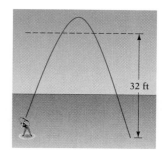

20. An object is projected into the air with an initial velocity of 64 ft/s. At what times will the object be on the ground? Use the equation $h = 64t - 16t^2$, where h is the height, in feet, above the ground after t seconds.

SUPPLEMENTAL EXERCISES 11.5

Solve.

21. The sum of the squares of four consecutive integers is 86. Find the four integers.

22. A tin can has a volume of 1500 cm^3 and a height of 15 cm. Find the radius of the tin can. Round to the nearest hundredth.

23. Find the radius of a right circular cone that has a volume of 800 cm^3 and a height of 12 cm. Round to the nearest hundredth.

24. The hypotenuse of a right triangle is $\sqrt{13}$ cm. One leg is 1 cm shorter than twice the length of the other leg. Find the lengths of the legs of the right triangle.

25. The radius of a large pizza is 1 in. less than twice the radius of a small pizza. The difference between the areas of the two pizzas is 33π in^2. Find the radius of the large pizza.

26. The perimeter of a rectangular garden is 54 ft. The area of the garden is 180 ft^2. Find the length and width of the garden.

Project in Mathematics

Geometric Construction of Completing the Square

Completing the square as a method of solving a quadratic equation has been known for centuries. The Persian mathematician al-Khwarizmi used this method in a textbook written around A.D. 825. The method was very geometric. That is, al-Khwarizmi literally completed a square. To understand how this method works, consider the following geometric shapes: A square whose area is x^2, a rectangle whose area is x, and another square whose area is 1.

Now consider the expression $x^2 + 6x$. From our discussion in this chapter, to complete the square, we would add $\left(\frac{1}{2} \cdot 6\right)^2 = 3^2 = 9$ to the expression. Here is the geometric construction that al-Khwarizmi used.

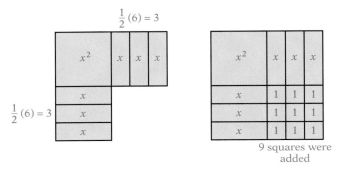

Note that it is necessary to add 9 squares to the figure to "complete the square." One of the difficulties of using a geometric method such as this is that it cannot easily be extended to $x^2 - 6x$. There is no way to draw an area of $-6x$! That limitation did not bother al-Khwarizmi much. Negative numbers were not a significant part of mathematics until well into the 13th century.

Solve.

1. Show how al-Khwarizmi would have completed the square for $x^2 + 4x$.

2. Show how al-Khwarizmi would have completed the square for $x^2 + 10x$.

3. Do the geometric constructions you drew for Exercises 1 and 2 correspond to the algebraic method shown in this chapter?

Focus on Problem Solving

Algebraic Manipulation and Graphing Techniques

Problem solving is often easier when we have both algebraic manipulation and graphing techniques at our disposal. Solving quadratic equations and graphing quadratic equations in two variables are used here to solve problems involving profit.

A company's revenue, R, is the total amount of money the company earned by selling its products. The cost, C, is the total amount of money the company spent to manufacture and sell its products. A company's profit, P, is the difference between the revenue and cost: $P = R - C$. A company's revenue and cost may be represented by equations.

A company manufactures and sells woodstoves. The total monthly cost, in dollars, to produce n woodstoves is $C = 30n + 2000$. Write a variable expression for the company's monthly profit if the revenue, in dollars, obtained from selling all n woodstoves is $R = 150n - 0.4n^2$.

Substitute the variable expressions for revenue and cost into the equation $P = R - C$.

$$P = R - C$$
$$P = (150n - 0.4n^2) - (30n + 2000)$$

Simplify the right side of the equation.

$$P = 150n - 0.4n^2 - 30n - 2000$$
$$P = -0.4n^2 + 120n - 2000$$

The company's monthly profit is $P = -0.4n^2 + 120n - 2000$.

How many woodstoves must the company manufacture and sell in order to make a profit of $6000 a month?

Substitute 6000 for P in the profit equation.

$$P = -0.4n^2 + 120n - 2000$$
$$6000 = -0.4n^2 + 120n - 2000$$

Write the equation in standard form.

$$0 = -0.4n^2 + 120n - 8000$$

$-0.4 = -\dfrac{4}{10}$. Multiply each side of the equation by $-\dfrac{10}{4}$.

$$0 = n^2 - 300n + 20{,}000$$

Factor.

$$0 = (n - 100)(n - 200)$$
$$n = 100 \quad n = 200$$

The company will make a monthly profit of $6000 if either 100 woodstoves are manufactured and sold or 200 woodstoves are manufactured and sold.

The graph of the profit equation $P = -0.4n^2 + 120n - 2000$ is shown at the right. The highest point on the graph is the point at which the maximum profit can be made. For this example, a maximum profit of $7000 occurs when the company manufactures and sells 150 woodstoves.

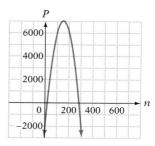

Solve.

1. The total cost, in dollars, for a company to produce and sell n guitars per month is $C = 240n + 1200$. The company's revenue, in dollars, from selling all n guitars is $R = 400n - 2n^2$.

   a. How many guitars must the company produce and sell each month in order to make a monthly profit of $1200?

   b. Graph the profit equation. What is the maximum monthly profit the company can make?

2. A company's total monthly cost, in dollars, for manufacturing and selling n videotapes per month is $C = 35n + 2000$. The company's revenue, in dollars, from selling all n videotapes is $R = 175n - 0.2n^2$.

   a. How many videotapes must be produced and sold each month in order for the company to make a monthly profit of $18,000?

   b. Graph the profit equation. How many videotapes must the company produce and sell in order to make the maximum monthly profit?

Chapter Summary

Key Words

A *quadratic equation* is an equation that can be written in the form $ax^2 + bx + c = 0$, $a \neq 0$. A quadratic equation is also called a *second-degree equation*. (Objective 11.1.1)

The equation $4x^2 - 3x + 6 = 0$ is a quadratic equation. In this equation, $a = 4$, $b = -3$, and $c = 6$.

A quadratic equation is in *standard form* when the polynomial is in descending order and equal to zero. (Objective 11.1.1)

The quadratic equation $5x^2 + 3x - 1 = 0$ is in standard form.

Adding to a binomial the constant term that makes it a perfect-square trinomial is called *completing the square*. (Objective 11.2.1)

Adding to $x^2 - 4x$ the constant term 4 results in a perfect-square trinomial:
$$x^2 - 4x + 4 = (x - 2)(x - 2) = (x - 2)^2$$

The graph of an equation of the form $y = ax^2 + bx + c$ is a *parabola*. The *vertex* is the lowest point on a parabola that opens up or the highest point on a parabola that opens down. (Objective 11.4.1)

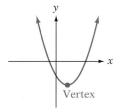

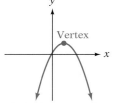

Parabola that opens up   Parabola that opens down

Essential Rules and Procedures

Solving a Quadratic Equation by Factoring
Write the equation in standard form, factor the left side of the equation, apply the Principle of Zero Products, solve for the variable. (Objective 11.1.1)

$$x^2 + 2x = 15$$
$$x^2 + 2x - 15 = 0$$
$$(x - 3)(x + 5) = 0$$
$$x - 3 = 0 \qquad x + 5 = 0$$
$$x = 3 \qquad\qquad x = -5$$

Principle of Taking the Square Root of Each Side of an Equation
If $x^2 = a$, then $x = \pm\sqrt{a}$.

This principle is used to solve quadratic equations by taking square roots. (Objective 11.1.2)

$$3x^2 - 48 = 0$$
$$3x^2 = 48$$
$$x^2 = 16$$
$$\sqrt{x^2} = \pm\sqrt{16}$$
$$x = \pm 4$$

Solving a Quadratic Equation by Completing the Square
When the quadratic equation is in the form $x^2 + bx = c$, add to each side of the equation the term that completes the square on $x^2 + bx$. Factor the perfect-square trinomial, and write it as the square of a binomial. Take the square root of each side of the equation and solve for x.   (Objective 11.2.1)

$$x^2 + 6x = 5$$
$$x^2 + 6x + 9 = 5 + 9$$
$$(x + 3)^2 = 14$$
$$\sqrt{(x + 3)^2} = \pm\sqrt{14}$$
$$x + 3 = \pm\sqrt{14}$$
$$x = -3 \pm \sqrt{14}$$

The Quadratic Formula
The solutions of $ax^2 + bx + c = 0$, $a \neq 0$, are
$$x = \frac{-b \pm \sqrt{b^2 - 4ac}}{2a}$$
(Objective 11.3.1)

$$3x^2 = x + 5$$
$$3x^2 - x - 5 = 0$$
$$x = \frac{-b \pm \sqrt{b^2 - 4ac}}{2a}$$
$$= \frac{-(-1) \pm \sqrt{(-1)^2 - 4(3)(-5)}}{2(3)}$$
$$= \frac{1 \pm \sqrt{1 + 60}}{6} = \frac{1 \pm \sqrt{61}}{6}$$

Graphing a Quadratic Equation in Two Variables
Find several solutions of the equation. Graph the ordered-pair solutions on a rectangular coordinate system. Draw a parabola through the points.   (Objective 11.4.1)

$$y = x^2 - x - 2$$

x	y
0	-2
1	-2
-1	0
2	0
-2	4
3	4

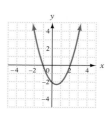

Chapter Review Exercises

1. Solve: $b^2 - 16 = 0$

2. Solve: $x^2 - x - 3 = 0$

3. Solve: $x^2 - 3x - 5 = 0$

4. Solve: $49x^2 = 25$

5. Graph: $y = -\frac{1}{4}x^2$

6. Graph: $y = -3x^2$

7. Solve: $x^2 = 10x - 2$

8. Solve: $6x(x + 1) = x - 1$

9. Solve: $4y^2 + 9 = 0$

10. Solve: $5x^2 + 20x + 12 = 0$

11. Solve: $x^2 - 4x + 1 = 0$

12. Solve: $x^2 - x = 30$

13. Solve: $6x^2 + 13x - 28 = 0$

14. Solve: $x^2 = 40$

15. Solve: $3x^2 - 4x = 1$

16. Solve: $x^2 - 2x - 10 = 0$

17. Solve: $x^2 - 12x + 27 = 0$

18. Solve: $(x - 7)^2 = 81$

19. Graph: $y = 2x^2 + 1$

20. Graph: $y = \frac{1}{2}x^2 - 1$

21. Solve: $(y + 4)^2 - 25 = 0$

22. Solve: $4x^2 + 16x = 7$

23. Solve: $24x^2 + 34x + 5 = 0$

24. Solve: $x^2 = 4x - 8$

25. Solve: $x^2 - 5 = 8x$

26. Solve: $25(2x^2 - 2x + 1) = (x + 3)^2$

27. Solve: $\left(x - \frac{1}{2}\right)^2 = \frac{9}{4}$

28. Solve: $16x^2 = 30x - 9$

29. Solve: $x^2 + 7x = 3$

30. Solve: $12x^2 + 10 = 29x$

31. Solve: $4(x-3)^2 = 20$

32. Solve: $x^2 + 8x - 3 = 0$

33. Graph: $y = x^2 - 3x$

34. Graph: $y = x^2 - 4x + 3$

35. Solve: $(x+9)^2 = x + 11$

36. Solve: $(x-2)^2 - 24 = 0$

37. Solve: $x^2 + 6x + 12 = 0$

38. Solve: $x^2 + 6x - 2 = 0$

39. Solve: $18x^2 - 52x = 6$

40. Solve: $2x^2 + 5x = 1$

41. Graph: $y = -x^2 + 4x - 5$

42. Solve: $x^2 + 3 = 9x$

43. Solve: $2x^2 + 5 = 7x$

44. Graph: $y = 4 - x^2$

45. It took an air balloon 1 h more to fly 60 mi against the wind than it took to fly 60 mi with the wind. The rate of the wind was 5 mph. Find the rate of the air balloon in calm air.

46. The height of a triangle is 2 m more than twice the length of the base. The area of the triangle is 20 m². Find the height of the triangle and the length of the base.

47. The sum of the squares of two consecutive positive odd integers is thirty-four. Find the two integers.

48. In 5 h, two campers rowed 12 mi down a stream and then rowed back to their campsite. The rate of the stream's current was 1 mph. Find the rate of the boat in still water.

49. An object is thrown into the air with an initial velocity of 32 ft/s. At what times will the object be 12 ft above the ground? Use the equation $h = 32t - 16t^2$, where h is the height, in feet, above the ground after t seconds.

50. A smaller drain takes 8 h longer to empty a tank than does a larger drain. Working together, the drains can empty the tank in 3 h. How long would it take each drain, working alone, to empty the tank?

Chapter Test

1. Solve: $3(x + 4)^2 - 60 = 0$

2. Solve: $2x^2 + 8x = 3$

3. Solve: $3x^2 + 7x = 20$

4. Solve: $x^2 - 3x = 6$

5. Solve: $x^2 + 4x - 16 = 0$

6. Graph: $y = x^2 + 2x - 4$

7. Solve: $x^2 + 4x + 2 = 0$

8. Solve: $x^2 + 3x = 8$

9. Solve: $2x^2 - 5x - 3 = 0$

10. Solve: $2x^2 - 6x + 1 = 0$

11. Solve: $3x^2 - x = 1$

12. Solve: $2(x - 5)^2 = 36$

13. Solve: $x^2 - 6x - 5 = 0$

14. Solve: $x^2 - 5x = 1$

15. Solve: $x^2 - 5x = 2$

16. Solve: $6x^2 - 17x = -5$

17. Solve: $x^2 + 3x - 7 = 0$

18. Solve: $2x^2 - 4x - 5 = 0$

19. Solve: $2x^2 - 3x - 2 = 0$

20. Graph: $y = x^2 - 2x - 3$

21. Solve: $3x^2 - 2x = 3$

22. The length of a rectangle is 2 ft less than twice the width. The area of the rectangle is 40 ft². Find the length and width of the rectangle.

23. It took a motorboat 1 h more to travel 60 mi against the current than it took to go 60 mi with the current. The rate of the current was 1 mph. Find the rate of the boat in calm water.

24. The sum of the squares of three consecutive odd integers is 83. Find the middle odd integer.

25. A jogger ran 7 mi at a constant rate and then reduced the rate by 3 mph. An additional 8 mi was run at the reduced rate. The total time spent jogging the 15 mi was 3 h. Find the rate for the last 8 mi.

Cumulative Review Exercises

1. Simplify: $2x - 3[2x - 4(3 - 2x) + 2] - 3$

2. Solve: $-\frac{3}{5}x = -\frac{9}{10}$

3. Solve: $2x - 3(4x - 5) = -3x - 6$

4. Solve: $2x - 3(2 - 3x) > 2x - 5$

5. Find the x- and y-intercepts of the line $4x - 3y = 12$.

6. Find the equation of the line that contains the point whose coordinates are $(-3, 2)$ and has slope $-\frac{4}{3}$.

7. Find the domain and range of the relation $\{(-2, -8), (-1, -1), (0, 0), (1, 1), (2, 8)\}$. Is the relation a function?

8. Evaluate $f(x) = -3x + 10$ at $x = -9$.

9. Graph: $y = \frac{1}{4}x - 2$

10. Graph the solution set of $2x - 3y > 6$.

11. Solve by substitution: $3x - y = 5$
$$y = 2x - 3$$

12. Solve by the addition method: $3x + 2y = 2$
$$5x - 2y = 14$$

13. Simplify: $\dfrac{(2a^{-2}b)^2}{-3a^{-5}b^4}$

14. Divide: $(x^2 - 8) \div (x - 2)$

15. Factor: $4y(x - 4) - 3(x - 4)$

16. Factor: $3x^3 + 2x^2 - 8x$

17. Divide: $\dfrac{3x^2 - 6x}{4x - 6} \div \dfrac{2x^2 + x - 6}{6x^2 - 24x}$

18. Subtract: $\dfrac{x}{2(x - 1)} - \dfrac{1}{(x - 1)(x + 1)}$

19. Simplify: $\dfrac{1 - \dfrac{7}{x} + \dfrac{12}{x^2}}{2 - \dfrac{1}{x} - \dfrac{15}{x^2}}$

20. Solve: $\dfrac{x}{x + 6} = \dfrac{3}{x}$

21. Multiply: $(\sqrt{a} - \sqrt{2})(\sqrt{a} + \sqrt{2})$

22. Solve: $3 = 8 - \sqrt{5x}$

23. Solve: $2x^2 - 7x = -3$

24. Solve: $3(x - 2)^2 = 36$

25. Solve: $3x^2 - 4x - 5 = 0$

26. Graph: $y = x^2 - 3x - 2$

27. In a certain state, the sales tax is $7\frac{1}{4}\%$. The sales tax on a chemistry text-book is $2.61. Find the cost of the textbook before the tax is added.

28. Find the cost per pound of a mixture made from 20 lb of cashews that cost $3.50 per pound and 50 lb of peanuts that cost $1.75 per pound.

29. A stock investment of 100 shares paid a dividend of $215. At this rate, how many additional shares are required to earn a dividend of $752.50?

30. A 720 mi trip from one city to another takes 3 h when a plane is flying with the wind. The return trip, against the wind, takes 4.5 h. Find the rate of the plane in calm air and the rate of the wind.

31. A student received a 70, a 91, an 85, and a 77 on four tests in a math class. What scores on the fifth test will enable the student to receive a minimum of 400 points?

32. A guy wire is attached to a point 30 m above the ground on a telephone pole that is perpendicular to the ground. The wire is anchored to the ground at a point 10 m from the base of the pole. Find the length of the guy wire. Round to the nearest hundredth.

33. The sum of the squares of three consecutive odd integers is 155. Find the middle odd integer.

Final Exam

1. Evaluate: $-|-3|$

2. Subtract: $-15 - (-12) - 3$

3. Write $\frac{1}{8}$ as a percent.

4. Simplify: $-2^4 \cdot (-2)^4$

5. Simplify: $-7 - \frac{12 - 15}{2 - (-1)} \cdot (-4)$

6. Evaluate $\frac{a^2 - 3b}{2a - 2b^2}$ when $a = 3$ and $b = -2$.

7. Simplify: $6x - (-4y) - (-3x) + 2y$

8. Multiply: $(-15z)\left(-\frac{2}{5}\right)$

9. Simplify: $-2[5 - 3(2x - 7) - 2x]$

10. Solve: $20 = -\frac{2}{5}x$

11. Solve: $4 - 2(3x + 1) = 3(2 - x) + 5$

12. Find 19% of 80.

13. Solve: $4 - x \geq 7$

14. Solve: $2 - 2(y - 1) \leq 2y - 6$

15. Find the slope of the line that contains the points whose coordinates are $(-1, -3)$ and $(2, -1)$.

16. Find the equation of the line that contains the point whose coordinates are $(3, -4)$ and has slope $-\frac{2}{3}$.

17. Graph the line with slope $-\frac{1}{2}$ and y-intercept $(0, -3)$.

18. Graph: $f(x) = \frac{2}{3}x - 4$

19. Find the range of the function given by the equation $f(x) = -x + 5$ if the domain is $\{-6, -3, 0, 3, 6\}$.

20. Graph the solution set of $3x - 2y \geq 6$.

21. Solve by substitution: $y = 4x - 7$
$y = 2x + 5$

22. Solve by the addition method: $4x - 3y = 11$
$2x + 5y = -1$

23. Subtract: $(2x^2 - 5x + 1) - (5x^2 - 2x - 7)$

24. Simplify: $(-3xy^3)^4$

25. Multiply: $(3x^2 - x - 2)(2x + 3)$

26. Simplify: $\dfrac{(-2x^2y^3)^3}{(-4x^{-1}y^4)^2}$

27. Simplify: $(4x^{-2}y)^3(2xy^{-2})^{-2}$

28. Divide: $\dfrac{12x^3y^2 - 16x^2y^2 - 20y^2}{4xy^2}$

29. Divide: $(5x^2 - 2x - 1) \div (x + 2)$

30. Write 0.000000039 in scientific notation.

31. Factor: $2a(4 - x) - 6(x - 4)$

32. Factor: $x^2 - 5x - 6$

33. Factor: $2x^2 - x - 3$

34. Factor: $6x^2 - 5x - 6$

35. Factor: $8x^3 - 28x^2 + 12x$

36. Factor: $25x^2 - 16$

37. Factor: $75y - 12x^2y$

38. Solve: $2x^2 = 7x - 3$

39. Multiply: $\dfrac{2x^2 - 3x + 1}{4x^2 - 2x} \cdot \dfrac{4x^2 + 4x}{x^2 - 2x + 1}$

40. Subtract: $\dfrac{5}{x + 3} - \dfrac{3x}{2x - 5}$

41. Simplify: $\dfrac{x - \dfrac{3}{2x - 1}}{1 - \dfrac{2}{2x - 1}}$

42. Solve: $\dfrac{5x}{3x - 5} - 3 = \dfrac{7}{3x - 5}$

43. Solve $a = 3a - 2b$ for a.

44. Simplify: $\sqrt{49x^6}$

45. Add: $2\sqrt{27a} + 8\sqrt{48a}$

46. Simplify: $\dfrac{\sqrt{3}}{\sqrt{5} - 2}$

47. Solve: $\sqrt{x + 4} - \sqrt{x - 1} = 1$

48. Solve: $(x - 3)^2 = 7$

49. Solve: $4x^2 - 2x - 1 = 0$

50. Graph: $y = x^2 - 4x + 3$

51. Translate and simplify "the sum of twice a number and three times the difference between the number and two."

52. Because of depreciation, the value of an office machine is now $2400. This is 80% of its original value. Find the original value of the machine.

53. A coin bank contains quarters and dimes. There are three times as many dimes as quarters. The total value of the coins in the bank is $11. Find the number of dimes in the bank.

54. One angle of a triangle is 10° more than the measure of the second angle. The third angle is 10° more than the measure of the first angle. Find the measure of each angle of the triangle.

55. The manufacturer's cost for a laser printer is $900. The manufacturer sells the printer for $1485. Find the markup rate.

56. An investment of $3000 is made at an annual simple interest rate of 8%. How much additional money must be invested at 11% so that the total interest earned is 10% of the total investment?

57. A grocer mixes 4 lb of peanuts that cost $2 per pound with 2 lb of walnuts that cost $5 per pound. What is the cost per pound of the resulting mixture?

58. A pharmacist mixes 20 L of a solution that is 60% acid with 30 L of a solution that is 20% acid. What is the percent concentration of acid in the resulting mixture?

59. A small plane flew at a constant rate for 1 h. The pilot then doubled the plane's speed. An additional 1.5 h was flown at the increased speed. If the entire flight was 860 km, how far did the plane travel during the first hour?

60. With the current, a motorboat travels 50 mi in 2.5 h. Against the current, it takes twice as long to travel 50 mi. Find the rate of the boat in calm water and the rate of the current.

61. The length of a rectangle is 5 m more than the width. The area of the rectangle is 50 m². Find the dimensions of the rectangle.

62. A paint formula requires 2 oz of dye for every 15 oz of base paint. How many ounces of dye are required for 120 oz of base paint?

63. It takes a chef 1 h to prepare a dinner. The chef's apprentice can prepare the same dinner in 1.5 h. How long would it take the chef and the apprentice, working together, to prepare the dinner?

64. The hypotenuse of a right triangle measures 14 cm. One leg of the triangle measures 8 cm. Find the length of the other leg of the triangle. Round to the nearest tenth.

65. It took a plane $\frac{1}{2}$ h more to fly 500 mi against the wind than it took to fly the same distance with the wind. The rate of the plane in calm air is 225 mph. Find the rate of the wind.

APPENDIX: Guidelines for Using Graphing Calculators

Texas Instruments *TI-82*

To evaluate an expression

a. Press the $\boxed{Y=}$ key. A menu showing y_1 through y_8 will be displayed vertically with the cursor on y_1. Press $\boxed{CLEAR}$, if necessary, to delete an unwanted expression.

b. Input the expression to be evaluated. For example, to input the expression $-3a^2b - 4c$, use the following keystrokes:

$\boxed{Y=}$ $\boxed{CLEAR}$ $\boxed{(-)}$ 3 $\boxed{ALPHA}$ A $\boxed{\wedge}$ 2 $\boxed{ALPHA}$ B $\boxed{-}$ 4 $\boxed{ALPHA}$ C $\boxed{2nd}$ QUIT

Note the difference between the keys for a *negative* sign $\boxed{(-)}$ and a *minus* sign $\boxed{-}$.

c. Store the value of each variable that will be used in the expression. For example, to evaluate the expression above when $a = 3$, $b = -2$, and $c = -4$, use the following keystrokes:

3 $\boxed{STO\triangleright}$ $\boxed{ALPHA}$ A $\boxed{ENTER}$ $\boxed{(-)}$ 2 $\boxed{STO\triangleright}$ $\boxed{ALPHA}$ B $\boxed{ENTER}$ $\boxed{(-)}$ 4 $\boxed{STO\triangleright}$ $\boxed{ALPHA}$ C
$\boxed{ENTER}$

These steps store the value of each variable.

d. Press $\boxed{2nd}$ Y-VARS $\boxed{1}$ $\boxed{1}$ $\boxed{ENTER}$. The value of the expression, y_1, for the given values is displayed, in this case, $y_1 = 70$.

To graph a function

a. Press the $\boxed{Y=}$ key. A menu showing y_1 through y_8 will be displayed vertically with the cursor on y_1. Press $\boxed{CLEAR}$, if necessary, to delete an unwanted expression.

b. Input the expression for each function that is to be graphed. Press $\boxed{X,T,\theta}$ to input x. For example, to input $y = \dfrac{1}{2}x - 3$, use the following keystrokes:

$\boxed{Y=}$ $\boxed{(}$ 1 $\boxed{\div}$ 2 $\boxed{)}$ $\boxed{X,T,\theta}$ $\boxed{-}$ 3

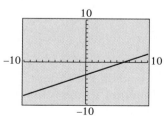

c. Set the domain and range by pressing $\boxed{WINDOW}$. Enter the values for the minimum x-value (Xmin), the maximum x-value (Xmax), the distance between tick marks on the x-axis (Xscl), the minimum y-value (Ymin), the maximum y-value (Ymax), and the distance between tick marks on the y-axis (Yscl). Now press $\boxed{GRAPH}$. For the graph shown at the left, Xmin $= -10$, Xmax $= 10$, Xscl $= 1$, Ymin $= -10$, Ymax $= 10$, and Yscl $= 1$. This is called the standard viewing rectangle. Pressing $\boxed{ZOOM}$ $\boxed{6}$ is a quick way to set the calculator to the standard viewing rectangle. *Note:* This will also immediately graph the function in that window.

d. Press the $\boxed{Y=}$ key. The equal sign has a black rectangle around it. This indicates that the function is *active* and will be graphed when the $\boxed{GRAPH}$ key is pressed. A function is deactivated by using the arrow keys. Move the cursor

A1

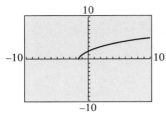

over the equal sign and press $\boxed{\text{ENTER}}$. When the cursor is moved to the right, the black rectangle will not be present and that equation will not be active.

e. Graphing some radical equations requires special care. To graph the function $y = \sqrt{2x + 3}$, enter the following keystrokes:

$\boxed{\text{Y=}}$ $\boxed{\text{2nd}}$ $\sqrt{}$ $\boxed{(}$ 2 $\boxed{\text{X,T,}\theta}$ $\boxed{+}$ 3 $\boxed{)}$

The graph is shown at the left.

To display the *x*-coordinates of rectangular coordinates as integers

a. Set the viewing window as follows: Xmin = −47, Xmax = 47, Xscl = 10, Ymin = −32, Ymax = 32, Yscl = 10.

b. Graph the function and use the TRACE feature. Press $\boxed{\text{TRACE}}$ and then move the cursor with the $\boxed{\triangleleft}$ and $\boxed{\triangleright}$ keys. The values of *x* and $y = f(x)$ displayed on the bottom of the screen are the coordinates of a point on the graph.

To display the *x*-coordinates of rectangular coordinates in tenths

a. Set the viewing window as follows: $\boxed{\text{ZOOM}}$ $\boxed{4}$

b. Graph the function and use the TRACE feature. Press $\boxed{\text{TRACE}}$ and then move the cursor with the $\boxed{\triangleleft}$ and $\boxed{\triangleright}$ keys. The values of *x* and $y = f(x)$ displayed on the bottom of the screen are the coordinates of a point on the graph.

To evaluate a function for a given value of *x*, or to produce a pair of rectangular coordinates

a. Input the equation; for example, input $y_1 = 2x^3 - 3x + 2$.

b. Press $\boxed{\text{2nd}}$ $\boxed{\text{QUIT}}$.

c. Input a value for *x*; for example, to input 3 press 3 $\boxed{\text{STO}\triangleright}$ $\boxed{\text{X,T,}\theta}$ $\boxed{\text{ENTER}}$.

d. Press $\boxed{\text{2nd}}$ $\boxed{\text{Y-VARS}}$ $\boxed{1}$ $\boxed{1}$ $\boxed{\text{ENTER}}$. The value of the function, y_1, for the given *x*-value is shown; in this case, $y_1 = 47$. An ordered pair of the function is (3, 47).

e. Repeat steps (c)–(d) to produce as many pairs as desired. The TABLE feature of the *TI-82* can also be used to determine ordered pairs.

Zoom Features of the *TI-82*

To zoom in or out on a graph

a. Here are two methods of using ZOOM. The first method uses the built-in features of the calculator. Move the cursor to a point on the graph that is of interest. Press $\boxed{\text{ZOOM}}$. The ZOOM menu will appear. Press $\boxed{2}$ $\boxed{\text{ENTER}}$ to zoom in on the graph by the amount shown under the SET FACTORS menu. The center of the new graph is the location at which you placed the cursor. Press $\boxed{3}$ $\boxed{\text{ENTER}}$ to zoom out on the graph by the amount under the SET FACTORS menu. (The SET FACTORS menu is accessed by pressing $\boxed{\text{ZOOM}}$ $\boxed{\triangleright}$ $\boxed{4}$.)

b. The second method uses the ZBOX option under the ZOOM menu. To use this method, press ZOOM 1 . A cursor will appear on the graph. Use the arrow keys to move the cursor to a portion of the graph that is of interest. Press ENTER . Now use the arrow keys to draw a box around the portion of the graph you wish to see. Press ENTER . The portion of the graph defined by the box will be drawn.

c. Pressing ZOOM 6 resets the window to the standard 10 × 10 viewing window.

Solving Equations with the *TI-82*

This discussion is based on the fact that the solution of an equation can be related to the x-intercepts of a graph. For instance, the solutions of the equation $x^2 = x + 1$ are the x-intercepts of the graph of $y = x^2 - x - 1$.

To solve $x^2 = x + 1$, rewrite the equation with all terms on one side. The equation is now $x^2 - x - 1 = 0$. Think of this equation as $y_1 = x^2 - x - 1$. The x-intercepts of the graph of y_1 are the solutions of the equation $x^2 = x + 1$.

a. Enter $y = x^2 - x - 1$.

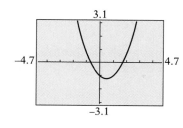

b. Graph the equation. You may need to adjust the viewing window so that the x-intercepts are visible.

c. Press 2nd CALC 2 .

d. Move the cursor to a point on the curve that is to the left of an x-intercept. Press ENTER .

e. Move the cursor to a point on the curve that is to the right of an x-intercept. Press ENTER .

f. Press ENTER .

g. A root is shown as an x-coordinate on the bottom of the screen; in this case, one root is approximately -0.62, and the other root is approximately 1.62. The SOLVE feature under the MATH menu can also be used to find solutions of equations.

Solving Systems of Equations in Two Variables with the *TI-82*

To solve a system of equations

a. Solve each equation for y.

b. Enter the first equation as y_1. For instance, let $y_1 = x^2 - 1$.

c. Enter the second equation as y_2. For instance, let $y_2 = 1 - \frac{1}{2}x$.

d. Graph both equations. (*Note:* The points of intersection must appear on the screen. It may be necessary to adjust the viewing window so that the point(s) of intersection are displayed.)

e. Press 2nd CALC 5 .

f. Move the cursor to the left of the first point of intersection. Press ENTER .

g. Move the cursor to the right of the first point of intersection. Press ENTER .

h. Press ENTER .

i. The first point of intersection is approximately $(-1.68, 1.84)$.

j. Repeat this procedure for each point of intersection.

Finding Minimum or Maximum Values of a Function with the *TI-82*

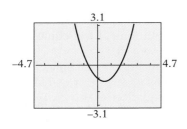

a. Enter the function into y_1. The equation $y_1 = x^2 - x - 1$ is used here.

b. Graph the equation. You may need to adjust the viewing window so that the maximum or minimum points are visible.

c. Press 2nd CALC 3 to determine a minimum value or press 2nd CALC 4 to determine a maximum value.

d. Move the cursor to a point on the curve that is to the left of the minimum (maximum). Press ENTER .

e. Move the cursor to a point on the curve that is to the right of the minimum (maximum). Press ENTER .

f. Press ENTER .

g. The minimum (maximum) is shown as the y-coordinate on the bottom of the screen; in this case the minimum value is -1.25 when $x = 0.5$.

CASIO *fx-7700GB*

To evaluate an expression

a. For example, to input the expression $-3a^2b - 4c$, use the keystrokes shown below. Note the difference between the keys for a *negative* sign $(-)$ and a *minus* sign ⬚. To enter $(-)$, press SHIFT $(-)$.

SHIFT $(-)$ 3 ALPHA A x^y 2 ALPHA B ⬚ 4 ALPHA C SHIFT F MEM F1
1 EXE

The number 1 entered here can be any number from 1 to 6.

b. Store the value of each variable that will be used in the expression. For example, to evaluate the expression above when $a = 3$, $b = -2$, and $c = -4$, use the following keystrokes:

These steps store the values of each variable.

c. To evaluate the expression, recall the expression from the function menu. Press SHIFT F MEM F2 1 EXE . The value of the expression is displayed as 70.

To graph a function

a. Ensure the calculator is in graphics mode. Press $\boxed{\texttt{MODE}}\ \boxed{1}\ \boxed{\texttt{MODE}}\ \boxed{+}\ \boxed{\texttt{MODE}}$ $\boxed{\texttt{SHIFT}}\ \boxed{+}$.

b. To graph $y = \dfrac{1}{2}x - 3$, use the following keystrokes:

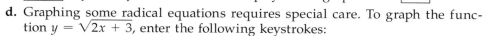

c. Set the domain and range by pressing $\boxed{\texttt{Range}}$. Enter the values for the minimum x-value (Xmin), the maximum x-value (Xmax), the distance between tick marks on the x-axis (Xscl), the minimum y-value (Ymin), the maximum y-value (Ymax), and the distance between tick marks on the y-axis (Yscl). Press $\boxed{\texttt{EXE}}$ after each entry. For the graph shown at the left, Xmin $= -10$, Xmax $= 10$, Xscl $= 1$, Ymin $= -10$, Ymax $= 10$, and Yscl $= 1$. Press the $\boxed{\texttt{Range}}$ key until you return to the display of the graph. Press $\boxed{\texttt{EXE}}$.

d. Graphing some radical equations requires special care. To graph the function $y = \sqrt{2x + 3}$, enter the following keystrokes:

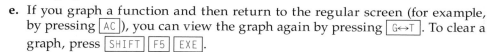

The graph is shown at the left.

e. If you graph a function and then return to the regular screen (for example, by pressing $\boxed{\texttt{AC}}$), you can view the graph again by pressing $\boxed{\texttt{G}\leftrightarrow\texttt{T}}$. To clear a graph, press $\boxed{\texttt{SHIFT}}\ \boxed{\texttt{F5}}\ \boxed{\texttt{EXE}}$.

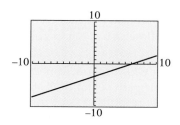

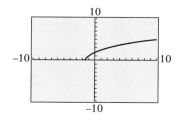

To display the x-coordinates of rectangular coordinates as integers

a. Set the viewing window as follows: Xmin $= -47$, Xmax $= 47$, Xscl $= 10$, Ymin $= -31$, Ymax $= 31$, Yscl $= 10$.

b. Graph the function and use the TRACE feature. Press $\boxed{\texttt{F1}}$ and then use the cursor keys to move along the graph. The values of x and $y = f(x)$ displayed on the bottom of the screen are the coordinates of a point on the graph.

To display the x-coordinates of rectangular coordinates in tenths

a. Set the viewing window as follows: $\boxed{\texttt{Range}}\ \boxed{\texttt{F1}}\ \boxed{\texttt{Range}}\ \boxed{\texttt{Range}}$

b. Graph the function and use the TRACE feature. Press $\boxed{\texttt{F1}}$ and then use the cursor keys to move along the graph.

To produce a pair of rectangular coordinates

a. Input the function; for example, input $3x - 4$.

b. Press $\boxed{\texttt{SHIFT}}\ \boxed{\texttt{F}\,\texttt{MEM}}\ \boxed{\texttt{F1}}\ \boxed{1}\ \boxed{\texttt{AC}}$ to store the function in f_1.

c. Input any value for x; for example, input 3 using the keystrokes $3\ \boxed{\rightarrow}\ \boxed{\texttt{X,}\theta\texttt{,T}}$ $\boxed{\texttt{EXE}}$ to store the value in x.

d. Press $\boxed{\texttt{SHIFT}}\ \boxed{\texttt{F}\,\texttt{MEM}}\ \boxed{\texttt{F2}}\ \boxed{1}\ \boxed{\texttt{EXE}}$ to find the corresponding function value for the stored x-value; in this example you should get 5. The point is $(3, 5)$.

e. Repeat steps (c)–(d) to produce as many pairs as desired.

Zoom Features of the CASIO *fx-7700GB*

To zoom in or out on a graph

a. Here are two methods of using ZOOM. The first method uses the built-in features of the calculator. Press [F1]. This activates the cursor. Use the arrow keys to move the cursor to a portion of the graph that is of interest. Press [F2]. The ZOOM menu will appear. Press [F3] to zoom in on the graph by the amount shown in FCT (factor). The center of the new graph is the location at which you placed the cursor. Press [F2] followed by [F4] to zoom out on the graph by the amount shown in FCT.

b. The second method uses the BOX option under the ZOOM menu. If necessary, press [G↔T] to view the graph. Press [F2]. Press [F1] to select BOX. Use the arrow keys to move the cursor to a portion of the graph that is of interest. Press [EXE]. Now use the arrow keys to draw a box around the portion of the graph you wish to see. Press [EXE]. The portion of the graph defined by the box will be drawn.

Solving Equations with the CASIO *fx-7700GB*

This discussion is based on the fact that the solution of an equation can be related to the x-intercepts of a graph. For instance, the solutions of the equation $x^2 - x = 1$ are the x-intercepts of the graph of $y = x^2 - x - 1$, which are the zeros of f.

To solve $x^2 = x + 1$, rewrite the equation with all terms on one side. The equation is now $x^2 - x - 1 = 0$. Think of this equation as $y_1 = x^2 - x - 1$. The x-intercepts of the graph of y_1 are the solutions of the equation $x^2 = x + 1$.

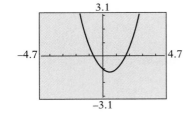

a. Press [Graph], enter $y = x^2 - x - 1$, and press [EXE]. You may need to adjust the viewing window so that the x-intercepts are visible. Note that if you adjust the viewing window after graphing the function, the equation must be reentered.

b. Press [F1] and move the cursor to an approximate x-intercept.

c. Press [F2] [F3].

d. Repeat steps (c) and (d) until you can approximate the x-intercepts to the desired degree of accuracy. The roots are approximately -0.62 and 1.62.

Solving Systems of Equations in Two Variables with the CASIO *fx-7700GB*

To solve a system of equations

a. Solve each equation for y.

b. Enter the first equation as f_1. For instance, let $y = x^2 - 1$.

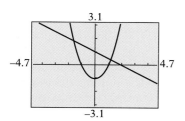

c. Enter the second equation as f_2. For instance, let $y = 1 - \frac{1}{2}x$.

d. Graph both equations. (*Note:* The points of intersection must appear on the screen. It may be necessary to adjust the viewing window so that the point(s) of intersection are displayed.)

e. Press F1 and then move the cursor to an approximate intersection point.

f. Press F2 F3.

g. Repeat steps (e) and (f) until you can approximate the point of intersection to the desired degree of accuracy. The first point of intersection is approximately $(-1.68, 1.84)$.

Finding Minimum or Maximum Values of a Function with the CASIO *fx-7700GB*

To find the minimum (maximum) value of a function

a. Graph the function as the equation $y = x^2 - x - 1$.

b. Press F1 and then move the cursor to the approximate minimum (maximum).

c. Press F2 F3.

d. Repeat steps (b) and (c) until you can approximate the minimum (maximum) value to the desired degree of accuracy. The minimum (maximum) value is shown as the y-coordinate on the bottom of the screen; in this case the minimum value is -1.25 when $x = 0$.

SHARP EL-9300

To evaluate an expression

a. The SOLVER mode of the calculator is used to evaluate expressions. To enter SOLVER mode, press 2ndF SOLVER CL. The expression $-3a^2b - 4c$ must be entered as the equation $-3a^2b - 4c = t$. The letter t can be any letter other than one used in the expression. When entering an expression in SOLVER mode, the variables appear on the screen in lower case. Use the following keystrokes to input $-3a^2b - 4c = t$:

(−) 3 ALPHA A a^b 2 ▷ ALPHA B − 4 ALPHA C ALPHA = ALPHA T ENTER

Note the difference between the keys for a *negative* sign (−) and a *minus* sign −.

b. After you press ENTER, the variables used in the equation will be displayed on the screen. To evaluate the expression for $a = 3$, $b = -2$, and $c = -4$, input each value, pressing ENTER after each number. When the cursor moves to t, press ENTER. A small window will appear. Press ENTER. In this case you will see "$t = 70$" on the screen.

c. Pressing ENTER again will allow you to evaluate the expression for new values of a, b, and c. Press ⊞ to return to normal operation.

To graph a function

a. Press the ⟋ key. The screen will show Y1=.

b. Input the expression for a function that is to be graphed. Press $\boxed{\text{X}/\theta/\text{T}}$ to enter x. For example, to input $y = \dfrac{1}{2}x - 3$, use the following keystrokes:

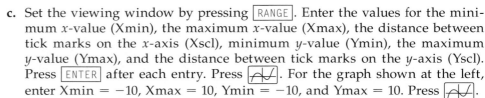

c. Set the viewing window by pressing $\boxed{\text{RANGE}}$. Enter the values for the minimum x-value (Xmin), the maximum x-value (Xmax), the distance between tick marks on the x-axis (Xscl), minimum y-value (Ymin), the maximum y-value (Ymax), and the distance between tick marks on the y-axis (Yscl). Press $\boxed{\text{ENTER}}$ after each entry. Press $\boxed{\sim}$. For the graph shown at the left, enter Xmin $= -10$, Xmax $= 10$, Ymin $= -10$, and Ymax $= 10$. Press $\boxed{\sim}$.

d. Press $\boxed{\text{EQTN}}$ to return to the equation. The equal sign has a black rectangle around it. This indicates that the function is *active* and will be graphed when the $\boxed{\sim}$ key is pressed. A function is deactivated by using the arrow keys. Move the cursor over the equal sign and press $\boxed{\text{ENTER}}$. When the cursor is moved to the right, the black rectangle will not be present and that equation will not be active.

e. Graphing some radical equations requires special care. To graph the function $y = \sqrt{2x + 3}$, enter the following keystrokes:

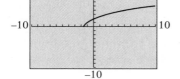

The graph is shown at the left.

To display the x-coordinates of rectangular coordinates as integers

a. Set the viewing window as follows: Xmin $= -47$, Xmax $= 47$, Xscl $= 10$, Ymin $= -31$, Ymax $= 31$, Yscl $= 10$.

b. Graph the function and use the right and left arrow keys to trace the function. The values of x and $y = f(x)$ displayed on the bottom of the screen are the coordinates of a point on the graph.

To display the x-coordinates of rectangular coordinates in tenths

a. Set the viewing window as follows: Xmin $= -4.7$, Xmax $= 4.7$, Xscl $= 1$, Ymin $= -3.1$, Ymax $= 3.1$, Yscl $= 1$. This is accomplished by pressing $\boxed{\text{RANGE}}$ $\boxed{\text{MENU}}$ A $\boxed{\text{ENTER}}$ $\boxed{\text{RANGE}}$.

b. Graph the function and use the arrow keys to move along the graph of the function. The coordinates are displayed at the bottom of the screen.

To evaluate a function for a given value of x, or to produce a pair of rectangular coordinates

a. Enter SOLVER mode. Press $\boxed{\text{2ndF}}$ SOLVER $\boxed{\text{CL}}$.

b. Input the expression; for instance, input $x^3 - 4x^2 + 1 = y$. Press $\boxed{\text{ENTER}}$.

c. Move the cursor to the x variable (if it is not already there). Input any value for x; for example, input 3 and then press $\boxed{\text{ENTER}}$. The cursor will now be over y. Press $\boxed{\text{ENTER}}$ twice to evaluate the function. In this case, the value is -8. An ordered pair of the function is $(3, -8)$.

d. Repeat step (c) to produce as many pairs as desired.

Zoom Features of the SHARP EL-9300

To zoom in or out on a graph

a. Here are two methods of using ZOOM. The first method uses the built-in features of the calculator. Move the cursor to a point on the graph that is of interest. Press `ZOOM`. The ZOOM menu will appear. Press `2` to zoom in on the graph by the amount shown by FACTOR. The center of the new graph is the location at which you placed the cursor. Press `ZOOM` `3` to zoom out on the graph by the amount shown in FACTOR.

b. The second method uses the BOX option under the ZOOM menu. To use this method, press `ZOOM` `1`. A cursor will appear on the graph. Use the arrow keys to move the cursor to a portion of the graph that is of interest. Press `ENTER`. Use the arrow keys to draw a box around the portion of the graph you wish to see. Press `ENTER`.

Solving Equations or Systems of Equations in Two Variables with the SHARP EL-9300

a. The x-intercept, y-intercept, and the point of intersection of the two graphs can be determined by using the JUMP command. Graph the functions of interest. Using the arrow keys, place the cursor on the graph of one of the functions. Press `2ndF` JUMP; the JUMP menu will appear. Press `1` to jump to the intersection of two graphs, press `4` to jump to the x-intercept, or press `5` to jump to the y-intercept. If there is more than one intercept or intersection, pressing `2ndF` JUMP again will allow you to find the remaining points. *Important:* The intersection must be a point in the viewing window.

Finding Maximum and Minimum Values of a Function with the SHARP EL-9300

a. The maximum and minimum values of a function can be determined by using the JUMP command. Graph the function of interest. Press `2ndF` JUMP; the JUMP menu will appear. Press `2` to jump to the minimum value or press `3` to jump to the maximum value of the function. If there is more than one minimum or maximum, pressing `2ndF` JUMP again will allow you to find the remaining points. *Important:* The minimum or maximum must be a point in the viewing window.

b. The minimum (maximum) value is shown as the y-coordinate on the bottom of the screen. For the equation $y_1 = x^2 - x - 1$, the minimum value is -1.25 when $x = 0$.

SOLUTIONS to Chapter 1 Problems

SECTION 1.1 *pages 3 – 6*

Problem 1 $A = \{1, 2, 3, 4\}$

Problem 2 $-5 < -1$

$-1 = -1$

$5 > -1$

The element 5 is greater than -1.

Problem 3 **A.** 9 **B.** -62

Problem 4 **A.** $|-5| = 5$ **B.** $-|-9| = -9$

SECTION 1.2 *pages 10 – 16*

Problem 1 **A.** $-162 + 98 = -64$

B. $-154 + (-37) = -191$

C. $-36 + 17 + (-21) = -19 + (-21)$
$= -40$

Problem 2 $-8 - 14 = -8 + (-14)$
$= -22$

Problem 3 $4 - (-3) - 12 - (-7) - 20$
$= 4 + 3 + (-12) + 7 + (-20)$
$= 7 + (-12) + 7 + (-20)$
$= -5 + 7 + (-20)$
$= 2 + (-20)$
$= -18$

Problem 4 **A.** $-38 \cdot 51 = -1938$

B. $-7(-8)(9)(-2) = 56(9)(-2)$
$= 504(-2)$
$= -1008$

Problem 5 **A.** $(-135) \div (-9) = 15$

B. $\dfrac{84}{-6} = -14$

C. $-\dfrac{36}{-12} = -(-3) = 3$

Problem 6

Strategy To find the average daily high temperature:
- Add the seven temperature readings.
- Divide by 7.

Solution $-5 + (-6) + 3 + 0 + (-4) + (-7) + (-2) = -11 + 3 + 0 + (-4) + (-7) + (-2)$
$= -8 + 0 + (-4) + (-7) + (-2)$
$= -8 + (-4) + (-7) + (-2)$
$= -12 + (-7) + (-2)$
$= -19 + (-2)$
$= -21$

$-21 \div 7 = -3$

The average daily high temperature was $-3°C$.

SECTION 1.3   *pages 22 – 29*

Problem 1

$$\begin{array}{r} 0.16 \\ 25\overline{)4.00} \\ -2\ 5 \\ \hline 1\ 50 \\ -1\ 50 \\ \hline 0 \end{array}$$

$$\frac{4}{25} = 0.16$$

Problem 2

$$\begin{array}{r} 0.444 \\ 9\overline{)4.000} \\ -3\ 6 \\ \hline 40 \\ -36 \\ \hline 40 \\ -36 \\ \hline 4 \end{array}$$

$$\frac{4}{9} = 0.\overline{4}$$

Problem 3 $125\% = 125\left(\dfrac{1}{100}\right) = \dfrac{125}{100} = 1\dfrac{1}{4}$

$125\% = 125(0.01) = 1.25$

Problem 4 $16\dfrac{2}{3}\% = 16\dfrac{2}{3}\left(\dfrac{1}{100}\right) = \dfrac{50}{3}\left(\dfrac{1}{100}\right) = \dfrac{1}{6}$

Problem 5 $6.08\% = 6.08(0.01) = 0.0608$

Problem 6   **A.** $0.043 = 0.043(100\%) = 4.3\%$

B. $2.57 = 2.57(100\%) = 257\%$

Problem 7 $\dfrac{5}{9} = \dfrac{5}{9}(100\%) = \dfrac{500}{9}\% \approx 55.6\%$

Problem 8 $\dfrac{9}{16} = \dfrac{9}{16}(100\%) = \dfrac{900}{16}\% = 56\dfrac{1}{4}\%$

Problem 9   Prime factorization of 9 and 12:

$9 = 3 \cdot 3 \qquad 12 = 2 \cdot 2 \cdot 3$

$\text{LCM} = 2 \cdot 2 \cdot 3 \cdot 3 = 36$

$$\frac{5}{9} - \frac{11}{12} = \frac{20}{36} - \frac{33}{36} = \frac{20}{36} + \frac{-33}{36} = \frac{20 + (-33)}{36} = \frac{-13}{36} = -\frac{13}{36}$$

Problem 10 $-\dfrac{7}{8} - \dfrac{5}{6} + \dfrac{1}{2} = -\dfrac{21}{24} - \dfrac{20}{24} + \dfrac{12}{24} = \dfrac{-21}{24} + \dfrac{-20}{24} + \dfrac{12}{24} = \dfrac{-21 + (-20) + 12}{24} = \dfrac{-29}{24} = -\dfrac{29}{24}$

Problem 11

$$\begin{array}{r} 3.097 \\ 4.9 \\ +\ 3.09 \\ \hline 11.087 \end{array}$$

Problem 12

$$\begin{array}{r} 67.910 \\ -16.127 \\ \hline 51.783 \end{array}$$

$16.127 - 67.91 = -51.783$

Problem 13 $-\dfrac{7}{12} \cdot \dfrac{9}{14} = -\dfrac{7 \cdot 9}{12 \cdot 14} = -\dfrac{\overset{1}{\cancel{7}} \cdot \overset{1}{\cancel{3}} \cdot 3}{2 \cdot 2 \cdot \cancel{3} \cdot 2 \cdot \cancel{7}} = -\dfrac{3}{8}$

Problem 14 $-\dfrac{3}{8} \div \left(-\dfrac{5}{12}\right) = \dfrac{3}{8} \cdot \dfrac{12}{5} = \dfrac{3 \cdot 12}{8 \cdot 5} = \dfrac{3 \cdot \overset{1}{\cancel{2}} \cdot \overset{1}{\cancel{2}} \cdot 3}{\cancel{2} \cdot \cancel{2} \cdot 2 \cdot 5} = \dfrac{9}{10}$

Problem 15
$$\begin{array}{r} 5.44 \\ \times\ \ \ 3.8 \\ \hline 4352 \\ 1632 \\ \hline 20.672 \end{array}$$

$(-5.44)(3.8) = -20.672$

Problem 16
$$\begin{array}{r} 0.231 \\ 1.7\overline{)0.3{,}940} \\ -3\ 4\ \ \ \ \\ \hline 54\ \ \ \\ -51\ \ \ \\ \hline 30\ \\ -17\ \\ \hline 13 \end{array}$$

$-0.394 \div 1.7 \approx -0.23$

SECTION 1.4   *pages 34 – 38*

Problem 1
$(-5)^3 = (-5)(-5)(-5) = 25(-5) = -125$

$-5^3 = -(5 \cdot 5 \cdot 5) = -(25 \cdot 5) = -125$

Problem 2
$(-3)^3 = (-3)(-3)(-3) = 9(-3) = -27$

$(-3)^4 = (-3)(-3)(-3)(-3) = 9(-3)(-3) = -27(-3) = 81$

Problem 3
$(3^3)(-2)^3 = (3 \cdot 3 \cdot 3) \cdot (-2)(-2)(-2) = 27 \cdot (-8) = -216$

$\left(-\dfrac{2}{5}\right)^2 = \left(-\dfrac{2}{5}\right)\left(-\dfrac{2}{5}\right) = \dfrac{2 \cdot 2}{5 \cdot 5} = \dfrac{4}{25}$

Problem 4
$$\begin{aligned} 36 \div (8-5)^2 - (-3)^2 \cdot 2 &= 36 \div (3)^2 - (-3)^2 \cdot 2 \\ &= 36 \div 9 - 9 \cdot 2 \\ &= 4 - 9 \cdot 2 \\ &= 4 - 18 \\ &= -14 \end{aligned}$$

Problem 5
$$\begin{aligned} 27 \div 3^2 + (-3)^2 \cdot 4 \\ = 27 \div 9 + 9 \cdot 4 \\ = 3 + 9 \cdot 4 \\ = 3 + 36 \\ = 39 \end{aligned}$$

Problem 6
$$\begin{aligned} 4 - 3[4 - 2(6-3)] \div 2 \\ = 4 - 3[4 - 2(3)] \div 2 \\ = 4 - 3[4 - 6] \div 2 \\ = 4 - 3(-2) \div 2 \\ = 4 - (-6) \div 2 \\ = 4 - (-3) \\ = 7 \end{aligned}$$

SOLUTIONS to Chapter 2 Problems

SECTION 2.1   *pages 51 – 54*

Problem 1 -4

Problem 2
$$\begin{aligned} 2xy + y^2 \\ 2(-4)(2) + (2)^2 &= 2(-4)(2) + 4 \\ &= -8(2) + 4 \\ &= -16 + 4 \\ &= -12 \end{aligned}$$

Problem 3 $\dfrac{a^2 + b^2}{a + b}$

$\dfrac{(5)^2 + (-3)^2}{5 + (-3)} = \dfrac{25 + 9}{5 + (-3)}$

$\qquad\qquad = \dfrac{34}{2}$

$\qquad\qquad = 17$

Problem 4 $x^3 - 2(x + y) + z^2$

$(2)^3 - 2[2 + (-4)] + (-3)^2$

$\quad = (2)^3 - 2(-2) + (-3)^2$

$\quad = 8 - 2(-2) + 9$

$\quad = 8 + 4 + 9$

$\quad = 12 + 9$

$\quad = 21$

Problem 5 $V = \dfrac{1}{3}\pi r^2 h$

$V = \dfrac{1}{3}\pi(4.5)^2(9.5)$ ▶ $r = \dfrac{1}{2}d = \dfrac{1}{2}(9) = 4.5$

$V = \dfrac{1}{3}\pi(20.25)(9.5)$

$V \approx 201.5$

The volume is approximately 201.5 cm³.

SECTION 2.2   *pages 56 – 63*

Problem 1 $7 + (-7) = 0$

Problem 2   The Associative Property of Addition

Problem 3   **A.** $9x + 6x = (9 + 6)x = 15x$ **B.** $-4y - 7y = [-4 + (-7)]y = -11y$

Problem 4   **A.** $3a - 2b + 5a = 3a + 5a - 2b$ **B.** $x^2 - 7 + 9x^2 - 14 = x^2 + 9x^2 - 7 - 14$

$\qquad\qquad\qquad\quad = (3a + 5a) - 2b$ $\qquad\qquad\qquad\qquad = (x^2 + 9x^2) + (-7 - 14)$

$\qquad\qquad\qquad\quad = 8a - 2b$ $\qquad\qquad\qquad\qquad = 10x^2 - 21$

Problem 5   **A.** $-7(-2a) = [-7(-2)]a$ **B.** $-\dfrac{5}{6}(-30y^2) = \left[-\dfrac{5}{6}(-30)\right]y^2$ **C.** $(-5x)(-2) = (-2)(-5x)$

$\qquad\qquad\quad = 14a$ $\qquad\qquad\qquad\qquad = 25y^2$ $\qquad\qquad\qquad = [-2(-5)]x$

$\qquad\qquad\qquad\qquad\qquad\qquad\qquad\qquad\qquad\qquad\qquad = 10x$

Problem 6   **A.** $7(4 + 2y) = 7(4) + 7(2y)$ **B.** $-(5x - 12) = -1(5x - 12)$

$\qquad\qquad\qquad = 28 + 14y$ $\qquad\qquad\qquad\quad = -1(5x) - (-1)(12)$

$\qquad\qquad\qquad\qquad\qquad\qquad\qquad\qquad = -5x + 12$

C. $(3a - 1)5 = (3a)(5) - (1)(5)$ **D.** $-3(6a^2 - 8a + 9) = -3(6a^2) - (-3)(8a) + (-3)(9)$

$\qquad\qquad = 15a - 5$ $\qquad\qquad\qquad\qquad\qquad = -18a^2 + 24a - 27$

Problem 7 $7(x - 2y) - 3(-x - 2y)$

$\quad = 7x - 14y + 3x + 6y$

$\quad = 10x - 8y$

Problem 8 $3y - 2[x - 4(2 - 3y)]$

$\quad = 3y - 2[x - 8 + 12y]$

$\quad = 3y - 2x + 16 - 24y$

$\quad = -2x - 21y + 16$

SECTION 2.3   *pages 67 – 71*

Problem 1   **A.** 18 <u>less than</u> the <u>cube of</u> x **B.** y <u>decreased by</u> the <u>sum</u> of z and 9

$\qquad\qquad x^3 - 18$ $\qquad\qquad y - (z + 9)$

C. the <u>difference between</u> the <u>square</u> of q and the <u>sum</u> of r and t

$\qquad q^2 - (r + t)$

Problem 2   the unknown number: n
the square of the number: n^2
the product of five and the square of
   the number: $5n^2$

$5n^2 + n$

Problem 3   the unknown number: n
twice the number: $2n$
the sum of seven and twice the number: $7 + 2n$

$3(7 + 2n)$

Problem 4   the unknown number: n
the difference between twice the number and 17: $2n - 17$

$$n - (2n - 17) = n - 2n + 17$$
$$= -n + 17$$

Problem 5   the unknown number: n

three-fourths of the number: $\dfrac{3}{4}n$

one-fifth of the number: $\dfrac{1}{5}n$

$$\frac{3}{4}n + \frac{1}{5}n = \frac{15}{20}n + \frac{4}{20}n$$
$$= \frac{19}{20}n$$

Problem 6   the time required by the newer model: t
the time required by the older model is
   twice the time required for the
   newer model: $2t$

Problem 7   the length of one piece: L
the length of the second piece: $6 - L$

SOLUTIONS to Chapter 3 Problems

SECTION 3.1   *pages 87–94*

Problem 1
$$5 - 4x = 8x + 2$$

$$\begin{array}{c|c} 5 - 4\left(\dfrac{1}{4}\right) & 8\left(\dfrac{1}{4}\right) + 2 \\ 5 - 1 & 2 + 2 \\ & 4 = 4 \end{array}$$

Yes, $\dfrac{1}{4}$ is a solution.

Problem 2
$$10x - x^2 = 3x - 10$$

$$\begin{array}{c|c} 10(5) - (5)^2 & 3(5) - 10 \\ 50 - 25 & 15 - 10 \\ & 25 \neq 5 \end{array}$$

No, 5 is not a solution.

Problem 3
$$x - \frac{1}{3} = -\frac{3}{4}$$
$$x - \frac{1}{3} + \frac{1}{3} = -\frac{3}{4} + \frac{1}{3}$$
$$x + 0 = -\frac{9}{12} + \frac{4}{12}$$
$$x = -\frac{5}{12}$$

Check: $x - \dfrac{1}{3} = -\dfrac{3}{4}$

$$\begin{array}{c|c} -\dfrac{5}{12} - \dfrac{1}{3} & -\dfrac{3}{4} \\ -\dfrac{3}{4} & = -\dfrac{3}{4} \end{array}$$

The solution is $-\dfrac{5}{12}$.

Problem 4
$$-8 = 5 + x$$
$$-8 - 5 = 5 - 5 + x$$
$$-13 = x$$

The solution is -13.

Problem 5
$$-\frac{2x}{5} = 6$$
$$\left(-\frac{5}{2}\right)\left(-\frac{2}{5}x\right) = \left(-\frac{5}{2}\right)(6)$$
$$1x = -15$$
$$x = -15$$

The solution is -15.

Problem 6
$$6x = 10$$
$$\frac{6x}{6} = \frac{10}{6}$$
$$x = \frac{5}{3}$$

The solution is $\frac{5}{3}$.

Check: $$\frac{6x = 10}{6\left(\frac{5}{3}\right)\ \Big|\ 10}$$
$$10 = 10$$

Problem 7
$$4x - 8x = 16$$
$$-4x = 16$$
$$\frac{-4x}{-4} = \frac{16}{-4}$$
$$x = -4$$

The solution is -4.

Problem 8
$$PB = A$$
$$P(60) = 27$$
$$60P = 27$$
$$\frac{60P}{60} = \frac{27}{60}$$
$$P = 0.45$$

27 is 45% of 60.

Problem 9

Strategy   To find the percent of the questions answered correctly, solve the basic percent equation using $B = 80$ and $A = 72$. The percent is unknown.

Solution
$$PB = A$$
$$P(80) = 72$$
$$80P = 72$$
$$\frac{80P}{80} = \frac{72}{80}$$
$$P = 0.9$$

90% of the questions were answered correctly.

Problem 10

Strategy   To find how many gallons are used efficiently, solve the basic percent equation using $B = 15$ and $P = 32\% = 0.32$. The amount is unknown.

Solution
$$PB = A$$
$$0.32(15) = A$$
$$4.8 = A$$

Out of 15 gal of gasoline, 4.8 gal are used efficiently.

SECTION 3.2   *pages 100 – 107*

Problem 1
$$5x + 7 = 10$$
$$5x + 7 - 7 = 10 - 7$$
$$5x = 3$$
$$\frac{5x}{5} = \frac{3}{5}$$
$$x = \frac{3}{5}$$

The solution is $\frac{3}{5}$.

Problem 2
$$11 = 11 + 3x$$
$$11 - 11 = 11 - 11 + 3x$$
$$0 = 3x$$
$$\frac{0}{3} = \frac{3x}{3}$$
$$0 = x$$

The solution is 0.

Problem 3

$$\frac{5}{2} - \frac{2}{3}x = \frac{1}{2}$$

$$6\left(\frac{5}{2} - \frac{2}{3}x\right) = 6\left(\frac{1}{2}\right)$$

$$6\left(\frac{5}{2}\right) - 6\left(\frac{2}{3}x\right) = 3$$

$$15 - 4x = 3$$

$$15 - 15 - 4x = 3 - 15$$

$$-4x = -12$$

$$\frac{-4x}{-4} = \frac{-12}{-4}$$

$$x = 3$$

The solution is 3.

Problem 4

$$5x + 4 = 6 + 10x$$

$$5x - 10x + 4 = 6 + 10x - 10x$$

$$-5x + 4 = 6$$

$$-5x + 4 - 4 = 6 - 4$$

$$-5x = 2$$

$$\frac{-5x}{-5} = \frac{2}{-5}$$

$$x = -\frac{2}{5}$$

The solution is $-\frac{2}{5}$.

Problem 5

$$5x - 4(3 - 2x) = 2(3x - 2) + 6$$

$$5x - 12 + 8x = 6x - 4 + 6$$

$$13x - 12 = 6x + 2$$

$$13x - 6x - 12 = 6x - 6x + 2$$

$$7x - 12 = 2$$

$$7x - 12 + 12 = 2 + 12$$

$$7x = 14$$

$$\frac{7x}{7} = \frac{14}{7}$$

$$x = 2$$

The solution is 2.

Problem 6

$$-2[3x - 5(2x - 3)] = 3x - 8$$

$$-2[3x - 10x + 15] = 3x - 8$$

$$-2[-7x + 15] = 3x - 8$$

$$14x - 30 = 3x - 8$$

$$14x - 3x - 30 = 3x - 3x - 8$$

$$11x - 30 = -8$$

$$11x - 30 + 30 = -8 + 30$$

$$11x = 22$$

$$\frac{11x}{11} = \frac{22}{11}$$

$$x = 2$$

The solution is 2.

Problem 7

Strategy

The total cost was $8000, the unit cost was $15, and the fixed costs were $2000. Therefore, $T = 8000$, $U = 15$, and $F = 2000$. To find the number of units made, replace each of the variables with its given value, and solve for N.

Solution

$$T = U \cdot N + F$$

$$8000 = 15N + 2000$$

$$8000 - 2000 = 15N + 2000 - 2000$$

$$6000 = 15N$$

$$\frac{6000}{15} = \frac{15N}{15}$$

$$400 = N$$

The number of units made was 400.

Problem 8

Strategy

The lever is 14 ft long, so $d = 14$. One force is 6 ft from the fulcrum, so $x = 6$. The one force is 40 lb, so $F_1 = 40$. To find the force when the system balances, replace the variables F_1, x, and d in the lever system equation with the given values, and solve for F_2.

Solution

$$F_1x = F_2(d - x)$$

$$40(6) = F_2(14 - 6)$$

$$240 = F_2(8)$$

$$240 = 8F_2$$

$$\frac{240}{8} = \frac{8F_2}{8}$$

$$30 = F_2$$

A 30 lb force must be applied to the other end.

SECTION 3.3   *pages 113 – 119*

Problem 1   The solution set is the numbers greater than -2.

Problem 2

$$x + 2 < -2$$
$$x + 2 - 2 < -2 - 2$$
$$x < -4$$

Problem 3

$$5x + 3 > 4x + 5$$
$$5x - 4x + 3 > 4x - 4x + 5$$
$$x + 3 > 5$$
$$x + 3 - 3 > 5 - 3$$
$$x > 2$$

Problem 4

$$3x < 9$$
$$\frac{3x}{3} < \frac{9}{3}$$
$$x < 3$$

Problem 5

$$-\frac{3}{4}x \geq 18$$
$$-\frac{4}{3}\left(-\frac{3}{4}x\right) \leq -\frac{4}{3}(18)$$
$$x \leq -24$$

Problem 6

$$5 - 4x > 9 - 8x$$
$$5 - 4x + 8x > 9 - 8x + 8x$$
$$5 + 4x > 9$$
$$5 - 5 + 4x > 9 - 5$$
$$4x > 4$$
$$\frac{4x}{4} > \frac{4}{4}$$
$$x > 1$$

Problem 7

$$8 - 4(3x + 5) \leq 6(x - 8)$$
$$8 - 12x - 20 \leq 6x - 48$$
$$-12 - 12x \leq 6x - 48$$
$$-12 - 12x - 6x \leq 6x - 6x - 48$$
$$-12 - 18x \leq -48$$
$$-12 + 12 - 18x \leq -48 + 12$$
$$-18x \leq -36$$
$$\frac{-18x}{-18} \geq \frac{-36}{-18}$$
$$x \geq 2$$

SOLUTIONS to Chapter 4 Problems

SECTION 4.1   *pages 135 – 137*

Problem 1   the unknown number: n

nine less than twice a number	is	five times the sum of the number and twelve

$$2n - 9 = 5(n + 12)$$
$$2n - 9 = 5n + 60$$
$$2n - 5n - 9 = 5n - 5n + 60$$
$$-3n - 9 = 60$$
$$-3n - 9 + 9 = 60 + 9$$
$$-3n = 69$$
$$\frac{-3n}{-3} = \frac{69}{-3}$$
$$n = -23$$

The number is -23.

Problem 2

Strategy   To find the number of carbon atoms, write and solve an equation using n to represent the number of carbon atoms.

Solution

| eight | represents | twice the number of carbon atoms |

$$8 = 2n$$
$$\frac{8}{2} = \frac{2n}{2}$$
$$4 = n$$

There are 4 carbon atoms in a butane molecule.

Problem 3

Strategy To find the number of 10-speed bicycles made, write and solve an equation using n to represent the number of 10-speed bicycles and $160 - n$ to represent the number of 3-speed bicycles.

Solution

| four times the number of 3-speed bicycles made | equals | 30 less than the number of 10-speed bicycles made |

$$4(160 - n) = n - 30$$
$$640 - 4n = n - 30$$
$$640 - 4n - n = n - n - 30$$
$$640 - 5n = -30$$
$$640 - 640 - 5n = -30 - 640$$
$$-5n = -670$$
$$\frac{-5n}{-5} = \frac{-670}{-5}$$
$$n = 134$$

There are 134 10-speed bicycles made each day.

SECTION 4.2 *pages 142 – 145*

Problem 1

Strategy
- First consecutive integer: n
  Second consecutive integer: $n + 1$
  Third consecutive integer: $n + 2$
- The sum of the three integers is -12.

Solution
$$n + (n + 1) + (n + 2) = -12$$
$$3n + 3 = -12$$
$$3n = -15$$
$$n = -5$$

$$n + 1 = -5 + 1 = -4$$
$$n + 2 = -5 + 2 = -3$$

The three consecutive integers are -5, -4, and -3.

Problem 2

Strategy
- Number of dimes: x
  Number of nickels: $5x$
  Number of quarters: $x + 6$

Coin	Number	Value	Total value
Dime	x	10	$10x$
Nickel	$5x$	5	$5(5x)$
Quarter	$x + 6$	25	$25(x + 6)$

- The sum of the total values of each denomination of coin equals the total value of all the coins (630 cents).

Solution

$$10x + 5(5x) + 25(x + 6) = 630$$
$$10x + 25x + 25x + 150 = 630$$
$$60x + 150 = 630$$
$$60x = 480$$
$$x = 8$$

$$5x = 5(8) = 40$$
$$x + 6 = 8 + 6 = 14$$

The bank contains 8 dimes, 40 nickels, and 14 quarters.

SECTION 4.3   *pages 149 – 151*

Problem 1

Strategy
- Width of the rectangle: W
  Length of the rectangle: $W + 3$
- Use the equation for the perimeter of a rectangle.

Solution

$$2L + 2W = P$$
$$2(W + 3) + 2W = 42$$
$$2W + 6 + 2W = 42$$
$$4W + 6 = 42$$
$$4W = 36$$
$$W = 9$$

The width of the rectangle is 9 m.

Problem 2

Strategy
- Measure of the second angle: x
  Measure of the first angle: $2x$
  Measure of the third angle: $x - 8$
- Use the equation $A + B + C = 180°$.

Solution

$$x + 2x + (x - 8) = 180$$
$$4x - 8 = 180$$
$$4x = 188$$
$$x = 47$$

$$2x = 2(47) = 94$$
$$x - 8 = 47 - 8 = 39$$

The measure of the first angle is 94°.
The measure of the second angle is 47°.
The measure of the third angle is 39°.

SECTION 4.4   *pages 154 – 157*

Problem 1

Strategy   Given: $C = \$60$
$S = \$90$
Unknown markup rate: r
Use the equation $S = C + rC$.

Solution

$$S = C + rC$$
$$90 = 60 + 60r$$
$$30 = 60r$$
$$0.5 = r$$

The markup rate is 50%.

Problem 2

Strategy   Given: $r = 40\% = 0.40$
$S = \$133$
Unknown cost: C
Use the equation $S = C + rC$.

Solution

$$S = C + rC$$
$$133 = C + 0.4C$$
$$133 = 1.40C$$
$$95 = C$$

The cost is $95.

Problem 3

Strategy

Given: $R = \$29.80$
$S = \$22.35$
Unknown discount rate: r
Use the equation $S = R - rR$.

Solution

$$S = R - rR$$
$$22.35 = 29.80 - 29.80r$$
$$-7.45 = -29.80r$$
$$0.25 = r$$

The discount rate is 25%.

Problem 4

Strategy

Given: $S = \$43.50$
$r = 25\% = 0.25$
Unknown regular price: R
Use the equation $S = R - rR$.

Solution

$$S = R - rR$$
$$43.50 = R - 0.25R$$
$$43.50 = 0.75R$$
$$58 = R$$

The regular price is $58.

SECTION 4.5   *pages 160 – 162*

Problem 1

Strategy

■ Additional amount: x

	Principal	·	Rate	=	Interest
Amount at 7%	2500	·	0.07	=	0.07(2500)
Amount at 10%	x	·	0.10	=	0.10x
Amount at 9%	2500 + x	·	0.09	=	0.09(2500 + x)

■ The sum of the interest earned by the two investments equals the interest earned on the total investment.

Solution

$$0.07(2500) + 0.10x = 0.09(2500 + x)$$
$$175 + 0.10x = 225 + 0.09x$$
$$175 + 0.01x = 225$$
$$0.01x = 50$$
$$x = 5000$$

$5000 more must be invested at 10%.

SECTION 4.6   *pages 166 – 170*

Problem 1

Strategy

■ Pounds of $.75 fertilizer: x

	Amount	Cost	Value
$.90 fertilizer	20	$.90	0.90(20)
$.75 fertilizer	x	$.75	0.75x
$.85 fertilizer	20 + x	$.85	0.85(20 + x)

■ The sum of the values before mixing equals the value after mixing.

Solution $0.90(20) + 0.75x = 0.85(20 + x)$
$18 + 0.75x = 17 + 0.85x$
$18 - 0.10x = 17$
$-0.10x = -1$
$x = 10$

10 lb of the $.75 fertilizer must be added.

Problem 2

Strategy   ■ Pure orange juice is 100% orange juice. 100% = 1.00
Amount of pure orange juice: x

	Amount	Percent	Quantity
Pure orange juice	x	1.00	$1.00x$
Fruit drink	5	0.10	$5(0.10)$
Orange drink	$x + 5$	0.25	$0.25(x + 5)$

■ The sum of the quantities before mixing is equal to the quantity after mixing.

Solution $1.00x + 5(0.10) = 0.25(x + 5)$
$1.00x + 0.5 = 0.25x + 1.25$
$0.75x + 0.5 = 1.25$
$0.75x = 0.75$
$x = 1$

To make the orange drink, 1 qt of pure orange juice is added to the fruit drink.

SECTION 4.7   *pages 175 – 177*

Problem 1

Strategy   ■ Rate of the first train: r
Rate of the second train: $2r$

	Rate	Time	Distance
First train	r	3	$3r$
Second train	$2r$	3	$3(2r)$

■ The sum of the distances traveled by each train equals 306 mi.

Solution $3r + 3(2r) = 306$
$3r + 6r = 306$
$9r = 306$
$r = 34$

$2r = 2(34) = 68$

The first train is traveling at 34 mph.
The second train is traveling at 68 mph.

Problem 2

Strategy   ■ Time spent flying out: t
Time spent flying back: $7 - t$

	Rate	Time	Distance
Out	120	t	$120t$
Back	90	$7 - t$	$90(7 - t)$

■ The distance out equals the distance back.

Solution $120t = 90(7 - t)$
$120t = 630 - 90t$
$210t = 630$
$t = 3$ (The time out was 3 h.)

The distance $= 120t = 120(3) = 360$ mi.

The parcel of land was 360 mi away.

SECTION 4.8   *pages 181 – 182*

Problem 1

Strategy   To find the minimum selling price, write and solve an inequality using S to represent the selling price.

Solution
$$340 < 0.70S$$
$$\frac{340}{0.70} < \frac{0.70S}{0.70}$$
$$485.71429 < S$$

The minimum selling price is $485.72.

Problem 2

Strategy   To find the maximum number of miles:
- ■ Write an expression for the cost of each car, using x to represent the number of miles driven during the week.
- ■ Write and solve an inequality.

Solution

cost of a Company A car	is less than	cost of a Company B car

$$9(7) + 0.10x < 12(7) + 0.08x$$
$$63 + 0.10x < 84 + 0.08x$$
$$63 + 0.10x - 0.08x < 84 + 0.08x - 0.08x$$
$$63 + 0.02x < 84$$
$$63 - 63 + 0.02x < 84 - 63$$
$$0.02x < 21$$
$$\frac{0.02x}{0.02} < \frac{21}{0.02}$$
$$x < 1050$$

The maximum number of miles is 1049.

SOLUTIONS to Chapter 5 Problems

SECTION 5.1   *pages 201 – 204*

Problem 1

Problem 2 $A(4, 2)$, $B(-3, 4)$, $C(-3, 0)$, $D(0, 0)$

Problem 3

Strategy   Graph the ordered pairs on a rectangular coordinate system where the reading on the horizontal axis represents the age of the car in years and the reading on the vertical axis represents the price paid for the car in hundreds of dollars.

Solution

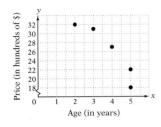

SECTION 5.2   *pages 208 – 216*

Problem 1 $y = -\dfrac{1}{2}x - 3$

$$\begin{array}{c|c} -4 & -\dfrac{1}{2}(2) - 3 \\ & -1 - 3 \\ & -4 \\ \hline -4 = -4 \end{array}$$

Yes, $(2, -4)$ is a solution of $y = -\dfrac{1}{2}x - 3$.

Problem 2 $y = -\dfrac{1}{4}x + 1$

$$= -\dfrac{1}{4}(4) + 1$$
$$= -1 + 1$$
$$= 0$$

The ordered-pair solution is $(4, 0)$.

Problem 3

Problem 4

Problem 5
$$5x - 2y = 10$$
$$5x - 5x - 2y = -5x + 10$$
$$-2y = -5x + 10$$
$$\dfrac{-2y}{-2} = \dfrac{-5x + 10}{-2}$$
$$y = \dfrac{-5x}{-2} + \dfrac{10}{-2}$$
$$y = \dfrac{5}{2}x - 5$$

Problem 6   **A.** $5x - 2y = 10$
$$-2y = -5x + 10$$
$$y = \dfrac{5}{2}x - 5$$

B. $x - 3y = 9$
$$-3y = -x + 9$$
$$y = \dfrac{1}{3}x - 3$$

Problem 7 x-intercept: $4x - y = 4$
$$4x - 0 = 4$$
$$4x = 4$$
$$x = 1$$

y-intercept: $4x - y = 4$
$$4(0) - y = 4$$
$$-y = 4$$
$$y = -4$$

The x-intercept is $(1, 0)$.   The y-intercept is $(0, -4)$.

Problem 8 A.   B.

SECTION 5.3 *pages 220 – 226*

Problem 1 A. $m = \dfrac{y_2 - y_1}{x_2 - x_1} = \dfrac{3 - 2}{1 - (-1)} = \dfrac{1}{2}$ B. $m = \dfrac{y_2 - y_1}{x_2 - x_1} = \dfrac{-5 - 2}{4 - 1} = \dfrac{-7}{3}$

The slope is $\dfrac{1}{2}$.   The slope is $-\dfrac{7}{3}$.

C. $m = \dfrac{y_2 - y_1}{x_2 - x_1} = \dfrac{7 - 3}{2 - 2} = \dfrac{4}{0}$ D. $m = \dfrac{y_2 - y_1}{x_2 - x_1} = \dfrac{-3 - (-3)}{-5 - 1} = \dfrac{0}{-6} = 0$

The slope is undefined.   The slope is 0.

Problem 2 $m = \dfrac{8650 - 6100}{1 - 4} = \dfrac{2550}{-3} = -850$

A slope of -850 means that the value of the car is decreasing at a rate of \$850 per year.

Problem 3 y-intercept $= (0, b) = (0, -1)$ **Problem 4** Solve the equation for y.

$m = -\dfrac{1}{4}$

$$x - 2y = 4$$
$$-2y = -x + 4$$
$$y = \dfrac{1}{2}x - 2$$

y-intercept $= (0, b) = (0, -2)$

$m = \dfrac{1}{2}$

SECTION 5.4 *pages 229 – 231*

Problem 1 $y = \dfrac{3}{2}x + b$ **Problem 2** $m = \dfrac{2}{5}$ $(x_1, y_1) = (5, 4)$

$-2 = \dfrac{3}{2}(4) + b$

$-2 = 6 + b$

$-8 = b$

$y = \dfrac{3}{2}x - 8$

$y - y_1 = m(x - x_1)$

$y - 4 = \dfrac{2}{5}(x - 5)$

$y - 4 = \dfrac{2}{5}x - 2$

$y = \dfrac{2}{5}x + 2$

SECTION 5.5   *pages 235 – 240*

Problem 1   The domain is {1}.
The range is {0, 1, 2, 3, 4}.
There are ordered pairs with the same
first coordinate and different second
coordinates. The relation is not a function.

Problem 2 $f(x) = -5x + 1$
$f(2) = -5(2) + 1$
$f(2) = -10 + 1$
$f(2) = -9$

The ordered pair $(2, -9)$ is an element
of the function.

Problem 3 $f(x) = 4x - 3$
$f(-5) = 4(-5) - 3 = -20 - 3 = -23$
$f(-3) = 4(-3) - 3 = -12 - 3 = -15$
$f(-1) = 4(-1) - 3 = -4 - 3 = -7$
$f(1) = 4(1) - 3 = 4 - 3 = 1$

The range is $\{-23, -15, -7, 1\}$.
The ordered pairs $(-5, -23)$, $(-3, -15)$,
$(-1, -7)$, and $(1, 1)$ belong to the function.

Problem 4 $f(x) = -\dfrac{1}{2}x - 3$

$y = -\dfrac{1}{2}x - 3$

Problem 5   a. $d = 40t$
$f(t) = 40t$

b.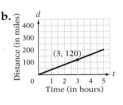

c. The ordered pair (3, 120) means that in 3 h, the
car travels a distance of 120 mi.

SECTION 5.6   *pages 246 – 248*

Problem 1 $x - 3y < 2$
$x - x - 3y < -x + 2$
$-3y < -x + 2$
$\dfrac{-3y}{-3} > \dfrac{-x + 2}{-3}$

$y > \dfrac{1}{3}x - \dfrac{2}{3}$

Problem 2 $x < 3$

SOLUTIONS to Chapter 6 Problems

SECTION 6.1   *pages 267–269*

Problem 1

$$2x - 5y = 8 \qquad\qquad -x + 3y = -5$$

$2(-1) - 5(-2)$	8		$-(-1) + 3(-2)$	-5
$-2 + 10$	8		$1 + (-6)$	-5
$8 = 8$			$-5 = -5$	

Yes, $(-1, -2)$ is a solution of the system of equations.

Problem 2   **A.**

The solution is $(-3, 2)$.

B.

The lines are parallel and therefore do not intersect. The system of equations is inconsistent and has no solution.

SECTION 6.2   *pages 272–275*

Problem 1   (1) $\qquad 7x - y = 4$
(2) $\qquad 3x + 2y = 9$

$$7x - y = 4$$
$$-y = -7x + 4 \qquad\qquad$$ ► Solve equation (1) for y.
$$y = 7x - 4$$

$$3x + 2y = 9$$
$$3x + 2(7x - 4) = 9 \qquad\qquad$$ ► Substitute $7x - 4$ for y in equation (2).
$$3x + 14x - 8 = 9$$
$$17x - 8 = 9$$
$$17x = 17$$
$$x = 1$$

$$7x - y = 4$$
$$7(1) - y = 4 \qquad\qquad$$ ► Substitute the value of x in equation (1).
$$7 - y = 4$$
$$-y = -3$$
$$y = 3$$

The solution is $(1, 3)$.

Problem 2

$$3x - y = 4$$
$$y = 3x + 2$$

$$3x - y = 4$$
$$3x - (3x + 2) = 4$$
$$3x - 3x - 2 = 4$$
$$-2 = 4$$

The system of equations is inconsistent and has no solution.

Problem 3

$$y = -2x + 1$$
$$6x + 3y = 3$$

$$6x + 3y = 3$$
$$6x + 3(-2x + 1) = 3$$
$$6x - 6x + 3 = 3$$
$$3 = 3$$

The system of equations is dependent. The solutions are the ordered pairs that satisfy the equation $y = -2x + 1$.

SECTION 6.3   *pages 278 – 281*

Problem 1   (1) $x - 2y = 1$
(2) $2x + 4y = 0$

$$2(x - 2y) = 2 \cdot 1 \qquad \blacktriangleright \text{Eliminate } y.$$
$$2x + 4y = 0$$

$$2x - 4y = 2$$
$$2x + 4y = 0$$
$$\qquad 4x = 2 \qquad \blacktriangleright \text{Add the equations.}$$
$$x = \frac{2}{4} = \frac{1}{2}$$

$$2x + 4y = 0 \qquad \blacktriangleright \text{Replace } x \text{ in equation (2).}$$
$$2\left(\frac{1}{2}\right) + 4y = 0$$
$$1 + 4y = 0$$
$$4y = -1$$
$$y = -\frac{1}{4}$$

The solution is $\left(\dfrac{1}{2}, -\dfrac{1}{4}\right)$.

Problem 2   (1) $2x - 3y = 4$
(2) $-4x + 6y = -8$

$$2(2x - 3y) = 2 \cdot 4 \qquad \blacktriangleright \text{Eliminate } y.$$
$$-4x + 6y = -8$$

$$4x - 6y = 8$$
$$-4x + 6y = -8$$
$$0 + 0 = 0 \qquad \blacktriangleright \text{Add the equations.}$$
$$0 = 0$$

The system of equations is dependent. The solutions are the ordered pairs that satisfy the equation $2x - 3y = 4$.

SECTION 6.4   *pages 283 – 287*

Problem 1

Strategy   ■ Rate of the current: c
Rate of the canoeist in calm water: r

	Rate	Time	Distance
With current	$r + c$	3	$3(r + c)$
Against current	$r - c$	4	$4(r - c)$

■ The distance traveled with the current is 24 mi.
The distance traveled against the current is 24 mi.

Solution $3(r + c) = 24$ $\dfrac{3(r + c)}{3} = \dfrac{24}{3}$ $r + c = 8$

$4(r - c) = 24$ $\dfrac{4(r - c)}{4} = \dfrac{24}{4}$ $r - c = 6$

$$2r = 14$$
$$r = 7$$

$$r + c = 8$$
$$7 + c = 8$$
$$c = 1$$

The rate of the current is 1 mph.
The rate of the canoeist in calm water is 7 mph.

Problem 2

Strategy   ▬ The number of dimes in the first bank: d
The number of quarters in the first bank: q

First bank

	Number	Value	Total value
Dimes	d	10	$10d$
Quarters	q	25	$25q$

Second bank

	Number	Value	Total value
Dimes	$2d$	10	$20d$
Quarters	$\dfrac{1}{2}q$	25	$\dfrac{25}{2}q$

▬ The total value of the coins in the first bank is $3.90.
The total value of the coins in the second bank is $3.30.

Solution $10d + 25q = 390$ $10d + 25q = 390$ $10d + 25q = 390$

$20d + \dfrac{25}{2}q = 330$ $-2\left(20d + \dfrac{25}{2}q\right) = -2(330)$ $-40d - 25q = -660$

$$-30d = -270$$
$$d = 9$$

$$10d + 25q = 390$$
$$10(9) + 25q = 390$$
$$90 + 25q = 390$$
$$25q = 300$$
$$q = 12$$

There are 9 dimes and 12 quarters in the first bank.

SOLUTIONS to Chapter 7 Problems

SECTION 7.1   *pages 303 – 305*

Problem 1 $2x^2 + 4x - 3$
$\underline{5x^2 - 6x}$
$7x^2 - 2x - 3$

Problem 2 $(-4x^2 - 3xy + 2y^2) + (3x^2 - 4y^2)$
$= (-4x^2 + 3x^2) - 3xy + (2y^2 - 4y^2)$
$= -x^2 - 3xy - 2y^2$

Problem 3   The opposite of $2y^2 - xy + 5x^2$ is $-2y^2 + xy - 5x^2$.

$$\begin{array}{r} 8y^2 - 4xy + x^2 \\ -2y^2 + xy - 5x^2 \\ \hline 6y^2 - 3xy - 4x^2 \end{array}$$

Problem 4
$$\begin{aligned} (-3a^2 - 4a + 2) &- (5a^3 + 2a - 6) \\ &= (-3a^2 - 4a + 2) + (-5a^3 - 2a + 6) \\ &= -5a^3 - 3a^2 + (-4a - 2a) + (2 + 6) \\ &= -5a^3 - 3a^2 - 6a + 8 \end{aligned}$$

SECTION 7.2   *pages 307–309*

Problem 1 $(3x^2)(6x^3) = (3 \cdot 6)(x^2 \cdot x^3) = 18x^5$

Problem 2 $(-3xy^2)(-4x^2y^3) = [(-3)(-4)](x \cdot x^2)(y^2 \cdot y^3) = 12x^3y^5$

Problem 3 $(3x)(2x^2y)^3 = (3x)(2^3x^6y^3) = (3x)(8x^6y^3) = (3 \cdot 8)(x \cdot x^6)y^3 = 24x^7y^3$

SECTION 7.3   *pages 311–315*

Problem 1   **A.** $(-2y + 3)(-4y) = 8y^2 - 12y$ **B.** $-a^2(3a^2 + 2a - 7) = -3a^4 - 2a^3 + 7a^2$

Problem 2
$$\begin{array}{r} 2y^3 + 2y^2 - 3 \\ 3y - 1 \\ \hline -2y^3 - 2y^2 + 3 \\ 6y^4 + 6y^3 - 9y \\ \hline 6y^4 + 4y^3 - 2y^2 - 9y + 3 \end{array}$$

Problem 3
$$\begin{array}{r} 3x^3 - 2x^2 + x - 3 \\ 2x + 5 \\ \hline 15x^3 - 10x^2 + 5x - 15 \\ 6x^4 - 4x^3 + 2x^2 - 6x \\ \hline 6x^4 + 11x^3 - 8x^2 - x - 15 \end{array}$$

Problem 4 $(4y - 5)(3y - 3) = 12y^2 - 12y - 15y + 15 = 12y^2 - 27y + 15$

Problem 5 $(3a + 2b)(3a - 5b) = 9a^2 - 15ab + 6ab - 10b^2 = 9a^2 - 9ab - 10b^2$

Problem 6 $(2a + 5c)(2a - 5c) = (2a)^2 - (5c)^2 = 4a^2 - 25c^2$

Problem 7 $(3x + 2y)^2 = (3x)^2 + 2(3x)(2y) + (2y)^2 = 9x^2 + 12xy + 4y^2$

Problem 8

Strategy   To find the area in terms of x, replace the variables L and W in the equation $A = LW$ with the given values, and solve for A.

Solution
$$\begin{aligned} A &= LW \\ A &= (x + 7)(x - 4) \\ A &= x^2 - 4x + 7x - 28 \\ A &= x^2 + 3x - 28 \end{aligned}$$

The area is $(x^2 + 3x - 28)$ m².

Problem 9

Strategy   To find the area of the triangle in terms of x, replace the variables b and h in the equation $A = \frac{1}{2}bh$ with the given values, and solve for A.

Solution
$$\begin{aligned} A &= \frac{1}{2}bh \\ A &= \frac{1}{2}(x + 3)(4x - 6) \\ A &= \frac{1}{2}(4x^2 + 6x - 18) \\ A &= 2x^2 + 3x - 9 \end{aligned}$$

The area is $(2x^2 + 3x - 9)$ cm².

SECTION 7.4   *pages 320–327*

Problem 1 $\dfrac{2^{-2}}{2^3} = 2^{-2-3} = 2^{-5} = \dfrac{1}{2^5} = \dfrac{1}{32}$

Problem 2 $\dfrac{b^8}{a^{-5}b^6} = a^5b^2$

Problem 3   **A.** $\dfrac{12x^{-8}y^4}{-16xy^{-3}} = -\dfrac{3x^{-9}y^7}{4} = -\dfrac{3y^7}{4x^9}$ **B.** $(-3ab)(2a^3b^{-2})^{-3} = (-3ab)(2^{-3}a^{-9}b^6) = -\dfrac{3a^{-8}b^7}{2^3} = -\dfrac{3b^7}{8a^8}$

Problem 4   **A.** $57{,}000{,}000{,}000 = 5.7 \cdot 10^{10}$ **B.** $0.000000017 = 1.7 \cdot 10^{-8}$

Problem 5   **A.** $5 \cdot 10^{12} = 5{,}000{,}000{,}000{,}000$ **B.** $4.0162 \cdot 10^{-9} = 0.0000000040162$

Problem 6   **A.** $(2.4 \cdot 10^{-9})(1.6 \cdot 10^3) = 3.84 \cdot 10^{-6}$ **B.** $\dfrac{5.4 \cdot 10^{-2}}{1.8 \cdot 10^{-4}} = 3 \cdot 10^2$

Problem 7
$$\dfrac{4x^3y + 8x^2y^2 - 4xy^3}{2xy} = \dfrac{4x^3y}{2xy} + \dfrac{8x^2y^2}{2xy} - \dfrac{4xy^3}{2xy}$$
$$= 2x^2 + 4xy - 2y^2$$

Problem 8
$$\dfrac{24x^2y^2 - 18xy + 6y}{6xy} = \dfrac{24x^2y^2}{6xy} - \dfrac{18xy}{6xy} + \dfrac{6y}{6xy}$$
$$= 4xy - 3 + \dfrac{1}{x}$$

Problem 9

$$
\begin{array}{r}
x^2 + 2x + 2 \\
x - 2 \overline{)\, x^3 + 0x^2 - 2x - 4} \\
\underline{x^3 - 2x^2} \\
2x^2 - 2x \\
\underline{2x^2 - 4x} \\
2x - 4 \\
\underline{2x - 4} \\
0
\end{array}
$$

$$(x^3 - 2x - 4) \div (x - 2) = x^2 + 2x + 2$$

SOLUTIONS to Chapter 8 Problems

SECTION 8.1   *pages 343 – 345*

Problem 1
$4x^6y = 2 \cdot 2 \cdot x^6 \cdot y$
$18x^2y^6 = 2 \cdot 3 \cdot 3 \cdot x^2 \cdot y^6$
$\text{GCF} = 2 \cdot x^2 \cdot y = 2x^2y$

Problem 2   **A.**
$14a^2 = 2 \cdot 7 \cdot a^2$
$21a^4b = 3 \cdot 7 \cdot a^4 \cdot b$
The GCF is $7a^2$.

$14a^2 - 21a^4b = 7a^2(2) + 7a^2(-3a^2b)$
$ = 7a^2(2 - 3a^2b)$

    B.
$6x^4y^2 = 2 \cdot 3 \cdot x^4 \cdot y^2$
$9x^3y^2 = 3 \cdot 3 \cdot x^3 \cdot y^2$
$12x^2y^4 = 2 \cdot 2 \cdot 3 \cdot x^2 \cdot y^4$
The GCF is $3x^2y^2$.

$6x^4y^2 - 9x^3y^2 + 12x^2y^4$
$= 3x^2y^2(2x^2) + 3x^2y^2(-3x) + 3x^2y^2(4y^2)$
$= 3x^2y^2(2x^2 - 3x + 4y^2)$

Problem 3 $a(b - 7) + b(b - 7) = (b - 7)(a + b)$

Problem 4 $3y(5x - 2) - 4(2 - 5x) = 3y(5x - 2) + 4(5x - 2) = (5x - 2)(3y + 4)$

Problem 5 $y^5 - 5y^3 + 4y^2 - 20 = (y^5 - 5y^3) + (4y^2 - 20)$
$ = y^3(y^2 - 5) + 4(y^2 - 5) = (y^2 - 5)(y^3 + 4)$

SECTION 8.2   *pages 348 – 351*

Problem 1

Factors of 15	Sum
−1, −15	−16
−3, −5	−8

$$x^2 - 8x + 15 = (x - 3)(x - 5)$$

Problem 2

Factors of −18	Sum
1, −18	−17
−1, 18	17
2, −9	−7
−2, 9	7
3, −6	−3
−3, 6	3

$$x^2 + 3x - 18 = (x + 6)(x - 3)$$

Problem 3   The GCF is $3b$.

$$3a^2b - 18ab - 81b = 3b(a^2 - 6a - 27)$$

Factors of −27	Sum
1, −27	−26
−1, 27	26
3, −9	−6
−3, 9	6

$$3a^2b - 18ab - 81b = 3b(a + 3)(a - 9)$$

Problem 4   The GCF is 4.

$$4x^2 - 40xy + 84y^2$$
$$= 4(x^2 - 10xy + 21y^2)$$

Factors of 21	Sum
−1, −21	−22
−3, −7	−10

$$4x^2 - 40xy + 84y^2 = 4(x - 3y)(x - 7y)$$

SECTION 8.3   *pages 354 – 360*

Problem 1

Factors of 6	Factors of 5
1, 6	−1, −5
2, 3	

Trial Factors	Middle Term
$(x - 1)(6x - 5)$	$-5x - 6x = -11x$

$$6x^2 - 11x + 5 = (x - 1)(6x - 5)$$

Problem 2

Factors of 8	Factors of −15
1, 8	1, −15
2, 4	−1, 15
	3, −5
	−3, 5

Trial Factors	Middle Term
$(x + 1)(8x - 15)$	$-15x + 8x = -7x$
$(x - 1)(8x + 15)$	$15x - 8x = 7x$
$(x + 3)(8x - 5)$	$-5x + 24x = 19x$
$(x - 3)(8x + 5)$	$5x - 24x = -19x$
$(2x + 1)(4x - 15)$	$-30x + 4x = -26x$
$(2x - 1)(4x + 15)$	$30x - 4x = 26x$
$(2x + 3)(4x - 5)$	$-10x + 12x = 2x$
$(2x - 3)(4x + 5)$	$10x - 12x = -2x$
$(8x + 1)(x - 15)$	$-120x + x = -119x$
$(8x - 1)(x + 15)$	$120x - x = 119x$
$(8x + 3)(x - 5)$	$-40x + 3x = -37x$
$(8x - 3)(x + 5)$	$40x - 3x = 37x$
$(4x + 1)(2x - 15)$	$-60x + 2x = -58x$
$(4x - 1)(2x + 15)$	$60x - 2x = 58x$
$(4x + 3)(2x - 5)$	$-20x + 6x = -14x$
$(4x - 3)(2x + 5)$	$20x - 6x = 14x$

$$8x^2 + 14x - 15 = (4x - 3)(2x + 5)$$

Problem 3

Factors of 24	Factors of -1
1, 24	1, -1
2, 12	
3, 8	
4, 6	

Trial Factors	Middle Term
$(1 - y)(24 + y)$	$y - 24y = -23y$
$(2 - y)(12 + y)$	$2y - 12y = -10y$
$(3 - y)(8 + y)$	$3y - 8y = -5y$
$(4 - y)(6 + y)$	$4y - 6y = -2y$

$24 - 2y - y^2 = (4 - y)(6 + y)$

Problem 4

The GCF is $2a^2$.

$4a^2b^2 - 30a^2b + 14a^2$
$\quad = 2a^2(2b^2 - 15b + 7)$

Factors of 2	Factors of 7
1, 2	$-1, -7$

Trial Factors	Middle Term
$(b - 1)(2b - 7)$	$-7b - 2b = -9b$
$(b - 7)(2b - 1)$	$-b - 14b = -15b$

$4a^2b^2 - 30a^2b + 14a^2$
$\quad = 2a^2(b - 7)(2b - 1)$

Problem 5

$a \cdot c = -14$
$-1(14) = -14, \quad -1 + 14 = 13$

$2a^2 + 13a - 7$
$\quad = 2a^2 - a + 14a - 7$
$\quad = (2a^2 - a) + (14a - 7)$
$\quad = a(2a - 1) + 7(2a - 1)$
$\quad = (2a - 1)(a + 7)$

Problem 6

$a \cdot c = -12$
$1(-12) = -12, \quad 1 - 12 = -11$

$4a^2 - 11a - 3$
$\quad = 4a^2 + a - 12a - 3$
$\quad = (4a^2 + a) - (12a + 3)$
$\quad = a(4a + 1) - 3(4a + 1)$
$\quad = (4a + 1)(a - 3)$

Problem 7

The GCF is $5x$.

$15x^3 + 40x^2 - 80x = 5x(3x^2 + 8x - 16)$
$3(-16) = -48$
$-4(12) = -48, \quad -4 + 12 = 8$

$3x^2 + 8x - 16$
$\quad = 3x^2 - 4x + 12x - 16$
$\quad = (3x^2 - 4x) + (12x - 16)$
$\quad = x(3x - 4) + 4(3x - 4)$
$\quad = (3x - 4)(x + 4)$

$15x^3 + 40x^2 - 80x = 5x(3x^2 + 8x - 16)$
$\qquad\qquad\qquad\quad = 5x(3x - 4)(x + 4)$

SECTION 8.4   *pages 363 – 366*

Problem 1   **A.** $25a^2 - b^2 = (5a)^2 - b^2 = (5a + b)(5a - b)$

              B. $6x^2 - 1$ is nonfactorable over the integers.

              C. $n^8 - 36 = (n^4)^2 - 6^2 = (n^4 + 6)(n^4 - 6)$

Problem 2 $n^4 - 81 = (n^2 + 9)(n^2 - 9) = (n^2 + 9)(n + 3)(n - 3)$

Problem 3   **A.** $16y^2 = (4y)^2, 1 = 1^2$

$(4y + 1)^2 = (4y)^2 + 2(4y)(1) + (1)^2$
$= 16y^2 + 8y + 1$

$16y^2 + 8y + 1 = (4y + 1)^2$

B. $x^2 = (x)^2, 36 = 6^2$

$(x + 6)^2 = x^2 + 2(x)(6) + 6^2$
$= x^2 + 12x + 36$

The polynomial is not a perfect square.

$x^2 + 14x + 36$ is nonfactorable over the integers.

Problem 4   **A.** $12x^3 - 75x = 3x(4x^2 - 25) = 3x(2x + 5)(2x - 5)$

B. $a^2b - 7a^2 - b + 7 = (a^2b - 7a^2) - (b - 7)$
$= a^2(b - 7) - (b - 7)$
$= (b - 7)(a^2 - 1)$
$= (b - 7)(a + 1)(a - 1)$

C. $4x^3 + 28x^2 - 120x = 4x(x^2 + 7x - 30) = 4x(x + 10)(x - 3)$

SECTION 8.5   *pages 369 – 372*

Problem 1

$2x^2 - 50 = 0$
$2(x^2 - 25) = 0$
$x^2 - 25 = 0$
$(x + 5)(x - 5) = 0$

$x + 5 = 0 \qquad x - 5 = 0$
$x = -5 \qquad\quad x = 5$

The solutions are −5 and 5.

Problem 2

$(x + 2)(x - 7) = 52$
$x^2 - 5x - 14 = 52$
$x^2 - 5x - 66 = 0$
$(x - 11)(x + 6) = 0$

$x - 11 = 0 \qquad x + 6 = 0$
$x = 11 \qquad\quad x = -6$

The solutions are −6 and 11.

Problem 3

Strategy   First positive integer: n
Second positive integer: $n + 1$
Square of the first positive integer: n^2
Square of the second positive integer: $(n + 1)^2$

The sum of the squares of the two integers is 85.

Solution $n^2 + (n + 1)^2 = 85$
$n^2 + n^2 + 2n + 1 = 85$
$2n^2 + 2n - 84 = 0$
$2(n^2 + n - 42) = 0$
$n^2 + n - 42 = 0$
$(n + 7)(n - 6) = 0$

$n + 7 = 0 \qquad n - 6 = 0$
$n = -7 \qquad\quad n = 6$

−7 is not a positive integer.

$n + 1 = 6 + 1 = 7$

The two integers are 6 and 7.

Problem 4

Strategy   Width $= x$
Length $= 2x + 3$

The area of the rectangle is 90 m². Use the equation $A = LW$.

Solution $A = LW$
$90 = (2x + 3)x$
$90 = 2x^2 + 3x$
$0 = 2x^2 + 3x - 90$
$0 = (2x + 15)(x - 6)$

$2x + 15 = 0 \qquad x - 6 = 0$
$2x = -15 \qquad\quad x = 6$
$x = -\dfrac{15}{2}$

The width cannot be a negative number.

$2x + 3 = 2(6) + 3 = 12 + 3 = 15$

The width is 6 m. The length is 15 m.

SOLUTIONS to Chapter 9 Problems

SECTION 9.1   *pages 387–390*

Problem 1   **A.** $\dfrac{6x^5y}{12x^2y^3} = \dfrac{\overset{1}{\cancel{2}}\cdot\overset{1}{\cancel{3}}\cdot x^5y}{\underset{1}{\cancel{2}}\cdot 2\cdot\underset{1}{\cancel{3}}\cdot x^2y^3} = \dfrac{x^3}{2y^2}$

B. $\dfrac{x^2+2x-24}{16-x^2} = \dfrac{\overset{-1}{\cancel{(x-4)}}(x+6)}{\underset{1}{\cancel{(4-x)}}(4+x)} = -\dfrac{x+6}{x+4}$

Problem 2   **A.** $\dfrac{12x^2+3x}{10x-15}\cdot\dfrac{8x-12}{9x+18}$

$= \dfrac{3x(4x+1)}{5(2x-3)}\cdot\dfrac{4(2x-3)}{9(x+2)}$

$= \dfrac{\overset{1}{\cancel{3}}x(4x+1)\cdot 2\cdot 2\overset{1}{\cancel{(2x-3)}}}{5\underset{1}{\cancel{(2x-3)}}\cdot\underset{1}{\cancel{3}}\cdot 3(x+2)}$

$= \dfrac{4x(4x+1)}{15(x+2)}$

B. $\dfrac{x^2+2x-15}{9-x^2}\cdot\dfrac{x^2-3x-18}{x^2-7x+6}$

$= \dfrac{(x-3)(x+5)}{(3-x)(3+x)}\cdot\dfrac{(x+3)(x-6)}{(x-1)(x-6)}$

$= \dfrac{\overset{-1}{\cancel{(x-3)}}(x+5)\cdot\overset{1}{\cancel{(x+3)}}\overset{1}{\cancel{(x-6)}}}{\underset{1}{\cancel{(3-x)}}\underset{1}{\cancel{(3+x)}}\cdot(x-1)\underset{1}{\cancel{(x-6)}}}$

$= -\dfrac{x+5}{x-1}$

Problem 3   **A.** $\dfrac{a^2}{4bc^2-2b^2c}\div\dfrac{a}{6bc-3b^2}$

$= \dfrac{a^2}{4bc^2-2b^2c}\cdot\dfrac{6bc-3b^2}{a}$

$= \dfrac{a^2\cdot 3b\overset{1}{\cancel{(2c-b)}}}{2bc\underset{1}{\cancel{(2c-b)}}\cdot a} = \dfrac{3a}{2c}$

B. $\dfrac{3x^2+26x+16}{3x^2-7x-6}\div\dfrac{2x^2+9x-5}{x^2+2x-15}$

$= \dfrac{3x^2+26x+16}{3x^2-7x-6}\cdot\dfrac{x^2+2x-15}{2x^2+9x-5}$

$= \dfrac{\overset{1}{\cancel{(3x+2)}}(x+8)}{\underset{1}{\cancel{(3x+2)}}\cancel{(x-3)}}\cdot\dfrac{\overset{1}{\cancel{(x+5)}}\overset{1}{\cancel{(x-3)}}}{(2x-1)\underset{1}{\cancel{(x+5)}}}$

$= \dfrac{x+8}{2x-1}$

SECTION 9.2   *pages 394–396*

Problem 1 $8uv^2 = 2\cdot 2\cdot 2\cdot u\cdot v\cdot v$
$12uw = 2\cdot 2\cdot 3\cdot u\cdot w$

LCM $= 2\cdot 2\cdot 2\cdot 3\cdot u\cdot v\cdot v\cdot w$
  $= 24uv^2w$

Problem 2 $m^2-6m+9 = (m-3)(m-3)$
$m^2-2m-3 = (m+1)(m-3)$

LCM $= (m-3)(m-3)(m+1)$

Problem 3   The LCM is $36xy^2z$.

$\dfrac{x-3}{4xy^2} = \dfrac{x-3}{4xy^2}\cdot\dfrac{9z}{9z} = \dfrac{9xz-27z}{36xy^2z}$

$\dfrac{2x+1}{9y^2z} = \dfrac{2x+1}{9y^2z}\cdot\dfrac{4x}{4x} = \dfrac{8x^2+4x}{36xy^2z}$

Problem 4 $\dfrac{2x}{25-x^2} = \dfrac{2x}{-(x^2-25)} = -\dfrac{2x}{x^2-25}$

The LCM is $(x-5)(x+5)(x+2)$.

$\dfrac{x+4}{x^2-3x-10} = \dfrac{x+4}{(x+2)(x-5)}\cdot\dfrac{x+5}{x+5} = \dfrac{x^2+9x+20}{(x+2)(x-5)(x+5)}$

$\dfrac{2x}{25-x^2} = -\dfrac{2x}{(x-5)(x+5)}\cdot\dfrac{x+2}{x+2} = -\dfrac{2x^2+4x}{(x+2)(x-5)(x+5)}$

SECTION 9.3   *pages 398 – 402*

Problem 1   **A.** $\dfrac{3}{xy} + \dfrac{12}{xy} = \dfrac{3 + 12}{xy} = \dfrac{15}{xy}$

B. $\dfrac{2x^2}{x^2 - x - 12} - \dfrac{7x + 4}{x^2 - x - 12} = \dfrac{2x^2 - (7x + 4)}{x^2 - x - 12} = \dfrac{2x^2 - 7x - 4}{x^2 - x - 12} = \dfrac{(2x + 1)(x - 4)}{(x + 3)(x - 4)} = \dfrac{2x + 1}{x + 3}$

Problem 2   **A.** The LCM of the denominators is 24y.

$$\dfrac{z}{8y} - \dfrac{4z}{3y} + \dfrac{5z}{4y} = \dfrac{z}{8y} \cdot \dfrac{3}{3} - \dfrac{4z}{3y} \cdot \dfrac{8}{8} + \dfrac{5z}{4y} \cdot \dfrac{6}{6} = \dfrac{3z}{24y} - \dfrac{32z}{24y} + \dfrac{30z}{24y} = \dfrac{3z - 32z + 30z}{24y} = \dfrac{z}{24y}$$

B. The LCM is $x - 2$.

$$\dfrac{5x}{x - 2} - \dfrac{3}{2 - x} = \dfrac{5x}{x - 2} \cdot \dfrac{1}{1} - \dfrac{3}{-(x - 2)} \cdot \dfrac{-1}{-1} = \dfrac{5x}{x - 2} - \dfrac{-3}{x - 2} = \dfrac{5x - (-3)}{x - 2} = \dfrac{5x + 3}{x - 2}$$

C. The LCM is $y - 7$.

$$y + \dfrac{5}{y - 7} = \dfrac{y}{1} + \dfrac{5}{y - 7} = \dfrac{y}{1} \cdot \dfrac{y - 7}{y - 7} + \dfrac{5}{y - 7} = \dfrac{y^2 - 7y}{y - 7} + \dfrac{5}{y - 7} = \dfrac{y^2 - 7y + 5}{y - 7}$$

Problem 3   **A.** The LCM is $(3x - 1)(x + 4)$.

$$\dfrac{4x}{3x - 1} - \dfrac{9}{x + 4} = \dfrac{4x}{3x - 1} \cdot \dfrac{x + 4}{x + 4} - \dfrac{9}{x + 4} \cdot \dfrac{3x - 1}{3x - 1}$$

$$= \dfrac{4x^2 + 16x}{(3x - 1)(x + 4)} - \dfrac{27x - 9}{(3x - 1)(x + 4)}$$

$$= \dfrac{(4x^2 + 16x) - (27x - 9)}{(3x - 1)(x + 4)}$$

$$= \dfrac{4x^2 + 16x - 27x + 9}{(3x - 1)(x + 4)} = \dfrac{4x^2 - 11x + 9}{(3x - 1)(x + 4)}$$

B. The LCM is $(x + 5)(x - 5)$.

$$\dfrac{2x - 1}{x^2 - 25} + \dfrac{2}{5 - x} = \dfrac{2x - 1}{(x + 5)(x - 5)} + \dfrac{2}{-(x - 5)} \cdot \dfrac{-1(x + 5)}{-1(x + 5)}$$

$$= \dfrac{2x - 1}{(x + 5)(x - 5)} + \dfrac{-2(x + 5)}{(x + 5)(x - 5)}$$

$$= \dfrac{(2x - 1) + (-2)(x + 5)}{(x + 5)(x - 5)}$$

$$= \dfrac{2x - 1 - 2x - 10}{(x + 5)(x - 5)}$$

$$= \dfrac{-11}{(x + 5)(x - 5)} = -\dfrac{11}{(x + 5)(x - 5)}$$

SECTION 9.4   *pages 405 – 406*

Problem 1   **A.** The LCM of 3, x, 9, and x^2 is $9x^2$.

$$\dfrac{\dfrac{1}{3} - \dfrac{1}{x}}{\dfrac{1}{9} - \dfrac{1}{x^2}} = \dfrac{\dfrac{1}{3} - \dfrac{1}{x}}{\dfrac{1}{9} - \dfrac{1}{x^2}} \cdot \dfrac{9x^2}{9x^2} = \dfrac{\dfrac{1}{3} \cdot 9x^2 - \dfrac{1}{x} \cdot 9x^2}{\dfrac{1}{9} \cdot 9x^2 - \dfrac{1}{x^2} \cdot 9x^2} = \dfrac{3x^2 - 9x}{x^2 - 9} = \dfrac{3x(x - 3)}{(x - 3)(x + 3)} = \dfrac{3x}{x + 3}$$

B. The LCM of x and x^2 is x^2.

$$\frac{1 + \dfrac{4}{x} + \dfrac{3}{x^2}}{1 + \dfrac{10}{x} + \dfrac{21}{x^2}} = \frac{1 + \dfrac{4}{x} + \dfrac{3}{x^2}}{1 + \dfrac{10}{x} + \dfrac{21}{x^2}} \cdot \frac{x^2}{x^2}$$

$$= \frac{1 \cdot x^2 + \dfrac{4}{x} \cdot x^2 + \dfrac{3}{x^2} \cdot x^2}{1 \cdot x^2 + \dfrac{10}{x} \cdot x^2 + \dfrac{21}{x^2} \cdot x^2}$$

$$= \frac{x^2 + 4x + 3}{x^2 + 10x + 21}$$

$$= \frac{(x + 1)\overset{1}{\cancel{(x + 3)}}}{\underset{1}{\cancel{(x + 3)}}(x + 7)} = \frac{x + 1}{x + 7}$$

SECTION 9.5   *pages 409 – 413*

Problem 1   The LCM of 3 and $3x$ is $3x$.

$$x + \frac{1}{3} = \frac{4}{3x}$$

$$3x\left(x + \frac{1}{3}\right) = 3x\left(\frac{4}{3x}\right)$$

$$3x \cdot x + \frac{3x}{1} \cdot \frac{1}{3} = \frac{3x}{1} \cdot \frac{4}{3x}$$

$$3x^2 + x = 4$$

$$3x^2 + x - 4 = 0$$

$$(3x + 4)(x - 1) = 0$$

$$3x + 4 = 0 \qquad x - 1 = 0$$

$$3x = -4 \qquad x = 1$$

$$x = -\frac{4}{3}$$

Both $-\dfrac{4}{3}$ and 1 check as solutions.

The solutions are $-\dfrac{4}{3}$ and 1.

Problem 2   The LCM is $x + 2$.

$$\frac{5x}{x + 2} = 3 - \frac{10}{x + 2}$$

$$\frac{(x + 2)}{1} \cdot \frac{5x}{x + 2} = \frac{(x + 2)}{1}\left(3 - \frac{10}{x + 2}\right)$$

$$5x = (x + 2)3 - 10$$

$$5x = 3x + 6 - 10$$

$$5x = 3x - 4$$

$$2x = -4$$

$$x = -2$$

-2 does not check as a solution.
The equation has no solution.

Problem 3   **A.**

$$\frac{2}{x + 3} = \frac{6}{5x + 5}$$

$$(x + 3)(5x + 5)\frac{2}{x + 3} = (x + 3)(5x + 5)\frac{6}{5x + 5}$$

$$(5x + 5)2 = (x + 3)6$$

$$10x + 10 = 6x + 18$$

$$4x + 10 = 18$$

$$4x = 8$$

$$x = 2$$

The solution is 2.

B.
$$\frac{5}{2x-3} = \frac{10}{x+3}$$
$$(x+3)(2x-3)\frac{5}{2x-3} = (x+3)(2x-3)\frac{10}{x+3}$$
$$(x+3)5 = (2x-3)10$$
$$5x+15 = 20x-30$$
$$-15x+15 = -30$$
$$-15x = -45$$
$$x = 3$$

The solution is 3.

Problem 4

Strategy   To find the total area that 270 ceramic tiles will cover, write and solve a proportion using x to represent the number of square feet that 270 tiles will cover.

Solution
$$\frac{4}{9} = \frac{x}{270}$$
$$270\left(\frac{4}{9}\right) = 270\left(\frac{x}{270}\right)$$
$$120 = x$$

A 120 ft² area can be tiled using 270 ceramic tiles.

Problem 5

Strategy   To find the additional amount of medication required for a 180 lb adult, write and solve a proportion using x to represent the additional medication. Then $3 + x$ is the total amount required for a 180 lb adult.

Solution
$$\frac{120}{3} = \frac{180}{3+x}$$
$$\frac{40}{1} = \frac{180}{3+x}$$
$$(3+x)\cdot 40 = (3+x)\cdot\frac{180}{3+x}$$
$$120 + 40x = 180$$
$$40x = 60$$
$$x = 1.5$$

1.5 additional ounces are required for a 180 lb adult.

SECTION 9.6   *pages 418 – 420*

Problem 1   **A.**
$$s = \frac{A+L}{2}$$
$$2\cdot s = 2\left(\frac{A+L}{2}\right)$$
$$2s = A+L$$
$$2s-A = A-A+L$$
$$2s-A = L$$

B.
$$S = a+(n-1)d$$
$$S = a+nd-d$$
$$S-a = a-a+nd-d$$
$$S-a = nd-d$$
$$S-a+d = nd-d+d$$
$$S-a+d = nd$$
$$\frac{S-a+d}{d} = \frac{nd}{d}$$
$$\frac{S-a+d}{d} = n$$

Problem 2

$$S = C + rC$$
$$S = (1 + r)C$$
$$\frac{S}{1 + r} = \frac{(1 + r)C}{1 + r}$$
$$\frac{S}{1 + r} = C$$

SECTION 9.7   *pages 422 – 426*

Problem 1

Strategy
- Time for one printer to complete the job: t

	Rate	Time	Part
1st printer	$\frac{1}{t}$	3	$\frac{3}{t}$
2nd printer	$\frac{1}{t}$	5	$\frac{5}{t}$

- The sum of the parts of the task completed must equal 1.

Solution

$$\frac{3}{t} + \frac{5}{t} = 1$$
$$t\left(\frac{3}{t} + \frac{5}{t}\right) = t \cdot 1$$
$$3 + 5 = t$$
$$8 = t$$

Working alone, one printer takes 8 h to print the payroll.

Problem 2

Strategy
- Rate sailing across the lake: r
  Rate sailing back: $2r$

	Distance	Rate	Time
Across	6	r	$\frac{6}{r}$
Back	6	$2r$	$\frac{6}{2r}$

- The total time for the trip was 3 h.

Solution

$$\frac{6}{r} + \frac{6}{2r} = 3$$
$$2r\left(\frac{6}{r} + \frac{6}{2r}\right) = 2r(3)$$
$$2r \cdot \frac{6}{r} + 2r \cdot \frac{6}{2r} = 6r$$
$$12 + 6 = 6r$$
$$18 = 6r$$
$$3 = r$$

The rate across the lake was 3 km/h.

SOLUTIONS to Chapter 10 Problems

SECTION 10.1   *pages 445 – 448*

Problem 1 $-5\sqrt{32} = -5\sqrt{2^5} = -5\sqrt{2^4 \cdot 2} = -5\sqrt{2^4}\sqrt{2} = -5 \cdot 2^2\sqrt{2} = -20\sqrt{2}$

Problem 2 $\sqrt{216} = \sqrt{2^3 \cdot 3^3} = \sqrt{2^2 \cdot 3^2(2 \cdot 3)} = \sqrt{2^2 \cdot 3^2}\sqrt{2 \cdot 3} = 2 \cdot 3\sqrt{2 \cdot 3} = 6\sqrt{6}$

Problem 3 $\sqrt{y^{19}} = \sqrt{y^{18} \cdot y} = \sqrt{y^{18}}\sqrt{y} = y^9\sqrt{y}$

Problem 4   **A.** $\sqrt{45b^7} = \sqrt{3^2 \cdot 5 \cdot b^7} = \sqrt{3^2 b^6(5 \cdot b)} = \sqrt{3^2 b^6}\sqrt{5b} = 3b^3\sqrt{5b}$

        B. $3a\sqrt{28a^9b^{18}} = 3a\sqrt{2^2 \cdot 7 \cdot a^9 b^{18}} = 3a\sqrt{2^2 a^8 b^{18}(7a)} = 3a\sqrt{2^2 a^8 b^{18}}\sqrt{7a} = 3a \cdot 2a^4 b^9\sqrt{7a} = 6a^5 b^9\sqrt{7a}$

Problem 5 $\sqrt{25(a + 3)^2} = \sqrt{5^2(a + 3)^2} = 5(a + 3) = 5a + 15$

SECTION 10.2   *pages 451 – 452*

Problem 1   **A.** $9\sqrt{3} + 3\sqrt{3} - 18\sqrt{3} = -6\sqrt{3}$

           B. $2\sqrt{50} - 5\sqrt{32} = 2\sqrt{2 \cdot 5^2} - 5\sqrt{2^5} = 2\sqrt{5^2}\sqrt{2} - 5\sqrt{2^4}\sqrt{2}$

$$= 2 \cdot 5\sqrt{2} - 5 \cdot 2^2\sqrt{2} = 10\sqrt{2} - 20\sqrt{2} = -10\sqrt{2}$$

Problem 2   **A.** $y\sqrt{28y} + 7\sqrt{63y^3} = y\sqrt{2^2 \cdot 7y} + 7\sqrt{3^2 \cdot 7 \cdot y^3}$

$$= y\sqrt{2^2}\sqrt{7y} + 7\sqrt{3^2 \cdot y^2}\sqrt{7y}$$

$$= y \cdot 2\sqrt{7y} + 7 \cdot 3y\sqrt{7y}$$

$$= 2y\sqrt{7y} + 21y\sqrt{7y} = 23y\sqrt{7y}$$

           B. $2\sqrt{27a^5} - 4a\sqrt{12a^3} + a^2\sqrt{75a}$

$$= 2\sqrt{3^3 \cdot a^5} - 4a\sqrt{2^2 \cdot 3 \cdot a^3} + a^2\sqrt{3 \cdot 5^2 \cdot a}$$

$$= 2\sqrt{3^2 \cdot a^4}\sqrt{3a} - 4a\sqrt{2^2 \cdot a^2}\sqrt{3a} + a^2\sqrt{5^2}\sqrt{3a}$$

$$= 2 \cdot 3a^2\sqrt{3a} - 4a \cdot 2a\sqrt{3a} + a^2 \cdot 5\sqrt{3a}$$

$$= 6a^2\sqrt{3a} - 8a^2\sqrt{3a} + 5a^2\sqrt{3a} = 3a^2\sqrt{3a}$$

SECTION 10.3   *pages 455 – 458*

Problem 1 $\sqrt{5a}\,\sqrt{15a^3b^4}\,\sqrt{3b^5} = \sqrt{225a^4b^9} = \sqrt{3^2 5^2 a^4 b^9} = \sqrt{3^2 5^2 a^4 b^8}\,\sqrt{b} = 3 \cdot 5a^2b^4\sqrt{b} = 15a^2b^4\sqrt{b}$

Problem 2 $\sqrt{5x}(\sqrt{5x} - \sqrt{25y}) = \sqrt{5^2 x^2} - \sqrt{5^3 xy} = \sqrt{5^2 x^2} - \sqrt{5^2}\,\sqrt{5xy} = 5x - 5\sqrt{5xy}$

Problem 3 $(3\sqrt{x} - \sqrt{y})(5\sqrt{x} - 2\sqrt{y}) = 15(\sqrt{x})^2 - 6\sqrt{xy} - 5\sqrt{xy} + 2(\sqrt{y})^2 = 15x - 11\sqrt{xy} + 2y$

Problem 4 $(2\sqrt{x} + 7)(2\sqrt{x} - 7) = (2\sqrt{x})^2 - 7^2 = 4x - 49$

Problem 5   **A.** $\dfrac{\sqrt{15x^6y^7}}{\sqrt{3x^7y^9}} = \sqrt{\dfrac{15x^6y^7}{3x^7y^9}} = \sqrt{\dfrac{5}{xy^2}} = \dfrac{\sqrt{5}}{\sqrt{xy^2}} = \dfrac{\sqrt{5}}{y\sqrt{x}} = \dfrac{\sqrt{5}}{y\sqrt{x}} \cdot \dfrac{\sqrt{x}}{\sqrt{x}} = \dfrac{\sqrt{5x}}{xy}$

           B. $\dfrac{\sqrt{y}}{\sqrt{y} + 3} = \dfrac{\sqrt{y}}{\sqrt{y} + 3} \cdot \dfrac{\sqrt{y} - 3}{\sqrt{y} - 3} = \dfrac{y - 3\sqrt{y}}{y - 9}$

           C. $\dfrac{5 + \sqrt{y}}{1 - 2\sqrt{y}} = \dfrac{5 + \sqrt{y}}{1 - 2\sqrt{y}} \cdot \dfrac{1 + 2\sqrt{y}}{1 + 2\sqrt{y}} = \dfrac{5 + 10\sqrt{y} + \sqrt{y} + 2y}{1 - 4y} = \dfrac{5 + 11\sqrt{y} + 2y}{1 - 4y}$

SECTION 10.4   *pages 461 – 467*

Problem 1

$$\sqrt{4x} + 3 = 7$$
$$\sqrt{4x} = 4$$
$$(\sqrt{4x})^2 = 4^2$$
$$4x = 16$$
$$x = 4$$

Check:
$$\sqrt{4x} + 3 = 7$$

$\sqrt{4 \cdot 4} + 3$	7
$\sqrt{4^2} + 3$	7
$4 + 3$	7
$7 = 7$	

The solution is 4.

Problem 2   **A.**

$$\sqrt{3x - 2} - 5 = 0$$
$$\sqrt{3x - 2} = 5$$
$$(\sqrt{3x - 2})^2 = 5^2$$
$$3x - 2 = 25$$
$$3x = 27$$
$$x = 9$$

Check:
$$\sqrt{3x - 2} - 5 = 0$$

$\sqrt{3 \cdot 9 - 2} - 5$	0
$\sqrt{27 - 2} - 5$	0
$\sqrt{25} - 5$	0
$\sqrt{5^2} - 5$	0
$5 - 5$	0
$0 = 0$	

The solution is 9.

B. $\sqrt{4x - 7} + 5 = 0$

$\sqrt{4x - 7} = -5$

$(\sqrt{4x - 7})^2 = (-5)^2$

$4x - 7 = 25$

$4x = 32$

$x = 8$

There is no solution.

Check: $\sqrt{4x - 7} + 5 = 0$

$\begin{array}{c|c} \sqrt{4 \cdot 8 - 7} + 5 & 0 \\ \sqrt{32 - 7} + 5 & 0 \\ \sqrt{25} + 5 & 0 \\ 5 + 5 & 0 \end{array}$

$10 \neq 0$

Problem 3 $\sqrt{x} + \sqrt{x + 9} = 9$

$\sqrt{x} = 9 - \sqrt{x + 9}$

$(\sqrt{x})^2 = (9 - \sqrt{x + 9})^2$

$x = 81 - 18\sqrt{x + 9} + (x + 9)$

$18\sqrt{x + 9} = 90$

$\sqrt{x + 9} = 5$

$(\sqrt{x + 9})^2 = 5^2$

$x + 9 = 25$

$x = 16$

Check: $\sqrt{x} + \sqrt{x + 9} = 9$

$\begin{array}{c|c} \sqrt{16} + \sqrt{16 + 9} & 9 \\ 4 + \sqrt{25} & 9 \\ 4 + 5 & 9 \end{array}$

$9 = 9$

The solution is 16.

Problem 4

Strategy To find the distance, use the Pythagorean Theorem. The hypotenuse is the length of the ladder. One leg is the distance from the bottom of the ladder to the base of the building. The distance along the building from the ground to the top of the ladder is the unknown leg.

Solution $a = \sqrt{c^2 - b^2}$

$a = \sqrt{(12)^2 - (5)^2}$

$a = \sqrt{144 - 25}$

$a = \sqrt{119}$

$a \approx 10.91$

The distance is 10.91 ft.

Problem 5

Strategy To find the length of the pendulum, replace T in the equation with the given value and solve for L.

Solution
$$T = 2\pi\sqrt{\frac{L}{32}}$$

$$1.5 = 2\pi\sqrt{\frac{L}{32}}$$

$$\frac{1.5}{2\pi} = \sqrt{\frac{L}{32}}$$

$$\left(\frac{1.5}{2\pi}\right)^2 = \left(\sqrt{\frac{L}{32}}\right)^2$$

$$\left(\frac{1.5}{2\pi}\right)^2 = \frac{L}{32}$$

$$32\left(\frac{1.5}{2\pi}\right)^2 = L \qquad \blacktriangleright \text{Use the } \pi \text{ key on your calculator.}$$

$$1.82 \approx L$$

The length of the pendulum is 1.82 ft.

SOLUTIONS to Chapter 11 Problems

SECTION 11.1 *pages 483 – 487*

Problem 1

$$\frac{3y^2}{2} + y - \frac{1}{2} = 0$$

$$2\left(\frac{3y^2}{2} + y - \frac{1}{2}\right) = 2(0)$$

$$3y^2 + 2y - 1 = 0$$

$$(3y - 1)(y + 1) = 0$$

$$3y - 1 = 0 \qquad y + 1 = 0$$

$$3y = 1 \qquad\qquad y = -1$$

$$y = \frac{1}{3}$$

The solutions are $\frac{1}{3}$ and -1.

Problem 2   **A.** $4x^2 - 96 = 0$

$$4x^2 = 96$$

$$x^2 = 24$$

$$\sqrt{x^2} = \pm\sqrt{24}$$

$$x = \pm 2\sqrt{6}$$

The solutions are $2\sqrt{6}$ and $-2\sqrt{6}$.

B. $x^2 + 81 = 0$

$$x^2 = -81$$

$$\sqrt{x^2} = \pm\sqrt{-81} \quad \text{(Not a real number)}$$

The equation has no real number solution.

Problem 3

$$(x + 5)^2 = 20$$

$$\sqrt{(x + 5)^2} = \pm\sqrt{20}$$

$$x + 5 = \pm 2\sqrt{5}$$

$$x + 5 = 2\sqrt{5} \qquad\qquad x + 5 = -2\sqrt{5}$$

$$x = -5 + 2\sqrt{5} \qquad\qquad x = -5 - 2\sqrt{5}$$

The solutions are $-5 + 2\sqrt{5}$ and $-5 - 2\sqrt{5}$.

SECTION 11.2 *pages 489 – 493*

Problem 1   **A.** $3x^2 - 6x - 2 = 0$

$$3x^2 - 6x = 2$$

$$\frac{1}{3}(3x^2 - 6x) = \frac{1}{3} \cdot 2$$

$$x^2 - 2x = \frac{2}{3}$$

Complete the square.

$$x^2 - 2x + 1 = \frac{2}{3} + 1$$

$$(x - 1)^2 = \frac{5}{3}$$

$$\sqrt{(x - 1)^2} = \pm\sqrt{\frac{5}{3}}$$

$$x - 1 = \pm\frac{\sqrt{15}}{3}$$

$$x - 1 = \frac{\sqrt{15}}{3} \qquad\qquad x - 1 = -\frac{\sqrt{15}}{3}$$

$$x = 1 + \frac{\sqrt{15}}{3} \qquad\qquad x = 1 - \frac{\sqrt{15}}{3}$$

$$= \frac{3 + \sqrt{15}}{3} \qquad\qquad = \frac{3 - \sqrt{15}}{3}$$

The solutions are $\dfrac{3 + \sqrt{15}}{3}$ and $\dfrac{3 - \sqrt{15}}{3}$.

B. $y^2 - 6y + 10 = 0$
$$y^2 - 6y = -10$$

Complete the square.
$$y^2 - 6y + 9 = -10 + 9$$
$$(y - 3)^2 = -1$$
$$\sqrt{(y - 3)^2} = \pm\sqrt{-1}$$

$\sqrt{-1}$ is not a real number.
The equation has no real number solution.

Problem 2 $x^2 + 8x + 8 = 0$
$$x^2 + 8x = -8$$
$$x^2 + 8x + 16 = -8 + 16$$
$$(x + 4)^2 = 8$$
$$\sqrt{(x + 4)^2} = \pm\sqrt{8}$$
$$x + 4 = \pm 2\sqrt{2}$$

$x + 4 = 2\sqrt{2}$	$x + 4 = -2\sqrt{2}$
$x = -4 + 2\sqrt{2}$	$x = -4 - 2\sqrt{2}$
$\approx -4 + 2(1.414)$	$\approx -4 - 2(1.414)$
$\approx -4 + 2.828$	$\approx -4 - 2.828$
≈ -1.172	≈ -6.828

The solutions are approximately -1.172 and -6.828.

SECTION 11.3   *pages 495 – 498*

Problem 1   **A.** $3x^2 + 4x - 4 = 0$

$a = 3, b = 4, c = -4$

$$x = \frac{-(4) \pm \sqrt{(4)^2 - 4(3)(-4)}}{2 \cdot 3}$$
$$= \frac{-4 \pm \sqrt{16 + 48}}{6}$$
$$= \frac{-4 \pm \sqrt{64}}{6} = \frac{-4 \pm 8}{6}$$

$x = \dfrac{-4 + 8}{6}$	$x = \dfrac{-4 - 8}{6}$
$= \dfrac{4}{6} = \dfrac{2}{3}$	$= \dfrac{-12}{6} = -2$

The solutions are $\dfrac{2}{3}$ and -2.

B. $x^2 + 2x = 1$
$$x^2 + 2x - 1 = 0$$

$a = 1, b = 2, c = -1$

$$x = \frac{-(2) \pm \sqrt{(2)^2 - 4(1)(-1)}}{2 \cdot 1}$$
$$= \frac{-2 \pm \sqrt{4 + 4}}{2}$$
$$= \frac{-2 \pm \sqrt{8}}{2}$$
$$= \frac{-2 \pm 2\sqrt{2}}{2}$$
$$= \frac{2(-1 \pm \sqrt{2})}{2} = -1 \pm \sqrt{2}$$

The solutions are $-1 + \sqrt{2}$ and $-1 - \sqrt{2}$.

C. $z^2 + 2z + 6 = 0$

$a = 1, b = 2, c = 6$

$$z = \frac{-2 \pm \sqrt{2^2 - 4(1)(6)}}{2(1)}$$
$$= \frac{-2 \pm \sqrt{4 - 24}}{2} = \frac{-2 \pm \sqrt{-20}}{2}$$

$\sqrt{-20}$ is not a real number.
The equation has no real number solution.

SECTION 11.4   *pages 501 – 503*

Problem 1   **A.** $a = 1$. a is positive.
The parabola opens up.

B. $a = -1$. a is negative.
The parabola opens down.

SECTION 11.5   *pages 507 – 509*

Problem 1

Strategy
  ■ This is a geometry problem.
  ■ Width of the rectangle: W
     Length of the rectangle: $W + 3$
  ■ Use the equation $A = LW$.

Solution

$A = LW$
$40 = (W + 3)W$
$40 = W^2 + 3W$
$0 = W^2 + 3W - 40$
$0 = (W + 8)(W - 5)$

$W + 8 = 0 \qquad W - 5 = 0$
$\qquad W = -8 \qquad\qquad W = 5$

The solution -8 is not possible.
The width is 5 m.

ANSWERS to Chapter 1 Odd-Numbered Exercises

CONCEPT REVIEW 1.1 *pages 6 – 7*

1. Sometimes true   **3.** Never true   **5.** Sometimes true   **7.** Always true   **9.** Always true

SECTION 1.1 *pages 7 – 10*

1. >   **3.** <   **5.** >   **7.** >   **9.** <   **11.** >   **13.** >   **15.** <   **17.** $\{1, 2, 3, 4, 5, 6, 7, 8\}$
19. $\{1, 2, 3, 4, 5, 6, 7, 8\}$ **21.** $\{-6, -5, -4, -3, -2, -1\}$ **23.** 5   **25.** $-23, -18$ **27.** 21, 37
29. $-52, -46, 0$ **31.** $-17, 0, 4, 29$ **33.** 5, 6, 7, 8, 9   **35.** $-10, -9, -8, -7, -6, -5$ **37.** -22
39. 31   **41.** 168   **43.** -630 **45.** 18   **47.** -49 **49.** 16   **51.** 12   **53.** -29 **55.** -14
57. 0   **59.** -34 **61. a.** $8, 5, 2, -1, -3$ **b.** $8, 5, 2, 1, 3$ **63.** >   **65.** <   **67.** >   **69.** <
71. $-19, -|-8|, |-5|, 6$ **73.** $-22, -(-3), |-14|, |-25|$ **75. a.** 5°F with a 20 mph wind feels colder.   **b.** -15°F
with a 20 mph wind feels colder.   **77.** $-4, 4$ **79.** $-3, 11$ **81.** Negative   **83.** 0   **85.** True

CONCEPT REVIEW 1.2 *page 17*

1. Sometimes true   **3.** Always true   **5.** Always true   **7.** Never true   **9.** Never true

SECTION 1.2 *pages 17 – 21*

1. -11 **3.** -9 **5.** -3 **7.** 1   **9.** -5 **11.** -30 **13.** 9   **15.** 1   **17.** -10 **19.** -28
21. -392 **23.** 8   **25.** -7 **27.** 7   **29.** -2 **31.** -28 **33.** -13 **35.** 6   **37.** -9 **39.** 2
41. -138 **43.** -8 **45.** 42   **47.** -20 **49.** -16 **51.** 25   **53.** 0   **55.** -72 **57.** -102
59. 140   **61.** -70 **63.** 162   **65.** 120   **67.** 36   **69.** 192   **71.** -108 **73.** 0   **75.** -2 **77.** 8
79. 0   **81.** -9 **83.** -9 **85.** 9   **87.** -24 **89.** -12 **91.** -13 **93.** -18 **95.** 19   **97.** 26
99. 11   **101.** -13 **103.** 13   **105.** The temperature is 3°C.   **107.** The difference is 14°C.   **109.** The
difference is 399°C.   **111.** The difference in elevation is 5662 m.   **113.** The difference in elevation is 6028 m.
115. The difference in elevation is 9248 m.   **117.** The average daily high temperature is -3°C.   **119.** The
temperature dropped 100°F.   **121.** The difference is $4247 million.   **123.** The difference is $6618 million.
125. The score is 93.   **127.** 17   **129.** 3   **131.** $-4, -9, -14$ **133.** $1, -0.25, 0.0625$ **135.** 5436
137. b   **139.** a   **141.** No. $10 - (-8) = 18$

CONCEPT REVIEW 1.3 *page 30*

1. Never true   **3.** Sometimes true   **5.** Always true   **7.** Never true   **9.** Always true

SECTION 1.3 *pages 30 – 34*

1. $0.\overline{3}$ **3.** 0.25   **5.** 0.4   **7.** $0.1\overline{6}$ **9.** 0.125   **11.** $0.\overline{2}$ **13.** $0.\overline{45}$ **15.** $0.58\overline{3}$ **17.** $0.2\overline{6}$
19. 0.5625   **21.** 0.24   **23.** 0.225   **25.** $0.68\overline{1}$ **27.** $0.458\overline{3}$ **29.** $0.\overline{15}$ **31.** $0.0\overline{81}$ **33.** $\frac{3}{4}$, 0.75
35. $\frac{1}{2}$, 0.5   **37.** $\frac{16}{25}$, 0.64   **39.** $1\frac{3}{4}$, 1.75   **41.** $\frac{19}{100}$, 0.19   **43.** $\frac{1}{20}$, 0.05   **45.** $4\frac{1}{2}$, 4.5   **47.** $\frac{2}{25}$, 0.08
49. $\frac{1}{9}$ **51.** $\frac{5}{16}$ **53.** $\frac{1}{200}$ **55.** $\frac{1}{16}$ **57.** 0.073   **59.** 0.158   **61.** 0.0915   **63.** 0.1823   **65.** 15%
67. 5%   **69.** 17.5%   **71.** 115%   **73.** 0.8%   **75.** 6.5%   **77.** 54%   **79.** 33.3%   **81.** 44.4%

83. 250% **85.** $37\frac{1}{2}\%$ **87.** $35\frac{5}{7}\%$ **89.** 125% **91.** $155\frac{5}{9}\%$ **93.** $\frac{13}{12}$ **95.** $-\frac{5}{24}$ **97.** $-\frac{19}{24}$

99. $\frac{5}{26}$ **101.** $\frac{7}{24}$ **103.** 0 **105.** $-\frac{7}{16}$ **107.** $\frac{11}{24}$ **109.** 1 **111.** $\frac{11}{8}$ **113.** $\frac{169}{315}$ **115.** -38.8

117. -6.192 **119.** 13.355 **121.** 4.676 **123.** -10.03 **125.** -37.19 **127.** -17.5 **129.** 853.2594

131. $-\frac{3}{8}$ **133.** $\frac{1}{10}$ **135.** $\frac{15}{64}$ **137.** $\frac{3}{2}$ **139.** $-\frac{8}{9}$ **141.** $\frac{2}{3}$ **143.** 4.164 **145.** 4.347

147. -4.028 **149.** -2.22 **151.** 12.26448 **153.** 0.75 **155.** -2060.55 **157.** 0.09 **159.** Natural number, integer, positive integer, rational number, real number **161.** Rational number, real number

163. Irrational number, real number **165.** The cost is \$1.24. **167.** $\frac{17}{99} = 0.\overline{17}$; $\frac{45}{99} = 0.\overline{45}$; $\frac{73}{99} = 0.\overline{73}$; $\frac{83}{99} = 0.\overline{83}$;

$\frac{33}{99} = 0.\overline{33} = 0.\overline{3}$, yes; $\frac{1}{99} = 0.\overline{01}$, yes **169.** Top row: $-\frac{1}{6}$, 0; second row: $-\frac{1}{2}$; third row: $\frac{1}{3}, \frac{1}{2}$ **171.** $0.70x$

173. $a = 2, b = 3, c = 6$

CONCEPT REVIEW 1.4 *pages 38 – 39*

1. Never true **3.** Never true **5.** Always true **7.** Never true **9.** Never true

---

SECTION 1.4 *pages 39 – 40*

1. 36 **3.** -49 **5.** 9 **7.** 81 **9.** $\frac{1}{4}$ **11.** 0.09 **13.** 12 **15.** 0.216 **17.** -12 **19.** 16

21. -864 **23.** -1008 **25.** 3 **27.** $-77{,}760$ **29.** 9 **31.** 12 **33.** 1 **35.** 7 **37.** -36

39. 13 **41.** 4 **43.** 15 **45.** -1 **47.** 4 **49.** 0.51 **51.** 1.7 **53.** $\frac{17}{48}$ **55.** $<$ **57.** $>$

59. 100 **61.** 120 **63.** It will take the computer 17 s. **65.** 6 **67.** 9

CHAPTER REVIEW EXERCISES *pages 45 – 47*

1. $\{1, 2, 3, 4, 5, 6\}$ (Obj. 1.1.1) **2.** 62.5% (Obj. 1.3.2) **3.** -4 (Obj. 1.1.2) **4.** 4 (Obj. 1.2.2) **5.** 17 (Obj. 1.2.4) **6.** $0.\overline{7}$ (Obj. 1.3.1) **7.** -5.3578 (Obj. 1.3.4) **8.** 8 (Obj. 1.4.2) **9.** 4 (Obj. 1.1.2)

10. -14 (Obj. 1.2.2) **11.** 67.2% (Obj. 1.3.2) **12.** $\frac{159}{200}$ (Obj. 1.3.2) **13.** -9 (Obj. 1.2.4) **14.** 0.85 (Obj. 1.3.1) **15.** $-\frac{1}{2}$ (Obj. 1.3.4) **16.** 9 (Obj. 1.4.2) **17.** $<$ (Obj. 1.1.1) **18.** -16 (Obj. 1.2.1)

19. 90 (Obj. 1.2.3) **20.** -6.881 (Obj. 1.3.3) **21.** $-5, -3$ (Obj. 1.1.1) **22.** 0.07 (Obj. 1.3.2) **23.** 12 (Obj. 1.4.1) **24.** $>$ (Obj. 1.1.1) **25.** -3 (Obj. 1.2.1) **26.** -48 (Obj. 1.2.3) **27.** $\frac{1}{15}$ (Obj. 1.3.3)

28. -108 (Obj. 1.4.1) **29.** 277.8% (Obj. 1.3.2) **30.** 2.4 (Obj. 1.3.2) **31.** 2 (Obj. 1.1.2)

32. -14 (Obj. 1.2.2) **33.** -8 (Obj. 1.2.4) **34.** 0.35 (Obj. 1.3.1) **35.** $-\frac{7}{6}$ (Obj. 1.3.4) **36.** 12 (Obj. 1.4.2) **37.** 3 (Obj. 1.1.2) **38.** 18 (Obj. 1.2.2) **39.** $288\frac{8}{9}\%$ (Obj. 1.3.2) **40.** **a.** $12, 8, 1, -7$

b. $12, 8, 1, 7$ (Obj. 1.1.2) **41.** 12 (Obj. 1.2.4) **42.** $0.\overline{63}$ (Obj. 1.3.1) **43.** -11.5 (Obj. 1.3.4)

44. 8 (Obj. 1.4.2) **45.** $>$ (Obj. 1.1.1) **46.** -8 (Obj. 1.2.1) **47.** 72 (Obj. 1.2.3) **48.** $-\frac{11}{24}$ (Obj. 1.3.3) **49.** $-17, -9, 0, 4$ (Obj. 1.1.1) **50.** 0.2% (Obj. 1.3.2) **51.** -4 (Obj. 1.4.1)

52. 3.561 (Obj. 1.3.3) **53.** -17 (Obj. 1.1.2) **54.** -27 (Obj. 1.2.2) **55.** $-\frac{1}{10}$ (Obj. 1.3.4)

56. $<$ (Obj. 1.1.1) **57.** 0.72 (Obj. 1.3.1) **58.** -128 (Obj. 1.4.1) **59.** 7.5% (Obj. 1.3.2)

60. $54\frac{2}{7}\%$ (Obj. 1.3.2) **61.** -18 (Obj. 1.2.1) **62.** 0 (Obj. 1.2.3) **63.** 9 (Obj. 1.4.2)

64. $\frac{7}{8}$ (Obj. 1.3.3) **65.** 16 (Obj. 1.2.4) **66.** $<$ (Obj. 1.1.1) **67.** -3 (Obj. 1.4.1) **68.** -6 (Obj. 1.2.1) **69.** 300 (Obj. 1.2.3) **70.** $\{-3, -2, -1\}$ (Obj. 1.1.1) **71.** The temperature is 8°C. (Obj. 1.2.5) **72.** The average low temperature was -2°C. (Obj. 1.2.5) **73.** The difference was 13°C. (Obj. 1.2.5) **74.** The temperature is -6°C. (Obj. 1.2.5) **75.** The difference is 714°C. (Obj. 1.2.5)

CHAPTER TEST   *pages 47–48*

1. $\frac{11}{20}$ (Obj. 1.3.2) **2.** $-6, -8$ (Obj. 1.1.1) **3.** -3 (Obj. 1.2.2) **4.** 0.85 (Obj. 1.3.1) **5.** $-\frac{1}{14}$ (Obj. 1.3.4) **6.** -15 (Obj. 1.2.4) **7.** $-\frac{8}{3}$ (Obj. 1.4.1) **8.** 2 (Obj. 1.2.1) **9.** $\{1, 2, 3, 4, 5, 6\}$ (Obj. 1.1.1) **10.** 159% (Obj. 1.3.2) **11.** 29 (Obj. 1.1.2) **12.** $>$ (Obj. 1.1.1) **13.** $-\frac{23}{18}$ (Obj. 1.3.3) **14.** 258 (Obj. 1.2.3) **15.** 3 (Obj. 1.4.2) **16.** $\frac{5}{6}$ (Obj. 1.3.4) **17.** 23.1% (Obj. 1.3.2) **18.** 0.062 (Obj. 1.3.2) **19.** 14 (Obj. 1.2.2) **20.** $0.4\overline{3}$ (Obj. 1.3.1) **21.** -2.43 (Obj. 1.3.4) **22.** 15 (Obj. 1.2.4) **23.** 640 (Obj. 1.4.1) **24.** -12 (Obj. 1.2.1) **25.** -34 (Obj. 1.1.2) **26.** $>$ (Obj. 1.1.1) **27. a.** 17, 6, -5, -9 **b.** 17, 6, 5, 9 (Obj. 1.1.2) **28.** $69\frac{13}{23}$% (Obj. 1.3.2) **29.** -11.384 (Obj. 1.3.3) **30.** 160 (Obj. 1.2.3) **31.** -2 (Obj. 1.4.2) **32.** The temperature is 4°C. (Obj. 1.2.5) **33.** The average high temperature was -4°C. (Obj. 1.2.5)

ANSWERS to Chapter 2 Odd-Numbered Exercises

CONCEPT REVIEW 2.1   *page 54*

1. Always true **3.** Sometimes true **5.** Always true

SECTION 2.1   *pages 54–56*

1. $2x^2, 5x, \underline{-8}$ **3.** $-a^4, 6$ **5.** $\underline{7x^2y}, 6xy^2$ **7.** 1, -9 **9.** 1, -4, -1 **11.** 10 **13.** 32 **15.** 21
17. 16 **19.** -9 **21.** $\overline{3}$ **23.** -7 **25.** 13 **27.** -15 **29.** 41 **31.** 1 **33.** 5 **35.** 1
37. 57 **39.** 5 **41.** 12 **43.** 6 **45.** 10 **47.** 8 **49.** -3 **51.** -2 **53.** -22 **55.** 4
57. 20 **59.** 24 **61.** 4.96 **63.** -5.68 **65.** The volume is 25.8 in³. **67.** The area is 93.7 cm².
69. The volume is 484.9 m³. **71.** 24 **73.** $\frac{1}{2}$ **75.** 13 **77.** -1 **79.** 4 **81.** 81

CONCEPT REVIEW 2.2   *page 63*

1. Never true **3.** Always true **5.** Always true **7.** Never true

SECTION 2.2   *pages 64 – 67*

1. 2   **3.** 5   **5.** 6   **7.** −8   **9.** −4   **11.** The Inverse Property of Addition   **13.** The Commutative Property of Addition   **15.** The Associative Property of Addition   **17.** The Commutative Property of Multiplication   **19.** The Associative Property of Multiplication   **21.** $14x$ **23.** $5a$ **25.** $-6y$
27. $-3b - 7$ **29.** $5a$ **31.** $-2ab$ **33.** $5xy$ **35.** 0   **37.** $-\dfrac{5}{6}x$ **39.** $-\dfrac{1}{24}x^2$ **41.** $11x$ **43.** $7a$
45. $-14x^2$ **47.** $-x + 3y$ **49.** $17x - 3y$ **51.** $-2a - 6b$ **53.** $-3x - 8y$ **55.** $-4x^2 - 2x$ **57.** $12x$
59. $-21a$ **61.** $6y$ **63.** $8x$ **65.** $-6a$ **67.** $12b$ **69.** $-15x^2$ **71.** x^2 **73.** x **75.** n **77.** x
79. n **81.** $2x$ **83.** $-2x$ **85.** $-15a^2$ **87.** $6y$ **89.** $3y$ **91.** $-2x$ **93.** $-x - 2$ **95.** $8x - 6$
97. $-2a - 14$ **99.** $-6y + 24$ **101.** $35 - 21b$ **103.** $-9 + 15x$ **105.** $15x^2 + 6x$ **107.** $2y - 18$
109. $-15x - 30$ **111.** $-6x^2 - 28$ **113.** $-6y^2 + 21$ **115.** $-6a^2 + 7b^2$ **117.** $4x^2 - 12x + 20$
119. $-3y^2 + 9y + 21$ **121.** $-12a^2 - 20a + 28$ **123.** $12x^2 - 9x + 12$ **125.** $10x^2 - 20xy - 5y^2$
127. $-8b^2 + 6b - 9$ **129.** $a - 7$ **131.** $-11x + 13$ **133.** $-4y - 4$ **135.** $-2x - 16$ **137.** $18y - 51$
139. $a + 7b$ **141.** $6x + 28$ **143.** $5x - 75$ **145.** $4x - 4$ **147.** $38x - 63$ **149.** b **151.** $20x + y$
153. $-3x + y$ **155.** 0   **157.** $-a + b$ **159.** **a.** Yes **b.** No

CONCEPT REVIEW 2.3   *pages 71 – 72*

1. Never true   **3.** Never true   **5.** Always true   **7.** Never true

SECTION 2.3   *pages 72 – 77*

1. $19 - d$ **3.** $r - 12$ **5.** $28a$ **7.** $5(n - 7)$ **9.** $y - 3y$ **11.** $-6b$ **13.** $\dfrac{4}{p - 6}$ **15.** $\dfrac{x - 9}{2x}$
17. $-4s - 21$ **19.** $\dfrac{d + 8}{d}$ **21.** $\dfrac{3}{8}(t + 15)$ **23.** $w + \dfrac{7}{w}$ **25.** $d + (16d - 3)$ **27.** $\dfrac{n}{19}$ **29.** $n + 40$
31. $(n - 90)^2$ **33.** $\dfrac{4}{9}n + 20$ **35.** $n(n + 10)$ **37.** $7n + 14$ **39.** $\dfrac{12}{n + 2}$ **41.** $\dfrac{2}{n + 1}$ **43.** $60 - \dfrac{n}{50}$
45. $n^2 + 3n$ **47.** $(n + 3) + n^3$ **49.** $n^2 - \dfrac{1}{4}n$ **51.** $2n - \dfrac{7}{n}$ **53.** $n^3 - 12n$ **55.** $n + (n + 10); \ 2n + 10$
57. $n - (9 - n); \ 2n - 9$ **59.** $\dfrac{1}{5}n - \dfrac{3}{8}n; \ -\dfrac{7}{40}n$ **61.** $(n + 9) + 4; \ n + 13$ **63.** $2(3n + 40); \ 6n + 80$
65. $7(5n); \ 35n$ **67.** $17n + 2n; \ 19n$ **69.** $n + 12n; \ 13n$ **71.** $3(n^2 + 4); \ 3n^2 + 12$ **73.** $\dfrac{3}{4}(16n + 4); \ 12n + 3$
75. $16 - (n + 9); \ -n + 7$ **77.** $\dfrac{4n}{2} + 5; \ 2n + 5$ **79.** $6(n + 8); \ 6n + 48$ **81.** $7 - (n + 2); \ -n + 5$
83. $\dfrac{1}{3}(n + 6n); \ \dfrac{7n}{3}$ **85.** $(8 + n^3) + 2n^3; \ 3n^3 + 8$ **87.** $(n + 12) + (n - 6); \ 2n + 6$
89. $(n + 9) + (n - 20); \ 2n - 11$ **91.** $14 + (n - 3)10; \ 10n - 16$ **93.** $\dfrac{1}{2}j$ **95.** $4c$ **97.** $s + 25$
99. $\dfrac{1}{2}L - 3$ **101.** $0.32 + 0.23(w - 1)$ **103.** S and $12 - S$ **105.** g and $20 - g$ **107.** $640 + 24h$
109. even   **111.** even   **113.** even   **115.** even   **117.** odd   **119.** $2x$ **121.** $\dfrac{3}{5}x$

CHAPTER REVIEW EXERCISES   *pages 80 – 82*

1. y^2 (Obj. 2.2.2)   **2.** $3x$ (Obj. 2.2.3)   **3.** $-10a$ (Obj. 2.2.3)   **4.** $-4x + 8$ (Obj. 2.2.4)   **5.** $7x + 38$ (Obj. 2.2.5)   **6.** 16 (Obj. 2.1.1)   **7.** 9 (Obj. 2.2.1)   **8.** $36y$ (Obj. 2.2.3)   **9.** $6y - 18$ (Obj. 2.2.4)
10. $-3x + 21y$ (Obj. 2.2.5)   **11.** $-8x^2 + 12y^2$ (Obj. 2.2.4)   **12.** $5x$ (Obj. 2.2.2)   **13.** 22 (Obj. 2.1.1)
14. $2x$ (Obj. 2.2.3)   **15.** $15 - 35b$ (Obj. 2.2.4)   **16.** $-7x + 33$ (Obj. 2.2.5)   **17.** The Commutative Property of Multiplication (Obj. 2.2.1)   **18.** $24 - 6x$ (Obj. 2.2.4)   **19.** $5x^2$ (Obj. 2.2.2)   **20.** $-7x + 14$ (Obj. 2.2.5)

21. $-9y^2 + 9y + 21$ (Obj. 2.2.4)   **22.** $2x + y$ (Obj. 2.2.5)   **23.** 3 (Obj. 2.1.1)   **24.** $36y$ (Obj. 2.2.3)
25. $5x - 43$ (Obj. 2.2.5)   **26.** $2x$ (Obj. 2.2.3)   **27.** $-6x^2 + 21y^2$ (Obj. 2.2.4)   **28.** 6 (Obj. 2.1.1)
29. $-x + 6$ (Obj. 2.2.5)   **30.** $-5a - 2b$ (Obj. 2.2.2)   **31.** $-10x^2 + 15x - 30$ (Obj. 2.2.4)
32. $-9x - 7y$ (Obj. 2.2.2)   **33.** $6a$ (Obj. 2.2.3)   **34.** $17x - 24$ (Obj. 2.2.5)   **35.** $-2x - 5y$ (Obj. 2.2.2)
36. $30b$ (Obj. 2.2.3)   **37.** 21 (Obj. 2.2.1)   **38.** $-2x^2 + 4x$ (Obj. 2.2.2)   **39.** $-6x^2$ (Obj. 2.2.3)
40. $15x - 27$ (Obj. 2.2.5)   **41.** $-8a^2 + 3b^2$ (Obj. 2.2.4)   **42.** The Multiplication Property of Zero (Obj. 2.2.1)
43. $b - 7b$ (Obj. 2.3.1)   **44.** $n + 2n^2$ (Obj. 2.3.2)   **45.** $\dfrac{6}{n} - 3$ (Obj. 2.3.2)   **46.** $\dfrac{10}{y-2}$ (Obj. 2.3.1)
47. $8\left(\dfrac{2n}{16}\right); n$ (Obj. 2.3.3)   **48.** $4(2 + 5n); 8 + 20n$ (Obj. 2.3.3)   **49.** $s + 15$ (Obj. 2.3.4)
50. b and $20 - b$ (Obj. 2.3.4)

CHAPTER TEST   *pages 82 – 83*

1. $36y$ (Obj. 2.2.3)   **2.** $4x - 3y$ (Obj. 2.2.2)   **3.** $10n - 6$ (Obj. 2.2.5)   **4.** 2 (Obj. 2.1.1)

5. The Multiplication Property of One (Obj. 2.2.1)   **6.** $4x - 40$ (Obj. 2.2.4)   **7.** $\dfrac{1}{12}x^2$ (Obj. 2.2.2)

8. $4x$ (Obj. 2.2.3)   **9.** $-24y^2 + 48$ (Obj. 2.2.4)   **10.** 19 (Obj. 2.2.1)   **11.** 6 (Obj. 2.1.1)
12. $-3x + 13y$ (Obj. 2.2.5)   **13.** b (Obj. 2.2.2)   **14.** $78a$ (Obj. 2.2.3)   **15.** $3x^2 - 15x + 12$ (Obj. 2.2.4)

16. -32 (Obj. 2.1.1)   **17.** $37x - 5y$ (Obj. 2.2.5)   **18.** $\dfrac{n+8}{17}$ (Obj. 2.3.1)   **19.** $(a + b) - b^2$ (Obj. 2.3.1)

20. $n^2 + 11n$ (Obj. 2.3.2)   **21.** $20(n + 9); 20n + 180$ (Obj. 2.3.3)   **22.** $(n + 2) + (n - 3); 2n - 1$ (Obj. 2.3.3)

23. $n - \dfrac{1}{4}(2n); \dfrac{1}{2}n$ (Obj. 2.3.3)   **24.** $30d$ (Obj. 2.3.4)   **25.** L and $9 - L$ (Obj. 2.3.4)

CUMULATIVE REVIEW EXERCISES   *pages 83 – 84*

1. -7 (Obj. 1.2.1)   **2.** 5 (Obj. 1.2.2)   **3.** 24 (Obj. 1.2.3)   **4.** -5 (Obj. 1.2.4)   **5.** 1.25 (Obj. 1.3.1)

6. $\dfrac{3}{5}, 0.60$ (Obj. 1.3.2)   **7.** $\{-4, -3, -2, -1\}$ (Obj. 1.1.1)   **8.** 8% (Obj. 1.3.2)   **9.** $\dfrac{11}{48}$ (Obj. 1.3.3)

10. $\dfrac{5}{18}$ (Obj. 1.3.4)   **11.** $\dfrac{1}{4}$ (Obj. 1.3.4)   **12.** $\dfrac{8}{3}$ (Obj. 1.4.1)   **13.** -5 (Obj. 1.4.2)   **14.** $\dfrac{53}{48}$ (Obj. 1.4.2)

15. -8 (Obj. 2.1.1)   **16.** $5x^2$ (Obj. 2.2.2)   **17.** $-a - 12b$ (Obj. 2.2.2)   **18.** $3a$ (Obj. 2.2.3)
19. $20b$ (Obj. 2.2.3)   **20.** $20 - 10x$ (Obj. 2.2.4)   **21.** $6y - 21$ (Obj. 2.2.4)   **22.** $-6x^2 + 8y^2$ (Obj. 2.2.4)
23. $-8y^2 + 20y + 32$ (Obj. 2.2.4)   **24.** $-10x + 15$ (Obj. 2.2.5)   **25.** $5x - 17$ (Obj. 2.2.5)
26. $13x - 16$ (Obj. 2.2.5)   **27.** $6x + 29y$ (Obj. 2.2.5)   **28.** $6 - 12n$ (Obj. 2.3.2)
29. $5 + (n - 7); n - 2$ (Obj. 2.3.3)   **30.** $\dfrac{1}{2}s$ (Obj. 2.3.4)

ANSWERS to Chapter 3 Odd-Numbered Exercises

CONCEPT REVIEW 3.1   *page 94*

1. Sometimes true   **3.** Always true   **5.** Always true   **7.** Always true

---

SECTION 3.1   *pages 95 – 99*

1. Yes   **3.** No   **5.** No   **7.** Yes   **9.** Yes   **11.** Yes   **13.** No   **15.** No   **17.** No   **19.** Yes
21. No   **23.** No   **25.** 2   **27.** 15   **29.** 6   **31.** -6 **33.** 3   **35.** 0   **37.** -2 **39.** -7
41. -7 **43.** -12 **45.** 2   **47.** -5 **49.** 15   **51.** 9   **53.** 14   **55.** -1 **57.** 1   **59.** $-\dfrac{1}{2}$

61. $-\dfrac{3}{4}$ **63.** $\dfrac{1}{12}$ **65.** $-\dfrac{7}{12}$ **67.** $-\dfrac{5}{7}$ **69.** $-\dfrac{1}{3}$ **71.** 1.869 **73.** 0.884 **75.** 7 **77.** -7

79. -4 **81.** 5 **83.** 9 **85.** -8 **87.** 0 **89.** -7 **91.** 8 **93.** 12 **95.** 12 **97.** -18

99. -18 **101.** 6 **103.** -12 **105.** 3 **107.** -24 **109.** 9 **111.** $\dfrac{1}{3}$ **113.** $\dfrac{15}{7}$ **115.** 4 **117.** 3

119. 4.48 **121.** 2.06 **123.** -2.1 **125.** 24% **127.** 7.2 **129.** 400 **131.** 9 **133.** 25% **135.** 5

137. 200% **139.** 400 **141.** 7.7 **143.** 200 **145.** 400 **147.** 20 **149.** 80%

151. The expected profit is \$18,000. **153.** The 1994 commerical cost is 514% of the 1967 commercial cost.

155. 9000 seats were added to the stadium. **157.** A student must answer correctly $83\dfrac{1}{3}$% of the questions.

159. There are about 3000 accredited colleges and universities in the United States. **161.** 19% of the total number of students are in the fine arts college. **163.** 60% of the people surveyed preferred a drink that was not cola flavored. **165.** The total sales for January were \$22,250. **167.** -21 **169.** -27 **171.** 21

173. 22° **175.** The cost of dinner was \$45. **177.** The new value is two times the orginal value.

179. One possible answer is $5x = -10$.

CONCEPT REVIEW 3.2 *page 107*

1. Always true **3.** Never true **5.** Never true **7.** Always true

SECTION 3.2 *pages 108 – 113*

1. 3 **3.** 6 **5.** -1 **7.** 1 **9.** -9 **11.** 3 **13.** 3 **15.** 3 **17.** 2 **19.** 2 **21.** 4

23. $\dfrac{6}{7}$ **25.** $\dfrac{2}{3}$ **27.** -1 **29.** $\dfrac{3}{4}$ **31.** $\dfrac{1}{3}$ **33.** $-\dfrac{3}{4}$ **35.** $\dfrac{1}{3}$ **37.** $-\dfrac{1}{6}$ **39.** 1 **41.** 0

43. $\dfrac{13}{10}$ **45.** $\dfrac{2}{5}$ **47.** $-\dfrac{3}{2}$ **49.** 18 **51.** 8 **53.** -16 **55.** 25 **57.** 21 **59.** 15 **61.** -16

63. -21 **65.** $\dfrac{15}{2}$ **67.** $-\dfrac{18}{5}$ **69.** 2 **71.** 3 **73.** 1 **75.** 2 **77.** 3.95 **79.** -0.8 **81.** -11

83. 0 **85.** 3 **87.** 8 **89.** 2 **91.** -2 **93.** -3 **95.** 2 **97.** -2 **99.** -2 **101.** -7

103. 0 **105.** -2 **107.** -2 **109.** -2 **111.** 4 **113.** -3 **115.** 2 **117.** 10 **119.** 3

121. $\dfrac{3}{4}$ **123.** $\dfrac{2}{7}$ **125.** $-\dfrac{3}{4}$ **127.** 3 **129.** -14 **131.** 7 **133.** 3 **135.** 4 **137.** 2 **139.** 2

141. 2 **143.** -7 **145.** $\dfrac{4}{7}$ **147.** $\dfrac{1}{2}$ **149.** $-\dfrac{1}{3}$ **151.** $\dfrac{10}{3}$ **153.** $-\dfrac{1}{4}$ **155.** 0.5 **157.** 0

159. -1 **161.** The initial velocity is 8 ft/s. **163.** The depreciated value will be \$38,000 in 2 years.

165. The value of x is 11°. **167.** The approximate length of the humerus is 31.8 in. **169.** The first "4 min mile" was run in approximately 1952. **171.** The population P_2 is 51,000 people. **173.** The break-even point is 350 television sets. **175.** The break-even point is 1200 compact discs. **177.** No, the see-saw is not balanced.

179. A force of 25 lb must be applied to the other end of the lever. **181.** The fulcrum must be placed 10 ft from the 128 lb acrobat. **183.** No solution **185.** $-\dfrac{11}{4}$ **187.** 62 **189.** One possible answer is $3x - 6 = 2x - 2$.

CONCEPT REVIEW 3.3 *pages 119 – 120*

1. Always true **3.** Sometimes true **5.** Never true **7.** Always true **9.** Sometimes true

SECTION 3.3 *pages 120 – 123*

1. (number line -5 to 5) **3.** (number line -5 to 5) **5.** $x < 2$ (number line -5 to 5)

7. $x > 3$ (number line -5 to 5) **9.** $n \ge 3$ (number line -5 to 5) **11.** $x \le -4$ (number line -5 to 5)

13. $x \ge -1$ (number line -5 to 5) **15.** $y \ge -9$ **17.** $x < 12$ **19.** $x \ge 5$ **21.** $x < -11$ **23.** $x \le 10$

25. $x \ge -6$ **27.** $x > 2$ **29.** $d < -\dfrac{1}{6}$ **31.** $x \ge -\dfrac{31}{24}$ **33.** $x < \dfrac{5}{4}$ **35.** $x > \dfrac{5}{24}$ **37.** $x \le -1.2$

39. $x \le 0.70$ **41.** $x < -7.3$ **43.** $x \le -3$ **45.** $x > -2$

47. $x > 0$ **49.** $n \ge 4$ **51.** $x > -2$

53. $x < \dfrac{5}{3}$ **55.** $x \ge 5$ **57.** $x > -\dfrac{5}{2}$ **59.** $x \le -\dfrac{2}{3}$ **61.** $x < -18$ **63.** $x > -16$ **65.** $y \ge 6$

67. $x \ge -6$ **69.** $b \le 33$ **71.** $n < \dfrac{3}{4}$ **73.** $x \le -\dfrac{6}{7}$ **75.** $y \le \dfrac{5}{6}$ **77.** $x \le \dfrac{27}{28}$ **79.** $y > -\dfrac{3}{2}$

81. $x > -0.5$ **83.** $y \ge -0.8$ **85.** $x \le 4.2$ **87.** $x \le 5$ **89.** $x < -5.4$ **91.** $x < 4$ **93.** $x < -4$

95. $x \ge 1$ **97.** $x \le 5$ **99.** $x < 0$ **101.** $x < 20$ **103.** $x > 500$ **105.** $x > 2$ **107.** $x \le -5$

109. $y \le \dfrac{5}{2}$ **111.** $x < \dfrac{25}{11}$ **113.** $x > 11$ **115.** $n \le \dfrac{11}{18}$ **117.** $x \ge 6$ **119.** $x \le \dfrac{2}{5}$ **121.** $t < 1$

123. $n > \dfrac{7}{10}$ **125.** $\{1, 2\}$ **127.** $\{1, 2, 3\}$ **129.** $\{3, 4, 5\}$ **131.** $\{10, 11, 12, 13\}$ **133.**

135.

CHAPTER REVIEW EXERCISES   *pages 127–129*

1. No (Obj. 3.1.1)   **2.** 20 (Obj. 3.1.2)   **3.** -7 (Obj. 3.1.3)   **4.** 7 (Obj. 3.2.1)   **5.** 4 (Obj. 3.2.2)

6. $-\dfrac{1}{5}$ (Obj. 3.2.3)   **7.** 405 (Obj. 3.1.4)   **8.** 25 (Obj. 3.1.4)   **9.** 67.5% (Obj. 3.1.4)

10. (Obj. 3.3.1)   **11.** $x > 2$ (Obj. 3.3.1)

12. $x > -4$ (Obj. 3.3.2)   **13.** $x \ge -4$ (Obj. 3.3.3)   **14.** $x \ge 4$ (Obj. 3.3.3)

15. Yes (Obj. 3.1.1)   **16.** 2.5 (Obj. 3.1.2)   **17.** -49 (Obj. 3.1.3)   **18.** $\dfrac{1}{2}$ (Obj. 3.2.1)   **19.** $\dfrac{1}{3}$ (Obj. 3.2.2)

20. 10 (Obj. 3.2.3)   **21.** 16 (Obj. 3.1.4)   **22.** 125 (Obj. 3.1.4)   **23.** $16\dfrac{2}{3}\%$ (Obj. 3.1.4)

24. $x < -4$ (Obj. 3.3.1)   **25.** $x \le -2$ (Obj. 3.3.2)

26. -2 (Obj. 3.2.3)   **27.** $\dfrac{5}{6}$ (Obj. 3.1.2)   **28.** 20 (Obj. 3.1.3)   **29.** 6 (Obj. 3.2.1)   **30.** 0 (Obj. 3.2.3)

31. $x < 12$ (Obj. 3.3.3)   **32.** 15 (Obj. 3.1.4)   **33.** 5 (Obj. 3.2.1)   **34.** $x > 5$ (Obj. 3.3.3)

35. 4% (Obj. 3.1.4)   **36.** $x > -18$ (Obj. 3.3.3)   **37.** $x < \dfrac{1}{2}$ (Obj. 3.3.3)   **38.** The measure of the third angle is 110°. (Obj. 3.2.4)   **39.** A force of 24 lb must be applied to the other end of the lever. (Obj. 3.2.4) **40.** It takes 3 s for the velocity to increase from 4 ft/s to 100 ft/s. (Obj. 3.2.4)   **41.** The discount is $79.25. (Obj. 3.2.4)   **42.** This represents a 40% increase. (Obj. 3.1.4)   **43.** The depth is 80 ft. (Obj. 3.2.4) **44.** The fulcrum is 3 ft from the 25 lb force. (Obj. 3.2.4)   **45.** The length of the rectangle is 24 ft. (Obj. 3.2.4)

CHAPTER TEST   *pages 129–130*

1. -12 (Obj. 3.1.3)   **2.** $-\dfrac{1}{2}$ (Obj. 3.2.2)   **3.** -3 (Obj. 3.2.1)   **4.** No (Obj. 3.1.1)   **5.** $\dfrac{1}{8}$ (Obj. 3.1.2)

6. $-\dfrac{1}{3}$ (Obj. 3.2.3)   **7.** 5 (Obj. 3.2.1)   **8.** $\dfrac{1}{2}$ (Obj. 3.2.2)   **9.** -5 (Obj. 3.1.2)   **10.** -5 (Obj. 3.2.2)

11. $-\dfrac{40}{3}$ (Obj. 3.1.3)   **12.** $-\dfrac{22}{7}$ (Obj. 3.2.3)   **13.** 2 (Obj. 3.2.3)   **14.** $\dfrac{12}{11}$ (Obj. 3.2.3)

15. -3 (Obj. 3.2.3)   **16.** 125% (Obj. 3.1.4)   **17.** 40 (Obj. 3.1.4)   **18.** (Obj. 3.3.1)

19. $x \le -1$ (Obj. 3.3.1)   **20.** $x > -2$ (Obj. 3.3.2)

21. $x > \dfrac{1}{2}$ (Obj. 3.3.1)   **22.** $x \le -\dfrac{9}{2}$ (Obj. 3.3.3)   **23.** $x \ge -16$ (Obj. 3.3.2)   **24.** $x \le -\dfrac{22}{7}$ (Obj. 3.3.3)

25. $x \le -3$ (Obj. 3.3.3)   **26.** $x > 2$ (Obj. 3.3.3)   **27.** 4 (Obj. 3.2.2)   **28.** 24 (Obj. 3.1.4)

29. $x \ge 3$ (Obj. 3.3.2)   **30.** Yes (Obj. 3.1.1)   **31.** The astronaut would weigh 30 lb on the moon. (Obj. 3.1.4)   **32.** The final temperature of the water after mixing is 60°C. (Obj. 3.2.4)   **33.** The number of calculators produced was 200. (Obj. 3.2.4)

CUMULATIVE REVIEW EXERCISES   *pages 130–132*

1. 6   (Obj. 1.2.2)   **2.** -48 (Obj. 1.2.3)   **3.** $-\dfrac{19}{48}$ (Obj. 1.3.3)   **4.** -2 (Obj. 1.3.4)   **5.** 54   (Obj. 1.4.1)

6. 24   (Obj. 1.4.2)   **7.** 6   (Obj. 2.1.1)   **8.** $-17x$ (Obj. 2.2.2)   **9.** $-5a - 2b$ (Obj. 2.2.2)

10. $2x$ (Obj. 2.2.3)   **11.** $36y$ (Obj. 2.2.3)   **12.** $2x^2 + 6x - 4$ (Obj. 2.2.4)   **13.** $-4x + 14$ (Obj. 2.2.5)

14. $6x - 34$ (Obj. 2.2.5)   **15.** $\{-7, -6, -5, -4, -3, -2, -1\}$ (Obj. 1.1.1)   **16.** $87\dfrac{1}{2}\%$ (Obj. 1.3.2)

17. 3.42   (Obj. 1.3.2)   **18.** $\dfrac{5}{8}$ (Obj. 1.3.2)   **19.** Yes   (Obj. 3.1.1)   **20.** -5 (Obj. 3.1.2)

21. -25 (Obj. 3.1.3)   **22.** 3   (Obj. 3.2.1)   **23.** -3 (Obj. 3.2.2)   **24.** 13   (Obj. 3.2.3)

25. $x < -\dfrac{8}{9}$ (Obj. 3.3.2)   **26.** $x \geq 12$ (Obj. 3.3.3)   **27.** $x > 9$ (Obj. 3.3.3)   **28.** $8 - \dfrac{n}{12}$ (Obj. 2.3.2)

29. $n + (n + 2); 2n + 2$ (Obj. 2.3.3)   **30.** b and $35 - b$ (Obj. 2.3.4)

31. The length of the shorter piece: s; the length of the longer piece is four inches less than three times the length of the shorter piece: $3s - 4$ (Obj. 2.3.4)   **32.** 17% of the computer programmer's salary is deducted for income tax.   (Obj. 3.1.4)   **33.** The number of cameras produced was 250.   (Obj. 3.2.4)   **34.** The final temperature of the water after mixing is 60°C. (Obj. 3.2.4)   **35.** A force of 24 lb must be applied to the other end of the lever.   (Obj. 3.2.4)

ANSWERS to Chapter 4 Odd-Numbered Exercises

CONCEPT REVIEW 4.1   *page 137*

1. Always true   **3.** Always true   **5.** Never true

SECTION 4.1   *pages 137–141*

1. $n - 15 = 7; n = 22$ **3.** $7n = -21; n = -3$ **5.** $3n - 4 = 5; n = 3$ **7.** $4(2n + 3) = 12; n = 0$

9. $12 = 6(n - 3); n = 5$ **11.** $22 = 6n - 2; n = 4$ **13.** $4n + 7 = 2n + 3; n = -2$ **15.** $5n - 8 = 8n + 4;$ $n = -4$ **17.** $2(n - 25) = 3n; n = -50$ **19.** $3n = 2(20 - n); 8$ and 12   **21.** $3n + 2(18 - n) = 44; 8$ and 10

23. The original value was $16,000.   **25.** The rating of the second computer is 15 mips.   **27.** The number of three-point baskets was 15.   **29.** The amount of mulch is 15 lb.   **31.** The monthly payment is $117.75.

33. There were 5 h of labor required.   **35.** The two amounts deposited were $2000 and $3000.   **37.** The length is 80 ft. The width is 50 ft.   **39.** The perimeter of the larger square is 8 ft.   **41.** The width is 3 ft.

43. The two angles are 28° and 62°.   **45.** The two angles are 48° and 132°.   **47.** One-third of the container is filled at 3.39 P.M.   **49.** The phone call lasted 15 min.   **51.** There are 60 coins in the bank.

CONCEPT REVIEW 4.2   *page 145*

1. Always true   **3.** Always true   **5.** Never true

SECTION 4.2   *pages 146–149*

1. The integers are 17, 18, and 19.   **3.** The integers are 26, 28, and 30.   **5.** The integers are 17, 19, and 21.

7. The integers are 8 and 10.   **9.** The integers are 7 and 9.   **11.** The integers are $-9, -8,$ and -7.   **13.** The integers are 10, 12, and 14.   **15.** The integers are 4, 6, and 8.   **17.** There are 12 dimes and 15 quarters in the bank.   **19.** The executive bought eight 23¢ stamps and thirty-two 32¢ stamps.   **21.** There are five 29¢ stamps in the drawer.   **23.**   There are 28 quarters in the bank.   **25.**   There are 20 one-dollar bills and 6 five-dollar bills in the cash box.   **27.**   There are 11 pennies in the bank.   **29.**   There are twenty-seven 22¢ stamps in the collection.   **31.**   There are thirteen 3¢ stamps, eighteen 7¢ stamps, and nine 12¢ stamps in the collection. **33.**   There are eighteen 6¢ stamps, six 8¢ stamps, and twenty-four 15¢ stamps.   **35.**   The integers are −12, −10, −8, and −6.   **37.**   There is $6.70 in the bank.   **39.**   For any three consecutive odd integers, the sum of the first and third integers is twice the second integer.

CONCEPT REVIEW 4.3   *page 152*

1. Always true   **3.** Always true   **5.** Never true

SECTION 4.3   *pages 152 – 154*

1.   The length is 50 ft. The width is 25 ft.   **3.**   The length is 130 ft. The width is 52 ft.   **5.**   The lengths of the three sides are 13 ft, 12 ft, and 14 ft.   **7.**   The lengths of the three sides are 11 ft, 6 ft, and 13 ft.   **9.**   The lengths of the three sides are 60 ft, 60 ft, and 15 ft.   **11.**   The length is 13.5 m. The width is 12.8 m.   **13.**   The measures of the three angles are 36°, 36°, and 108°.   **15.**   The measures of the three angles are 32°, 58°, and 90°. **17.**   The measures of the three angles are 38°, 38°, and 104°.   **19.**   The measures of the three angles are 60°, 45°, and 75°.   **21.**   The measures of the three angles are 105°, 40°, and 35°.   **23.**   The measures of the three angles are 57°, 19°, and 104°.   **25.**   The length is 9 cm. The width is 3 cm.   **27.**   The length is 16x.

CONCEPT REVIEW 4.4   *page 157*

1. Always true   **3.** Never true   **5.** Always true

SECTION 4.4   *pages 157 – 160*

1.   The selling price is $35.   **3.**   The markup rate is 87.5%.   **5.**   The selling price is $196.   **7.**   The markup rate is 40%.   **9.**   The selling price is $111.60.   **11.**   The markup rate is 23.3%.   **13.**   The cost is $120. **15.**   The sale price is $71.25.   **17.**   The discount rate is 25%.   **19.**   The price is $24.50.   **21.**   The discount rate is 40%.   **23.**   The price is $218.50.   **25.**   The discount rate is 26%.   **27.**   The regular price is $310. **29.**   The markup is $18.   **31.**   The markup rate is 40%.   **33.**   The cost is $230.   **35.**   The regular price was $80. **37.**   No; the single discount that would give the same sale price is 28%.

CONCEPT REVIEW 4.5   *page 163*

1. Always true   **3.** Always true   **5.** Never true

SECTION 4.5   *pages 163 – 166*

1. $9000 was invested at 7%, and $6000 was invested at 6.5%.   **3.** $1500 was invested in the mutual fund. **5.** $200,000 was deposited at 10%, and $100,000 was deposited at 8.5%.   **7.**   The electrician has $3000 invested in bonds that earn 8% annual simple interest.   **9.** $2500 is invested at 11%.   **11.** $40,500 was invested at 8%, and $13,500 was invested at 12%.   **13.**   The total amount invested was $650,000.   **15.**   The total amount invested was $500,000.   **17.**   The amount of the research consultant's investment is $45,000.   **19.**   The total annual interest received was $3040.   **21.**   The value of the investment in three years is $3831.87.   **23a.**   By age 55, the couple should have $170,000 saved for retirement.

CONCEPT REVIEW 4.6   *pages 170 – 171*

1. Always true   **3.** Always true   **5.** Never true   **7.** Always true

SECTION 4.6 *pages 171 – 174*

1. 2 lb of diet supplement and 3 lb of vitamin supplement should be used. **3.** The cost is $6.98 per pound. **5.** 56 oz of the $4.30 alloy and 144 oz of the $1.80 alloy were used. **7.** The cost is $2.90 per pound. **9.** 10 kg of hard candy must be used. **11.** 30 lb of the $2.20 meat and 20 lb of the $4.20 meat were used. **13.** 8 kg of soil supplement must be used. **15.** 37 lb of almonds and 63 lb of walnuts were used. **17.** The cost per pound is $.70. **19.** 9.6 lb of lima beans must be used. **21.** 20 ml of the 13% solution and 30 ml of the 18% solution should be used. **23.** The concentration of silver is 50%. **25.** 30 lb of the 60% rye grass is used. **27.** The concentration of hydrocortisone is 0.74%. **29.** 100 ml of the 7% solution and 200 ml of the 4% solution are used. **31.** 100 oz of the 8% saline solution must be added. **33.** The concentration of the resulting alloy is 27%. **35.** 10 oz of pure bran flakes must be added. **37.** 3 lb of the 20% jasmine and 2 lb of the 15% jasmine are used. **39.** 1.2 oz of dried apricots must be added. **41.** The cost is $3.65 per ounce. **43.** 10 oz of water evaporated from the solution. **45.** 75 g of pure water must be added. **47.** 85 adults and 35 children attended the performance.

CONCEPT REVIEW 4.7 *pages 177 – 178*

1. Never true **3.** Never true **5.** Always true

SECTION 4.7 *pages 178 – 180*

1. The rate of the first plane is 105 mph. The rate of the second plane 130 mph. **3.** The second runner will overtake the first runner in 3 h. **5.** The car traveled at 40 mph for 0.5 h. **7.** The distance between the two airports is 300 mi. **9.** It took the campers $\frac{1}{3}$ h to canoe downstream. **11.** The jet overtakes the propeller-driven plane 570 mi from the starting point. **13.** The plane flew 2 h at 115 mph and 3 h at 125 mph. **15.** The sailboat traveled 33 mi in the first 3 h. **17.** The rate of the passenger train is 60 mph. The rate of the freight train is 45 mph. **19.** The joggers will meet 1 h after they begin. **21.** The rate of the cyclist is 13 mph. **23.** The rate of the car is 60 mph. **25.** The joggers will meet at 7:48 A.M. **27.** The cyclist's average speed is $13\frac{1}{3}$ mph.

CONCEPT REVIEW 4.8 *pages 182 – 183*

1. Never true **3.** Always true **5.** Never true

SECTION 4.8 *pages 183 – 186*

1. The smallest integer is 2. **3.** The amount of income tax to be paid is $3150 or more. **5.** The organization must collect more than 440 lb. **7.** The student must score 78 or better. **9.** The dollar amount in sales must be more than $5714. **11.** The dollar amount the agent expects to sell is $20,000 or less. **13.** A person must use this service for more than 60 min. **15.** 8 or less ounces must be added. **17.** The ski area is more than 38 mi away. **19.** The maximum number of miles you can drive is 166. **21.** The integers are 1, 3, and 5; or 3, 5, and 7. **23.** The crew can prepare 8 to 12 aircraft in this period of time. **25.** The minimum number is 467 calendars.

CHAPTER REVIEW EXERCISES   *pages 193 – 195*

1. $5n - 4 = 16$; $n = 4$ (Obj. 4.1.1)   **2.** The length of the shorter piece is 14 in.   (Obj. 4.1.2)   **3.** The two numbers are 8 and 13.   (Obj. 4.1.2)   **4.** $6(n + 3) = 2n - 10$; $n = -7$ (Obj. 4.1.1)   **5.** The discount rate is $33\frac{1}{3}$%.   (Obj. 4.4.2)   **6.** $8000 is invested at 6%, and $7000 is invested at 7%.   (Obj. 4.5.1)   **7.** The mixture costs $1.84 per pound.   (Obj. 4.6.1)   **8.** The rate of the motorcyclist is 45 mph.   (Obj. 4.7.1)   **9.** The measures of the three angles are 16°, 82°, and 82°.   (Obj. 4.3.2)   **10.** The markup rate is 70%.   (Obj. 4.4.1)   **11.** The concentration of butterfat is 14%.   (Obj. 4.6.2)   **12.** The length is 80 ft. The width is 20 ft.   (Obj. 4.3.1)   **13.** The maximum number of miles you can drive a Company A car is 359.   (Obj. 4.8.1)   **14.** The regular price is $32.   (Obj. 4.4.2)   **15.** 7 qt of cranberry juice and 3 qt of apple juice were used.   (Obj. 4.6.1)   **16.** The selling price is $595.   (Obj. 4.4.1)   **17.** The two numbers are 6 and 30.   (Obj. 4.1.2)   **18.** $5600 is deposited in the 12% account.   (Obj. 4.5.1)   **19.** The two integers are 9 and 10.   (Obj. 4.2.1)   **20.** One liter of pure water should be added.   (Obj. 4.6.2)   **21.** The lowest score the student can receive is 72.   (Obj. 4.8.1)   **22.** $15 = \frac{2}{3}n + 3$; $n = 18$ (Obj. 4.1.1)   **23.** The height is 993 ft.   (Obj. 4.1.2)   **24.** The measures of the three angles are 75°, 60°, and 45°.   (Obj. 4.3.2)   **25.** The ticket seller held 30 one-dollar bills.   (Obj. 4.2.2)   **26.** The length of the longer piece is 7 ft.   (Obj. 4.1.2)   **27.** The number of hours of consultation was 8.   (Obj. 4.1.2)   **28.** 5 dimes and 7 quarters were left for a tip.   (Obj. 4.2.2)   **29.** The cost is $671.25.   (Obj. 4.4.1)   **30.** They will meet after 4 min.   (Obj. 4.7.1)   **31.** The measures of the three sides are 8 in., 12 in., and 15 in.   (Obj. 4.3.1)   **32.** The integers are -17, -15, and -13.   (Obj. 4.2.1)   **33.** The maximum width is 11 ft.   (Obj. 4.8.1)

CHAPTER TEST   *pages 195 – 197*

1. $6n + 13 = 3n - 5$; $n = -6$ (Obj. 4.1.1)   **2.** $3n - 15 = 27$; $n = 14$ (Obj. 4.1.1)   **3.** The numbers are 8 and 10.   (Obj. 4.1.2)   **4.** The lengths are 6 ft and 12 ft.   (Obj. 4.1.2)   **5.** The cost is $200.   (Obj. 4.4.1)   **6.** The discount rate is 20%.   (Obj. 4.4.2)   **7.** 20 gal of the 15% solution must be added.   (Obj. 4.6.2)   **8.** The length is 14 m. The width is 5 m.   (Obj. 4.3.1)   **9.** The integers are 5, 7, and 9.   (Obj. 4.2.1)   **10.** Five or more residents are in the nursing home.   (Obj. 4.8.1)   **11.** $5000 is invested at 10%, and $2000 is invested at 15%.   (Obj. 4.5.1)   **12.** 8 lb of the $7 coffee and 4 lb of the $4 coffee should be used.   (Obj. 4.6.1)   **13.** The rate of the first plane is 225 mph. The rate of the second plane is 125 mph.   (Obj. 4.7.1)   **14.** The measures of the three angles are 48°, 33°, and 99°.   (Obj. 4.3.2)   **15.** There are 15 nickels and 35 quarters in the bank.   (Obj. 4.2.2)   **16.** The amounts to be deposited are $1400 at 6.75% and $1000 at 9.45%.   (Obj. 4.5.1)   **17.** The minimum length is 24 ft.   (Obj. 4.8.1)   **18.** The sale price is $79.20.   (Obj. 4.4.2)

CUMULATIVE REVIEW EXERCISES   *pages 197 – 198*

1. -12, -6 (Obj. 1.1.1)   **2.** 6   (Obj. 1.2.2)   **3.** $-\frac{1}{6}$ (Obj. 1.4.1)   **4.** $-\frac{11}{6}$ (Obj. 1.4.2)   **5.** -18 (Obj. 1.1.2)   **6.** -24 (Obj. 2.1.1)   **7.** $9x + 4y$ (Obj. 2.2.2)   **8.** $-12 + 8x + 20x^3$ (Obj. 2.2.4)   **9.** $4x + 4$ (Obj. 2.2.5)   **10.** $6x^2$ (Obj. 2.2.2)   **11.** No   (Obj. 3.1.1)   **12.** -3 (Obj. 3.2.1)   **13.** -15 (Obj. 3.1.3)   **14.** 3   (Obj. 3.2.1)   **15.** -10 (Obj. 3.2.3)   **16.** $\frac{2}{5}$ (Obj. 1.3.2)   **17.** $x \ge 4$ (Obj. 3.3.2)   **18.** $x \ge -3$ (Obj. 3.3.2)   **19.** $x > 18$ (Obj. 3.3.3)   **20.** $x \ge 4$ (Obj. 3.3.3)   **21.** 2.5%   (Obj. 1.3.2)   **22.** 12%   (Obj. 1.3.2)   **23.** 3   (Obj. 3.1.4)   **24.** 45   (Obj. 3.1.4)   **25.** $8n + 12 = 4n$; $n = -3$ (Obj. 4.1.1)   **26.** The area of the garage is 600 ft².   (Obj. 4.1.2)   **27.** The number of hours of labor is 5.   (Obj. 4.1.2)   **28.** 20% of the libraries had the reference book.   (Obj. 3.1.4)   **29.** The amount of money deposited is $2000.   (Obj. 4.5.1)   **30.** The markup rate is 75%.   (Obj. 4.4.1)   **31.** 60 g of the gold alloy must be used.   (Obj. 4.6.1)   **32.** 30 oz of pure water must be added.   (Obj. 4.6.2)   **33.** The measure of one of the equal angles is 47°.   (Obj. 4.3.2)   **34.** The middle even integer is 14.   (Obj. 4.2.1)   **35.** There are 7 dimes in the bank.   (Obj. 4.2.2)

ANSWERS to Chapter 5 Odd-Numbered Exercises

CONCEPT REVIEW 5.1 *page 204*

1. Always true   **3.** Never true   **5.** Never true   **7.** Always true

SECTION 5.1 *pages 205 – 208*

1.   **3.**   **5.**   **7.** $A(2, 3), B(4, 0), C(-4, 1), D(-2, -2)$

9. $A(-2, 5), B(3, 4), C(0, 0), D(-3, -2)$ **11.** **a.** $2; -4$ **b.** $1; -3$ **13.**

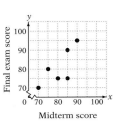

15. 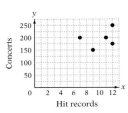   **17.** 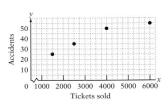   **19.** Quadrant IV   **21.** Quadrant III

23. 4 units   **25.** 2 units   **27.** 5 units   **29.** **a.** The signs of both coordinates are positive.   **b.** The sign of the first coordinate is negative. The sign of the second coordinate is positive.   **c.** The signs of both coordinates are negative.   **d.** The sign of the first coordinate is positive. The sign of the second coordinate is negative.

CONCEPT REVIEW 5.2 *page 216*

1. Never true   **3.** Always true   **5.** Never true   **7.** Never true   **9.** Always true

SECTION 5.2 *pages 216 – 220*

1. $m = 4, b = 1$ **3.** $m = \dfrac{5}{6}, b = \dfrac{1}{6}$ **5.** Yes   **7.** No   **9.** No   **11.** Yes   **13.** No   **15.** $(3, 7)$

17. $(6, 3)$ **19.** $(0, 1)$ **21.** $(-5, 0)$ **23.**   **25.** 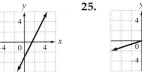   **27.**

29.   **31.**   **33.**   **35.**   **37.**

39.
41.
43.
45.
47.

49.
51.
53. $y = -3x + 10$
55. $y = 4x - 3$
57. $y = -\dfrac{3}{2}x + 3$

59. $y = \dfrac{2}{5}x - 2$
61. $y = -\dfrac{2}{7}x + 2$
63. $y = -\dfrac{1}{3}x + 2$
65.
67.

69.
71.
73.
75.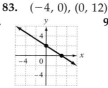
77. $(3, 0), (0, -3)$

79. $(3, 0), (0, -6)$
81. $(10, 0), (0, -2)$
83. $(-4, 0), (0, 12)$
85. $(0, 0), (0, 0)$
87. $(-6, 0), (0, 3)$

89.
91.
93.
95.
97.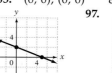

99.
101.
103.
105.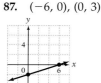
107.

109.
111. **a.** $y = -\dfrac{1}{2}x - 5$ **b.** $(-2, -4)$
113. 1 unit

CONCEPT REVIEW 5.3 *page 227*

1. Always true **3.** Never true **5.** Never true

SECTION 5.3 *pages 227–229*

1. -2 **3.** $\dfrac{1}{3}$ **5.** $-\dfrac{5}{2}$ **7.** $-\dfrac{1}{2}$ **9.** -1 **11.** Undefined **13.** 0 **15.** $-\dfrac{1}{3}$ **17.** 0 **19.** -5

21. Undefined **23.** $-\dfrac{2}{3}$ **25.** The slope is 0.8. After being connected, each minute of a transatlantic phone call costs an additional $.80. **27.** The slope is -500. The median family income in the United States has been decreasing $500 per year.

29. **31.** **33.** **35.** **37.**

39. **41.** **43.** **45.**

47. Increasing the coefficient of x increases the slope. **49.** Increasing the constant term increases the y-intercept. **51.** No, not all graphs of straight lines have a y-intercept. For example, $x = 2$ does not have a y-intercept.

CONCEPT REVIEW 5.4 *pages 231–232*

1. Always true **3.** Sometimes true **5.** Always true

SECTION 5.4 *pages 232–234*

1. $y = 2x + 2$ **3.** $y = -3x - 1$ **5.** $y = \dfrac{1}{3}x$ **7.** $y = \dfrac{3}{4}x - 5$ **9.** $y = -\dfrac{3}{5}x$ **11.** $y = \dfrac{1}{4}x + \dfrac{5}{2}$

13. $y = -\dfrac{2}{3}x - 7$ **15.** $y = 2x - 3$ **17.** $y = -2x - 3$ **19.** $y = \dfrac{2}{3}x$ **21.** $y = \dfrac{1}{2}x + 2$

23. $y = -\dfrac{3}{4}x - 2$ **25.** $y = \dfrac{3}{4}x + \dfrac{5}{2}$ **27.** $y = -\dfrac{4}{3}x - 9$ **29.** Yes; $y = -x + 6$ **31.** No **33.** 7

35. -1 **37.** $F = \dfrac{9}{5}C + 32$ **39.** $y = -\dfrac{2}{3}x + \dfrac{5}{3}$

CONCEPT REVIEW 5.5 *page 241*

1. Always true **3.** Never true **5.** Always true **7.** Always true

---

SECTION 5.5 *pages 241–246*

1. {(85, 75), (80, 70), (85, 70), (75, 80), (90, 80)}; no **3.** {(4, L), (6, W), (4, W), (2, W), (1, L), (6, W)}; no
5. {(11, 200), (9, 150), (7, 200), (12, 175), (12, 250)}; no **7.** Domain: {0, 2, 4, 6}; range: {0}; yes **9.** Domain: {2};
range: {2, 4, 6, 8}; no **11.** Domain: {0, 1, 2, 3}; range: {0, 1, 2, 3}; yes **13.** Domain: {−3, −2, −1, 1, 2, 3};
range: {−3, 2, 3, 4}; yes **15.** 40; (10, 40) **17.** −11; (−6, −11) **19.** 12; (−2, 12) **21.** $\dfrac{7}{2}; \left(\dfrac{1}{2}, \dfrac{7}{2}\right)$
23. 2; (−5, 2) **25.** 32; (−4, 32) **27.** Range: {−19, −13, −7, −1, 5}; the ordered pairs (−5, −19), (−3, −13),
(−1, −7), (1, −1), and (3, 5) belong to the function. **29.** Range: {1, 2, 3, 4, 5}; the ordered pairs (−4, 1), (−2, 2),
(0, 3), (2, 4), and (4, 5) belong to the function. **31.** Range: {6, 7, 15}; the ordered pairs (−3, 15), (−1, 7), (0, 6),
(1, 7), and (3, 15) belong to the function. **33.** **35.** 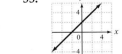 **37.**

39. **41.** **43.** **45.** **47.**

49.   **51.**   **53.**

55. a. $f(x) = 30,000 - 5000x$ **b.** 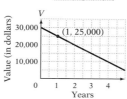   **c.** The value of the computer after 1 year is $25,000.

57. a. $f(m) = 0.18m + 50$ **b.** 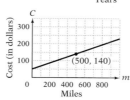   **c.** The cost to drive this car 500 mi is $140.

CONCEPT REVIEW 5.6   *page 249*

1. Never true   **3.** Always true   **5.** Always true

SECTION 5.6   *pages 249 – 250*

1.   **3.**   **5.**   **7.**   **9.**

11.   **13.**   **15.**   **17.** $y \geq 2x + 2$ **19.** $y > 2$

21.   **23.**

CHAPTER REVIEW EXERCISES   *pages 257 – 260*

1. $(3, 0)$ (Obj. 5.2.1)   **2.** $y = 3x - 1$ (Obj. 5.4.1/5.4.2)   **3.**   (Obj. 5.2.3)

4.   (Obj. 5.1.1)   **5.** 79 (Obj. 5.5.1)   **6.** 0 (Obj. 5.3.1)   **7.**   (Obj. 5.2.2)

8.   (Obj. 5.2.3)   **9.** $y = -\frac{2}{3}x + \frac{4}{3}$ (Obj. 5.4.1/5.4.2)   **10.** (2, 0), (0, −3) (Obj. 5.2.3)

11.   (Obj. 5.3.2)   **12.**   (Obj. 5.5.2)   **13.** $y = \frac{2}{3}x + 3$ (Obj. 5.4.1/5.4.2)

14. 2 (Obj. 5.3.1)   **15.** −4 (Obj. 5.5.1)   **16.** Domain: {−20, −10, 0, 10}; range: {−10, −5, 0, 5}; yes (Obj. 5.5.1)

17.   (Obj. 5.2.2)   **18.**   (Obj. 5.3.2)   **19.**   (Obj. 5.1.1)

20.   (Obj. 5.5.2)   **21.** (6, 0), (0, −4) (Obj. 5.2.3)   **22.** $y = 2x + 2$ (Obj. 5.4.1/5.4.2)

23.   (Obj. 5.2.3)   **24.**   (Obj. 5.6.1)   **25.** $-\frac{7}{5}$ (Obj. 5.3.1)   **26.** $y = \frac{1}{2}x + 2$

(Obj. 5.4.1/5.4.2)

27.   (Obj. 5.3.2)   **28.**   (Obj. 5.2.3)   **29.** (−2, −5) (Obj. 5.2.1)

30. $y = -3x + 2$ (Obj. 5.4.1/5.4.2)   **31.** 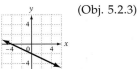 (Obj. 5.5.2)   **32.** 0 (Obj. 5.5.1)

33. Undefined (Obj. 5.3.1)   **34.** $y = \frac{1}{2}x - 2$ (Obj. 5.4.1/5.4.2)   **35.** Yes (Obj. 5.2.1)   **36.** (2, −1)

(Obj. 5.2.1)   **37.** (0, 0); (0, 0) (Obj. 5.2.3)   **38.** $y = 3$ (Obj. 5.4.1/5.4.2)   **39.**   (Obj. 5.6.1)

40. 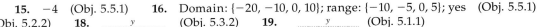 (Obj. 5.3.2)   **41.** $y = 3x - 4$ (Obj. 5.4.1/5.4.2)   **42.**   (Obj. 5.2.3)

43. Domain: {−10, −5, 5}; range: {−5, 0}; no (Obj. 5.5.1)   **44.** Range: {−53, −23, 7, 37, 67} (Obj. 5.5.1)

45. Range: {2, 3, 4, 5, 6}   **46.**   (Obj. 5.1.2)   **47.**   (Obj. 5.1.2)
(Obj. 5.5.1)

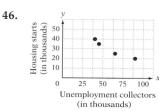

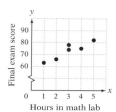

48. {(25, 4.8), (35, 3.5), (40, 2.1), (20, 5.5), (45, 1.0)}; yes   (Obj. 5.5.1)   **49. a.** $f(s) = 70s + 40,000$
b.   **c.** The contractor's cost to build a house of 1500 ft^2 is \$145,000.   (Obj. 5.5.2)

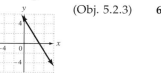

50. The slope is 58.5. The number of heart surgeries performed is increasing at a rate of 58.5 per year.   (Obj. 5.3.1)

CHAPTER TEST   *pages 261 – 263*

1. $y = -\dfrac{1}{3}x$ (Obj. 5.4.1/5.4.2)   **2.** $\dfrac{7}{11}$ (Obj. 5.3.1)   **3.** (8, 0); (0, −12)   (Obj. 5.2.3)   **4.** (9, −13)   (Obj. 5.2.1)

5.   (Obj. 5.2.3)   **6.**   (Obj. 5.2.2)   **7.** 10   (Obj. 5.5.1)   **8.** No   (Obj. 5.2.1)

    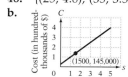

9.   (Obj. 5.3.2)   **10.**   (Obj. 5.6.1)   **11.**   (Obj. 5.1.1)

12.   (Obj. 5.3.2)   **13.** 0   (Obj. 5.3.1)   **14.** $y = -\dfrac{2}{5}x + 7$ (Obj. 5.4.1/5.4.2)

15. $y = 4x - 7$ (Obj. 5.4.1/5.4.2)   **16.** 13   (Obj. 5.5.1)   **17.**   (Obj. 5.2.2)

18.   (Obj. 5.6.1)   **19.**   (Obj. 5.2.3)   **20.**   (Obj. 5.3.2)

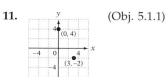

21.   (Obj. 5.5.2)   **22.**   (Obj. 5.5.2)   **23.** 25   (Obj. 5.5.1)

24. Domain: {8}; range: {0, 2, 4, 8}; no   (Obj. 5.5.1)   **25.** $y = \frac{3}{5}x - 3$ (Obj. 5.4.1/5.4.2)

26. {−22, −7, 5, 11, 26}   (Obj. 5.5.1)   **27.**   (Obj. 5.1.2)   **28. a.** $f(t) = 8t + 1000$

b.   **c.** The cost to manufacture 340 toasters is $3720.   (Obj. 5.5.2)

29. {(8.5, 64), (9.4, 68), (10.1, 76), (11.4, 87), (12.0, 92)}; yes   (Obj. 5.5.1)   **30.** The slope is 36. The cost of lumber has been increasing by $36 per month.   (Obj. 5.3.1)

CUMULATIVE REVIEW EXERCISES   *pages 263 – 264*

1. −12   (Obj. 1.4.2)   **2.** $-\frac{5}{8}$ (Obj. 2.1.1)   **3.** −17x + 28   (Obj. 2.2.5)   **4.** $\frac{3}{2}$ (Obj. 3.2.1)   **5.** 1

(Obj. 3.2.3)   **6.** $\frac{1}{15}$ (Obj. 1.3.2)   **7.** {1, 2, 3, 4, 5, 6, 7, 8}   (Obj. 1.1.1)   **8.** −4, 0, 5   (Obj. 1.1.1)

9. $a \geq -1$ (Obj. 3.3.3)   **10.** $y = \frac{4}{5}x - 3$ (Obj. 5.2.3)   **11.** (−2, −7)   (Obj. 5.2.1)   **12.** 0   (Obj. 5.3.1)

13. (4, 0); (0, 10)   (Obj. 5.2.3)   **14.** $y = -x + 5$ (Obj. 5.4.1/5.4.2)   **15.**   (Obj. 5.2.2)

16.   (Obj. 5.2.3)   **17.**   (Obj. 5.5.2)   **18.**   (Obj. 5.6.1)

19. Domain: {0, 1, 2, 3, 4}; range: {0, 1, 2, 3, 4}; yes   (Obj. 5.5.1)   **20.** −11   (Obj. 5.5.1)   **21.** {−7, −2, 3, 8, 13, 18}   (Obj. 5.5.1)   **22.** 7 and 17   (Obj. 4.1.2)   **23.** The fulcrum is 7 ft from the 80 lb force.   (Obj. 3.2.4)   **24.** The length of the first side is 22 ft.   (Obj. 4.3.1)   **25.** The sale price is $62.30.   (Obj. 4.4.2)

ANSWERS to Chapter 6 Odd-Numbered Exercises

CONCEPT REVIEW 6.1   *pages 269 – 270*

1. Always true   **3.** Sometimes true   **5.** Sometimes true

---

SECTION 6.1   *pages 270 – 272*

1. Yes   **3.** Yes   **5.** No   **7.** No   **9.** No   **11.** Yes   **13.** Yes   **15.** (4, 1)   **17.** (4, 1)   **19.** (4, 3)
21. (3, −2)   **23.** (2, −2)   **25.** The system of equations is inconsistent and has no solution.   **27.** The system of equations is dependent. The solutions are the ordered pairs that satisfy the equation $y = 2x - 2$.   **29.** (1, −4)
31. (0, 0)   **33.** The system of equations is inconsistent and has no solution.   **35.** (0, −2)   **37.** (1, −1)

39. $(2, 0)$ **41.** $(-3, -2)$ **43.** The system of equations is inconsistent and has no solution.
45. $y = 2$ **47.** $y = x$ **49. a.** Always true   **b.** Never true   **c.** Always true
$y = x + 4$ $y = -x + 2$

CONCEPT REVIEW 6.2   *page 275*

1. Always true   **3.** Never true   **5.** Never true

SECTION 6.2   *pages 276 – 277*

1. $(2, 1)$ **3.** $(4, 1)$ **5.** $(-1, 1)$ **7.** $(3, 1)$ **9.** $(1, 1)$ **11.** $(-1, 1)$ **13.** The system of equations is inconsistent and has no solution.   **15.** The system of equations is inconsistent and has no solution.
17. $\left(-\dfrac{3}{4}, -\dfrac{3}{4}\right)$ **19.** $\left(\dfrac{9}{5}, \dfrac{6}{5}\right)$ **21.** $(-7, -23)$ **23.** $(1, 1)$ **25.** $(2, 0)$ **27.** $(2, 1)$ **29.** $\left(\dfrac{9}{19}, -\dfrac{13}{19}\right)$
31. $(5, 7)$ **33.** $(1, 7)$ **35.** $\left(\dfrac{17}{5}, -\dfrac{7}{5}\right)$ **37.** $\left(-\dfrac{6}{11}, \dfrac{31}{11}\right)$ **39.** $(2, 3)$ **41.** $(0, 0)$ **43.** The system of
equations is dependent. The solutions are the ordered pairs that satisfy the equation $3x + y = 4$.   **45.** $\left(\dfrac{20}{17}, -\dfrac{15}{17}\right)$
47. $(5, 2)$ **49.** $(-17, -8)$ **51.** $\left(-\dfrac{5}{7}, \dfrac{13}{7}\right)$ **53.** $(3, -2)$ **55.** $\left(-\dfrac{11}{20}, \dfrac{23}{40}\right)$ **57.** $(3, -2)$ **59.** $(1, 7)$
61. 2 **63.** 2

CONCEPT REVIEW 6.3   *page 281*

1. Always true   **3.** Never true   **5.** Never true

SECTION 6.3   *pages 281 – 283*

1. $(5, -1)$ **3.** $(1, 3)$ **5.** $(1, 1)$ **7.** $(3, -2)$ **9.** The system of equations is dependent. The solutions are the ordered pairs that satisfy the equation $2x - y = 1$.   **11.** $(3, 1)$ **13.** The system of equations is inconsistent and has no solution.   **15.** $\left(-\dfrac{13}{17}, -\dfrac{24}{17}\right)$ **17.** $(2, 0)$ **19.** $(0, 0)$ **21.** $(5, -2)$ **23.** $\left(\dfrac{32}{19}, -\dfrac{9}{19}\right)$
25. $(3, 4)$ **27.** $(1, -1)$ **29.** The system of equations is dependent. The solutions are the ordered pairs that
satisfy the equation $5x + 15y = 20$.   **31.** $(3, 1)$ **33.** $(-1, 2)$ **35.** $(1, 1)$ **37.** $\left(\dfrac{1}{2}, -\dfrac{1}{2}\right)$ **39.** $\left(\dfrac{2}{3}, \dfrac{1}{9}\right)$
41. $\left(\dfrac{7}{25}, -\dfrac{1}{25}\right)$ **43.** $(5, 2)$ **45.** $(1, 4)$ **47.** $(1, 2)$ **49.** $A = 3, B = -1$ **51.** $A = 2$ **53. a.** $k \neq 1$
b. $k \neq \dfrac{3}{2}$ **c.** $k \neq 4$

CONCEPT REVIEW 6.4   *pages 287 – 288*

1. Always true   **3.** Never true   **5.** Never true

SECTION 6.4   *pages 288 – 291*

1. The rate of the plane in calm air was 400 mph. The rate of the wind was 50 mph.   **3.** The rate of the boat in calm water was 7 mph. The rate of the current was 3 mph.   **5.** The rate of the plane in calm air was 125 mph. The rate of the wind was 25 mph.   **7.** The rate of the plane in calm air was 100 mph. The rate of the wind was 20 mph.   **9.** The rate of the plane in calm air was 110 mph. The rate of the wind was 30 mph.   **11.** The rate of the plane in calm air was 180 km/h. The rate of the wind was 20 km/h.   **13.** The cost for one copy of the word processing program was $245.   **15.** The dividend per share of the oil company was $.25. The dividend per share of the movie company was $.45.   **17.** The team scored 21 two-point baskets and 15 three-point baskets.

19. There are 8 nickels and 10 quarters in the first bank.   **21.** There are 8 dimes and 10 quarters in the bank. **23.** The two angles measure 55° and 125°.   **25.** $6000 was invested in the 9% account, and $4000 was invested in the 8% account.   **27.** There are 0 dimes and 5 nickels, or 1 dime and 3 nickels, or 2 dimes and 1 nickel in the bank.

CHAPTER REVIEW EXERCISES   *pages 295 – 297*

1. $(-1, 1)$ (Obj. 6.2.1)   **2.** $(2, -3)$ (Obj. 6.1.1)   **3.** $(-3, 1)$ (Obj. 6.3.1)   **4.** $(1, 6)$ (Obj. 6.2.1)   **5.** $(1, 1)$ (Obj. 6.1.1)   **6.** $(1, -5)$ (Obj. 6.3.1)   **7.** $(3, 2)$ (Obj. 6.2.1)   **8.** The system of equations is dependent. The solutions are the ordered pairs that satisfy the equation $8x - y = 25$. (Obj. 6.3.1)   **9.** $(3, -1)$ (Obj. 6.3.1) **10.** $(1, -3)$ (Obj. 6.1.1)   **11.** $(-2, -7)$ (Obj. 6.3.1)   **12.** $(4, 0)$ (Obj. 6.2.1)   **13.** Yes (Obj. 6.1.1) **14.** $\left(-\dfrac{5}{6}, \dfrac{1}{2}\right)$ (Obj. 6.3.1)   **15.** $(-1, -2)$ (Obj. 6.3.1)   **16.** $(3, -3)$ (Obj. 6.1.1)   **17.** The system of equations is inconsistent and has no solution. (Obj. 6.3.1)   **18.** $\left(-\dfrac{1}{3}, \dfrac{1}{6}\right)$ (Obj. 6.2.1)   **19.** The system of equations is dependent. The solutions are the ordered pairs that satisfy the equation $y = 2x - 4$. (Obj. 6.1.1) **20.** The system of equations is dependent. The solutions are the ordered pairs that satisfy the equation $3x + y = -2$. (Obj. 6.3.1)   **21.** $(-2, -13)$ (Obj. 6.3.1)   **22.** The system of equations is dependent. The solutions are the ordered pairs that satisfy the equation $4x + 3y = 12$. (Obj. 6.2.1)   **23.** $(4, 2)$ (Obj. 6.1.1) **24.** The system of equations is inconsistent and has no solution. (Obj. 6.3.1)   **25.** $\left(\dfrac{1}{2}, -1\right)$ (Obj. 6.2.1) **26.** No (Obj. 6.1.1)   **27.** $\left(\dfrac{2}{3}, -\dfrac{1}{6}\right)$ (Obj. 6.3.1)   **28.** The system of equations is inconsistent and has no solution. (Obj. 6.2.1)   **29.** $\left(\dfrac{1}{3}, 6\right)$ (Obj. 6.2.1)   **30.** The system of equations is inconsistent and has no solution. (Obj. 6.1.1)   **31.** $(0, -1)$ (Obj. 6.3.1)   **32.** $(-1, -3)$ (Obj. 6.2.1)   **33.** $(0, -2)$ (Obj. 6.2.1) **34.** $(2, 0)$ (Obj. 6.3.1)   **35.** $(-4, 2)$ (Obj. 6.3.1)   **36.** $(1, 6)$ (Obj. 6.2.1)   **37.** The rate of the plane in calm air is 180 mph. The rate of the wind is 20 mph. (Obj. 6.4.1)   **38.** There were 60 adult tickets and 140 children's tickets sold. (Obj. 6.4.2)   **39.** The rate of the canoeist in calm water was 8 mph. The rate of the current was 2 mph. (Obj. 6.4.1)   **40.** There were 130 mailings requiring 32¢ in postage. (Obj. 6.4.2)   **41.** The rate of the boat in calm water was 14 km/h. The rate of the current was 2 km/h. (Obj. 6.4.1)   **42.** The customer purchased 4 compact discs at $15 each and 6 at $10 each. (Obj. 6.4.2)   **43.** The rate of the plane in calm air was 125 km/h. The rate of the wind was 15 km/h. (Obj. 6.4.1)   **44.** The rate of the paddle boat in calm water was 3 mph. The rate of the current was 1 mph. (Obj. 6.4.1)   **45.** There are 350 bushels of lentils and 200 bushels of corn in the silo. (Obj. 6.4.2)   **46.** The rate of the plane in calm air was 105 mph. The rate of the wind was 15 mph. (Obj. 6.4.1)   **47.** There are 4 dimes in the coin purse. (Obj. 6.4.2)   **48.** The rate of the plane in calm air was 325 mph. The rate of the wind was 25 mph. (Obj. 6.4.1)   **49.** The investor bought 1300 of the $6 shares and 200 of the $25 shares. (Obj. 6.4.2)   **50.** The rate of the sculling team in calm water was 9 mph. The rate of the current was 3 mph. (Obj. 6.4.1)

CHAPTER TEST   *page 298*

1. $(3, 1)$ (Obj. 6.2.1)   **2.** $(2, 1)$ (Obj. 6.3.1)   **3.** Yes (Obj. 6.1.1)   **4.** $(1, -1)$ (Obj. 6.2.1)   **5.** $\left(\dfrac{1}{2}, -1\right)$ (Obj. 6.3.1)   **6.** $(-2, 6)$ (Obj. 6.1.1)   **7.** The system of equations is inconsistent and has no solution. (Obj. 6.2.1)   **8.** $(2, -1)$ (Obj. 6.2.1)   **9.** $(2, -1)$ (Obj. 6.3.1)   **10.** $\left(\dfrac{22}{7}, -\dfrac{5}{7}\right)$ (Obj. 6.2.1)   **11.** $(1, -2)$ (Obj. 6.3.1)   **12.** No (Obj. 6.1.1)   **13.** $(2, 1)$ (Obj. 6.2.1)   **14.** $(2, -2)$ (Obj. 6.3.1)   **15.** $(2, 0)$ (Obj. 6.1.1) **16.** $(4, 1)$ (Obj. 6.2.1)   **17.** $(3, -2)$ (Obj. 6.3.1)   **18.** The system of equations is dependent. The solutions are the ordered pairs that satisfy the equation $3x + 6y = 2$. (Obj. 6.1.1)   **19.** $(1, 1)$ (Obj. 6.2.1)   **20.** $(4, -3)$ (Obj. 6.3.1)   **21.** $(-6, 1)$ (Obj. 6.2.1)   **22.** $(1, -4)$ (Obj. 6.3.1)   **23.** The rate of the plane in calm air is 100 mph. The rate of the wind is 20 mph. (Obj. 6.4.1)   **24.** The rate of the boat in calm water is 14 mph. The rate of the current is 2 mph. (Obj. 6.4.1)   **25.** There are 40 dimes and 30 nickels in the first bank. (Obj. 6.4.2)

CUMULATIVE REVIEW EXERCISES *pages 299 – 300*

1. $-8, -4$ (Obj. 1.1.1) **2.** {1, 2, 3, 4, 5, 6, 7, 8, 9, 10} (Obj. 1.1.1) **3.** 10 (Obj. 1.4.2) **4.** $40a - 28$ (Obj. 2.2.5) **5.** $\frac{3}{2}$ (Obj. 2.1.1) **6.** $-\frac{3}{2}$ (Obj. 3.1.3) **7.** $-\frac{7}{2}$ (Obj. 3.2.3) **8.** $-\frac{2}{9}$ (Obj. 3.2.3)
9. $x < -5$ (Obj. 3.3.3) **10.** $x \leq 3$ (Obj. 3.3.3) **11.** 24% (Obj. 3.1.4) **12.** $(4, 0), (0, -2)$ (Obj. 5.2.3)
13. $-\frac{7}{5}$ (Obj. 5.3.1) **14.** $y = -\frac{3}{2}x$ (Obj. 5.4.1/5.4.2) **15.** (Obj. 5.2.3)

16. (Obj. 5.2.2) **17.** (Obj. 5.6.1) **18.** (Obj. 5.5.2)

19. Domain: {−5, 0, 1, 5}; range: {5}; yes (Obj. 5.5.1) **20.** 3 (Obj. 5.5.1) **21.** Yes (Obj. 6.1.1)
22. $(-2, 1)$ (Obj. 6.2.1) **23.** $(0, 2)$ (Obj. 6.1.1) **24.** $(2, 1)$ (Obj. 6.3.1) **25.** {−22, −12, −2, 8, 18}
(Obj. 5.5.1) **26.** 200 cameras were produced during the month. (Obj. 3.2.4) **27.** $3750 should be invested
at 9.6%, and $5000 should be invested at 7.2%. (Obj. 4.5.1) **28.** The rate of the wind is 30 mph. (Obj. 6.4.1)
29. The rate of the boat in calm water is 10 mph. (Obj. 6.4.1) **30.** There are 40 dimes in the first bank.
(Obj. 6.4.2)

ANSWERS to Chapter 7 Odd-Numbered Exercises

CONCEPT REVIEW 7.1 *page 305*

1. Always true **3.** Never true **5.** Always true

SECTION 7.1 *pages 305 – 307*

1. $-2x^2 + 3x$ **3.** $y^2 - 8$ **5.** $5x^2 + 7x + 20$ **7.** $x^3 + 2x^2 - 6x - 6$ **9.** $2a^3 - 3a^2 - 11a + 2$
11. $5x^2 + 8x$ **13.** $7x^2 + xy - 4y^2$ **15.** $3a^2 - 3a + 17$ **17.** $5x^3 + 10x^2 - x - 4$ **19.** $3r^3 + 2r^2 - 11r + 7$
21. $-2x^3 + 3x^2 + 10x + 11$ **23.** $4x$ **25.** $3y^2 - 4y - 2$ **27.** $-7x - 7$ **29.** $4x^3 + 3x^2 + 3x + 1$
31. $y^3 - y^2 + 6y - 6$ **33.** $-y^2 - 13xy$ **35.** $2x^2 - 3x - 1$ **37.** $-2x^3 + x^2 + 2$ **39.** $3a^3 - 2$
41. $4y^3 - 2y^2 + 2y - 4$ **43.** Binomial **45.** Monomial **47.** No **49.** Yes **51.** Yes
53. $x^2 + x - 1$ **55.** $2x^3 + x^2 - 5x + 6$ **57.** $x^3 - x^2 + 2x + 6$ **59.** Yes; for example,
$(2x^3 + 3x - 4) + (-2x^3 + 5x^2 - 6) = 5x^2 + 3x - 10$

CONCEPT REVIEW 7.2 *page 309*

1. Never true **3.** Always true **5.** Never true

SECTION 7.2 *pages 309 – 311*

1. $2x^2$ **3.** $12x^2$ **5.** $6a^7$ **7.** x^3y^5 **9.** $-10x^9y$ **11.** x^7y^8 **13.** $-6x^3y^5$ **15.** x^4y^5z **17.** $a^3b^5c^4$
19. $-a^5b^8$ **21.** $-6a^5b$ **23.** $40y^{10}z^6$ **25.** $-20a^2b^3$ **27.** $x^3y^5z^3$ **29.** $-12a^{10}b^7$ **31.** 64 **33.** 4
35. -64 **37.** x^9 **39.** x^{14} **41.** x^4 **43.** $4x^2$ **45.** $-8x^6$ **47.** x^4y^6 **49.** $9x^4y^2$ **51.** $27a^8$
53. $-8x^7$ **55.** x^8y^4 **57.** a^4b^6 **59.** $16x^{10}y^3$ **61.** $-18x^3y^4$ **63.** $-8a^7b^5$ **65.** $-54a^9b^3$ **67.** $-72a^5b^5$

69. $32x^3$ **71.** $-3a^3b^3$ **73.** $13x^5y^5$ **75.** $27a^5b^3$ **77.** $-5x^7y^4$ **79.** a^{2n} **81.** a^{2n} **83.** True
85. False; $x^{2\cdot5}=x^{10}$ **87.** No; $2^{(3^2)}$ is larger. $(2^3)^2=8^2=64$, whereas $2^{(3^2)}=2^9=512$.   **89.** $x=y$ when n is odd.

CONCEPT REVIEW 7.3   *pages 315–316*

1. Always true   **3.** Sometimes true   **5.** Always true   **7.** Sometimes true

SECTION 7.3   *pages 316–320*

1. x^2-2x **3.** $-x^2-7x$ **5.** $3a^3-6a^2$ **7.** $-5x^4+5x^3$ **9.** $-3x^5+7x^3$ **11.** $12x^3-6x^2$
13. $6x^2-12x$ **15.** $-x^3y+xy^3$ **17.** $2x^4-3x^2+2x$ **19.** $2a^3+3a^2+2a$ **21.** $3x^6-3x^4-2x^2$
23. $-6y^4-12y^3+14y^2$ **25.** $-2a^3-6a^2+8a$ **27.** $6y^4-3y^3+6y^2$ **29.** $x^3y-3x^2y^2+xy^3$
31. x^3+4x^2+5x+2 **33.** $a^3-6a^2+13a-12$ **35.** $-2b^3+7b^2+19b-20$ **37.** $-6x^3+31x^2-41x+10$
39. $x^4-4x^3-3x^2+14x-8$ **41.** $15y^3-16y^2-70y+16$ **43.** $5a^4-20a^3-15a^2+62a-8$
45. $y^4+4y^3+y^2-5y+2$ **47.** x^2+4x+3 **49.** a^2+a-12 **51.** $y^2-5y-24$ **53.** $y^2-10y+21$
55. $2x^2+15x+7$ **57.** $3x^2+11x-4$ **59.** $4x^2-31x+21$ **61.** $3y^2-2y-16$ **63.** $9x^2+54x+77$
65. $21a^2-83a+80$ **67.** $15b^2+47b-78$ **69.** $2a^2+7ab+3b^2$ **71.** $6a^2+ab-2b^2$
73. $2x^2-3xy-2y^2$ **75.** $10x^2+29xy+21y^2$ **77.** $6a^2-25ab+14b^2$ **79.** $2a^2-11ab-63b^2$
81. $100a^2-100ab+21b^2$ **83.** $15x^2+56xy+48y^2$ **85.** $14x^2-97xy-60y^2$ **87.** $56x^2-61xy+15y^2$
89. y^2-25 **91.** $4x^2-9$ **93.** x^2+2x+1 **95.** $9a^2-30a+25$ **97.** $9x^2-49$ **99.** $4a^2+4ab+b^2$
101. $x^2-4xy+4y^2$ **103.** $16-9y^2$ **105.** $25x^2+20xy+4y^2$ **107.** The area is $(10x^2-35x)$ ft^2.
109. The area is $(18x^2+12x+2)$ in^2.   **111.** The area is $(4x^2+4x+1)$ km^2.   **113.** The area is $(4x^2+10x)$ m^2.
115. The area is $(\pi x^2+8\pi x+16\pi)$ cm^2.   **117.** The area is $(90x+2025)$ ft^2.   **119.** $4ab$
121. $9a^4-24a^3+28a^2-16a+4$ **123.** $24x^3-3x^2$ **125.** $x^{2n}+x^n$ **127.** $x^{2n}+2x^n+1$ **129.** x^2-1
131. x^4-1 **133.** x^6-1 **135.** 1024 **137.** $12x^2-x-20$ **139.** $7x^2-11x-8$

CONCEPT REVIEW 7.4   *page 327*

1. Never true   **3.** Always true   **5.** Never true   **7.** Never true

SECTION 7.4   *pages 327–331*

1. $\dfrac{1}{25}$ **3.** 64 **5.** $\dfrac{1}{27}$ **7.** 1 **9.** y^4 **11.** a^3 **13.** p^4 **15.** $2x^3$ **17.** $2k$ **19.** m^5n^2 **21.** $\dfrac{3r^2}{2}$
23. $-\dfrac{2a}{3}$ **25.** $\dfrac{1}{x^2}$ **27.** a^6 **29.** $\dfrac{4}{x^7}$ **31.** $5b^8$ **33.** $\dfrac{x^2}{3}$ **35.** 1 **37.** $\dfrac{1}{y^5}$ **39.** $\dfrac{1}{a^6}$ **41.** $\dfrac{1}{3x^3}$
43. $\dfrac{2}{3x^5}$ **45.** $\dfrac{y^4}{x^2}$ **47.** $\dfrac{2}{5m^3n^8}$ **49.** $\dfrac{1}{p^3q}$ **51.** $\dfrac{1}{2y^3}$ **53.** $\dfrac{7xz}{8y^3}$ **55.** $\dfrac{p^2}{2m^3}$ **57.** $-\dfrac{8x^3}{y^6}$ **59.** $\dfrac{9}{x^2y^4}$
61. $\dfrac{2}{x^4}$ **63.** $-\dfrac{5}{a^8}$ **65.** $-\dfrac{a^5}{8b^4}$ **67.** $\dfrac{10y^3}{x^4}$ **69.** $\dfrac{1}{2x^3}$ **71.** $\dfrac{3}{x^3}$ **73.** $\dfrac{1}{2x^2y^6}$ **75.** $-\dfrac{1}{6x^3}$ **77.** $2.37\cdot10^6$
79. $4.5\cdot10^{-4}$ **81.** $3.09\cdot10^5$ **83.** $6.01\cdot10^{-7}$ **85.** $710{,}000$ **87.** 0.000043 **89.** $671{,}000{,}000$
91. 0.00000713 **93.** $1.6\cdot10^{10}$ **95.** $5.98\cdot10^{24}$ **97.** $1.6\cdot10^{-19}$ **99.** $1\cdot10^{-12}$ **101.** $6.65\cdot10^{19}$
103. $3.22\cdot10^{-14}$ **105.** $3.6\cdot10^5$ **107.** $1.8\cdot10^{-18}$ **109.** $x+1$ **111.** $2a-5$ **113.** $3a+2$
115. $4b^2-3$ **117.** $x-2$ **119.** $-x+2$ **121.** x^2+3x-5 **123.** x^4-3x^2-1 **125.** $xy+2$
127. $-3y^3+5$ **129.** $3x-2+\dfrac{1}{x}$ **131.** $-3x+7-\dfrac{6}{x}$ **133.** $4a-5+6b$ **135.** $9x+6-3y$
137. $x+2$ **139.** $2x+1$ **141.** $2x-4$ **143.** $x+1+\dfrac{2}{x-1}$ **145.** $2x-1-\dfrac{2}{3x-2}$
147. $5x+12+\dfrac{12}{x-1}$ **149.** $a+3+\dfrac{4}{a+2}$ **151.** $y-6+\dfrac{26}{2y+3}$ **153.** $2x+5+\dfrac{8}{2x-1}$

155. $5a - 6 + \dfrac{4}{3a + 2}$ **157.** $3x - 5$ **159.** $x^2 + 2x + 3$ **161.** $x^2 - 3$ **163.** $\dfrac{3}{64}$

165. For 2^x: $\dfrac{1}{4}, \dfrac{1}{2}, 1, 2, 4$. For 2^{-x}: $4, 2, 1, \dfrac{1}{2}, \dfrac{1}{4}$.   **167.** 0.0625 **169.** xy^2 **171.** $\dfrac{x^6 z^6}{4y^5}$ **173.** 1 **175.** 0

177. True   **179.** True   **181.** $4y^2$ **183.** $x^3 - 4x^2 + 11x - 2$

CHAPTER REVIEW EXERCISES   *pages 336 – 338*

1. $21y^2 + 4y - 1$ (Obj. 7.1.1)   **2.** $-20x^3y^5$ (Obj. 7.2.1)   **3.** $-8x^3 - 14x^2 + 18x$ (Obj. 7.3.1)
4. $10a^2 + 31a - 63$ (Obj. 7.3.3)   **5.** $6x^2 - 7x + 10$ (Obj. 7.4.3)   **6.** $2x^2 + 3x - 8$ (Obj. 7.1.2)
7. -729 (Obj. 7.2.2)   **8.** $x^3 - 6x^2 + 7x - 2$ (Obj. 7.3.2)   **9.** $a^2 - 49$ (Obj. 7.3.4)   **10.** 1296 (Obj. 7.4.1)
11. $x - 6$ (Obj. 7.4.4)   **12.** $2x^3 + 9x^2 - 3x - 12$ (Obj. 7.1.1)   **13.** $18a^8b^6$ (Obj. 7.2.1)
14. $3x^4y - 2x^3y + 12x^2y$ (Obj. 7.3.1)   **15.** $8b^2 - 2b - 15$ (Obj. 7.3.3)   **16.** $-4y + 8$ (Obj. 7.4.3)
17. $13y^3 - 12y^2 - 5y - 1$ (Obj. 7.1.2)   **18.** 64 (Obj. 7.2.2)   **19.** $6y^3 + 17y^2 - 2y - 21$ (Obj. 7.3.2)
20. $4b^2 - 81$ (Obj. 7.3.4)   **21.** $\dfrac{b^6 c^2}{a^4}$ (Obj. 7.4.1)   **22.** $2y - 9$ (Obj. 7.4.4)   **23.** $3.97 \cdot 10^{-6}$ (Obj. 7.4.2)
24. $x^4 y^8 z^4$ (Obj. 7.2.1)   **25.** $-12y^5 + 4y^4 - 18y^3$ (Obj. 7.3.1)   **26.** $18x^2 - 48x + 24$ (Obj. 7.3.3)
27. 0.0000623 (Obj. 7.4.2)   **28.** $-7a^2 - a + 4$ (Obj. 7.1.2)   **29.** $9x^4 y^6$ (Obj. 7.2.2)
30. $12a^3 - 8a^2 - 9a + 6$ (Obj. 7.3.3)   **31.** $25y^2 - 70y + 49$ (Obj. 7.3.4)   **32.** $\dfrac{x^4 y^6}{9}$ (Obj. 7.4.1)
33. $x + 5 + \dfrac{4}{x + 12}$ (Obj. 7.4.4)   **34.** $240,000$ (Obj. 7.4.2)   **35.** $a^6 b^{11} c^9$ (Obj. 7.2.1)
36. $8a^3b^3 - 4a^2b^4 + 6ab^5$ (Obj. 7.3.1)   **37.** $6x^2 - 7xy - 20y^2$ (Obj. 7.3.3)   **38.** $4b^4 + 12b^2 - 1$ (Obj. 7.4.3)
39. $-4b^2 - 13b + 28$ (Obj. 7.1.2)   **40.** $100a^{15}b^{13}$ (Obj. 7.2.2)   **41.** $12b^5 - 4b^4 - 6b^3 - 8b^2 + 5$ (Obj. 7.3.2)
42. $36 - 25x^2$ (Obj. 7.3.4)   **43.** $\dfrac{2y^3}{x^3}$ (Obj. 7.4.1)   **44.** $a^2 - 2a + 6$ (Obj. 7.4.4)   **45.** $4b^3 - 5b^2 - 9b + 7$
(Obj. 7.1.1)   **46.** $-54a^{13}b^5c^7$ (Obj. 7.2.1)   **47.** $-18x^4 - 27x^3 + 63x^2$ (Obj. 7.3.1)   **48.** $30y^2 - 109y + 30$
(Obj. 7.3.3)   **49.** $9.176 \cdot 10^{12}$ (Obj. 7.4.2)   **50.** $-2y^2 + y - 5$ (Obj. 7.1.2)   **51.** $144x^{14}y^{18}z^{16}$ (Obj. 7.2.2)
52. $-12x^4 - 17x^3 - 2x^2 - 33x - 27$ (Obj. 7.3.2)   **53.** $64a^2 + 16a + 1$ (Obj. 7.3.4)   **54.** $\dfrac{2}{ab^6}$ (Obj. 7.4.1)
55. $b^2 + 5b + 2 + \dfrac{7}{b - 7}$ (Obj. 7.4.4)   **56.** The area is $(20x^2 - 35x)$ m². (Obj. 7.3.5)   **57.** The area is
$(25x^2 + 40x + 16)$ in². (Obj. 7.3.5)   **58.** The area is $(9x^2 - 4)$ ft². (Obj. 7.3.5)   **59.** The area is
$(\pi x^2 - 12\pi x + 36\pi)$ cm². (Obj. 7.3.5)   **60.** The area is $(15x^2 - 28x - 32)$ mi². (Obj. 7.3.5)

CHAPTER TEST   *pages 338 – 339*

1. $3x^3 + 6x^2 - 8x + 3$ (Obj. 7.1.1)   **2.** $-4x^4 + 8x^3 - 3x^2 - 14x + 21$ (Obj. 7.3.2)   **3.** $4x^3 - 6x^2$ (Obj. 7.3.1)
4. $-8a^6b^3$ (Obj. 7.2.2)   **5.** $-4x^6$ (Obj. 7.4.1)   **6.** $\dfrac{6b}{a}$ (Obj. 7.4.1)   **7.** $-5a^3 + 3a^2 - 4a + 3$ (Obj. 7.1.2)
8. $a^2 + 3ab - 10b^2$ (Obj. 7.3.3)   **9.** $4x^4 - 2x^2 + 5$ (Obj. 7.4.3)   **10.** $2x + 3 + \dfrac{2}{2x - 3}$ (Obj. 7.4.4)
11. $-6x^3y^6$ (Obj. 7.2.1)   **12.** $6y^4 - 9y^3 + 18y^2$ (Obj. 7.3.1)   **13.** $\dfrac{9}{x^3}$ (Obj. 7.4.1)   **14.** $4x^2 - 20x + 25$
(Obj. 7.3.4)   **15.** $10x^2 - 43xy + 28y^2$ (Obj. 7.3.3)   **16.** $x^3 - 7x^2 + 17x - 15$ (Obj. 7.3.2)   **17.** $\dfrac{a^4}{b^6}$ (Obj. 7.4.1)
18. $2.9 \cdot 10^{-5}$ (Obj. 7.4.2)   **19.** $16y^2 - 9$ (Obj. 7.3.4)   **20.** $3y^3 + 2y^2 - 10y$ (Obj. 7.1.2)   **21.** $9a^6b^4$
(Obj. 7.2.2)   **22.** $10a^3 - 39a^2 + 20a - 21$ (Obj. 7.3.2)   **23.** $9b^2 + 12b + 4$ (Obj. 7.3.4)   **24.** $-\dfrac{1}{4a^2b^5}$
(Obj. 7.4.1)   **25.** $4x + 8 + \dfrac{21}{2x - 3}$ (Obj. 7.4.4)   **26.** a^3b^7 (Obj. 7.2.1)   **27.** $a^2 + ab - 12b^2$ (Obj. 7.3.3)
28. 0.000000035 (Obj. 7.4.2)   **29.** The area is $(4x^2 + 12x + 9)$ m². (Obj. 7.3.5)   **30.** The area is
$(\pi x^2 - 10\pi x + 25\pi)$ in². (Obj. 7.3.5)

CUMULATIVE REVIEW EXERCISES   *pages 339 – 340*

1. $\dfrac{1}{144}$ (Obj. 1.3.3)   **2.** $\dfrac{25}{9}$ (Obj. 1.4.1)   **3.** $\dfrac{4}{5}$ (Obj. 1.4.2)   **4.** 87 (Obj. 1.1.2)   **5.** 0.775 (Obj. 1.3.1)

6. $-\dfrac{27}{4}$ (Obj. 2.1.1)   **7.** $-x - 4xy$ (Obj. 2.2.2)   **8.** $-12x$ (Obj. 2.2.3)   **9.** $-22x + 20$ (Obj. 2.2.5)

10. 8 (Obj. 2.2.1)   **11.** -18 (Obj. 3.1.3)   **12.** 16 (Obj. 3.2.2)   **13.** 12 (Obj. 3.2.3)   **14.** 80 (Obj. 3.2.2)

15. 24% (Obj. 3.1.4)   **16.** $x \geq -3$ (Obj. 3.3.3)   **17.** $-\dfrac{9}{5}$ (Obj. 5.3.1)   **18.** $y = -\dfrac{3}{2}x - \dfrac{3}{2}$

(Obj. 5.4.1/5.4.2)   **19.**   (Obj. 5.2.3)   **20.**   (Obj. 5.6.1)

21. Domain: $\{-8, -6, -4, -2\}$; range: $\{-7, -5, -2, 0\}$; yes (Obj. 5.5.1)   **22.** -2 (Obj. 5.5.1)
23. $(4, 1)$ (Obj. 6.2.1)   **24.** $(1, -4)$ (Obj. 6.3.1)   **25.** $5b^3 - 7b^2 + 8b - 10$ (Obj. 7.1.2)
26. $15x^3 - 26x^2 + 11x - 4$ (Obj. 7.3.2)   **27.** $20b^2 - 47b + 24$ (Obj. 7.3.3)   **28.** $25b^2 + 30b + 9$ (Obj. 7.3.4)
29. $-\dfrac{b^4}{4a}$ (Obj. 7.4.1)   **30.** $5y - 4 + \dfrac{1}{y}$ (Obj. 7.4.3)   **31.** $a - 7$ (Obj. 7.4.4)   **32.** $\dfrac{9y^2}{x^6}$ (Obj. 7.4.1)
33. $\{1, 9, 17, 21, 25\}$ (Obj. 5.5.1)   **34.** $5(n - 12)$; $5n - 60$ (Obj. 2.3.3)   **35.** $8n - 2n = 18$; $n = 3$ (Obj. 4.1.1)
36. The length is 15 m. The width is 6 m. (Obj. 4.3.1)   **37.** The selling price is \$43.20. (Obj. 4.4.1)
38. The concentration of orange juice is 28%. (Obj. 4.6.2)   **39.** The car overtakes the cyclist 25 mi from the
starting point. (Obj. 4.7.1)   **40.** The area is $(9x^2 + 12x + 4)$ ft^2. (Obj. 7.3.5)

ANSWERS to Chapter 8 Odd-Numbered Exercises

CONCEPT REVIEW 8.1   *pages 345 – 346*

1. Always true   **3.** Always true   **5.** Never true   **7.** Never true

SECTION 8.1   *pages 346 – 348*

1. x^3 **3.** xy^4 **5.** xy^4z^2 **7.** $7a^3$ **9.** 1   **11.** $3a^2b^2$ **13.** ab **15.** $2x$ **17.** $3x$
19. $5(a + 1)$ **21.** $8(2 - a^2)$ **23.** $4(2x + 3)$ **25.** $6(5a - 1)$ **27.** $x(7x - 3)$ **29.** $a^2(3 + 5a^3)$
31. $y(14y + 11)$ **33.** $2x(x^3 - 2)$ **35.** $2x^2(5x^2 - 6)$ **37.** $xy(x - y^2)$ **39.** $2a^5b + 3xy^3$ **41.** $6b^2(a^2b - 2)$
43. $2abc(3a + 2b)$ **45.** $9(2x^4y^2 - a^2b^2)$ **47.** $6x^3y^3(1 - 2x^3y^3)$ **49.** $x(x^2 - 3x - 1)$ **51.** $2(x^2 + 4x - 6)$
53. $b(b^2 - 5b - 7)$ **55.** $4(2y^2 - 3y + 8)$ **57.** $5y(y^2 - 4y + 2)$ **59.** $3y^2(y^2 - 3y - 2)$
61. $3y(y^2 - 3y + 8)$ **63.** $a^2(6a^3 - 3a - 2)$ **65.** $x^2(8y^2 - 4y + 1)$ **67.** $x^3y^3(4x^2y^2 - 8xy + 1)$
69. $(a + b)(x + 2)$ **71.** $(b + 2)(x - y)$ **73.** $(x - 2)(a - 5)$ **75.** $(y - 3)(b - 3)$ **77.** $(x - y)(a + 2)$
79. $(x + 4)(x^2 + 3)$ **81.** $(y + 2)(2y^2 + 3)$ **83.** $(a + 3)(b - 2)$ **85.** $(a - 2)(x^2 - 3)$ **87.** $(a - b)(3x - 2y)$
89. $(x - 3)(x + 4a)$ **91.** $(x - 5)(y - 2)$ **93.** $(7x + 2y)(3x - 7)$ **95.** $(2r + a)(a - 1)$ **97.** $(4x + 3y)(x - 4)$
99. $(2y - 3)(5xy + 3)$ **101. a.** $(x + 3)(2x + 5)$ **b.** $(2x + 5)(x + 3)$ **103. a.** $(a - b)(2a - 3b)$
b. $(2a - 3b)(a - b)$ **105.** 28   **107.** P doubles

CONCEPT REVIEW 8.2   *pages 351 – 352*

1. Never true   **3.** Always true   **5.** Never true

SECTION 8.2 *pages 352 – 354*

1. $(x + 1)(x + 2)$ **3.** $(x + 1)(x - 2)$ **5.** $(a + 4)(a - 3)$ **7.** $(a - 1)(a - 2)$ **9.** $(a + 2)(a - 1)$
11. $(b - 3)(b - 3)$ **13.** $(b + 8)(b - 1)$ **15.** $(y + 11)(y - 5)$ **17.** $(y - 2)(y - 3)$ **19.** $(z - 5)(z - 9)$
21. $(z + 8)(z - 20)$ **23.** $(p + 3)(p + 9)$ **25.** $(x + 10)(x + 10)$ **27.** $(b + 4)(b + 5)$ **29.** $(x + 3)(x - 14)$
31. $(b + 4)(b - 5)$ **33.** $(y + 3)(y - 17)$ **35.** $(p + 3)(p - 7)$ **37.** Nonfactorable over the integers
39. $(x - 5)(x - 15)$ **41.** $(x - 7)(x - 8)$ **43.** $(x + 8)(x - 7)$ **45.** $(a + 3)(a - 24)$ **47.** $(a - 3)(a - 12)$
49. $(z + 8)(z - 17)$ **51.** $(c + 9)(c - 10)$ **53.** $2(x + 1)(x + 2)$ **55.** $3(a + 3)(a - 2)$ **57.** $a(b + 5)(b - 3)$
59. $x(y - 2)(y - 3)$ **61.** $z(z - 3)(z - 4)$ **63.** $3y(y - 2)(y - 3)$ **65.** $3(x + 4)(x - 3)$ **67.** $5(z + 4)(z - 7)$
69. $2a(a + 8)(a - 4)$ **71.** $(x - 2y)(x - 3y)$ **73.** $(a - 4b)(a - 5b)$ **75.** $(x + 4y)(x - 7y)$
77. Nonfactorable over the integers **79.** $z^2(z - 5)(z - 7)$ **81.** $b^2(b - 10)(b - 12)$ **83.** $2y^2(y + 3)(y - 16)$
85. $x^2(x + 8)(x - 1)$ **87.** $4y(x + 7)(x - 2)$ **89.** $8(y - 1)(y - 3)$ **91.** $c(c + 3)(c + 10)$
93. $3x(x - 3)(x - 9)$ **95.** $(x - 3y)(x - 5y)$ **97.** $(a - 6b)(a - 7b)$ **99.** $(y + z)(y + 7z)$
101. $3y(x + 21)(x - 1)$ **103.** $3x(x + 4)(x - 3)$ **105.** $4z(z + 11)(z - 3)$ **107.** $(c + 4)(c + 5)$
109. $a^2(b - 5)(b - 9)$ **111.** $36, 12, -12, -36$ **113.** $22, 10, -10, -22$ **115.** $3, 4$ **117.** $5, 8, 9$ **119.** 2

CONCEPT REVIEW 8.3 *page 360*

1. Never true **3.** Sometimes true **5.** Always true

SECTION 8.3 *pages 360 – 363*

1. $(x + 1)(2x + 1)$ **3.** $(y + 3)(2y + 1)$ **5.** $(a - 1)(2a - 1)$ **7.** $(b - 5)(2b - 1)$ **9.** $(x + 1)(2x - 1)$
11. $(x - 3)(2x + 1)$ **13.** Nonfactorable over the integers **15.** $(2t - 1)(3t - 4)$ **17.** $(x + 4)(8x + 1)$
19. $(b - 4)(3b - 4)$ **21.** $(z - 14)(2z + 1)$ **23.** $(p + 8)(3p - 2)$ **25.** $(2x - 3)(3x - 4)$
27. $(b + 7)(5b - 2)$ **29.** $(2a - 3)(3a + 8)$ **31.** $(3t + 1)(6t - 5)$ **33.** $(3a + 7)(5a - 3)$
35. $(2y - 5)(4y - 3)$ **37.** $(2z + 3)(4z - 5)$ **39.** $(x + 5)(3x - 1)$ **41.** $(3x + 4)(4x + 3)$
43. Nonfactorable over the integers **45.** $(x + y)(3x - 2y)$ **47.** $(4 + z)(7 - z)$ **49.** $(1 - x)(8 + x)$
51. $3(x + 5)(3x - 4)$ **53.** $4(2x - 3)(3x - 2)$ **55.** $a^2(5a + 2)(7a - 1)$ **57.** $5(b - 7)(3b - 2)$
59. $2x(x + 1)(5x + 1)$ **61.** $2y(y - 4)(5y - 2)$ **63.** $yz(z + 2)(4z - 3)$ **65.** $b^2(4b + 5)(5b + 4)$
67. $xy(3x + 2)(3x + 2)$ **69.** $(t + 2)(2t - 5)$ **71.** $(p - 5)(3p - 1)$ **73.** $(3y - 1)(4y - 1)$
75. Nonfactorable over the integers **77.** $(3y + 1)(4y + 5)$ **79.** $(a + 7)(7a - 2)$ **81.** $(z + 2)(4z + 3)$
83. $(2p + 5)(11p - 2)$ **85.** $(y + 1)(8y + 9)$ **87.** $(2b - 3)(3b - 2)$ **89.** $(3b + 5)(11b - 7)$
91. $(3y - 4)(6y - 5)$ **93.** Nonfactorable over the integers **95.** $(2z - 5)(5z - 2)$ **97.** $(6z + 5)(6z + 7)$
99. $(2y - 3)(7y - 4)$ **101.** $(x + 6)(6x - 1)$ **103.** $3(x + 3)(4x - 1)$ **105.** $10(y + 1)(3y - 2)$
107. $x(x + 1)(2x - 5)$ **109.** $(a - 3b)(2a - 3b)$ **111.** $(y + z)(2y + 5z)$ **113.** $(1 + x)(18 - x)$
115. $4(9y + 1)(10y - 1)$ **117.** $8(t + 4)(2t - 3)$ **119.** $p(2p + 1)(3p + 1)$ **121.** $3(2z - 5)(5z - 2)$
123. $3a(2a + 3)(7a - 3)$ **125.** $y(3x - 5y)(3x - 5y)$ **127.** $xy(3x - 4y)(3x - 4y)$ **129.** $2y(y - 3)(4y - 1)$
131. $ab(a + 4)(a - 6)$ **133.** $5t(3 + 2t)(4 - t)$ **135.** $(y + 3)(2y + 1)$ **137.** $(2x - 1)(5x + 7)$
139. $7, 5, -7, -5$ **141.** $7, -7, 5, -5$ **143.** $11, -11, 7, -7$ **145.** $(x + 2)(x - 3)(x - 1)$

CONCEPT REVIEW 8.4 *pages 366 – 367*

1. Never true **3.** Sometimes true **5.** Always true **7.** Never true

SECTION 8.4 *pages 367 – 369*

1. $(x + 2)(x - 2)$ **3.** $(a + 9)(a - 9)$ **5.** $(2x + 1)(2x - 1)$ **7.** $(y + 1)^2$ **9.** $(a - 1)^2$ **11.** Nonfactorable
over the integers **13.** $(x^3 + 3)(x^3 - 3)$ **15.** $(5x + 1)(5x - 1)$ **17.** $(1 + 7x)(1 - 7x)$ **19.** $(x + y)^2$
21. $(2a + 1)^2$ **23.** $(8a - 1)^2$ **25.** Nonfactorable over the integers **27.** $(x^2 + y)(x^2 - y)$
29. $(3x + 4y)(3x - 4y)$ **31.** $(4b + 1)^2$ **33.** $(2b + 7)^2$ **35.** $(5a + 3b)^2$ **37.** $(xy + 2)(xy - 2)$
39. $(4 + xy)(4 - xy)$ **41.** $(2y - 9z)^2$ **43.** $(3ab - 1)^2$ **45.** $(m^2 + 16)(m + 4)(m - 4)$ **47.** $(x + 1)(9x + 4)$

49. $(y^4 + 9)(y^2 + 3)(y^2 - 3)$ **51.** $2(x + 3)(x - 3)$ **53.** $x^2(x + 7)(x - 5)$ **55.** $5(b + 3)(b + 12)$
57. Nonfactorable over the integers **59.** $2y(x + 11)(x - 3)$ **61.** $x(x^2 - 6x - 5)$ **63.** $3(y^2 - 12)$
65. $(2a + 1)(10a + 1)$ **67.** $y^2(x + 1)(x - 8)$ **69.** $5(a + b)(2a - 3b)$ **71.** $2(5 + x)(5 - x)$
73. $ab(3a - b)(4a + b)$ **75.** $2(x - 1)(a + b)$ **77.** $3a(2a - 1)^2$ **79.** $3(81 + a^2)$ **81.** $2a(2a - 5)(3a - 4)$
83. $(x - 2)(x + 1)(x - 1)$ **85.** $a(2a + 5)^2$ **87.** $3b(3a - 1)^2$ **89.** $6(4 + x)(2 - x)$
91. $(x + 2)(x - 2)(a + b)$ **93.** $x^2(x + y)(x - y)$ **95.** $2a(3a + 2)^2$ **97.** $b(2 - 3a)(1 + 2a)$
99. $(x - 5)(2 + x)(2 - x)$ **101.** $8x(3y + 1)^2$ **103.** $y^2(5 + x)(3 - x)$ **105.** $y(y + 3)(y - 3)$
107. $2x^2y^2(x + 1)(x - 1)$ **109.** $x^5(x^2 + 1)(x + 1)(x - 1)$ **111.** $2xy(3x - 2)(4x + 5)$ **113.** $x^2y^2(2x - 5)^2$
115. $(x - 2)^2(x + 2)$ **117.** $4xy^2(1 - x)(2 - 3x)$ **119.** $2ab(2a - 3b)(9a - 2b)$ **121.** $x^2y^2(1 - 3x)(5 + 4x)$
123. $(4x - 3 + y)(4x - 3 - y)$ **125.** $(x - 2 + y)(x - 2 - y)$ **127.** $12, -12$ **129.** $12, -12$ **131.** 9
133. 1 **135.** 4 **137.** 3 **139.** $(x + 2)(x^2 - 2x + 4)$ **141.** $(y - 3)(y^2 + 3y + 9)$
143. $(y + 4)(y^2 - 4y + 16)$ **145.** $(2x - 1)(4x^2 + 2x + 1)$

CONCEPT REVIEW 8.5   *pages 372 – 373*

1. Always true **3.** Always true **5.** Never true

SECTION 8.5   *pages 373 – 376*

1. $-3, -2$ **3.** $7, 3$ **5.** $0, 5$ **7.** $0, 9$ **9.** $0, -\dfrac{3}{2}$ **11.** $0, \dfrac{2}{3}$ **13.** $-2, 5$ **15.** $9, -9$ **17.** $\dfrac{7}{4}, -\dfrac{7}{4}$

19. $3, 5$ **21.** $8, -9$ **23.** $-2, 5$ **25.** $-\dfrac{1}{2}, 1$ **27.** $-\dfrac{2}{3}, -4$ **29.** $0, 7$ **31.** $-1, -4$ **33.** $2, 3$

35. $\dfrac{1}{2}, -4$ **37.** $\dfrac{1}{3}, 4$ **39.** $3, 9$ **41.** $9, -2$ **43.** $-1, -2$ **45.** $5, -9$ **47.** $4, -7$ **49.** $-2, -3$

51. $-8, -5$ **53.** $1, 3$ **55.** $-12, 5$ **57.** $3, 4$ **59.** $-\dfrac{1}{2}, -4$ **61.** The number is 7. **63.** The two
integers are 5 and 6. **65.** The integers are 15 and 16. **67.** The base is 18 ft. The height is 6 ft. **69.** The
length is 40 m. The width is 10 m. **71.** The length of a side of the original square was 6 cm. **73.** The
increase in area is 138.2 in². **75.** The dimensions are 4 in. by 7 in. **77.** The object will hit the ground 10 s
later. **79.** 12 consecutive natural numbers beginning with 1 will give a sum of 78. **81.** There are 8 teams in
the league. **83.** The ball will be 64 ft above the ground 2 s after it hits home plate. **85.** $-\dfrac{3}{2}, -5$

87. $9, -3$ **89.** $0, 9$ **91.** $-6, 3$ **93.** 48 or 3 **95.** The length is 10 cm. The width is 7 cm.

CHAPTER REVIEW EXERCISES   *pages 380 – 382*

1. $7y^3(2y^6 - 7y^3 + 1)$ (Obj. 8.1.1) **2.** $(a - 4)(3a + b)$ (Obj. 8.1.2) **3.** $(c + 2)(c + 6)$ (Obj. 8.2.1)
4. $a(a - 2)(a - 3)$ (Obj. 8.2.2) **5.** $(2x - 7)(3x - 4)$ (Obj. 8.3.1/8.3.2) **6.** $(y + 6)(3y - 2)$ (Obj. 8.3.1/8.3.2)
7. $(3a + 2)(6a - 5)$ (Obj. 8.3.1/8.3.2) **8.** $(ab + 1)(ab - 1)$ (Obj. 8.4.1) **9.** $4(y - 2)^2$ (Obj. 8.4.2)
10. $0, -\dfrac{1}{5}$ (Obj. 8.5.1) **11.** $3ab(4a + b)$ (Obj. 8.1.1) **12.** $(b - 3)(b - 10)$ (Obj. 8.2.1)
13. $(2x + 5)(5x + 2y)$ (Obj. 8.1.2) **14.** $3(a + 2)(a - 7)$ (Obj. 8.2.2) **15.** $n^2(n + 1)(n - 3)$ (Obj. 8.2.2)
16. Nonfactorable over the integers (Obj. 8.3.1/8.3.2) **17.** $(2x - 1)(3x - 2)$ (Obj. 8.3.1/8.3.2)
18. Nonfactorable over the integers (Obj. 8.4.1) **19.** $2, \dfrac{3}{2}$ (Obj. 8.5.1) **20.** $7(x + 1)(x - 1)$ (Obj. 8.4.2)
21. $x^3(3x^2 - 9x - 4)$ (Obj. 8.1.1) **22.** $(x - 3)(4x + 5)$ (Obj. 8.1.2) **23.** $(a + 7)(a - 2)$ (Obj. 8.2.1)
24. $(y + 9)(y - 4)$ (Obj. 8.2.1) **25.** $5(x - 12)(x + 2)$ (Obj. 8.2.2) **26.** $-3, 7$ (Obj. 8.5.1)
27. $(a + 2)(7a + 3)$ (Obj. 8.3.1/8.3.2) **28.** $(x + 20)(4x + 3)$ (Obj. 8.3.1/8.3.2) **29.** $(3y^2 + 5z)(3y^2 - 5z)$
(Obj. 8.4.1) **30.** $5(x + 2)(x - 3)$ (Obj. 8.2.2) **31.** $-\dfrac{3}{2}, \dfrac{2}{3}$ (Obj. 8.5.1) **32.** $2b(2b - 7)(3b - 4)$
(Obj. 8.3.1/8.3.2) **33.** $5x(x^2 + 2x + 7)$ (Obj. 8.1.1) **34.** $(x - 2)(x - 21)$ (Obj. 8.2.1) **35.** $(3a + 2)(a - 7)$
(Obj. 8.1.2) **36.** $(2x - 5)(4x - 9)$ (Obj. 8.3.1/8.3.2) **37.** $10x(a - 4)(a - 9)$ (Obj. 8.2.2)

38. $(a - 12)(2a + 5)$ (Obj. 8.3.1/8.3.2) **39.** $(3a - 5b)(7x + 2y)$ (Obj. 8.1.2) **40.** $(a^3 + 10)(a^3 - 10)$ (Obj. 8.4.1) **41.** $(4a + 1)^2$ (Obj. 8.4.1) **42.** $\frac{1}{4}, -7$ (Obj. 8.5.1) **43.** $10(a + 4)(2a - 7)$ (Obj. 8.3.1/8.3.2) **44.** $6(x - 3)$ (Obj. 8.1.1) **45.** $x^2y(3x^2 + 2x + 6)$ (Obj. 8.1.1) **46.** $(d - 5)(d + 8)$ (Obj. 8.2.1) **47.** $2(2x - y)(6x - 5)$ (Obj. 8.1.2) **48.** $4x(x + 1)(x - 6)$ (Obj. 8.2.2) **49.** $-2, 10$ (Obj. 8.5.1) **50.** $(x - 5)(3x - 2)$ (Obj. 8.3.1/8.3.2) **51.** $(2x - 11)(8x - 3)$ (Obj. 8.3.1/8.3.2) **52.** $(3x - 5)^2$ (Obj. 8.4.1) **53.** $(2y + 3)(6y - 1)$ (Obj. 8.3.1/8.3.2) **54.** $3(x + 6)^2$ (Obj. 8.4.2) **55.** The length is 100 yd. The width is 50 yd. (Obj. 8.5.2) **56.** The length is 100 yd. The width is 60 yd. (Obj. 8.5.2) **57.** The two integers are 4 and 5. (Obj. 8.5.2) **58.** The distance is 20 ft. (Obj. 8.5.2) **59.** The width of the resulting area is 15 ft. (Obj. 8.5.2) **60.** The length of a side of the original square was 20 ft. (Obj. 8.5.2)

CHAPTER TEST *page 382*

1. $3y^2(2x^2 + 3x + 4)$ (Obj. 8.1.1) **2.** $2x(3x^2 - 4x + 5)$ (Obj. 8.1.1) **3.** $(p + 2)(p + 3)$ (Obj. 8.2.1) **4.** $(x - 2)(a - b)$ (Obj. 8.1.2) **5.** $\frac{3}{2}, -7$ (Obj. 8.5.1) **6.** $(a - 3)(a - 16)$ (Obj. 8.2.1) **7.** $x(x + 5)(x - 3)$ (Obj. 8.2.2) **8.** $4(x + 4)(2x - 3)$ (Obj. 8.3.1/8.3.2) **9.** $(b + 6)(a - 3)$ (Obj. 8.1.2) **10.** $\frac{1}{2}, -\frac{1}{2}$ (Obj. 8.5.1) **11.** $(2x + 1)(3x + 8)$ (Obj. 8.3.1/8.3.2) **12.** $(x + 3)(x - 12)$ (Obj. 8.2.1) **13.** $2(b + 4)(b - 4)$ (Obj. 8.4.2) **14.** $(2a - 3b)^2$ (Obj. 8.4.1) **15.** $(p + 1)(x - 1)$ (Obj. 8.1.2) **16.** $5(x^2 - 9x - 3)$ (Obj. 8.1.1) **17.** Nonfactorable over the integers (Obj. 8.3.1/8.3.2) **18.** $(2x + 7y)(2x - 7y)$ (Obj. 8.4.1) **19.** $3, 5$ (Obj. 8.5.1) **20.** $(p + 6)^2$ (Obj. 8.4.1) **21.** $2(3x - 4y)^2$ (Obj. 8.4.2) **22.** $2y^2(y + 1)(y - 8)$ (Obj. 8.2.2) **23.** The length is 15 cm. The width is 6 cm. (Obj. 8.5.2) **24.** The length of the base is 12 in. (Obj. 8.5.2) **25.** The two integers are -13 and -12. (Obj. 8.5.2)

CUMULATIVE REVIEW EXERCISES *pages 383–384*

1. -8 (Obj. 1.2.2) **2.** -8.1 (Obj. 1.3.4) **3.** 4 (Obj. 1.4.2) **4.** -31 (Obj. 2.1.1) **5.** The Associative Property of Addition (Obj. 2.2.1) **6.** $18x^2$ (Obj. 2.2.3) **7.** $-6x + 24$ (Obj. 2.2.5) **8.** $\frac{2}{3}$ (Obj. 3.1.3) **9.** 5 (Obj. 3.2.3) **10.** $\frac{7}{4}$ (Obj. 3.2.2) **11.** 3 (Obj. 3.2.3) **12.** 35 (Obj. 3.1.4) **13.** $x \le -3$ (Obj. 3.3.3) **14.** $x < \frac{1}{5}$ (Obj. 3.3.3) **15.** (Obj. 5.2.2) **16.** (Obj. 5.5.2)

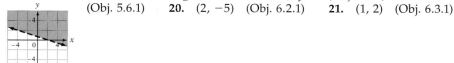

17. Domain: $\{-5, -3, -1, 1, 3\}$; range: $\{-4, -2, 0, 2, 4\}$; yes (Obj. 5.5.1) **18.** 61 (Obj. 5.5.1) **19.** (Obj. 5.6.1) **20.** $(2, -5)$ (Obj. 6.2.1) **21.** $(1, 2)$ (Obj. 6.3.1)

22. $3y^3 - 3y^2 - 8y - 5$ (Obj. 7.1.1) **23.** $-27a^{12}b^6$ (Obj. 7.2.2) **24.** $x^3 - 3x^2 - 6x + 8$ (Obj. 7.3.2) **25.** $4x + 8 + \frac{21}{2x - 3}$ (Obj. 7.4.4) **26.** $\frac{y^6}{x^{12}}$ (Obj. 7.4.1) **27.** $(a - b)(3 - x)$ (Obj. 8.1.2) **28.** $(x + 5y)(x - 2y)$ (Obj. 8.2.2) **29.** $2a^2(3a + 2)(a + 3)$ (Obj. 8.3.1/8.3.2) **30.** $(5a + 6b)(5a - 6b)$ (Obj. 8.4.1) **31.** $3(2x - 3y)^2$ (Obj. 8.4.2) **32.** $\frac{4}{3}, -5$ (Obj. 8.5.1) **33.** $\{-11, -7, -3, 1, 5\}$ (Obj. 5.5.1)

34. The average daily high temperature was $-3°C$. (Obj. 1.2.5) **35.** The length is 15 cm. The width is 6 cm. (Obj. 4.3.1) **36.** The pieces measure 4 ft and 6 ft. (Obj. 4.1.2) **37.** The maximum distance you can drive a Company A car is 359 mi. (Obj. 4.8.1) **38.** $6500 must be invested at an annual simple interest rate of 11%. (Obj. 4.5.1) **39.** The discount rate is 40%. (Obj. 4.4.2) **40.** The three integers are 10, 12, and 14. (Obj. 4.2.1)

ANSWERS to Chapter 9 Odd-Numbered Exercises

CONCEPT REVIEW 9.1 *page 391*

1. Always true **3.** Never true **5.** Always true **7.** Never true

SECTION 9.1 *pages 391 – 393*

1. $\dfrac{3}{4x}$ **3.** $\dfrac{1}{x+3}$ **5.** -1 **7.** $\dfrac{2}{3y}$ **9.** $-\dfrac{3}{4x}$ **11.** $\dfrac{a}{b}$ **13.** $-\dfrac{2}{x}$ **15.** $\dfrac{y-2}{y-3}$ **17.** $\dfrac{x+5}{x+4}$ **19.** $\dfrac{x+4}{x-3}$

21. $-\dfrac{x+2}{x+5}$ **23.** $\dfrac{2(x+2)}{x+3}$ **25.** $\dfrac{2x-1}{2x+3}$ **27.** $\dfrac{2}{3xy}$ **29.** $\dfrac{8xy^2ab}{3}$ **31.** $\dfrac{2}{9}$ **33.** $\dfrac{y^2}{x}$ **35.** $\dfrac{y(x+4)}{x(x+1)}$

37. $\dfrac{x^3(x-7)}{y^2(x-4)}$ **39.** $-\dfrac{y}{x}$ **41.** $\dfrac{x+3}{x+1}$ **43.** $\dfrac{x-5}{x+3}$ **45.** $-\dfrac{x+3}{x+5}$ **47.** $\dfrac{12x^4}{(x+1)(2x+1)}$ **49.** $-\dfrac{x+3}{x-12}$

51. $\dfrac{x+2}{x+4}$ **53.** $\dfrac{2x-5}{2x-1}$ **55.** $\dfrac{3x-4}{2x+3}$ **57.** $\dfrac{2xy^2ab^2}{9}$ **59.** $\dfrac{5}{12}$ **61.** $3x$ **63.** $\dfrac{y(x+3)}{x(x+1)}$ **65.** $\dfrac{x+7}{x-7}$

67. $-\dfrac{4ac}{y}$ **69.** $\dfrac{x-5}{x-6}$ **71.** 1 **73.** $-\dfrac{x+6}{x+5}$ **75.** $\dfrac{2x+3}{x-6}$ **77.** $\dfrac{4x+3}{2x-1}$ **79.** $\dfrac{(2x+5)(4x-1)}{(2x-1)(4x+5)}$

81. $-6, 1$ **83.** $-1, 1$ **85.** $3, -2$ **87.** -3 **89.** $\dfrac{4}{3}, -\dfrac{1}{2}$ **91.** $\dfrac{4a}{3b}$ **93.** $\dfrac{8}{9}$ **95.** $\dfrac{x-4}{y^4}$

97. $\dfrac{(x-2)(x-2)}{(x-4)(x+4)}$

CONCEPT REVIEW 9.2 *page 396*

1. Always true **3.** Never true **5.** Always true

SECTION 9.2 *pages 396 – 398*

1. $24x^3y^2$ **3.** $30x^4y^2$ **5.** $8x^2(x+2)$ **7.** $6x^2y(x+4)$ **9.** $40x^3(x-1)^2$ **11.** $4(x-3)^2$
13. $(2x-1)(2x+1)(x+4)$ **15.** $(x-7)^2(x+2)$ **17.** $(x+4)(x-3)$ **19.** $(x+5)(x-2)(x+7)$
21. $(x-7)(x-3)(x-5)$ **23.** $(x+2)(x+5)(x-5)$ **25.** $(2x-1)(x-3)(x+1)$
27. $(2x-5)(x-2)(x+3)$ **29.** $(x+3)(x-5)$ **31.** $(x+6)(x-3)$ **33.** $\dfrac{4x}{x^2}; \dfrac{3}{x^2}$ **35.** $\dfrac{4x}{12y^2}; \dfrac{3yz}{12y^2}$

37. $\dfrac{xy}{x^2(x-3)}; \dfrac{6x-18}{x^2(x-3)}$ **39.** $\dfrac{9x}{x(x-1)^2}; \dfrac{6x-6}{x(x-1)^2}$ **41.** $\dfrac{3x}{x(x-3)}; \dfrac{5}{x(x-3)}$ **43.** $\dfrac{3}{(x-5)^2}; -\dfrac{2x-10}{(x-5)^2}$

45. $\dfrac{3x}{x^2(x+2)}; \dfrac{4x+8}{x^2(x+2)}$ **47.** $\dfrac{x^2-6x+8}{(x+3)(x-4)}; \dfrac{x^2+3x}{(x+3)(x-4)}$ **49.** $\dfrac{3}{(x+2)(x-1)}; \dfrac{x^2-x}{(x+2)(x-1)}$

51. $\dfrac{5}{(2x-5)(x-2)}; \dfrac{x^2-3x+2}{(2x-5)(x-2)}$ **53.** $\dfrac{x^2-3x}{(x+3)(x-3)(x-2)}; \dfrac{2x^2-4x}{(x+3)(x-3)(x-2)}$ **55.** $-\dfrac{x^2-3x}{(x-3)^2(x+3)};$
$\dfrac{x^2+2x-3}{(x-3)^2(x+3)}$ **57.** $\dfrac{3x^2+12x}{(x-5)(x+4)}; \dfrac{x^2-5x}{(x-5)(x+4)}; -\dfrac{3}{(x-5)(x+4)}$ **59.** $\dfrac{300}{10^4}; \dfrac{5}{10^4}$ **61.** $\dfrac{b^2}{b}; \dfrac{5}{b}$ **63.** $\dfrac{y-1}{y-1}; \dfrac{y}{y-1}$
65. $\dfrac{x^2+1}{(x-1)^3}; \dfrac{x^2-1}{(x-1)^3}; \dfrac{x^2-2x+1}{(x-1)^3}$ **67.** $\dfrac{2b}{8(a+b)(a-b)}; \dfrac{a^2+ab}{8(a+b)(a-b)}$ **69.** $\dfrac{x-2}{(x+y)(x+2)(x-2)}; \dfrac{x+2}{(x+y)(x+2)(x-2)}$

CONCEPT REVIEW 9.3 *page 402*

1. Never true **3.** Never true **5.** Always true

SECTION 9.3 *pages 402 – 405*

1. $\dfrac{11}{y^2}$ **3.** $-\dfrac{7}{x+4}$ **5.** $\dfrac{8x}{2x+3}$ **7.** $\dfrac{5x+7}{x-3}$ **9.** $\dfrac{2x-5}{x+9}$ **11.** $\dfrac{-3x-4}{2x+7}$ **13.** $\dfrac{1}{x+5}$ **15.** $\dfrac{1}{x-6}$

17. $\dfrac{3}{2y-1}$ **19.** $\dfrac{1}{x-5}$ **21.** $\dfrac{4y+5x}{xy}$ **23.** $\dfrac{19}{2x}$ **25.** $\dfrac{5}{12x}$ **27.** $\dfrac{19x-12}{6x^2}$ **29.** $\dfrac{52y-35x}{20xy}$

31. $\dfrac{13x+2}{15x}$ **33.** $\dfrac{7}{24}$ **35.** $\dfrac{x+90}{45x}$ **37.** $\dfrac{x^2+2x+2}{2x^2}$ **39.** $\dfrac{2x^2+3x-10}{4x^2}$ **41.** $\dfrac{3y^2+8}{3y}$ **43.** $\dfrac{x^2+4x+4}{x+4}$

45. $\dfrac{4x+7}{x+1}$ **47.** $\dfrac{-3x^2+16x+2}{12x^2}$ **49.** $\dfrac{x^2-x+2}{x^2y}$ **51.** $\dfrac{16xy-12y+6x^2+3x}{12x^2y^2}$ **53.** $\dfrac{3xy-6y-2x^2-14x}{24x^2y}$

55. $\dfrac{9x+2}{(x-2)(x+3)}$ **57.** $\dfrac{2(x+23)}{(x-7)(x+3)}$ **59.** $\dfrac{2x^2-5x+1}{(x+1)(x-3)}$ **61.** $\dfrac{4x^2-34x+5}{(2x-1)(x-6)}$ **63.** $\dfrac{2a-5}{a-7}$

65. $\dfrac{4x+9}{(x+3)(x-3)}$ **67.** $\dfrac{-x+9}{(x+2)(x-3)}$ **69.** $\dfrac{14}{(x-5)^2}$ **71.** $\dfrac{-2(x+7)}{(x+6)(x-7)}$ **73.** $\dfrac{x-2}{2x}$ **75.** $\dfrac{x^2+x+4}{(x-2)(x+1)(x-1)}$

77. $\dfrac{x-4}{x-6}$ **79.** $\dfrac{2x+1}{x-1}$ **81.** 2 **83.** $\dfrac{b^2+b-7}{b+4}$ **85.** $\dfrac{4n}{(n-1)^2}$ **87.** $-\dfrac{3(x^2+8x+25)}{(x-3)(x+7)}$ **89.** 1

91. $\dfrac{x^2-x+2}{(x+5)(x+1)}$ **93.** $\dfrac{2}{3};\dfrac{3}{4};\dfrac{4}{5};\dfrac{50}{51};\dfrac{100}{101};\dfrac{1000}{1001}$ **95.** $\dfrac{6}{y}+\dfrac{7}{x}$ **97.** $\dfrac{2}{m^2n}+\dfrac{8}{mn^2}$

CONCEPT REVIEW 9.4 *page 407*

1. Sometimes true **3.** Never true **5.** Always true

SECTION 9.4 *pages 407 – 408*

1. $\dfrac{x}{x-3}$ **3.** $\dfrac{2}{3}$ **5.** $\dfrac{y+3}{y-4}$ **7.** $\dfrac{2(2x+13)}{5x+36}$ **9.** $\dfrac{3}{4}$ **11.** $\dfrac{x-2}{x+2}$ **13.** $\dfrac{x+2}{x+3}$ **15.** $\dfrac{x-6}{x+5}$ **17.** $-\dfrac{x-2}{x+1}$

19. $x-1$ **21.** $\dfrac{1}{2x-1}$ **23.** $\dfrac{x-3}{x+5}$ **25.** $\dfrac{x-7}{x-8}$ **27.** $\dfrac{2y-1}{2y+1}$ **29.** $\dfrac{x-2}{2x-5}$ **31.** $\dfrac{x-2}{x+1}$ **33.** $\dfrac{x-1}{x+4}$

35. $\dfrac{-x-1}{4x-3}$ **37.** $\dfrac{x+1}{2(5x-2)}$ **39.** $\dfrac{b+11}{4b-21}$ **41.** $\dfrac{5}{3}$ **43.** $-\dfrac{1}{x-1}$ **45.** $\dfrac{ab}{b+a}$ **47.** $\dfrac{y+4}{2(y-2)}$

CONCEPT REVIEW 9.5 *pages 413 – 414*

1. Always true **3.** Always true **5.** Sometimes true **7.** Always true

SECTION 9.5 *pages 414 – 418*

1. 1 **3.** $\dfrac{1}{4}$ **5.** 1 **7.** -3 **9.** $\dfrac{1}{2}$ **11.** 8 **13.** 5 **15.** -1 **17.** 5 **19.** No solution

21. 2, 4 **23.** $-\dfrac{3}{2}$, 4 **25.** 3 **27.** -1 **29.** 4 **31.** 15 **33.** 36 **35.** 113 **37.** -2 **39.** 4

41. 20,000 people voted in favor of the amendment. **43.** The office building requires 140 air vents. **45.** For 2 c of boiling water, 6 c of sugar are required. **47.** There are about 800 fish in the lake. **49.** 14 vials are required to treat 175 people. **51.** Yes, the shipment will be accepted. **53.** The distance between the two cities is 750 mi. **55.** For an annual yield of 1320 bushels, 10 additional acres must be planted. **57.** The recommended area of a window is 40 ft^2. **59.** To serve 70 people, 2 additional gallons are necessary.

61. There are 160 ml of water in 280 ml of soft drink. **63.** 1 **65.** 0, $-\dfrac{2}{3}$ **67.** The integers are -3 and -4.

69. The first person's share is $1.25 million. **71.** The cost would be $18.40.

CONCEPT REVIEW 9.6   *page 420*

1. Always true   **3.** Always true   **5.** Always true

SECTION 9.6   *pages 420 – 422*

1. $h = \dfrac{2A}{b}$ **3.** $t = \dfrac{d}{r}$ **5.** $T = \dfrac{PV}{nR}$ **7.** $L = \dfrac{P - 2W}{2}$ **9.** $b_1 = \dfrac{2A - hb_2}{h}$ **11.** $h = \dfrac{3V}{A}$ **13.** $S = C - Rt$

15. $P = \dfrac{A}{1 + rt}$ **17.** $w = \dfrac{A}{S + 1}$ **19. a.** $h = \dfrac{S - 2\pi r^2}{2\pi r}$ **b.** 5 in.   **c.** 4 in.   **21. a.** $r = \dfrac{S - C}{S}$

b. 20%   **c.** 30%

CONCEPT REVIEW 9.7   *pages 426 – 427*

1. Never true   **3.** Always true   **5.** Never true

SECTION 9.7   *pages 427 – 431*

1. It would take the experienced painter 6 h.   **3.** It would take 3 h to remove the earth.   **5.** It would take the computers 30 h.   **7.** It would take 6 min to cool the room.   **9.** It would take the second welder 15 h. **11.** It would take the other dock worker 10 min.   **13.** It would take the second mason 6 h.   **15.** It would take the second technician 3 h.   **17.** It would take the small unit $14\frac{2}{3}$ h.   **19.** It would have taken one of the welders 40 h.   **21.** The camper hiked at a rate of 5 mph.   **23.** The rate of the jogger was 8 mph. The rate of the cyclist was 20 mph.   **25.** The rate of the jet was 360 mph.   **27.** The rate of the second plane was 150 mph. **29.** The rate of the car was 48 mph.   **31.** The hiker walked at a rate of 3 mph.   **33.** The rate of the current is 2 mph.   **35.** The rate of the gulf stream is 6 mph.   **37.** The rate of the trucker was 55 mph.   **39.** The rate of the current is 5 mph.   **41.** It would take $1\frac{1}{19}$ h to fill the tank.   **43.** The amount of time spent traveling by canoe was 2 h.   **45.** The bus usually travels at a rate of 60 mph.

CHAPTER REVIEW EXERCISES   *pages 437 – 439*

1. $\dfrac{by^3}{6ax^2}$ (Obj. 9.1.2)   **2.** $\dfrac{22x - 1}{(3x - 4)(2x + 3)}$ (Obj. 9.3.2)   **3.** $x = -\dfrac{3}{4}y + 3$ (Obj. 9.6.1)   **4.** $\dfrac{2x^4}{3y^7}$ (Obj. 9.1.1)

5. $\dfrac{1}{x^2}$ (Obj. 9.1.3)   **6.** $\dfrac{x - 2}{3x - 10}$ (Obj. 9.4.1)   **7.** $72a^3b^5$ (Obj. 9.2.1)   **8.** $\dfrac{4x}{3x + 7}$ (Obj. 9.3.1)   **9.** $\dfrac{3x}{16x^2}$;

$\dfrac{10}{16x^2}$ (Obj. 9.2.2)   **10.** $\dfrac{x - 9}{x - 3}$ (Obj. 9.1.1)   **11.** $\dfrac{x - 4}{x + 3}$ (Obj. 9.1.3)   **12.** $\dfrac{2y - 3}{5y - 7}$ (Obj. 9.3.2)   **13.** $\dfrac{1}{x}$

(Obj. 9.1.2)   **14.** $\dfrac{1}{x + 3}$ (Obj. 9.3.1)   **15.** $15x^4(x - 7)^2$ (Obj. 9.2.1)   **16.** 10 (Obj. 9.5.1)   **17.** No solution

(Obj. 9.5.1)   **18.** $-\dfrac{4a}{5b}$ (Obj. 9.1.1)   **19.** $\dfrac{1}{x + 3}$ (Obj. 9.3.1)   **20.** $\dfrac{2x - 6}{(x + 3)(x - 3)}$; $\dfrac{7x + 21}{(x + 3)(x - 3)}$ (Obj. 9.2.2)

21. 5 (Obj. 9.5.1)   **22.** $\dfrac{x + 9}{4x}$ (Obj. 9.4.1)   **23.** $\dfrac{3x - 1}{x - 5}$ (Obj. 9.3.2)   **24.** $\dfrac{x - 3}{x + 3}$ (Obj. 9.1.3)

25. $-\dfrac{x + 6}{x + 3}$ (Obj. 9.1.1)   **26.** 2 (Obj. 9.5.1)   **27.** $x - 2$ (Obj. 9.4.1)   **28.** $\dfrac{x + 3}{x - 4}$ (Obj. 9.1.2)   **29.** 15

(Obj. 9.5.1)   **30.** 12 (Obj. 9.5.1)   **31.** $\dfrac{x}{x - 7}$ (Obj. 9.4.1)   **32.** $x = 2y + 15$ (Obj. 9.6.1)   **33.** $\dfrac{6b + 9a}{ab}$

(Obj. 9.3.2)   **34.** $(5x - 3)(2x - 1)(4x - 1)$ (Obj. 9.2.1)   **35.** $c = \dfrac{100m}{i}$ (Obj. 9.6.1)   **36.** 40 (Obj. 9.5.1)

37. 3 (Obj. 9.5.1)   **38.** $\dfrac{7x + 22}{60x}$ (Obj. 9.3.2)   **39.** $\dfrac{8a + 3}{4a - 3}$ (Obj. 9.1.2)   **40.** $\dfrac{3x^2 - x}{(6x - 1)(2x + 3)(3x - 1)}$;

$\dfrac{24x^3 - 4x^2}{(6x - 1)(2x + 3)(3x - 1)}$ (Obj. 9.2.2)   **41.** $\dfrac{2}{ab}$ (Obj. 9.3.1)   **42.** 62 (Obj. 9.5.1)   **43.** 6 (Obj. 9.5.1)

44. $\dfrac{b^3y}{10ax}$ (Obj. 9.1.3)   **45.** It would take the apprentice 6 h. (Obj. 9.7.1)   **46.** The spring would stretch 8 in. (Obj. 9.5.2)   **47.** The rate of the wind is 20 mph. (Obj. 9.7.2)   **48.** To make 10 gal of the garden spray will require 16 additional ounces. (Obj. 9.5.2)   **49.** Using both hoses, it would take 6 h to fill the pool. (Obj. 9.7.1)   **50.** The rate of the car is 45 mph. (Obj. 9.7.2)

CHAPTER TEST   *pages 439 – 440*

1. $\dfrac{x + 5}{x + 4}$ (Obj. 9.1.3)   **2.** $\dfrac{2}{x + 5}$ (Obj. 9.3.1)   **3.** $3(2x - 1)(x + 1)$ (Obj. 9.2.1)   **4.** -1 (Obj. 9.5.1)

5. $\dfrac{x + 1}{x^3(x - 2)}$ (Obj. 9.1.2)   **6.** $\dfrac{x - 3}{x - 2}$ (Obj. 9.4.1)   **7.** $\dfrac{3x + 6}{x(x - 2)(x + 2)}; \dfrac{x^2}{x(x - 2)(x + 2)}$ (Obj. 9.2.2)

8. $x = -\dfrac{5}{3}y - 5$ (Obj. 9.6.1)   **9.** 2 (Obj. 9.5.1)   **10.** $\dfrac{5}{(2x - 1)(3x + 1)}$ (Obj. 9.3.2)   **11.** 1 (Obj. 9.1.3)

12. $\dfrac{3}{x + 8}$ (Obj. 9.3.1)   **13.** $6x(x + 2)^2$ (Obj. 9.2.1)   **14.** $-\dfrac{x - 2}{x + 5}$ (Obj. 9.1.1)   **15.** 4 (Obj. 9.5.1)

16. $t = \dfrac{f - v}{a}$ (Obj. 9.6.1)   **17.** $\dfrac{2x^3}{3y^5}$ (Obj. 9.1.1)   **18.** $\dfrac{3}{(2x - 1)(x + 1)}$ (Obj. 9.3.2)   **19.** 3 (Obj. 9.5.1)

20. $\dfrac{x^3(x + 3)}{y(x + 2)}$ (Obj. 9.1.2)   **21.** $-\dfrac{3xy + 3y}{x(x + 1)(x - 1)}; \dfrac{x^2}{x(x + 1)(x - 1)}$ (Obj. 9.2.2)   **22.** $\dfrac{x + 3}{x + 5}$ (Obj. 9.4.1)   **23.** For 15 gal of water, 2 lb of additional salt will be required. (Obj. 9.5.2)   **24.** The rate of the wind is 20 mph. (Obj. 9.7.2)   **25.** It would take both pipes 6 min to fill the tank. (Obj. 9.7.1)

CUMULATIVE REVIEW EXERCISES   *pages 441 – 442*

1. -17 (Obj. 1.1.2)   **2.** -6 (Obj. 1.4.1)   **3.** $\dfrac{31}{30}$ (Obj. 1.4.2)   **4.** 21 (Obj. 2.1.1)   **5.** $5x - 2y$ (Obj. 2.2.2)   **6.** $-8x + 26$ (Obj. 2.2.4)   **7.** -20 (Obj. 3.2.1)   **8.** -12 (Obj. 3.2.3)   **9.** 10 (Obj. 3.1.4)

10. $x < \dfrac{9}{5}$ (Obj. 3.3.2)   **11.** $x \le 15$ (Obj. 3.3.3)   **12.** 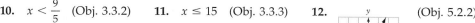 (Obj. 5.2.2)

13. (Obj. 5.5.2)   **14.** (Obj. 5.2.3)   **15.** (Obj. 5.6.1)

16. $\{-3, 5, 11, 19, 27\}$ (Obj. 5.5.1)   **17.** 37 (Obj. 5.5.1)   **18.** $\left(\dfrac{2}{3}, 3\right)$ (Obj. 6.2.1)   **19.** $\left(\dfrac{1}{2}, -1\right)$ (Obj. 6.3.1)

20. $-6x^4y^5$ (Obj. 7.2.1)   **21.** $a^{20}b^{15}$ (Obj. 7.2.2)   **22.** $\dfrac{a^3}{b^2}$ (Obj. 7.4.1)   **23.** $a^2 + ab - 12b^2$ (Obj. 7.3.3)

24. $3b^3 - b + 2$ (Obj. 7.4.3)   **25.** $x^2 + 2x + 4$ (Obj. 7.4.4)   **26.** $(4x + 1)(3x - 1)$ (Obj. 8.3.1/8.3.2)   **27.** $(y - 6)(y - 1)$ (Obj. 8.2.1)   **28.** $a(2a - 3)(a + 5)$ (Obj. 8.3.1/8.3.2)   **29.** $4(b + 5)(b - 5)$ (Obj. 8.4.2)

30. -3 and $\dfrac{5}{2}$ (Obj. 8.5.1)   **31.** $-\dfrac{x + 7}{x + 4}$ (Obj. 9.1.1)   **32.** 1 (Obj. 9.1.3)   **33.** $\dfrac{8}{(3x - 1)(x + 1)}$ (Obj. 9.3.2)

34. 1 (Obj. 9.5.1)   **35.** $a = \dfrac{f - v}{t}$ (Obj. 9.6.1)   **36.** $5x - 18 = -3; x = 3$ (Obj. 4.1.1)   **37.** $5000 must be invested at an annual simple interest rate of 11%. (Obj. 4.5.1)   **38.** There is 70% silver in the 120 g alloy. (Obj. 4.6.2)   **39.** The base is 10 in. The height is 6 in. (Obj. 8.5.2)   **40.** It would take both pipes 8 min to fill the tank. (Obj. 9.7.1)

ANSWERS to Chapter 10 Odd-Numbered Exercises

CONCEPT REVIEW 10.1 *pages 448 – 449*

1. Never true **3.** Always true **5.** Sometimes true

SECTION 10.1 *pages 449 – 450*

1. 4 **3.** 7 **5.** $4\sqrt{2}$ **7.** $2\sqrt{2}$ **9.** $18\sqrt{2}$ **11.** $10\sqrt{10}$ **13.** $\sqrt{15}$ **15.** $\sqrt{29}$ **17.** $-54\sqrt{2}$
19. $3\sqrt{5}$ **21.** 0 **23.** $48\sqrt{2}$ **25.** $\sqrt{105}$ **27.** 30 **29.** $30\sqrt{5}$ **31.** $5\sqrt{10}$ **33.** $4\sqrt{6}$ **35.** 18
37. 15.492 **39.** 16.971 **41.** 15.652 **43.** 18.762 **45.** x^3 **47.** $y^7\sqrt{y}$ **49.** a^{10} **51.** x^2y^2
53. $2x^2$ **55.** $2x\sqrt{6}$ **57.** $xy^3\sqrt{xy}$ **59.** $ab^5\sqrt{ab}$ **61.** $2x^2\sqrt{15x}$ **63.** $7a^2b^4$ **65.** $3x^2y^3\sqrt{2xy}$
67. $2x^5y^3\sqrt{10xy}$ **69.** $4a^4b^5\sqrt{5a}$ **71.** $8ab\sqrt{b}$ **73.** x^3y **75.** $8a^2b^3\sqrt{5b}$ **77.** $6x^2y^3\sqrt{3y}$ **79.** $4x^3y\sqrt{2y}$
81. $5a + 20$ **83.** $2x^2 + 8x + 8$ **85.** $x + 2$ **87.** $y + 1$ **89.** $0.05ab^2\sqrt{ab}$ **91.** $xy\sqrt{y+x}$
93. a. 1 **b.** 3 **c.** $3\sqrt{3}$ **95.** The length of a side of the square is 8.7 cm. **97.** $x \geq 0$
99. $x \geq -5$ **101.** $x \leq \dfrac{5}{2}$ **103.** All real numbers

CONCEPT REVIEW 10.2 *page 452*

1. Never true **3.** Always true **5.** Always true

SECTION 10.2 *pages 453 – 454*

1. $3\sqrt{2}$ **3.** $-\sqrt{7}$ **5.** $-11\sqrt{11}$ **7.** $10\sqrt{x}$ **9.** $-2\sqrt{y}$ **11.** $-11\sqrt{3b}$ **13.** $2x\sqrt{2}$ **15.** $-3a\sqrt{3a}$
17. $-5\sqrt{xy}$ **19.** $8\sqrt{5}$ **21.** $8\sqrt{2}$ **23.** $15\sqrt{2} - 10\sqrt{3}$ **25.** $\sqrt{x}$ **27.** $-12x\sqrt{3}$ **29.** $2xy\sqrt{x} - 3xy\sqrt{y}$
31. $-9x\sqrt{3x}$ **33.** $-13y^2\sqrt{2y}$ **35.** $4a^2b^2\sqrt{ab}$ **37.** $7\sqrt{2}$ **39.** $6\sqrt{x}$ **41.** $-3\sqrt{y}$ **43.** $-45\sqrt{2}$
45. $13\sqrt{3} - 12\sqrt{5}$ **47.** $32\sqrt{3} - 3\sqrt{11}$ **49.** $6\sqrt{x}$ **51.** $-34\sqrt{3x}$ **53.** $10a\sqrt{3b} + 10a\sqrt{5b}$
55. $-2xy\sqrt{3}$ **57.** $-7b\sqrt{ab} + 4a^2\sqrt{b}$ **59.** $3ab\sqrt{2a} - ab + 4ab\sqrt{3b}$ **61.** $8\sqrt{x+2}$ **63.** $2x\sqrt{2y}$
65. $2ab\sqrt{6ab}$ **67.** $-11\sqrt{2x+y}$ **69.** The perimeter is $(6\sqrt{3} + 2\sqrt{15})$ cm. **71.** The perimeter is 22.4 cm.
73. No

CONCEPT REVIEW 10.3 *page 459*

1. Always true **3.** Always true **5.** Sometimes true

SECTION 10.3 *pages 459 – 461*

1. 5 **3.** 6 **5.** x **7.** x^3y^2 **9.** $3ab^6\sqrt{2a}$ **11.** $12a^4b\sqrt{b}$ **13.** $2 - \sqrt{6}$ **15.** $x - \sqrt{xy}$
17. $5\sqrt{2} - \sqrt{5x}$ **19.** $4 - 2\sqrt{10}$ **21.** $x - 6\sqrt{x} + 9$ **23.** $3a - 3\sqrt{ab}$ **25.** $10abc$
27. $15x - 22y\sqrt{x} + 8y^2$ **29.** $x - y$ **31.** $10x + 13\sqrt{xy} + 4y$ **33.** 4 **35.** 7 **37.** 3 **39.** $x\sqrt{5}$
41. $\dfrac{a^2}{7}$ **43.** $\dfrac{\sqrt{3}}{3}$ **45.** $\sqrt{3}$ **47.** $\dfrac{\sqrt{3x}}{x}$ **49.** $\dfrac{\sqrt{2x}}{x}$ **51.** $\dfrac{3\sqrt{x}}{x}$ **53.** $\dfrac{2\sqrt{y}}{xy}$ **55.** $\dfrac{4\sqrt{b}}{7b}$ **57.** $\dfrac{5\sqrt{6}}{27}$
59. $\dfrac{\sqrt{y}}{3xy}$ **61.** $\dfrac{2\sqrt{3x}}{3y}$ **63.** $-\dfrac{\sqrt{2}+3}{7}$ **65.** $\dfrac{15 - 3\sqrt{5}}{20}$ **67.** $\dfrac{x\sqrt{y} + y\sqrt{x}}{x - y}$ **69.** $\dfrac{\sqrt{3xy}}{3y}$ **71.** $-2 - \sqrt{6}$
73. $-\dfrac{\sqrt{10}+5}{3}$ **75.** $\dfrac{x - 3\sqrt{x}}{x - 9}$ **77.** $-\dfrac{1}{2}$ **79.** 7 **81.** -5 **83.** $\dfrac{7}{4}$ **85.** $-\dfrac{42 - 26\sqrt{3}}{11}$
87. $\dfrac{14 - 9\sqrt{2}}{17}$ **89.** $\dfrac{a - 5\sqrt{a} + 4}{2a - 2}$ **91.** $\dfrac{y + 5\sqrt{y} + 6}{y - 9}$ **93.** -1.3 **95.** $-\dfrac{4}{9}$ **97.** $\dfrac{3}{2}$ **99. a.** True
b. True **c.** False; $x + 2\sqrt{x} + 1$ **d.** True **101.** No

CONCEPT REVIEW 10.4 *page 467*

1. Sometimes true **3.** Always true **5.** Sometimes true

SECTION 10.4 *pages 467–470*

1. 25 **3.** 144 **5.** 5 **7.** 16 **9.** 8 **11.** No solution **13.** 6 **15.** 24 **17.** -1 **19.** $-\dfrac{2}{5}$
21. $\dfrac{4}{3}$ **23.** 15 **25.** 5 **27.** 1 **29.** 1 **31.** 1 **33.** 2 **35.** No solution **37.** The number is 1.
39. The number is 11. **41.** The length of the hypotenuse is 10.30 cm. **43.** The length of the other leg is 9.75 ft. **45.** The periscope has to be 10.67 ft above the water. **47.** The pitcher's mound is less than halfway between home plate and second base. **49.** The length of the pendulum is 7.30 ft. **51.** The distance is 250 ft.
53. The bridge is 64 ft high. **55.** The height of the screen is 21.6 in. **57.** 5 **59.** 8 **61. a.** The perimeter is $12\sqrt{2}$ cm. **b.** The area is 12 cm². **63.** The area is 244.78 ft².

CHAPTER REVIEW EXERCISES *pages 476–477*

1. $-11\sqrt{3}$ (Obj. 10.2.1) **2.** $-x\sqrt{3}-x\sqrt{5}$ (Obj. 10.3.2) **3.** $x+8$ (Obj. 10.1.2) **4.** No solution (Obj. 10.4.1) **5.** 1 (Obj. 10.3.2) **6.** $7\sqrt{7}$ (Obj. 10.2.1) **7.** $12a+28\sqrt{ab}+15b$ (Obj. 10.3.1)
8. $7x^2+42x+63$ (Obj. 10.1.2) **9.** 12 (Obj. 10.1.1) **10.** 16 (Obj. 10.4.1) **11.** $4x\sqrt{5}$ (Obj. 10.2.1)
12. $5ab-7$ (Obj. 10.3.1) **13.** 3 (Obj. 10.4.1) **14.** $\sqrt{35}$ (Obj. 10.1.1) **15.** $7x^2y\sqrt{15xy}$ (Obj. 10.2.1)
16. $\dfrac{x}{3y^2}$ (Obj. 10.3.2) **17.** $9x-6\sqrt{xy}+y$ (Obj. 10.3.1) **18.** $a+4$ (Obj. 10.1.2) **19.** $20\sqrt{3}$ (Obj. 10.1.1)
20. $26\sqrt{3x}$ (Obj. 10.2.1) **21.** $\dfrac{8\sqrt{x}+24}{x-9}$ (Obj. 10.3.2) **22.** $3a\sqrt{2}+2a\sqrt{3}$ (Obj. 10.3.1) **23.** $9a^2\sqrt{2ab}$ (Obj. 10.1.2) **24.** $-6\sqrt{30}$ (Obj. 10.1.1) **25.** $-4a^2b^4\sqrt{5ab}$ (Obj. 10.2.1) **26.** $7x^2y^4$ (Obj. 10.3.2)
27. $\dfrac{3}{5}$ (Obj. 10.4.1) **28.** c^9 (Obj. 10.1.2) **29.** $15\sqrt{2}$ (Obj. 10.1.1) **30.** $18a\sqrt{5b}+5a\sqrt{b}$ (Obj. 10.2.1)
31. $\dfrac{16\sqrt{a}}{a}$ (Obj. 10.3.2) **32.** 8 (Obj. 10.4.1) **33.** $a^5b^3c^2$ (Obj. 10.3.1) **34.** $21\sqrt{70}$ (Obj. 10.1.1)
35. $2y^4\sqrt{6}$ (Obj. 10.1.2) **36.** 5 (Obj. 10.3.2) **37.** 5 (Obj. 10.4.1) **38.** $8y+10\sqrt{5y}-15$ (Obj. 10.3.1)
39. 99.499 (Obj. 10.1.1) **40.** 7 (Obj. 10.3.1) **41.** $25x^4\sqrt{6x}$ (Obj. 10.1.2) **42.** $-6x^3y^2\sqrt{2y}$ (Obj. 10.2.1)
43. $3a$ (Obj. 10.3.2) **44.** $20\sqrt{10}$ (Obj. 10.1.1) **45.** 20 (Obj. 10.4.1) **46.** $\dfrac{7a}{3}$ (Obj. 10.3.2)
47. 10 (Obj. 10.3.1) **48.** $6x^2y^2\sqrt{y}$ (Obj. 10.1.2) **49.** $36x^8y^5\sqrt{3xy}$ (Obj. 10.1.2) **50.** 3 (Obj. 10.4.1)
51. 20 (Obj. 10.1.1) **52.** $-2\sqrt{2x}$ (Obj. 10.2.1) **53.** 6 (Obj. 10.4.1) **54.** 3 (Obj. 10.3.1) **55.** The larger integer is 51. (Obj. 10.4.2) **56.** The explorer's weight on the surface of the earth is 144 lb. (Obj. 10.4.2)
57. The length of the other leg is 14.25 cm. (Obj. 10.4.2) **58.** The radius of the sharpest corner is 25 ft. (Obj. 10.4.2) **59.** The depth is 100 ft. (Obj. 10.4.2) **60.** The length of the guy wire is 26.25 ft. (Obj. 10.4.2)

CHAPTER TEST *pages 478–479*

1. $11x^4y$ (Obj. 10.1.2) **2.** $-5\sqrt{2}$ (Obj. 10.2.1) **3.** $6x^2\sqrt{xy}$ (Obj. 10.3.1) **4.** $3\sqrt{5}$ (Obj. 10.1.1)
5. $6x^3y\sqrt{2x}$ (Obj. 10.1.2) **6.** $6\sqrt{2y}+3\sqrt{2x}$ (Obj. 10.2.1) **7.** $y+8\sqrt{y}+15$ (Obj. 10.3.1) **8.** $\sqrt{2}$ (Obj. 10.3.2) **9.** 9 (Obj. 10.3.2) **10.** 11 (Obj. 10.4.1) **11.** 22.361 (Obj. 10.1.1) **12.** $4a^2b^5\sqrt{2ab}$ (Obj. 10.1.2) **13.** $a-\sqrt{ab}$ (Obj. 10.3.1) **14.** $4x^2y^2\sqrt{5y}$ (Obj. 10.3.1) **15.** $7ab\sqrt{a}$ (Obj. 10.3.2)
16. 25 (Obj. 10.4.1) **17.** $8x^6y^2\sqrt{3xy}$ (Obj. 10.1.2) **18.** $4ab\sqrt{2ab}-5ab\sqrt{ab}$ (Obj. 10.2.1) **19.** $a-4$ (Obj. 10.3.1) **20.** $-6-3\sqrt{5}$ (Obj. 10.3.2) **21.** 5 (Obj. 10.4.1) **22.** $3\sqrt{2}-x\sqrt{3}$ (Obj. 10.3.1)
23. $-6\sqrt{a}$ (Obj. 10.2.1) **24.** $6\sqrt{3}$ (Obj. 10.1.1) **25.** 7.937 (Obj. 10.1.1) **26.** $6ab\sqrt{a}$ (Obj. 10.3.2)

27. No solution   (Obj. 10.4.1)   28. The smaller integer is 40.   (Obj. 10.4.2)   29. The length of the pendulum is 5.07 ft.   (Obj. 10.4.2)   30. The ladder is 9.80 ft high on the building.   (Obj. 10.4.2)

CUMULATIVE REVIEW EXERCISES   *pages 479 – 480*

1. $-\dfrac{1}{12}$ (Obj. 1.4.2)   2. $2x + 18$ (Obj. 2.2.5)   3. $\dfrac{1}{13}$ (Obj. 3.2.3)   4. $x \le -\dfrac{9}{2}$ (Obj. 3.3.3)

5. $-\dfrac{4}{3}$ (Obj. 5.3.1)   6. $y = \dfrac{1}{2}x - 2$ (Obj. 5.4.1/5.4.2)   7. -18 (Obj. 5.5.1)

8.    (Obj. 5.5.2)   9.   (Obj. 5.6.1)   10. $(2, -1)$ (Obj. 6.1.1)

11. $(1, 1)$ (Obj. 6.2.1)   12. $(3, -2)$ (Obj. 6.3.1)   13. $\dfrac{6x^5}{y^3}$ (Obj. 7.4.1)   14. $-2b^2 + 1 - \dfrac{1}{3b^2}$ (Obj. 7.4.3)

15. $3x^2y^2(4x - 3y)$ (Obj. 8.1.1)   16. $(3b + 5)(3b - 4)$ (Obj. 8.3.1/8.3.2)   17. $2a(a - 3)(a - 5)$

(Obj. 8.3.1/8.3.2)   18. $\dfrac{1}{4(x + 1)}$ (Obj. 9.1.2)   19. $\dfrac{x - 5}{x - 3}$ (Obj. 9.4.1)   20. $\dfrac{x + 3}{x - 3}$ (Obj. 9.3.2)   21. $\dfrac{5}{3}$

(Obj. 9.5.1)   22. $38\sqrt{3a} - 35\sqrt{a}$ (Obj. 10.2.1)   23. 8   (Obj. 10.3.2)   24. 14   (Obj. 10.4.1)   25. The largest integer that satisfies the inequality is -33.   (Obj. 4.8.1)   26. The cost is \$24.50.   (Obj. 4.4.1)   27. 56 oz of pure water must be added.   (Obj. 4.6.2)   28. The two numbers are 8 and 13.   (Obj. 4.1.2)   29. It would take the small pipe 48 h to fill the tank.   (Obj. 9.7.1)   30. The smaller integer is 24.   (Obj. 10.4.2)

ANSWERS to Chapter 11 Odd-Numbered Exercises

CONCEPT REVIEW 11.1   *page 487*

1. Never true   3. Sometimes true   5. Always true

SECTION 11.1   *pages 487 – 489*

1. $a = 3, b = -4, c = 1$ 3. $a = 2, b = 0, c = -5$ 5. $a = 6, b = -3, c = 0$ 7. $x^2 - 3x - 8 = 0$
9. $x^2 - 16 = 0$ 11. $2x^2 + 12x + 13 = 0$ 13. -5 and 3   15. 1 and 3   17. -1 and -2 19. 3
21. 0 and $\dfrac{3}{2}$ 23. -2 and 5   25. $\dfrac{2}{3}$ and 1   27. -3 and $\dfrac{1}{3}$ 29. $-\dfrac{3}{2}$ and $\dfrac{4}{3}$ 31. -3 and $\dfrac{4}{5}$ 33. $\dfrac{1}{3}$

35. -4 and 4   37. $-\dfrac{2}{3}$ and $\dfrac{2}{3}$ 39. $\dfrac{2}{3}$ and 2   41. -6 and $\dfrac{3}{2}$ 43. $\dfrac{2}{3}$ 45. -3 and 6   47. 0 and 12

49. -6 and 6   51. -1 and 1   53. $-\dfrac{7}{2}$ and $\dfrac{7}{2}$ 55. $-\dfrac{2}{3}$ and $\dfrac{2}{3}$ 57. $-\dfrac{3}{4}$ and $\dfrac{3}{4}$ 59. There is no real number solution.   61. $2\sqrt{6}$ and $-2\sqrt{6}$ 63. -5 and 7   65. -7 and -3 67. -6 and 4   69. -9 and -1 71. 2 and 16   73. $-\dfrac{9}{2}$ and $-\dfrac{3}{2}$ 75. $-\dfrac{1}{3}$ and $\dfrac{7}{3}$ 77. $-\dfrac{12}{7}$ and $-\dfrac{2}{7}$ 79. $4 + 2\sqrt{5}$ and $4 - 2\sqrt{5}$ 81. There is no real number solution.   83. $\dfrac{1}{2} + \sqrt{6}$ and $\dfrac{1}{2} - \sqrt{6}$ 85. $-\dfrac{4}{3}$ and $\dfrac{8}{3}$

87. $\sqrt{2}$ and $-\sqrt{2}$ 89. 0 and 1   91. $\dfrac{\sqrt{ab}}{a}$ and $-\dfrac{\sqrt{ab}}{a}$ 93. The velocity is 10 m/s.

CONCEPT REVIEW 11.2 *page 494*

1. Always true **3.** Always true **5.** Never true

SECTION 11.2 *pages 494 – 495*

1. $x^2 + 12x + 36; (x + 6)^2$ **3.** $x^2 + 10x + 25; (x + 5)^2$ **5.** $x^2 - x + \dfrac{1}{4}; \left(x - \dfrac{1}{2}\right)^2$ **7.** -3 and 1

9. $-2 + \sqrt{3}$ and $-2 - \sqrt{3}$ **11.** There is no real number solution. **13.** $-3 + \sqrt{14}$ and $-3 - \sqrt{14}$

15. 2 **17.** $1 + \sqrt{2}$ and $1 - \sqrt{2}$ **19.** $\dfrac{-3 + \sqrt{13}}{2}$ and $\dfrac{-3 - \sqrt{13}}{2}$ **21.** -8 and 1 **23.** $-3 + \sqrt{5}$ and

$-3 - \sqrt{5}$ **25.** $4 + \sqrt{14}$ and $4 - \sqrt{14}$ **27.** 1 and 2 **29.** $\dfrac{3 + \sqrt{29}}{2}$ and $\dfrac{3 - \sqrt{29}}{2}$ **31.** $\dfrac{1 + \sqrt{5}}{2}$ and $\dfrac{1 - \sqrt{5}}{2}$

33. $\dfrac{5 + \sqrt{13}}{2}$ and $\dfrac{5 - \sqrt{13}}{2}$ **35.** $\dfrac{-1 + \sqrt{13}}{2}$ and $\dfrac{-1 - \sqrt{13}}{2}$ **37.** $-5 + 4\sqrt{2}$ and $-5 - 4\sqrt{2}$ **39.** $\dfrac{-3 + \sqrt{5}}{2}$ and

$\dfrac{-3 - \sqrt{5}}{2}$ **41.** $\dfrac{1 + \sqrt{17}}{2}$ and $\dfrac{1 - \sqrt{17}}{2}$ **43.** There is no real number solution. **45.** $\dfrac{1}{2}$ and 1 **47.** -3 and $\dfrac{1}{2}$

49. $\dfrac{1 + \sqrt{2}}{2}$ and $\dfrac{1 - \sqrt{2}}{2}$ **51.** $\dfrac{2 + \sqrt{5}}{2}$ and $\dfrac{2 - \sqrt{5}}{2}$ **53.** $3 + \sqrt{2}$ and $3 - \sqrt{2}$ **55.** $1 + \sqrt{13}$ and $1 - \sqrt{13}$

57. $\dfrac{3 + 3\sqrt{3}}{2}$ and $\dfrac{3 - 3\sqrt{3}}{2}$ **59.** $4 + \sqrt{7}$ and $4 - \sqrt{7}$ **61.** 1.193 and -4.193 **63.** 2.766 and -1.266

65. 0.152 and -1.652 **67.** -3 **69.** $6 + \sqrt{58}$ and $6 - \sqrt{58}$

CONCEPT REVIEW 11.3 *pages 498–499*

1. Always true **3.** Never true **5.** Never true

SECTION 11.3 *pages 499 – 500*

1. -7 and 1 **3.** -3 and 6 **5.** $1 + \sqrt{6}$ and $1 - \sqrt{6}$ **7.** $-3 + \sqrt{10}$ and $-3 - \sqrt{10}$ **9.** There is no real

number solution. **11.** $2 + \sqrt{13}$ and $2 - \sqrt{13}$ **13.** 0 and 1 **15.** $\dfrac{1 + \sqrt{2}}{2}$ and $\dfrac{1 - \sqrt{2}}{2}$ **17.** $-\dfrac{3}{2}$ and $\dfrac{3}{2}$

19. $\dfrac{3 + \sqrt{3}}{3}$ and $\dfrac{3 - \sqrt{3}}{3}$ **21.** $\dfrac{1 + \sqrt{10}}{3}$ and $\dfrac{1 - \sqrt{10}}{3}$ **23.** There is no real number solution. **25.** $\dfrac{4 + \sqrt{10}}{2}$ and

$\dfrac{4 - \sqrt{10}}{2}$ **27.** $-\dfrac{2}{3}$ and 3 **29.** $\dfrac{5 + \sqrt{73}}{6}$ and $\dfrac{5 - \sqrt{73}}{6}$ **31.** $\dfrac{1 + \sqrt{37}}{6}$ and $\dfrac{1 - \sqrt{37}}{6}$ **33.** $\dfrac{1 + \sqrt{41}}{5}$ and $\dfrac{1 - \sqrt{41}}{5}$

35. -3 and $\dfrac{4}{5}$ **37.** $-3 + 2\sqrt{2}$ and $-3 - 2\sqrt{2}$ **39.** $\dfrac{3 + 2\sqrt{6}}{2}$ and $\dfrac{3 - 2\sqrt{6}}{2}$ **41.** There is no real number

solution. **43.** $\dfrac{-1 + \sqrt{2}}{3}$ and $\dfrac{-1 - \sqrt{2}}{3}$ **45.** $-\dfrac{1}{2}$ and $\dfrac{2}{3}$ **47.** $\dfrac{-4 + \sqrt{5}}{2}$ and $\dfrac{-4 - \sqrt{5}}{2}$ **49.** $\dfrac{1 + 2\sqrt{3}}{2}$ and

$\dfrac{1 - 2\sqrt{3}}{2}$ **51.** $-\dfrac{3}{2}$ **53.** $\dfrac{-5 + \sqrt{2}}{3}$ and $\dfrac{-5 - \sqrt{2}}{3}$ **55.** $\dfrac{1 + \sqrt{19}}{3}$ and $\dfrac{1 - \sqrt{19}}{3}$ **57.** $\dfrac{5 + \sqrt{65}}{4}$ and $\dfrac{5 - \sqrt{65}}{4}$

59. $6 + \sqrt{11}$ and $6 - \sqrt{11}$ **61.** 5.690 and -3.690 **63.** 7.690 and -1.690 **65.** 4.589 and -1.089

67. 0.118 and -2.118 **69.** 1.105 and -0.905 **71.** $3 + \sqrt{13}$ and $3 - \sqrt{13}$ **73.** The ball hits the ground

after 4.81 s. **75. a.** False **b.** False **c.** False **d.** True

CONCEPT REVIEW 11.4 *page 503*

1. Always true **3.** Never true **5.** Never true

SECTION 11.4   *pages 503 – 506*

1. Up   **3.** Up   **5.** Up   **7.**   **9.**   **11.**   **13.**

15.   **17.**   **19.**   **21.**   **23.**

25.   **27.**   **29.**   **31.**   **33.**

35.   **37.** $y = x^2 - 8x + 15$ **39.** $y = 2x^2 - 12x + 22$ **41.** $(2, 0)$ and $(-2, 0)$

43. $(0, 0)$ and $(4, 0)$ **45.** $\left(-\dfrac{1}{2}, 0\right)$ and $(1, 0)$ **47.** $g(2) = -3$ **49.** $P(-3) = 38$ **51.**

53.   **55.**   **57.** 3   **59.** No

CONCEPT REVIEW 11.5   *page 509*

1. Never true   **3.** Always true   **5.** Always true

SECTION 11.5   *pages 509 – 511*

1. The length is 6 ft. The width is 4 ft.   **3.** The length is 100 ft. The width is 50 ft.   **5.** The integers are 7 and 9.   **7.** The integers are 5 and 7.   **9.** The integer is 0 or 2.   **11.** It would take the first computer 35 min and the second computer 14 min.   **13.** It would take the first engine 12 h and the second engine 6 h.
15. The rate of the plane in calm air is 100 mph.   **17.** The cyclist's rate during the first 150 mi was 50 mph.
19. The arrow will be 32 ft above the ground at 1 s and 2 s.   **21.** The integers are −6, −5, −4, and −3, or 3, 4, 5, and 6.   **23.** The radius is 7.98 cm.   **25.** The radius of the large pizza is 7 in.

CHAPTER REVIEW EXERCISES   *pages 515–518*

1. 4 and −4   (Obj. 11.1.2)   **2.** $\dfrac{1 + \sqrt{13}}{2}$ and $\dfrac{1 - \sqrt{13}}{2}$ (Obj. 11.2.1/11.3.1)   **3.** $\dfrac{3 + \sqrt{29}}{2}$ and $\dfrac{3 - \sqrt{29}}{2}$

(Obj. 11.2.1/11.3.1)   **4.** $\dfrac{5}{7}$ and $-\dfrac{5}{7}$ (Obj. 11.1.2)   **5.**   (Obj. 11.4.1)

6.   (Obj. 11.4.1)   **7.** $5 + \sqrt{23}$ and $5 - \sqrt{23}$ (Obj. 11.2.1/11.3.1)   **8.** $-\dfrac{1}{2}$ and $-\dfrac{1}{3}$ (Obj. 11.1.1)

9. There is no real number solution.   (Obj. 11.1.2)   **10.** $\dfrac{-10 + 2\sqrt{10}}{5}$ and $\dfrac{-10 - 2\sqrt{10}}{5}$ (Obj. 11.2.1/11.3.1)

11. $2 + \sqrt{3}$ and $2 - \sqrt{3}$ (Obj. 11.2.1/11.3.1)   **12.** 6 and −5   (Obj. 11.1.1)   **13.** $\dfrac{4}{3}$ and $-\dfrac{7}{2}$ (Obj. 11.1.1)

14. $2\sqrt{10}$ and $-2\sqrt{10}$ (Obj. 11.1.2)   **15.** $\dfrac{2 + \sqrt{7}}{3}$ and $\dfrac{2 - \sqrt{7}}{3}$ (Obj. 11.2.1/11.3.1)   **16.** $1 + \sqrt{11}$ and

$1 - \sqrt{11}$ (Obj. 11.2.1/11.3.1)   **17.** 3 and 9   (Obj. 11.1.1)   **18.** 16 and −2   (Obj. 11.1.2)

19.   (Obj. 11.4.1)   **20.**   (Obj. 11.4.1)   **21.** 1 and −9   (Obj. 11.1.2)

22. $\dfrac{-4 + \sqrt{23}}{2}$ and $\dfrac{-4 - \sqrt{23}}{2}$ (Obj. 11.2.1/11.3.1)   **23.** $-\dfrac{1}{6}$ and $-\dfrac{5}{4}$ (Obj. 11.1.1)   **24.** There is no real

number solution.   (Obj. 11.2.1/11.3.1)   **25.** $4 + \sqrt{21}$ and $4 - \sqrt{21}$ (Obj. 11.2.1/11.3.1)   **26.** $\dfrac{4}{7}$ (Obj. 11.1.1)

27. 2 and −1   (Obj. 11.1.2)   **28.** $\dfrac{3}{8}$ and $\dfrac{3}{2}$ (Obj. 11.1.1)   **29.** $\dfrac{-7 + \sqrt{61}}{2}$ and $\dfrac{-7 - \sqrt{61}}{2}$ (Obj. 11.2.1/11.3.1)

30. 2 and $\dfrac{5}{12}$ (Obj. 11.1.1)   **31.** $3 + \sqrt{5}$ and $3 - \sqrt{5}$ (Obj. 11.1.2)   **32.** $-4 + \sqrt{19}$ and $-4 - \sqrt{19}$

(Obj. 11.2.1/11.3.1)   **33.**   (Obj. 11.4.1)   **34.**   (Obj. 11.4.1)

35. −7 and −10   (Obj. 11.1.1)   **36.** $2 + 2\sqrt{6}$ and $2 - 2\sqrt{6}$ (Obj. 11.1.2)   **37.** There is no real number

solution.   (Obj. 11.2.1/11.3.1)   **38.** $-3 + \sqrt{11}$ and $-3 - \sqrt{11}$ (Obj. 11.2.1/11.3.1)   **39.** 3 and $-\dfrac{1}{9}$ (Obj. 11.1.1)

40. $\dfrac{-5 + \sqrt{33}}{4}$ and $\dfrac{-5 - \sqrt{33}}{4}$ (Obj. 11.2.1/11.3.1)   **41.**   (Obj. 11.4.1)

42. $\dfrac{9 + \sqrt{69}}{2}$ and $\dfrac{9 - \sqrt{69}}{2}$ (Obj. 11.2.1/11.3.1)   **43.** 1 and $\dfrac{5}{2}$ (Obj. 11.1.1)   **44.**   (Obj. 11.4.1)

45. The rate of the air balloon is 25 mph.   (Obj. 11.5.1)   **46.** The height is 10 m. The base is 4 m.   (Obj. 11.5.1)
47. The two integers are 3 and 5.   (Obj. 11.5.1)   **48.** The rate of the boat in still water is 5 mph.   (Obj. 11.5.1)
49. The object will be 12 ft above the ground at $\frac{1}{2}$ s and $1\frac{1}{2}$ s.   (Obj. 11.5.1)   **50.** It would take the smaller drain 12 h and the larger drain 4 h.   (Obj. 11.5.1)

CHAPTER TEST   *pages 518 – 519*

1. $-4 + 2\sqrt{5}$ and $-4 - 2\sqrt{5}$ (Obj. 11.1.2)   **2.** $\dfrac{-4 + \sqrt{22}}{2}$ and $\dfrac{-4 - \sqrt{22}}{2}$ (Obj. 11.2.1/11.3.1)   **3.** -4 and $\dfrac{5}{3}$

(Obj. 11.1.1)   **4.** $\dfrac{3 + \sqrt{33}}{2}$ and $\dfrac{3 - \sqrt{33}}{2}$ (Obj. 11.2.1/11.3.1)   **5.** $-2 + 2\sqrt{5}$ and $-2 - 2\sqrt{5}$ (Obj. 11.2.1/11.3.1)

6.   (Obj. 11.4.1)   **7.** $-2 + \sqrt{2}$ and $-2 - \sqrt{2}$ (Obj. 11.2.1/11.3.1)

8. $\dfrac{-3 + \sqrt{41}}{2}$ and $\dfrac{-3 - \sqrt{41}}{2}$ (Obj. 11.2.1/11.3.1)   **9.** $-\dfrac{1}{2}$ and 3   (Obj. 11.1.1)   **10.** $\dfrac{3 + \sqrt{7}}{2}$ and $\dfrac{3 - \sqrt{7}}{2}$

(Obj. 11.2.1/11.3.1)   **11.** $\dfrac{1 + \sqrt{13}}{6}$ and $\dfrac{1 - \sqrt{13}}{6}$ (Obj. 11.2.1/11.3.1)   **12.** $5 + 3\sqrt{2}$ and $5 - 3\sqrt{2}$ (Obj. 11.1.2)

13. $3 + \sqrt{14}$ and $3 - \sqrt{14}$ (Obj. 11.2.1/11.3.1)   **14.** $\dfrac{5 + \sqrt{29}}{2}$ and $\dfrac{5 - \sqrt{29}}{2}$ (Obj. 11.2.1/11.3.1)   **15.** $\dfrac{5 + \sqrt{33}}{2}$

and $\dfrac{5 - \sqrt{33}}{2}$ (Obj. 11.2.1/11.3.1)   **16.** $\dfrac{5}{2}$ and $\dfrac{1}{3}$ (Obj. 11.1.1)   **17.** $\dfrac{-3 + \sqrt{37}}{2}$ and $\dfrac{-3 - \sqrt{37}}{2}$ (Obj. 11.2.1/11.3.1)

18. $\dfrac{2 + \sqrt{14}}{2}$ and $\dfrac{2 - \sqrt{14}}{2}$ (Obj. 11.2.1/11.3.1)   **19.** $-\dfrac{1}{2}$ and 2   (Obj. 11.1.1)   **20.**   (Obj. 11.4.1)

21. $\dfrac{1 + \sqrt{10}}{3}$ and $\dfrac{1 - \sqrt{10}}{3}$ (Obj. 11.2.1/11.3.1)   **22.** The length is 8 ft. The width is 5 ft.   (Obj. 11.5.1)   **23.** The rate of the boat in calm water is 11 mph.   (Obj. 11.5.1)   **24.** The middle odd integer is 5 or -5.   (Obj. 11.5.1)
25. The rate for the last 8 mi was 4 mph.   (Obj. 11.5.1)

CUMULATIVE REVIEW EXERCISES   *pages 519 – 521*

1. $-28x + 27$ (Obj. 2.2.5)   **2.** $\dfrac{3}{2}$ (Obj. 3.1.3)   **3.** 3   (Obj. 3.2.3)   **4.** $x > \dfrac{1}{9}$ (Obj. 3.3.3)

5. $(3, 0)$ and $(0, -4)$ (Obj. 5.2.3)   **6.** $y = -\dfrac{4}{3}x - 2$ (Obj. 5.4.1/5.4.2)   **7.** Domain: $\{-2, -1, 0, 1, 2\}$;
range: $\{-8, -1, 0, 1, 8\}$; yes   (Obj. 5.5.1)   **8.** 37   (Obj. 5.5.1)   **9.**   (Obj. 5.2.2)

10.    (Obj. 5.6.1)   **11.** $(2, 1)$ (Obj. 6.2.1)   **12.** $(2, -2)$ (Obj. 6.3.1)   **13.** $-\dfrac{4a}{3b^2}$ (Obj. 7.4.1)

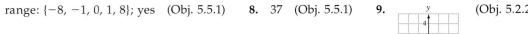

14. $x + 2 - \dfrac{4}{x - 2}$ (Obj. 7.4.4)   **15.** $(x - 4)(4y - 3)$ (Obj. 8.1.2)   **16.** $x(3x - 4)(x + 2)$ (Obj. 8.3.1/8.3.2)

17. $\dfrac{9x^2(x - 2)(x - 4)}{(2x - 3)^2(x + 2)}$ (Obj. 9.1.3)   **18.** $\dfrac{x + 2}{2(x + 1)}$ (Obj. 9.3.2)   **19.** $\dfrac{x - 4}{2x + 5}$ (Obj. 9.4.1)   **20.** -3 and 6

(Obj. 9.5.1)   **21.** $a - 2$ (Obj. 10.3.1)   **22.** 5 (Obj. 10.4.1)   **23.** $\dfrac{1}{2}$ and 3 (Obj. 11.1.1)   **24.** $2 + 2\sqrt{3}$ and

$2 - 2\sqrt{3}$ (Obj. 11.1.2)   **25.** $\dfrac{2 + \sqrt{19}}{3}$ and $\dfrac{2 - \sqrt{19}}{3}$ (Obj. 11.2.1/11.3.1)   **26.** 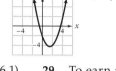   (Obj. 11.4.1)

27. The cost is \$36.   (Obj. 3.1.4)   **28.** The cost is \$2.25 per pound.   (Obj. 4.6.1)   **29.** To earn a dividend of \$752.50, 250 additional shares are required.   (Obj. 9.5.2)   **30.** The rate of the plane in calm air is 200 mph. The rate of the wind is 40 mph.   (Obj. 6.4.1)   **31.** The student must receive a score of 77 or better.   (Obj. 4.8.1)   **32.** The length of the guy wire is 31.62 m.   (Obj. 10.4.2)   **33.** The middle odd integer is -7 or 7.   (Obj. 11.5.1)

FINAL EXAM   *pages 521 – 524*

1. -3 (Obj. 1.1.2)   **2.** -6 (Obj. 1.2.2)   **3.** 12.5% (Obj. 1.3.2)   **4.** -256 (Obj. 1.4.1)

5. -11 (Obj. 1.4.2)   **6.** $-\dfrac{15}{2}$ (Obj. 2.1.1)   **7.** $9x + 6y$ (Obj. 2.2.2)   **8.** $6z$ (Obj. 2.2.3)

9. $16x - 52$ (Obj. 2.2.5)   **10.** -50 (Obj. 3.1.3)   **11.** -3 (Obj. 3.2.3)   **12.** 15.2 (Obj. 3.1.4)

13. $x \le -3$ (Obj. 3.3.3)   **14.** $y \ge \dfrac{5}{2}$ (Obj. 3.3.3)   **15.** $\dfrac{2}{3}$ (Obj. 5.3.1)   **16.** $y = -\dfrac{2}{3}x - 2$ (Obj. 5.4.1/5.4.2)

17.   (Obj. 5.3.2)   **18.**   (Obj. 5.5.2)   **19.** $\{-1, 2, 5, 8, 11\}$ (Obj. 5.5.1)

20.   (Obj. 5.6.1)   **21.** $(6, 17)$ (Obj. 6.2.1)   **22.** $(2, -1)$ (Obj. 6.3.1)

23. $-3x^2 - 3x + 8$ (Obj. 7.1.2)   **24.** $81x^4y^{12}$ (Obj. 7.2.2)   **25.** $6x^3 + 7x^2 - 7x - 6$ (Obj. 7.3.2)

26. $-\dfrac{x^8y}{2}$ (Obj. 7.4.1)   **27.** $\dfrac{16y^7}{x^8}$ (Obj. 7.4.1)   **28.** $3x^2 - 4x - \dfrac{5}{x}$ (Obj. 7.4.3)   **29.** $5x - 12 + \dfrac{23}{x + 2}$

(Obj. 7.4.4)   **30.** $3.9 \cdot 10^{-8}$ (Obj. 7.4.2)   **31.** $2(4 - x)(a + 3)$ (Obj. 8.1.2)   **32.** $(x - 6)(x + 1)$ (Obj. 8.2.1)

33. $(2x - 3)(x + 1)$ (Obj. 8.3.1/8.3.2)   **34.** $(3x + 2)(2x - 3)$ (Obj. 8.3.1/8.3.2)   **35.** $4x(2x - 1)(x - 3)$

(Obj. 8.3.1/8.3.2)   **36.** $(5x + 4)(5x - 4)$ (Obj. 8.4.1)   **37.** $3y(5 + 2x)(5 - 2x)$ (Obj. 8.4.2)   **38.** $\dfrac{1}{2}$ and 3

(Obj. 8.5.1/11.1.1)   **39.** $\dfrac{2(x + 1)}{x - 1}$ (Obj. 9.1.2)   **40.** $\dfrac{-3x^2 + x - 25}{(2x - 5)(x + 3)}$ (Obj. 9.3.2)   **41.** $x + 1$ (Obj. 9.4.1)

42. 2 (Obj. 9.5.1)   **43.** $a = b$ (Obj. 9.6.1)   **44.** $7x^3$ (Obj. 10.1.2)   **45.** $38\sqrt{3a}$ (Obj. 10.2.1)

46. $\sqrt{15} + 2\sqrt{3}$ (Obj. 10.3.2)   **47.** 5 (Obj. 10.4.1)   **48.** $3 + \sqrt{7}$ and $3 - \sqrt{7}$ (Obj. 11.1.2)   **49.** $\dfrac{1 + \sqrt{5}}{4}$

and $\dfrac{1 - \sqrt{5}}{4}$ (Obj. 11.2.1/11.3.1)   **50.** 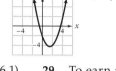   (Obj. 11.4.1)   **51.** $2x + 3(x - 2)$; $5x - 6$ (Obj. 2.3.3)

52. The original value of the machine was \$3000.   (Obj. 3.1.4)   **53.** There are 60 dimes in the bank.   (Obj. 4.2.2)

54. The angles measure 60°, 50°, and 70°.   (Obj. 4.3.2)   **55.** The markup rate is 65%.   (Obj. 4.4.1)   **56.** $6000 must be invested at 11%.   (Obj. 4.5.1)   **57.** The cost is $3 per pound.   (Obj. 4.6.1)   **58.** The percent concentration is 36%.   (Obj. 4.6.2)   **59.** The plane traveled 215 km during the first hour.   (Obj. 4.7.1)   **60.** The rate of the boat is 15 mph. The rate of the current is 5 mph.   (Obj. 6.4.1)   **61.** The length is 10 m. The width is 5 m.   (Obj. 8.5.2)   **62.** 16 oz of dye are required.   (Obj. 9.5.2)   **63.** It would take them 0.6 h to prepare the dinner.   (Obj. 9.7.1)   **64.** The length of the other leg is 11.5 cm.   (Obj. 10.4.2)   **65.** The rate of the wind is 25 mph.   (Obj. 11.5.1)

GLOSSARY

abscissa The first number in an ordered pair. It measures a horizontal distance and is also called the *x*-coordinate. (Section 5.1)

absolute value of a number Distance of a number from zero on the number line. (Section 1.1)

acute angle An angle that measures between 0° and 90°. (Section 4.3)

addend In addition, one of the numbers being added. (Section 1.2)

addition method Method of finding an exact solution of a system of linear equations wherein we use the Addition Property of Equations. (Section 6.3)

additive inverses Numbers that are the same distance from zero on the number line, but on opposite sides; also called opposites. (Section 2.2)

algebraic fraction A fraction in which the numerator and denominator are polynomials. (Section 9.1)

analytic geometry Geometry in which a coordinate system is used to study the relationships between variables. (Section 5.1)

axes The two number lines that form a rectangular coordinate system; also called coordinate axes. (Section 5.1)

base In exponential notation, the factor that is taken the number of times shown by the exponent. (Section 1.4)

binomial A polynomial of two terms. (Section 7.1)

clearing denominators Removing denominators from an equation that contains fractions by multiplying each side of the equation by the LCM of the denominators. (Sections 3.1/9.5)

combining like terms Using the Distributive Property to add the coefficients of like variable terms. (Section 2.2)

completing the square Adding to a binomial the constant term that makes it a perfect-square trinomial. (Section 11.2)

complex fraction A fraction whose numerator or denominator contains one or more fractions. (Section 9.4)

conjugates Binomial expressions that differ only in the sign of a term. The expressions $a + b$ and $a - b$ are conjugates. (Section 10.3)

consecutive even integers Even integers that follow one another in order. (Section 4.2)

consecutive integers Integers that follow one another in order. (Section 4.2)

consecutive odd integers Odd integers that follow one another in order. (Section 4.2)

constant term A term that includes no variable part; also called a constant. (Section 2.1)

coordinate axes The two number lines that form a rectangular coordinate system; also simply called axes. (Section 5.1)

coordinates of a point The numbers in the ordered pair that is associated with the point. (Section 5.1)

cost The price that a business pays for a product. (Section 4.4)

degree of a polynomial in one variable The largest exponent that appears on the variable. (Section 7.1)

dependent system of equations A system of equations that has an infinite number of solutions. (Section 6.1)

dependent variable In a function, the variable whose value depends on the value of another variable, known as the independent variable. (Section 5.5)

descending order The terms of a polynomial in one variable are arranged in descending order when the exponents of the variable decrease from left to right. (Section 7.1)

discount The amount by which a retailer reduces the regular price of a product for a promotional sale. (Section 4.4)

discount rate   The percent of the regular price that the discount represents. (Section 4.4)

domain   The set of first coordinates of the ordered pairs in a relation. (Section 5.5)

element of a set   One of the objects in a set. (Section 1.1)

equation   A statement of the equality of two mathematical expressions. (Section 3.1)

evaluating the function   Replacing x in $f(x)$ with some value and then simplifying the numerical expression that results. (Section 5.5)

evaluating the variable expression   Replacing each variable by its value and then simplifying the resulting numerical expression. (Section 2.1)

even integer   An integer that is divisible by 2. (Section 4.2)

exponent   In exponential notation, the elevated number that indicates how many times the factor occurs in the multiplication. (Section 1.4)

exponential form   The expression 2^5 is in exponential form. Compare *factored form*. (Section 1.4)

factor a polynomial   To write the polynomial as a product of other polynomials. (Section 8.1)

factor a trinomial of the form $ax^2 + bx + c$ To express the trinomial as the product of two binomials. (Section 8.2)

factored form   The expression $2 \cdot 2 \cdot 2 \cdot 2 \cdot 2$ is in factored form. Compare *exponential form*. (Section 1.4)

factors   In multiplication, the numbers that are multiplied. (Section 1.2)

FOIL   A method of finding the product of two binomials; the letters stand for First, Outer, Inner, and Last. (Section 7.3)

formula   An equation that states rules about measurements. (Section 9.6)

fraction in simplest form   A fraction in which there are no common factors other than 1 in the numerator and the denominator. (Sections 1.3/9.1)

function   A relation in which no two ordered pairs that have the same first coordinate have different second coordinates. (Section 5.5)

functional notation   A function designated by $f(x)$, which is the value of the function at x. (Section 5.5)

graph of a relation   The graph of the ordered pairs that belong to the relation. (Section 5.5)

graph of an equation in two variables   A graph of the ordered-pair solutions of an equation. (Section 5.2)

graph of an integer   A heavy dot directly above that number on the number line. (Section 1.1)

graph of an ordered pair   The dot drawn at the coordinates of the point in the plane. (Section 5.1)

graphing a point in the plane   Placing a dot at the location given by the ordered pair; also called plotting a point in the plane. (Section 5.1)

greater than   A number a is greater than another number b, written $a > b$, if a is to the right of b on the number line. (Section 1.1)

greater than or equal to   The sumbol $\geq$ means "is greater than or equal to." (Section 1.1)

greatest common factor   The greatest common factor (GCF) of two or more integers is the greatest integer that is a factor of all the integers. The greatest common factor of two or more monomials is the product of the GCF of the coefficients and the common variable factors. (Section 8.1)

half-plane   The solution set of an inequality in two variables. (Section 5.6)

hypotenuse   In a right triangle, the side opposite the 90° angle. (Section 10.4)

inconsistent system of equations   A system of equations that has no solution. (Section 6.1)

independent system of equations   A system of equations that has one solution. (Section 6.1)

independent variable   In a function, the variable that varies independently and whose value determines the value of the dependent variable. (Section 5.5)

inequality   An expression that contains the symbol $>$, $<$, $\geq$, or $\leq$. (Section 3.3)

integers The numbers ..., −3, −2, −1, 0, 1, 2, 3, ... (Section 1.1)

irrational number The square root of a number that is not a perfect square. The decimal representation of an irrational number never repeats or terminates and can only be approximated. (Sections 1.3/10.1)

isosceles triangle A triangle that has two equal angles and two equal sides. (Section 4.3)

least common denominator The smallest number that is a multiple of each denominator in question. (Section 1.3)

least common multiple (LCM) The LCM of two or more numbers is the smallest number that contains the prime factorization of each number. (Section 9.2)

legs In a right triangle, the sides opposite the acute angles. (Section 10.4)

less than A number a is less than another number b, written $a < b$, if a is to the left of b on the number line. (Section 1.1)

less than or equal to The symbol $\leq$ means "is less than or equal to." (Section 1.1)

like terms Terms of a variable expression that have the same variable part. (Section 2.2)

linear equation in two variables An equation of the form $y = mx + b$, where m is the coefficient of x and b is a constant; also called a linear function. (Section 5.2)

linear function An equation of the form $y = mx + b$, where m is the coefficient of x and b is a constant; also called a linear equation in two variables. (Section 5.5)

literal equation An equation that contains more than one variable. (Section 9.6)

markup The difference between selling price and cost. (Section 4.4)

markup rate The percent of the retailer's cost that the markup represents. (Section 4.4)

monomial A number, a variable, or a product of numbers and variables; a polynomial of one term. (Section 7.1)

multiplicative inverse of a number The reciprocal of the number. (Section 2.2)

natural numbers The numbers 1, 2, 3, ... (Section 1.1)

negative integers The integers to the left of zero on the number line. (Section 1.1)

negative slope A property of a line that slants downward to the right. (Section 5.3)

nonfactorable over the integers A polynomial that does not factor using only integers. (Section 8.2)

numerical coefficient The number part of a variable term. When the numerical coefficient is 1 or −1, the 1 is usually not written. (Section 2.1)

odd integer An integer that is not divisible by 2. (Section 4.2)

opposite of a polynomial The polynomial created when the sign of each term of the original polynomial is changed. (Section 7.1)

opposites Numbers that are the same distance from zero on the number line, but on opposite sides; also called additive inverses. (Section 1.1)

ordered pair A pair of numbers, such as (a, b), that can be used to identify a point in the plane determined by the axes of a rectangular coordinate system. (Section 5.1)

ordinate The second number in an ordered pair. It measures a vertical distance and is also called the y-coordinate. (Section 5.1)

origin The point of intersection of the two coordinate axes that form a rectangular coordinate system. (Section 5.1)

parabola The graph of a quadratic equation in two variables. (Section 11.4)

percent Parts of 100. (Section 1.3)

perfect square The square of an integer. (Section 10.1)

perfect-square trinomial A trinomial that is a product of a binomial and itself. (Section 8.4)

perimeter The distance around a geometric figure. (Section 4.3)

plane A flat surface determined by the intersection of two lines. (Section 5.1)

plotting a point in the plane   Placing a dot at the location given by the ordered pair; also called graphing a point in the plane. (Section 5.1)

polynomial   A variable expression in which the terms are monomials. (Section 7.1)

positive integers   The integers to the right of zero on the number line. (Section 1.1)

positive slope   A property of a line that slants upward to the right. (Section 5.3)

principal square root   The positive square root of a number. (Section 10.1)

product   In multiplication, the result of multiplying two numbers. (Section 1.2)

proportion   An equation that states the equality of two ratios or rates. (Section 9.5)

quadrant   One of the four regions into which the two axes of a rectangular coordinate system divide the plane. (Section 5.1)

quadratic equation   An equation of the form $ax^2 + bx + c = 0$, where a, b, and c are constants and a is not equal to zero; also called a second-degree equation. (Sections 8.5/11.1)

quadratic equation in two variables   An equation of the form $y = ax^2 + bx + c$, where a is not equal to zero. (Section 11.4)

quadratic function   A quadratic function is given by $f(x) = ax^2 + bx + c$, where a is not equal to zero. (Section 11.4)

quotient   In division, the result of dividing two numbers. (Section 1.2)

radical equation   An equation that contains a variable expression under a radical sign. (Section 10.4)

radical sign   The symbol $\sqrt{}$, which is used to indicate the positive, or principal, square root of a number. (Section 10.1)

radicand   In a radical expression, the expression under the radical sign. (Section 10.1)

range   The set of second coordinates of the ordered pairs in a relation. (Section 5.5)

rate   The quotient of two quantities that have different units. (Section 9.5)

rate of work   That part of a task that is completed in one unit of time. (Section 9.7)

ratio   The quotient of two quantities that have the same unit. (Section 9.5)

rational number   A number that can be written in the form a/b, where a and b are integers and b is not equal to zero. (Section 1.3)

rationalizing the denominator   The procedure used to remove a radical from the denominator of a fraction. (Section 10.3)

real numbers   The rational numbers and the irrational numbers. (Section 1.3)

reciprocal   Interchanging the numerator and denominator of a rational number yields that number's reciprocal. A fraction with the numerator and denominator interchanged. (Sections 2.2/9.1)

rectangular coordinate system   System formed by two number lines, one horizontal and one vertical, that intersect at the zero point of each line. (Section 5.1)

relation   Any set of ordered pairs. (Section 5.5)

repeating decimal   A decimal that is formed when dividing the numerator of a fraction by its denominator, in which one or more of the digits in the decimal repeat infinitely. (Section 1.3)

right angle   An angle whose measure is 90°. (Section 4.3)

right triangle   A triangle that includes an angle of 90°. (Section 4.3)

roster method   Method of writing a set by enclosing a list of the elements in braces. (Section 1.1)

scatter diagram   A graph of collected data as points in a coordinate system. (Section 5.1)

scientific notation   Notation in which a number is expressed as the product of two factors, one a number between 1 and 10 and the other a power of 10. (Section 7.4)

second-degree equation An equation of the form $ax^2 + bx + c = 0$, where a is not equal to zero; also called a quadratic equation. (Section 11.1)

selling price The price for which a business sells a product to a customer. (Section 4.4)

set A collection of objects. (Section 1.1)

slope of a line A measure of the slant of a line. The symbol for slope is m. (Section 5.3)

slope–intercept form The slope–intercept form of an equation of a straight line is $y = mx + b$. (Section 5.3)

solution of a linear equation in two variables An ordered pair whose coordinates make the equation a true statement. (Section 5.2)

solution of a system of equations in two variables An ordered pair that is a solution of each equation of the system. (Section 6.1)

solution of an equation A number that, when substituted for the variable, results in a true equation. (Section 3.1)

solution set of an inequality A set of numbers, each element of which, when substituted for the variable, results in a true inequality. (Section 3.3)

solving an equation Finding a solution of the equation. (Section 3.1)

square root A square root of a positive number x is a number a for which $a^2 = x$. (Section 10.1)

standard form A quadratic equation is in standard form when the polynomial is in descending order and equal to zero. $ax^2 + bx + c = 0$ is in standard form. (Sections 8.5/11.1)

substitution method An algebraic method of finding an exact solution of a system of linear equations. (Section 6.2)

sum In addition, the total of the numbers added. (Section 1.2)

system of equations Equations that are considered together. (Section 6.1)

terminating decimal A decimal that is formed when dividing the numerator of a fraction by its denominator and the remainder is zero. (Section 1.3)

terms of a variable expression The addends of the expression. (Section 2.1)

trinomial A polynomial of three terms. (Section 7.1)

undefined slope A property of a vertical line. (Section 5.3)

uniform motion The motion of a moving object whose speed and direction do not change. (Sections 4.7/9.7)

variable A letter of the alphabet used to stand for a quantity that is unknown or that can change. (Sections 1.1/2.1)

variable expression An expression that contains one or more variables. (Section 2.1)

variable term A term composed of a numerical coefficient and a variable part. (Section 2.1)

vertex The lowest point on a parabola that opens up; the highest point on a parabola that opens down. (Section 11.4)

x-coordinate The abscissa in an ordered pair. (Section 5.1)

x-intercept The point at which a graph crosses the x-axis. (Section 5.2)

xy-coordinate system A rectangular coordinate system in which the horizontal axis is labeled x and the vertical axis is labeled y. (Section 5.1)

y-coordinate The ordinate in an ordered pair. (Section 5.1)

y-intercept The point at which a graph crosses the y-axis. (Section 5.2)

zero slope A property of a horizontal line. (Section 5.3)

INDEX

Table of Properties

Properties of Real Numbers

The Associative Property of Addition

If a, b, and c are real numbers, then $(a + b) + c = a + (b + c)$.

The Associative Property of Multiplication

If a, b, and c are real numbers, then $(a \cdot b) \cdot c = a \cdot (b \cdot c)$.

The Commutative Property of Addition

If a and b are real numbers, then $a + b = b + a$.

The Commutative Property of Multiplication

If a and b are real numbers, then $a \cdot b = b \cdot a$.

The Addition Property of Zero

If a is a real number, then $a + 0 = 0 + a = a$.

The Multiplication Property of One

If a is a real number, then $a \cdot 1 = 1 \cdot a = a$.

The Multiplication Property of Zero

If a is a real number, then $a \cdot 0 = 0 \cdot a = 0$.

The Inverse Property of Multiplication

If a is a real number and $a \neq 0$, then

$$a \cdot \frac{1}{a} = \frac{1}{a} \cdot a = 1.$$

The Inverse Property of Addition

If a is a real number, then $a + (-a) = (-a) + a = 0$.

Distributive Property

If a, b, and c are real numbers, then $a(b + c) = ab + ac$ or $(b + c)a = ba + ca$.

Properties of Equations

Addition Property of Equations

The same number or variable term can be added to each side of an equation without changing the solution of the equation.

Multiplication Property of Equations

Each side of an equation can be multiplied by the same nonzero number without changing the solution of the equation.

Properties of Inequalities

Addition Property of Inequalities

If $a > b$, then $a + c > b + c$.
If $a < b$, then $a + c < b + c$.

Multiplication Property of Inequalities

If $a > b$ and $c > 0$, then $ac > bc$.
If $a < b$ and $c > 0$, then $ac < bc$.
If $a > b$ and $c < 0$, then $ac < bc$.
If $a < b$ and $c < 0$, then $ac > bc$.

Properties of Exponents

If m and n are integers, then $x^m \cdot x^n = x^{m+n}$.

If m and n are integers, then $(x^m)^n = x^{mn}$.

If $x \neq 0$, then $x^0 = 1$.

If m and n are integers and $x \neq 0$, then $\dfrac{x^m}{x^n} = x^{m-n}$.

If m, n, and p are integers, then $(x^m \cdot y^n)^p = x^{mp}y^{np}$.

If n is a positive integer and $x \neq 0$, then

$$x^{-n} = \frac{1}{x^n} \text{ and } \frac{1}{x^{-n}} = x^n.$$

If m, n, and p are integers and $y \neq 0$, then $\left(\dfrac{x^m}{y^n}\right)^p = \dfrac{x^{mp}}{y^{np}}$.